CONOVER and IMAN • Introduction to Modern Business Statistics
CORNELL • Experiments with Mixtures: Designs, Models and The Analysis of Mixture Data
COX • Planning of Experiments
DANIEL • Biostatistics: A Foundation for Analysis in the Health Sciences, *Third Edition*
DANIEL • Applications of Statistics to Industrial Experimentation
DANIEL and WOOD • Fitting Equations to Data: Computer Analysis of Multifactor Data, *Second Edition*
DAVID • Order Statistics, *Second Edition*
DAVISON • Multidimensional Scaling
DEMING • Sample Design in Business Research
DILLON and GOLDSTEIN • Multivariate Analysis: Methods and Applications
DODGE • Analysis of Experiments with Missing Data
DODGE and ROMIG • Sampling Inspection Tables, *Second Edition*
DOWDY and WEARDEN • Statistics for Research
DRAPER and SMITH • Applied Regression Analysis, *Second Edition*
DUNN • Basic Statistics: A Primer for the Biomedical Sciences, *Second Edition*
DUNN and CLARK • Applied Statistics: Analysis of Variance and Regression
ELANDT-JOHNSON and JOHNSON • Survival Models and Data Analysis
FLEISS • Statistical Methods for Rates and Proportions, *Second Edition*
FLEISS • The Design and Analysis of Clinical Experiments
FOX • Linear Statistical Models and Related Methods
FRANKEN, KÖNIG, ARNDT, and SCHMIDT • Queues and Point Processes
GALAMBOS • The Asymptotic Theory of Extreme Order Statistics
GIBBONS, OLKIN, and SOBEL • Selecting and Ordering Populations: A New Statistical Methodology
GNANADESIKAN • Methods for Statistical Data Analysis of Multivariate Observations
GOLDBERGER • Econometric Theory
GOLDSTEIN and DILLON • Discrete Discriminant Analysis
GREENBERG and WEBSTER • Advanced Econometrics: A Bridge to the Literature
GROSS and CLARK • Survival Distributions: Reliability Applications in the Biomedical Sciences
GROSS and HARRIS • Fundamentals of Queueing Theory, *Second Edition*
GUPTA and PANCHAPAKESAN • Multiple Decision Procedures: Theory and Methodology of Selecting and Ranking Populations
GUTTMAN, WILKS, and HUNTER • Introductory Engineering Statistics, *Third Edition*
HAHN and SHAPIRO • Statistical Models in Engineering
HALD • Statistical Tables and Formulas
HALD • Statistical Theory with Engineering Applications
HAND • Discrimination and Classification
HILDEBRAND, LAING, and ROSENTHAL • Prediction Analysis of Cross Classifications
HOAGLIN, MOSTELLER and TUKEY • Exploring Data Tables, Trends and Shapes
HOAGLIN, MOSTELLER, and TUKEY • Understanding Robust and Exploratory Data Analysis
HOEL • Elementary Statistics, *Fourth Edition*
HOEL and JESSEN • Basic Statistics for Business and Economics, *Third Edition*
HOGG and KLUGMAN • Loss Distributions

(*continued on back*)

Robust Statistics

Robust Statistics

The Approach Based
on Influence Functions

FRANK R. HAMPEL
ETH, Zürich, Switzerland

ELVEZIO M. RONCHETTI
Princeton University, Princeton, New Jersey

PETER J. ROUSSEEUW
Delft University of Technology, Delft, The Netherlands

WERNER A. STAHEL
ETH, Zürich, Switzerland

John Wiley & Sons
New York • Chichester • Brisbane • Toronto • Singapore

Library of Congress Cataloging in Publication Data:
Main entry under title:

Robust statistics.

(Wiley series in probability and mathematical
statistics. Probability and mathematical statistics,
ISSN 0271-6356)
    Includes index.
    1. Robust statistics.   I. Hampel, Frank R.,
1941–   . II. Series.
QA276.R618   1985   519.5   85-9428
ISBN 0-471-82921-8

Printed in the United States of America

10 9 8 7 6 5 4 3 2 1

To our families and friends

Non omnia possumus omnes

Preface

Statistics is the art and science of extracting useful information from empirical data. An effective way for conveying the information is to use parametric stochastic models. After some models had been used for more than two centuries, R. A. Fisher multiplied the number of useful models and derived statistical procedures based on them in the 1920s. His work laid the foundation for what is the most widely used statistical approach in today's sciences.

This "classical approach" is founded on stringent stochastic models, and before long it was noticed that the real world does not behave as nicely as described by their assumptions. In addition, the good performance and the valid application of the procedures require strict adherence to the assumptions. Consequently, nonparametric statistics emerged as a field of research, and some of its methods, such as the Wilcoxon test, became widely popular in applications. The basic principle was to make as few assumptions about the data as possible and still get the answer to a specific question like "Is there a difference?" While some problems of this kind did find very satisfactory solutions, parametric models continued to play an outstanding role because of their capacity to describe the information contained in a data set more completely, and because they are useful in a wider range of applications, especially in more complex situations.

Robust statistics combines the virtues of both approaches. Parametric models are used as vehicles of information, and procedures that do not depend critically on the assumptions inherent in these models are implemented.

Most applied statisticians have avoided the worst pitfalls of relying on untenable assumptions. Almost unavoidably, they have identified grossly aberrant observations and have corrected or discarded them before analyzing the data by classical statistical procedures. Formal tests for such outliers have been devised, and the growing field of diagnostics, both formal and informal, is based on this approach. While the combination of diagnosis, corrective action, and classical treatment is a robust procedure, there are other methods that have better performance.

Such a statement raises the question of how performance should be characterized. There is clearly a need for theoretical concepts to treat this problem. More specifically, we will have to discuss what violations of assumptions might arise in practice, how they can be conceptualized, and how the sensitivity—or its opposite, the robustness—of statistical procedures to these deviations should be measured. But a new theory may not only be used to describe the existing procedures under new aspects, it also suggests new methods that are superior in light of the theory, and we will extensively describe such new procedures.

The first theoretic approach to robust statistics was introduced by Huber in his famous paper published in 1964 in the *Annals of Mathematical Statistics*. He identified neighborhoods of a stochastic model which are supposed to contain the "true" distribution that generates the data. Then he found the estimator that behaves optimal over the whole neighborhood in a minimax spirit. In the following decade, he extended his basic idea and found a second approach. His book on the subject (Huber, 1981) made this fundamental work available to a wider audience.

The present book treats a different theoretic approach to robust statistics, originated by Hampel (1968, 1971, 1974). In order to describe the basic idea, let us draw on an analogy with the discussion of analytic functions in calculus. If you focus on a particular value of the function's argument, you can approximate the function by the tangent in this point. You should not forget, though, that the linearization may be useful at most up to the vicinity of the nearest singularity. Analogously, we shall introduce the influence function of an estimator or a test, which corresponds to the first derivative, and the breakdown point, which takes the place of the distance to the nearest singularity. We shall then derive optimal procedures with respect to the "infinitesimal" behavior as characterized by the influence function.

The approach is related to Huber's minimax approach, but it is mathematically simpler. It is not confined to models with invariance structure, as the minimax approach is. Instead, it can be used generally whenever maximum likelihood estimators and likelihood ratio tests make sense.

The intention of the book is to give a rather comprehensive account of the approach based on influence functions, covering both generalities and applications to some of the most important statistical models. It both introduces basic ideas of robust statistics and leads to research areas. It may serve as a textbook; the exercises range from simple questions to research problems. The book also addresses an audience interested in devising robust methods for applied problems. The mathematical background required includes the basics of calculus, linear algebra, and mathematical statistics. Occasionally, we allude to more high-powered mathematics.

Chapter 1 provides extensive background information and motivation for robust statistics in general, as well as for the specific approach of this book. Except for a few paragraphs, it may be read profitably even without the background just mentioned.

Chapters 2 through 7 deal with the mathematical formulation of the results. Chapter 2 treats the estimation problem in the case of a single parameter and introduces the basic concepts and results, and Chapter 3 does the same for the respective testing problem. Chapter 4 extends the approach to the estimation of a general finite-dimensional parameter and gives detailed advice as to how it may be applied in any decent particular problem at hand. Chapters 5 and 6 elaborate on the application of the approach to the estimation of covariance matrices and regression parameters, respectively, including possible variants and some related work on these problems. The testing problem in regression is treated in Chapter 7.

Chapter 8 gives an outlook on open problems and some complementary topics, including common misunderstandings of robust statistics. Most of it can be read in connection with Chapter 1 after skimming Chapter 2.

Two decades have passed since the fundamental concepts of robust statistics have been formulated. We feel that it is high time for the new procedures to find wide practical application. Computer programs are now available for the most widely used models, regression and covariance matrix estimation. We hope that by presenting this general and yet quite simple approach as well as the basic concepts, we help to get everyday's robust methods going.

HOW TO READ THE BOOK

The material covered in this book ranges from basic mathematical structures to refinements and extensions, from philosophical remarks to applied examples, and from basic concepts to remarks on rather specific work on special problems. For the reader's convenience, complementary parts are marked by an asterisk.

Different readers will direct their main interests to different parts of the book. A chart given before the table of contents indicates possible paths. The most important definitions of robustness notions can be found in Sections 2.1b–c, 2.2, 2.5a, 3.2a, 3.5, and 4.2a–b, and the most important mathematical results are probably in Sections 2.4a, 2.6c, 4.3a–b, and 6.3b.

The applied statistician will be particularly interested in the examples and methods such as those discussed in Sections 1.1d, 2.0, 2.1d, 2.1e, 2.6c, 3.1, 3.2c, 4.2d, 5.3c, 6.2, 6.3b, 6.4a, 7.3d, and 7.5d, and in the techniques implemented in the computer programs for robust regression and covariance matrices mentioned in Subsection 6.4c. Finally, he or she should become aware of the largely open problem of long-range dependence described in Section 8.1.

ACKNOWLEDGMENTS

J. W. Tukey, P. J. Huber, and C. Daniel have had decisive influences, directly or indirectly, on our view of robust statistics and data analysis.

H. Künsch has generously contributed a section.

Many colleagues have supported us by their encouragement and stimulating comments on the manuscript, in particular R. Carroll, B. Clarke, D. Donoho, A. Durbin, C. A. Field, I. Olkin, E. Spjøtvoll, G. S. Watson, and R. E. Welsch. Very valuable comments have been given by two anonymous referees.

The book developed as a result of a short course given at the 15th European Meeting of Statisticians in Palermo, Italy, in 1982; the course was proposed to us by the Chairman of the Programme Committee, J. Oosterhoff.

M. Karrer spent many overtime hours typing portions of the manuscript. Other portions were typed by P. Flemming and N. Zuidervaart. M. Mächler and J. Beran have done computations and searched for references. The second author has been partially supported by ARO contract #DAAG29-82-K-0178, and the third author by the Belgian National Science Foundation.

Last but not least we wish to thank our wives for their patience and support.

FRANK R. HAMPEL
ELVEZIO M. RONCHETTI
PETER J. ROUSSEEUW
WERNER A. STAHEL

Zürich, Switzerland
Princeton, New Jersey
Delft, The Netherlands
Zürich, Switzerland
March 1985

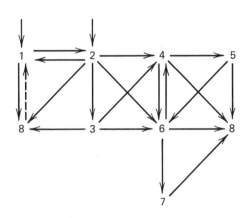

POSSIBLE PATHS THROUGH THE BOOK

Quick overview   Chapters 1, 2, parts of 8, and introductions to 3–7

One-parameter models   Chapters 1, 2, 3

Multiparameter models   Chapters 1, 2, 4, and 5

Regression $\begin{cases} \text{estimation} \\ \text{testing} \end{cases}$ Chapters 1, 2, 4, and 6
 Chapters 1, 2, 3, 6, and 7

Contents

      5.5a. Breakdown Point of M-Estimators, 296
     *5.5b. Breakdown at the Edge, 299
      5.5c. An Estimator with Breakdown Point $\frac{1}{2}$, 300

Exercises and Problems   303

6. LINEAR MODELS: ROBUST ESTIMATION   307

    6.1. Introduction   307
          6.1a. Overview, 307
          6.1b. The Model and the Classical Least-
                 Squares Estimates, 308

    6.2. Huber-Estimators   311

    6.3. M-Estimators for Linear Models   315
          6.3a. Definition, Influence Function, and
                 Sensitivities, 315
          6.3b. Most B-Robust and Optimal B-Robust
                 Estimators, 318
          6.3c. The Change-of-Variance Function; Most V-Robust
                 and Optimal V-Robust Estimators, 323

    6.4. Complements   328
          6.4a. Breakdown Aspects, 328
         *6.4b. Asymptotic Behavior of Bounded Influence
                 Estimators, 331
          6.4c. Computer Programs, 335
         *6.4d. Related Approaches, 337

Exercises and Problems   338

7. LINEAR MODELS: ROBUST TESTING   342

    7.1. Introduction   342
          7.1a. Overview, 342
          7.1b. The Test Problem in Linear Models, 343

    7.2. A General Class of Tests for Linear Models   345
          7.2a. Definition of τ-Test, 345
          7.2b. Influence Function and Asymptotic
                 Distribution, 347

Robust Statistics

Introduction and Motivation

1.1. THE PLACE AND AIMS OF ROBUST STATISTICS

1.1a. What Is Robust Statistics?

Robust statistics, in a loose, nontechnical sense, is concerned with the fact that many assumptions commonly made in statistics (such as normality, linearity, independence) are at most approximations to reality. One reason is the occurrence of gross errors, such as copying or keypunch errors. They usually show up as outliers, which are far away from the bulk of the data, and are dangerous for many classical statistical procedures. The outlier problem is well known and probably as old as statistics, and any method for dealing with it, such as subjective rejection or any formal rejection rule, belongs to robust statistics in this broad sense. Other reasons for deviations from idealized model assumptions include the empirical character of many models and the approximate character of many theoretical models. Thus, whenever the central limit theorem is invoked—be it together with the hypothesis of elementary errors to explain the frequent empirical fit of the normal distribution to measurement errors, be it for the normal approximation of, for example, a binomial or Poisson statistic, or be it that the data considered are themselves sums or means of more basic data, as happens frequently in sample surveys—in all these cases, the central limit theorem, being a limit theorem, can at most suggest approximate normality for real data.

Given this situation, the problem with the theories of classical parametric statistics (be they frequentist or Bayesian) is that they derive optimal procedures under exact parametric models, but say nothing about their behavior when the models are only approximately valid. Neither does nonparametric statistics specifically address this situation. It may be—and

1

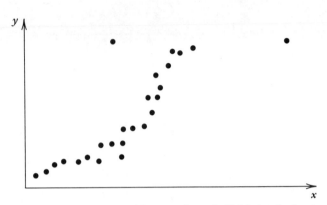

Figure 1. Various deviations from model assumptions. Artificial sketch, showing ordinary outlier, outlying leverage point, nonnormality, nonlinearity, and serially correlated errors. (The latter two are partly confounded, as are nonnormality and heteroscedasticity.)

often turns out to be the case—that classically optimum procedures (and even some "nonparametric" procedures) behave quite poorly already under slight violations of the strict model assumptions.

While the problem of robustness probably goes back to the prehistory of statistics and while a number of eminent statisticians (such as Newcomb, K. Pearson, Gosset, Jeffreys, and E. S. Pearson) were clearly aware of it, it has only been in recent decades that attempts have been made to formalize the problem beyond limited and ad hoc measures towards a theory of robustness. The reasons for this late start are not quite clear, although to some extent progress in the mathematical sciences facilitated the task, and modern computing power made the replacement of hitherto mainly subjective methods by formal methods both feasible and urgent—very urgent indeed, because it is hardly possible any more to detect outliers by subjective methods in the large and complicated data sets (such as multiple regressions) processed nowadays (cf. also Fig. 3). Another major reason was probably the growing general awareness of the need for robust procedures, due to the work by E. S. Pearson, G. E. P. Box, and J. W. Tukey, among others.

There exist now, in fact, a great variety of approaches towards the robustness problem (cf. the references cited in Subsection 1.1b): some consider rather general and abstract notions of stability; others study different topologies or other mathematical aspects related to robustness for the sake of their mathematical problems; many try to broaden the scope to some kind of nonparametric statistics, perhaps still using concepts like symmetry (including work on "adaptive" estimation); a widespread prag-

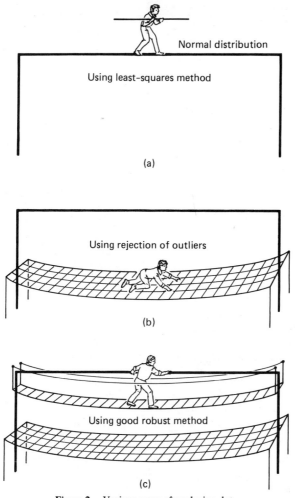

Figure 2. Various ways of analyzing data.

matic approach is to replace a given parametric model by another one, in particular to enlarge it to a "supermodel" by adding more parameters; there are now many Monte Carlo studies on a variety of hypothetical data distributions, and also some studies of real data; and research on rejection of outliers continues along traditional lines. It is quite possible that some of this work will give important impulses in the future, beyond the range of its immediate value. However, the emphasis in much of this work seems to lie

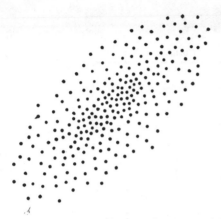

Figure 3. Ninety-nine views (two-dimensional projections; qualitatively) of a point cloud with an outlier in 20-dimensional space.

either on purely mathematical problems, or on rather ad hoc practical solutions, or on isolated pieces of empirical knowledge.

We shall, therefore, concentrate on some approaches which try to capture the essential features of real-life robustness problems within a mathematical framework (Huber, 1964, 1965; Hampel, 1968). For this aim, new statistical concepts had to be developed which describe the behavior of statistical procedures not only under strict parametric models, but also in neighborhoods of such models. There are several aims of such a theory of robustness. Its concepts provide a deeper insight into and more clarity about the vaguely defined empirical robustness problem; in fact, the main aspects of this problem can only now be defined mathematically. These definitions lead to new optimality questions and other tasks for the mathematical statistician. The theory also provides a framework for judging and comparing the existing statistical procedures under the aspect of robustness. For example, it throws much new light on the old problem of rejection of outliers and on the traditional methods used for this purpose. And finally, it develops and suggests new robust procedures, both better ones than those already known, and new ones for situations where none was known. While for the philosopher of statistics and the mathematical statistician such a theory of robustness is valuable in its own right, for the applied statistician it provides both a new enlarged frame of thinking and the mutual stimulation of reevaluation of existing procedures and development of new procedures.

In particular, the data analyst finds a formal framework for questions such as: Are the data unanimous in their message, or do different parts of the data tell different stories? In this case, what does the majority of the data say? Which minorities behave differently, and how? What is the influence of different parts of the data on the final result? Which data are of

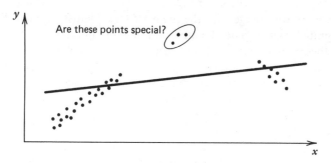

(a) Least-squares fit: average opinion of all points (noisy)

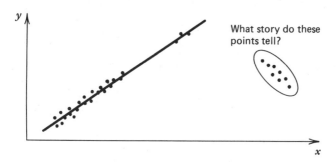

(b) Highly robust fit: clear opinion of majority of points

Figure 4.   Which fit do we want?

crucial importance, either for model choice or for the final results, and should be examined with special care? What could be the effects of gross errors on the results? How many gross errors can be tolerated by the design, model, and method used? What methods provide the greatest safety, and what methods are both fairly safe and fairly efficient? How much redundance has to be built into a design under specific circumstances and for specific purposes? What is a more realistic confidence interval for absolute constants, as opposed to contrasts? How safe are the results if, generally speaking, the model assumptions hold only approximately? Not all of these questions have an immediate answer, but many do, and all are in the background of the motivation of robustness theory.

    At present, the sharp focus on a hitherto neglected aspect of statistical theory and the sheer magnitude of the necessary groundwork for providing the needed concepts and methodology, strongly suggest the development of a specific theory of robustness. But it is hoped that in the future the theory of robustness will merge with the mainstream of statistics and will become an integrated part of general statistical theory and practice.

The theory of robustness is not just a superfluous mathematical decoration. It plays an essential role in organizing and reducing information about the behavior of statistical procedures to a managable form (just as, on a different level, statistics reduces data to the essentials). Compare also the stimulating discussion of the role of mathematical theory in statistics in Huber (1975b). And while Monte Carlo studies and numerical examples can be very useful and are even necessary to a limited extent, it is regrettable to see how many wasteful and superfluous studies have been and are still being undertaken, only because of lack of theoretical insight and understanding. As a great scientist once said: "There is nothing more practical than a good theory."

Of course, there are also some dangers in a theory of robustness, as in any theory related to the real world. One danger is that paradigms outlive their usefulness and even stifle creative new research (cf. Kuhn, 1962). However, at the time of this writing, we are still at the stage where the new paradigms of robustness theory are only slowly understood and accepted, and many fruitless discussions result from mere ignorance and lack of understanding of the new theory, and lack of awareness about which old problems have in fact already been clarified and solved. Another danger is that robustness theory is used too dogmatically and that it is taken too literally in applications, rather than as providing mere guidance. (Cf. also the first paragraph of the introduction of Hampel, 1973a for some critical remarks.) A more specific danger is the possible confusion between the informal and formal use of the word "robust," after some kinds of qualitative and quantitative "robustness" have been defined. Usually, the context should make clear what is meant, or else specifications like "qualitative robustness," "Π-robustness," "B-robustness," and so on. It does not seem feasible anymore to replace the term "robustness theory" by some fantasy term, as has been suggested, and as is common in pure mathematics; but it is good to keep in mind that "robustness theory," as the term is used here, really tries to capture essential features of the vague data-analytic robustness problem, and, on the other hand, that it can never capture all aspects.

Returning to the question "What is robust statistics?", we can perhaps say:

In a broad informal sense, robust statistics is a body of knowledge, partly formalized into "theories of robustness," relating to deviations from idealized assumptions in statistics.

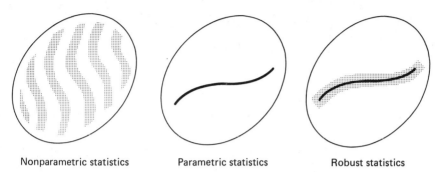

Nonparametric statistics   Parametric statistics   Robust statistics

Figure 5. The space of all probability distributions (usually of infinite dimension) on some sample space.

Nonparametric statistics allows "all" possible probability distributions and reduces the ignorance about them only by one or a few dimensions. Classical parametric statistics allows only a (very "thin") low-dimensional subset of all probability distributions—the parametric model, but provides the redundance necessary for efficient data reduction. Robust statistics allows a full (namely full-dimensional) neighborhood of a parametric model, thus being more realistic and yet, apart from some slight "fuzziness," providing the same advantages as a strict parametric model.

It gains much of its special appeal, but also much of its intrinsic tension, from the close connection between data-analytic problems and mathematical theory. As all of the present theories of robustness consider deviations from the various assumptions of parametric models, we might also say, in a more formal sense, and in a slightly more restricted way:

> *Robust statistics, as a collection of related theories, is the statistics of approximate parametric models.*

It is thus an extension of classical parametric statistics, taking into account that parametric models are only approximations to reality. It supplements all classical fields of statistics which use parametric models, by adding the aspect of robustness. It studies the behavior of statistical procedures, not only under strict parametric models, but also both in smaller and in larger neighborhoods of such parametric models. It describes the behavior in such neighborhoods in a lucid way by means of new robustness concepts; it helps to develop new, more robust procedures; and it defines optimal robust procedures which are optimal in certain precisely specified ways.

It may be noted that every specialist may see robustness theories under a different angle. For the functional analyst, they are strongly concerned with norms of derivatives of nonlinear functionals. For the decision theorist, they

lead in part to vector-valued risk functions, by adding robustness requirements to some classical requirement such as efficiency. For the Fisherian
statistician, they will be shown to be the natural extension of classical
Fisherian statistics. For the data analyst, they provide new tools and
concepts for the outlier problem, thereby clarifying some confusions, and
they suggest a broader outlook and new methods to be fruitfully discussed.
For engineers and many applied mathematicians, they are the sensitivity
analysis and perturbation theory of statistical procedures. Overall, and in
analogy with, for example, the stability aspects of differential equations or
of numerical computations, robustness theories can be viewed as stability
theories of statistical inference.

1.1b. The Relation to Some Other Key Words in Statistics

Robust statistics in the broad, vague, informal sense obviously encompasses
rejection of outliers, although this field seems to lead an isolated life of its
own, and only in recent years do some specialists for the rejection of outliers
appreciate the close natural relationship (cf. Barnett and Lewis, 1978).
These ties should become very beneficial, because robustness theories, as
well as some empirical robustness studies, have a lot to say about rejection
of outliers, including subjective rejection (cf. Section 1.4).

  Robust statistics in the broad sense also encompasses the problem of the
violation of the independence assumption, apart from systematic errors the
most dangerous violation of usual statistical assumptions. The effects, even
of weak unsuspected serial correlations, especially on confidence intervals
based on large samples, are most dramatic, already in series as short as
about 100 observations. However, because of the enormous mathematical
and also philosophical difficulties, no full-fledged theory has been developed
so far, only some rather pragmatic attempts for a preliminary practical
solution. While there have been a number of studies on short-term, quickly
decaying serial correlations, as in ARMA processes, which already pose
quite a serious problem, the main practical problem seems to lie with
long-term correlations, as in so-called self-similar processes. The approach
which seems to be most promising at present lies on a philosophical level
which has long been surpassed in the theories of robustness against outliers:
It consists in simply augmenting the parametric model by adding—in the
simplest form—one more parameter, namely the single correlation parameter of a self-similar process, and by estimating it and trying to adjust
confidence intervals and standard errors for its effect (cf. Section 8.1).

  Robust statistics is often confused with, or at least located close to
nonparametric statistics, although it has nothing to do with it directly. The

theories of robustness consider neighborhoods of parametric models and thus clearly belong to parametric statistics. Even if the term is used in a very vague sense, robust statistics considers the effects of only approximate fulfillment of assumptions, while nonparametric statistics makes rather weak but nevertheless strict assumptions (such as continuity of distribution or independence). The confusion stems from several facts. A number of statisticians work in both fields. Also, a number of nonparametric procedures happen to be very robust in certain ways and thus can also be used as robust procedures for parametric models. Perhaps even more important, one of the sources of nonparametric statistics seems to have been an overreaction against the demonstrated nonrobustness of some classical parametric procedures, resulting in abandonment of parametric models altogether. Nevertheless, nonparametric statistics has its own, rightful and important place in statistics. And even though robustness theories as such cannot be applied in a nonparametric situation, in which no parametric model is used, the concepts of robustness theories still provide valuable insight into the behavior of nonparametric methods (cf. also Subsection 1.2a and Hampel, 1983a).

There are no close connections either with adaptive (or asymptotically everywhere efficient) procedures, which are nonparametric in spirit, or with the study of procedures for all symmetric (or symmetric unimodal) distributions, which, on the one hand, contain the artificial restriction of symmetry and are far from exhausting a neighborhood of a model distribution, and, on the other hand, always contain members too distant from a given model distribution. Contrary to a widespread belief, there is no restriction to symmetry in robustness theory, except in the pure form of one out of three major approaches (see Section 1.3 and Subsection 8.2a); and the vaguely defined attempt to do well for a broad class of symmetric distributions, without special regard to a model distribution, is again more nonparametric than robust in spirit.

The augmentation of a given parametric model by adding, for example, a shape parameter or a transformation parameter may often lead to robust procedures, but the new "supermodel" is, strictly speaking, as unrealistic as the old one, only more complicated, and often much more arbitrary, and there is nothing in the supermodel itself to prove that procedures derived from it are robust, the supermodel still being far too thin in the space of all probability distributions.

It also has to be stressed again that the choice of the neighborhood of a parametric model is very important; it should be statistically meaningful, not just mathematically convenient. In particular, it should take into account the types of deviations from ideal models that occur in practice.

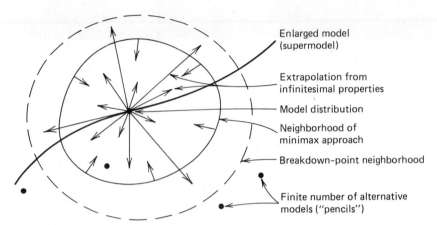

Figure 6.   Cross-section through space of all probability distributions (with parametric model) at a model distribution.

Robustness is also a relevant concept in mild and pragmatic forms of Bayesian theory. Here one needs not only robustness against deviations from the strict parametric model, as in frequentist theory, but also against changes of the a priori distribution of the parameters. A puristic Bayesian theory would put prior distributions on everything, including models and full neighborhoods of models, but this approach seems to be very hard to carry through.

The quantitative study of a finite number of parametric models (sometimes called "pencils" in the location-scale case) has a place in providing quantitative examples and checks for the accuracy of the extrapolations from the model by means of robustness theory; by itself, however, it consists only of isolated, rather arbitrary examples of numerical behavior without a connecting interpretation.

In view of the fact that the word "robust" has become fashionable and is being used in many different senses by different authors, it seems necessary to state what is meant and what is not meant by the robustness theory treated in this book, in order to avoid or at least reduce the great confusion that seems to be rather widespread. The point of view in this book is similar to that of Huber in that full neighborhoods of distinguished parametric models are studied; and while Huber developed two major theoretical approaches, namely the minimax approach to robust estimation and the capacities approach to robust testing and confidence intervals (cf. Huber, 1981), this book contains a third approach which we shall call the approach based on influence functions, or the infinitesimal approach for short, although it also considers an important global measure of robustness.

To facilitate the study of more or less loosely related work in various directions, especially in those directions mentioned in Subsections 1.1a and 1.1b, we shall give the names of some researchers which together with the few references given (and not cited already elsewhere) in this book may be helpful in tracking down further specific literature: Bickel, Jurečková, Serfling, Staudte (general mathematical results); Bednarski, Buja, Strassen, Wolf (cf. also Huber, 1973b and Huber and Strassen, 1973) (Choquet capacities); Bednarek-Kozek, Meshalkin, Zieliński, Zolotarev (general stability, characterizations, etc.); Hogg, Takeuchi (adaptivity); R. Beran, Millar (Hellinger distance, shrinking neighborhoods, etc.); Berger, Box, de Finetti, Dempster, Ramsay, A. F. M. Smith (Bayesian robustness); Bickel and Herzberg, Gastwirth and Rubin, Kariya, Portnoy (short-term serial correlations, cf. also Section 8.1); Anscombe, N. A. Campbell, Denby and Mallows, Hoaglin, Pregibon, Rey (1983) (with further references), Sacks and Ylvisaker (cf. also Huber 1970, 1979, additional work by Mosteller and Tukey, and many others) (various application-oriented work).

1.1c. The Aims of Robust Statistics

The main aims of robust statistics are:

(i) To describe the structure best fitting the bulk of the data.

(ii) To identify deviating data points (outliers) or deviating substructures for further treatment, if desired.

In the case of unbalanced data structures, as typically in regression, it is also necessary

(iii) To identify and give a warning about highly influential data points ("leverage points").

Furthermore, since not only the assumed marginal distributions but also the assumed correlation structures (in particular, the independence assumption) are only approximations, another task of robust statistics is

(iv) To deal with unsuspected serial correlations, or more generally, with deviations from the assumed correlation structures.

Some more detailed explanations seem in order.

For (i), we tentatively assume a parametric model and then try to do as well as we can with estimating and testing the parameters of the model, taking explicitly into account that the model may be distorted and that a minority of the data may not belong to the model at all. The inference is conditional, given that we are still willing to keep the model as an approximation for the majority of the data; it is safe in that it is influenced only to a limited extent by any minority of the data. The maximum permitted percentage of the minority, such that it has only limited influence, will be measured by the concept of "breakdown point" (cf. Hampel, 1968, 1971, 1974); often it is 50% so that indeed any majority can overrule any minority, as in the case of the sample median.

What is needed in addition is, of course, a decision whether to keep the assumed parametric model—for example, testing of the model and selection of the best-fitting model(s). Testing of a model within a larger model, as common in linear models, reduces to testing whether certain parameters can be assumed to equal zero, and thus falls into the framework of present robust testing theory. By contrast, the general question of "robust goodness of fit" apparently has not yet found systematic treatment, although there are a few scattered attempts (cf., e.g., Ylvisaker, 1977).

For (ii), the residuals from a robust fit automatically show outliers and the proper random variability of the "good" data, much clearer than for example residuals from least squares which tend to smear the effect of outliers over many data points, and where outliers blow up the residual mean-squared error, again making their detection more difficult. While in small or balanced data sets it is still possible to detect outliers by careful visual inspection of the data, this becomes virtually impossible with large, high-dimensional and unbalanced data sets, as are often fed into the computer nowadays. Even a careful analysis of residuals may not always show all deviating data points, or it may require much more time and ingenuity by the data analyst than is normally available. Some of the formal rules for detection of outliers are surprisingly unreliable; even the frequently used maximum Studentized residual can only tolerate up to about 10% outliers in the simplest situation before it breaks down. But this is already a result of robustness theory which will be discussed later in greater detail (Section 1.4).

Given the need for reliable and fast identification of outliers, and the means to accomplish such, the next question is how to treat the outliers which have been found. In principle they should be studied separately. This is in particular true if the outliers occur in certain patterns which may indicate unsuspected phenomena or a better model (cf. the examples in Daniel, 1976). Not all outliers are "bad" data caused by gross errors;

sometimes they are the most valuable data points of the whole set. Even if they are gross errors, it is sometimes possible to correct them, by going back to a more original data set, or by making guesses about their cause, for example, inadvertent interchange of two values. Even the mere frequencies and perhaps sizes of gross errors may be valuable information about the reliability of the data under consideration. Only if a model is well established as a useful approximation in routine investigations, and if there is only interest in the majority of the data and no interest in any outliers whatsoever, is the automatic rejection of outliers without further treatment justified. Even then, there are better methods than full acceptance or sudden, "hard" rejection of outliers beyond a certain point, in particular those which allow for a region of doubt whether an observation can be considered an outlier or not.

For (iii), while for many balanced and nearly balanced designs it is possible to find very robust methods which are almost fully efficient under the ideal parametric model, the situation changes if a few data points, obtained under highly informative conditions, have a dominating influence on the fit. Perhaps the simplest situation of this kind is simple linear regression, with one x_i far away from the others. This situation is not dealt with by merely robustifying with respect to the residuals (as in Huber, 1973a). However, we now have a conflict: If we downweight such a leverage point and it is a proper observation, we lose a lot of efficiency; on the other hand, if we do not downweight it and it is a gross error, it will completely spoil our result. Perhaps the best advice that we can give at present is to run two robust regressions, one with and one without downweighting of the leverage points, and then to compare the results. If they disagree beyond random variability, then either some leverage points are deviant, or the model is wrong. But even a very classical statistician who still uses only least squares, but wants to do more than a very superficial job, will want to know which are the most influential data points and will have a special look at them. Thus the influence of position in factor space is a valuable diagnostic tool even if not imbedded in a formal robustness theory.

In fact, the field of regression diagnostics (cf. Belsley et al., 1980 or Cook and Weisberg, 1982) also has the identification of leverage points as one of its main aims, and it is not surprising that various specific regression diagnostics correspond to (and are finite sample versions of) influence of position, influence of residual or their product, the total influence, as used in robust regression (cf. Section 6.2). They are, however, frequently evaluated at the least-squares procedure, rather than at a robust procedure, and this can cause masking effects and can make the identification of more than a few special values very cumbersome.

Really extreme leverage points, with practically no redundance from the other data, should be avoided in the design, if possible; sometimes, when they only later turn out to be crucial, one can still replicate them (exactly or approximately) after the main experiment, thus providing the redundance necessary for reliable judgment. There are still enough problems with moderate leverage points, especially if also gross errors are feared.

For (iv), in the past, formal robustness theories have concentrated on deviations in distribution shape (e.g., deviations from normality), in particular on outliers which are a danger for point estimation, confidence intervals, and tests with all sample sizes, but which can be fairly easily dealt with. However, not only do outliers occur rather frequently, but experience tends to show that even in cases where one would expect independent observations, as in astronomy, physics, and chemistry, there seem to be small but long-ranging correlations, of a kind which can hardly be detected individually, but which tend to pile up if summed over many lags and which therefore greatly endanger confidence intervals and tests based on large samples. The nominal confidence intervals derived under the independence assumption will be far too short if such persistent semi-systematic fluctuations are present.

The main effects of such semi-constant errors, as they are also called, can be modeled by means of self-similar processes. Thus, a possible program to attack this problem, which is still in the process of being carried out, is to estimate the correlation parameter of an approximating self-similar process, and then to use this parameter in order to correct the length of confidence intervals (cf. also Section 8.1 and Hampel, 1979).

1.1d. An Example

In order to illuminate some of the points made in Subsection 1.1c from the point of view of practice, we shall discuss a small example from the analysis of variance. In particular, we intend to show that in this example modern robust estimation techniques lead to the same results as a careful subjective analysis of the residuals [which is often cumbersome, virtually impossible in large and complicated designs, usually less efficient than modern robust techniques (cf. Section 1.4), and apparently often neglected even when it could be done fairly easily].

The example is a two-way analysis of variance for lettuce given in Cochran and Cox (1957) on p. 164 (Section 5.26, Table 5.6) of the second edition, and it is further discussed in Daniel (1976) on p. 38ff. The data are the numbers of lettuce plants emerging after application of two fertilizers on three levels each. We shall try to fit a main-effects model. The data after

Table 1. Data Minus Grand Mean of 353 and Least-Squares Residuals and Effects of 3^2 on Lettuce[a]

Data Minus 353			Least-Squares Residuals and Effects			
96	60	−27	6.3	20.3	−26.7	43
56	5	−62	9.7	8.7	−18.3	−0.3
−12	−75	−41	−16	−29	45	−42.7
			46.7	−3.3	−43.3	0

[a] Data from Cochran and Cox (1957, p. 164) and Daniel (1976, p. 39).

subtraction of their grand mean of 353 (as in Table 4.2 of Daniel, 1976) and the residuals and main effects as estimated by least squares are given in Table 1.

Daniel (1976, p. 40) argues that cell (3, 3) seems to contain an outlier and notes that treating it as a missing value for estimation would give it a residual of 101.25. He says that "...if it could be done tactfully, we would ask the experimenters whether by any chance the reported 312 (coded to −41) could have been misrecorded by 100 . . . ." He shows that the remaining eight observations fit a main-effects model with considerably decreased mean-squared error and with no hint of any special interactions. Later (p. 43) he remarks: "If this value is actually an error, and not a permanent interaction, then we do the experimenter no favor by reporting biased main effects and a spurious linear-by-linear interaction."

We shall now apply some formal robust methods. First, we ought to analyze the structure of the data. Clearly, the data set as given would be too small to study any serial correlations, even if some (spatial or other) ordering structure were known. The data are fully balanced for our main-effects model, so there are no leverage points (and the three major types of robust regression, named after Huber, Mallows, and Schweppe, cf. Section 6.3, all coincide). There are five independent parameters and only 4 degrees of freedom for error. Without additional knowledge, it would be possible to identify at most two or three outliers not in the same row or column (cf., e.g., Hampel, 1975, Section 3.1).

The following computations were done both with a pocket calculator and with a preliminary version of the robust regression package ROBSYS (cf. Subsection 6.4c) developed by A. Marazzi and A. Randriamiharisoa at the University of Lausanne, and tested with and applied to our data set by M. Mächler (ETH, Zurich).

Table 2.   Least Absolute Deviations Solution for the 3^2 on Lettuce

L_1-Fit			L_1-Residuals and Effects			
96	45	−22	0	15	−5	49.33
56	5	−62	0	0	0	9.33
−12	−63	−130	0	−12	89	−58.67
			56.33	5.33	−61.67	−9.67

As an intermediate step, we first compute a highly robust starting value for our vector of estimated effects. Because of the complete balance in the data, we can use for this purpose here the L_1-solution, which minimizes the sum of the absolute deviations (but in general the L_1-solution is *not* robust, cf. Subsection 6.4a). The results are given in Table 2. (The fact that five residuals vanish exactly reflects a peculiarity of the L_1-method.)

We note one very large residual (relative to the three other nonzero ones), clearly more outstanding than with the L_2- (least-squares) fit. Otherwise the L_1-solution, with its five enforced zeros, is not clear in the details and also rather inefficient.

We use it as starting point for an estimator which can reject outliers "smoothly" (cf. Section 1.4), namely the "2-4-8-estimator" (Hampel, 1974, Section 5.4, and Hampel, 1980, Sections 3.3.6 and 3.4) which is a very simple default version of a "three-part redescending M-estimator," with parameters $a = 2$, $b = 4$, and $r = 8$ as multiples of the "median deviation" (cf. Subsection 2.3a, Example 4, and Subsection 2.6a, Example 1, for the definitions in the one-parameter case; Subsections 4.2d and 4.4a for location and scale; and Section 6.2 for the generalization to regression). In effect, the 2-4-8-estimator is a weighted least-squares estimator with data-dependent weights (to be determined iteratively), which are $\equiv 1$ (like for ordinary least squares) for data with small residuals, $\equiv 0$ for clear outliers, and varying continuously between 1 and 0 for data with intermediate residuals.

For starting the iterations from the L_1-solution, we also need a robust scale estimate. (The mean deviation is *not* robust.) Normally, we would use the median deviation (equals the median of the absolute values of the residuals), but because of the five enforced zeros, which make the median deviation also equal to zero, we need a custom-tailored scale estimate. If it should tolerate up to two outliers (cf. above), it has to depend on the third- and fourth-largest absolute residual only. For simplicity, we take the mean of the two, namely 8.5, in place of the median deviation, without bothering about a correction factor. The exact choice, within a reasonable range, does

Table 3.   Solution with 2-4-8-Estimator and Iterated Median Deviation Scale, Starting from L_1 with Scale = 8.5, for the 3^2 on Lettuce

Fit			Residuals and Effects			
100.92	50.92	−22.83	−4.9167	9.0833	−4.1667	54.25
57.58	7.58	−66.17	−1.5833	−2.5833	4.1667	10.9167
−18.50	−68.50	−142.25	6.5	−6.5	101.25	−65.1667
			57.9167	7.9167	−65.8333	−11.25

not matter. (As it turns out, all starting values down to at least five, the smallest nonzero absolute residual, and up to ∞ lead to the same solution.)

The solution to which the iterations converge is given in Table 3.

The final median deviation is 4.92. With the 2-4-8-estimator, data with absolute residuals below two median deviations receive full weight and are treated as in least squares, while those with absolute residuals above eight median deviations are rejected completely. Table 3 shows that cell $(3, 3)$ is rejected completely, while all others receive full weight. We thus obtain exactly Daniel's (1976, p. 39) solution after his treating $(3, 3)$ as a missing value, or the least-squares solution for the remaining eight observations. (Only the scale estimate is somewhat different. In general, the solutions will not be exactly the same, of course, since some data might be just partly downweighted by the robust estimator, but they will be very similar.)

The next step is again the interpretation. We notice eight small residuals and one huge outlier conspicuously close to 100. Either there is a rather peculiar interaction concentrated in one cell, or the value is a gross error, most likely misrecorded by 100. In either case, the "robust" fit is a tight main-effects model for eight values and one distant outlier, and for informed judgment one should at least known it in addition to the least-squares fit where the outlier, though still visible for the experienced eye, is smeared over the whole table and blows up the mean-squared error considerably (from about 80 to about 1200).

Some readers who know already something about robust regression will wonder about other possible solutions. There are not so many possibilities. Using ROBSYS, the 2-4-8-estimator with iterated median deviation scale converged towards four different fixed points, depending on the starting value for scale, with 1, 2, 3, and 4 "outliers" rejected, respectively. The first and last solutions are those of Tables 3 and 2, respectively. The second solution almost rejects cell $(1, 2)$ in addition, with residual 18.36 (and all other absolute residuals except $(3, 3)$ below 4) and median deviation 2.91, and the third solution rejects cells $(1, 2)$ and $(3, 2)$ in addition, with residuals 17.5 and -10.75 respectively, and median deviation 1.25. The residual in $(3, 3)$ goes from 101.25 via 97.27 and 91.5 to 89, hence is always clearly sticking out. Also the effects do not change much and are always relatively far away from the least-squares effects. In comparing the four solutions, we shall most likely choose the one of Table 3, unless we have special reasons to mistrust cell $(1, 2)$ or even also $(3, 2)$. But also in view of the random variability of the estimates, the exact choice usually does not matter very much; it is a general experience that in the presence of outliers all reasonable robust estimates are relatively close together, but usually clearly separated from the least-squares solution.

Those who are still curious might consider yet other robust estimators. It is clear that 2-4-8-estimators with arbitrary fixed scale will yield intermediate solutions (even intermediate up to least squares, for increasing scale). Other redescending estimators will yield similar solutions. Regression with the "Huber-estimator" with "Proposal 2" scale and corner point c (cf. Example 2 in Subsection 4.2d, and Section 6.2) converged to the least-squares solution for c down to about 0.82, since Proposal 2 is not so resistant against outliers; and for smaller c tending to zero we obtain another continuous transition to the L_1-solution. The Huber-estimator with iterated median deviation scale converges to the L_1-solution because of the large number of parameters involved. Other estimators with monotone instead of redescending score functions will behave similar to Huber-estimators. They all can dampen the influence of the outlier on the fit, most strongly in the L_1-solution, but they cannot reject it completely. Nevertheless, one "monotone" robust starting value, such as a "weighted" L_1-estimator in general, is needed for safe identification of outliers before a redescending estimator can be applied.

1.2. WHY ROBUST STATISTICS?

1.2a. The Role of Parametric Models

The use of parametric models (such as the classes of normal, lognormal, exponential, Poisson distributions) is deeply entrenched in statistical practice. The reason for this is that they provide an approximate description of a complete set of data by means of a qualitative information (namely, which model is used) and very few simple numbers (the approximations or estimates for the parameters of the model), and that they, together with the parameter values, provide a complete and easily described stochastic model for the generation of the observed data and other fictive or future observations. They thus fulfill one of the main aims of statistics, namely data reduction or data condensation (perhaps from thousands or millions of numbers to only two; cf. also Fisher, 1925/70, p. 1), and they also allow the application of methods of probability theory to the complete description of the whole data set.

Put somewhat differently, parametric models allow the separation of the full information of a data set into pure structure and pure random variability, or "cosmos" and "chaos."

By contrast, nonparametric statistics does not make any assumptions about a low-dimensional class of distributions which is to describe the data fully, and leaving aside goodness-of-fit tests, which may actually be used to test parametric models, it considers only specific aspects of a data set, not its structure as a whole. We may learn, for example, that a certain functional of a distribution—or even a certain distribution—is or is not significantly different from another one, without knowing what the distribution is. Nonparametric statistics has a rightful and important place in statistics, but in general it does not exhaust the information contained in the data.

A source of confusion is the fact that the same statistics, like arithmetic mean or median, can be used in a parametric or a nonparametric way. In the former case, they estimate the parameters which fix the full distribution of the data; in the latter case, they look merely at one particular aspect of the data, no matter what their distribution is otherwise. It is a big difference whether I am told that a certain waiting time distribution has a median of three hours—which leaves me with all conceivable distributions compatible with this fact—or whether I am told that the half-life period of an exponentially distributed waiting time is three hours. Only parametric models provide the redundance necessary for a data description which is both simple and complete.

In some way intermediate between parametric and nonparametric models are "smoothing" models, such as data description by running means and medians and smoothing splines, and including so-called nonparametric regression and nonparametric density estimation. While the conditional expectation or density in a single point is sometimes studied in a truly nonparametric limit, the description of a complete function by a (small to moderate) finite number of estimated values still requires some additional smoothness property which provides the necessary redundance. Also fully adaptive estimation (which basically estimates a density f or a score function f'/f) has these features. In the limit of increasingly many data, it is also possible to arrive at a complete data description with these procedures, either by using more and more descriptive points or by using more and more smoothness properties; but such descriptions are not as simple, short, and elegant and easily amenable to the use of probability theory as those provided by parametric models.

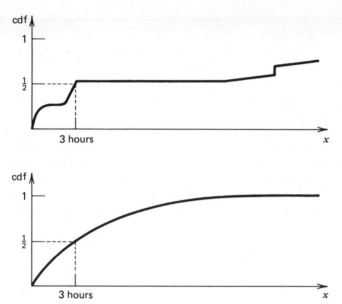

Figure 1. Two cumulative waiting-time distributions with median three hours. There are many more possibilities if no parametric model is specified.

1.2b. Types of Deviations from Parametric Models

This subsection describes the main aspects of the approximate character of parametric models.

It is tempting to forget over the beauty and elegance of the theory of parametric models that they are only approximations to reality. Thus, for some time during the last century, "everyone believed in the normal distribution, the mathematicians because they thought it was an experimental fact, the experimenters because they thought is was a mathematical theorem" (Lippmann according to Poincaré; cf. Stigler, 1975). But the central limit theorem only tells us about an imaginary limit, given certain assumptions. It does not tell us how far we are still away from that limit or whether, indeed, the assumptions are fulfilled; and any empirical check can at most prove statistically that the true distribution is within a certain neighborhood of a model distribution, while it can never prove that it coincides exactly with a model distribution. Moreover, large data sets of high quality show significant deviations from normality in cases which should be prime examples for the normal law of errors (cf. below).

A tacit hope in ignoring deviations from ideal models was that they would not matter; that statistical procedures which were optimal under the strict model would still be approximately optimal under the approximate model. Unfortunately, it turned out that this hope was often drastically wrong; even mild deviations often have much larger effects than were anticipated by most statisticians (cf. Tukey, 1960, inserts).

What are the types of deviations from strict parametric models? How strongly do they affect the behavior of classically optimal procedures? Are there any better, "more robust" procedures? These are some of the questions to be answered in the sequel.

We may distinguish four main types of deviations from strict parametric models:

(i)   The occurrence of gross errors.

(ii) Rounding and grouping.

(iii) The model may have been conceived as an approximation anyway, for example, by virtue of the central limit theorem.

(iv) Apart from the distributional assumptions, the assumption of independence (or of some other specific correlation structure) may only be approximately fulfilled.

For (i), gross errors or blunders are occasional values where something went wrong. Common examples are mistakes in copying or computation. The effects on the data are usually much bigger than those of the permanently acting causes of random variability, thus resulting in outliers (namely values which deviate from the pattern set by the majority of the data). Some gross errors will be hidden among the "good" data and will usually be harmless. On the other hand, outliers can also be caused by transient phenomena and by part of the data not fitting the same model. Outliers are the most obvious, the most discussed, and, if untreated, often the most dangerous type of deviation from parametric models and are, therefore, treated in greater detail in separate subsections (Subsections 1.2c and 1.4).

For the moment, it suffices to note that: (1) a single huge unnoticed gross error can spoil a statistical analysis completely (as in the case of least squares); (2) several percent gross errors are rather common; and (3) modern robust techniques can deal with outliers relatively easily, even better than classical methods for objective or subjective rejection of outliers.

For (ii), all data have only a limited accuracy and are thus basically discrete; morever, they are often rounded, grouped, or classified even more coarsely. There can also be small systematic but localized inaccuracies in the measuring scale. These local granularity effects appear to be the most harmless type of deviation from idealized models, and in fact they can often be neglected, but they should not be forgotten. There are several situations in which they even play a prominent role: very coarse classification so that a continuous distribution would be a very bad approximation; study of locally determined quantities such as density estimation or even quantile estimation; and models with infinite Fisher information, such as the rectangular or exponential shift family, which would allow "superefficient" estimation of their parameter.

Let us consider this last situation in greater detail. Kempthorne (1966) has shown that the superefficiency of these estimators, namely a variance going down like n^{-2} instead of n^{-1}, disappears as soon as the finite accuracy of every number is taken into account. The actual dependence of the variance of maximum likelihood and related estimators on n shows a transition from an n^{-2} curve to an n^{-1} curve as soon as the discreteness becomes noticeable with increasing n. There are other robustness considerations (namely the requirement of qualitative robustness; see Subsection 1.3f) also leading to estimators with the usual rate of convergence, even if infinite accuracy were in fact attainable. This is one of the instances where robustness theory eliminates pathological features in classical parametric statistics, in this case superefficiency caused by the model.

The local properties of granularity can be studied by means of the concept of "influence function" (see Subsection 1.3d and Chapter 2); the most relevant single number connected with this problem is the "local-shift sensitivity" (see Subsection 2.1c). There has been little systematic study, though, of robustness against granularity effects.

As a side remark, one may note that it becomes intuitively evident that estimators with infinite local-shift sensitivity, like median, quantiles, and hard rejection rules, are more susceptible to local effects than estimators which smooth over the data locally. Interestingly, the arithmetic mean has the lowest local-shift sensitivity among all location estimators and is in this specific sense the "most robust" estimator, although it is drastically nonrobust in other, much more important senses.

For (iii), large sets of measurement data of very high quality, which to all knowledge contain no gross errors, still tend to show small but noticeable deviations from the normal model. This had been known, but consciously ignored, already by Bessel (cf. Bessel, 1818; Bessel and Baeyer, 1838), shortly after the invention of the method of least squares. Also later, eminent statisticians, including K. Pearson (1902), Student (1927) and Jeffreys (1939/61, Chapter 5.7) collected high-quality samples and found that they are usually longer-tailed than the normal distribution. A recent collection of examples can be found in Romanowski and Green (1965) and Romanowski (1970). These authors also noted some slight skewness. (Cf. also the examples cited in Scheffé, 1959, p. 333.)

In his very interesting discussion, Jeffreys (1939/61) analyzes nine long series of measurements, including six series of pseudoastronomical data by K. Pearson with about 500 observations each, made for a different statistical purpose, and two series of residuals from latitude observations of about 5000 observations each. He tried to fit symmetric Pearson curves and found that seven of the nine series were longer-tailed than the normal, five significantly so, but only one was significantly shorter-tailed than the normal. However, he also noted (as had Pearson) that there were often strong serial correlations in the Pearson series (of supposedly independent observations!), and he discovered that there was a high, almost perfect correlation between the short-tailedness and the serial correlation (!). This means the less long-tailed the data (down to the normal or even shorter-tailed than the normal), the more one has to worry about serial correlations which invalidate standard errors, confidence intervals, and the like; closeness to normality may not be a reason for joy but a reason for even deeper suspicion. The Pearson series, which were obtained by single observers (including "K. P." himself) under highly uniform conditions, were fitted by type II and by t- (type VII) distributions with down to 8 degrees of freedom; extrapolation to the case of independence (using the correlation with the serial correlation) yields about $4\frac{1}{2}$ degrees of freedom. Jeffreys, somewhat more cautiously, surmises that independent data under such homogeneous conditions would be approximated by t-distributions with 5–9 (or $4\frac{1}{2}$–10) degrees of freedom. The latitude series, whose correlation structure apparently was not investigated, but which represent actual measurement series under nonuniform conditions, are fitted by t-distributions with $3\frac{1}{2}$ and $4\frac{1}{2}$ degrees of freedom. If they also show correlations, as might be suspected, an extrapolated independent series would need still fewer degrees of freedom. This, by the way, is in accordance with Huber (orally) who also found t_3 a suitable example for what high-quality data can look like.

Some critics of the Princeton robustness study (Andrews et al., 1972) have maintained that short-tailed alternatives to the normal deserve the same attention as long-tailed alternatives. They certainly deserve some attention, but much less than the long-tailed models, for two reasons: they are much less dangerous and are, overall, much less frequent. In fact, it seems at present that they occur mainly in three situations: either the data have a very limited range (or other a priori reasons for suggesting a tendency towards shorter tails); or the data have been prematurely and strongly cleaned of all suspected outliers (often at a very early stage and unknown to the statistician), rendering an artificially truncated distribution; or the short-tailedness may be connected with strong serial correlations, as in some of the Pearson series. Besides, robustness theory also gives extrapolations and qualitative results for short-tailed distributions.

The situation for models other than the normal distribution (and with it, implicitly, the two-parameter lognormal) has received far less attention, not surprisingly since the normal plays such a central role in statistics and probability theory; but the phenomena can be surmised to be similar. Thus it is known that the model of the exponential distribution for waiting time of failures, which can be derived under the assumption of lack of memory, often does not quite fit for very small and very large waiting times though it may be very good in the middle. There is no basic difference to the normal model; again, general robustness theory could be applied in principle without any problem, only the numerical and computational details would have to be worked out.

In complex designs, model features such as linearity and additivity of effects often hold in good approximation, but not exactly. Deviations of this type are also covered to some extent by the general robustness theories to be described, although more specific studies for particular features and classes of models would be very desirable. Outliers in the analysis of variance and the question of necessary redundancy in the presence of gross errors are briefly and concisely discussed in Hampel (1975, Chapter 3.1; cf. also Subsection 8.4b). Huber (1975a) has shown that slight deviations from linearity in, say, a simple linear regression situation are highly relevant also for the choice of the design of an experiment and may lead to optimal robust designs quite different from those of the classical theory of optimal designs, but closer to common sense and widespread practice. These and related questions should receive more attention in the future than they have in the past.

A model feature which leads up to consideration of the correlation structure and which is also frequently assumed in first approximation, is homoscedasticity or constancy of error variance. Ideally, heteroscedastic data should be weighted properly, but if the error variance ratios are not known and are so irregular that they cannot be inferred from the data, one way to safeguard against data with large error variance (which tend to behave like random outliers) is to downweight data with large (standardized) residuals appropriately, since they are likely to have large error variances. Now this can be achieved by suitable robust estimators. On the other hand, if the sizes of the residuals show a structure, such as an increase with increasing fit, it may be possible to fit a better model which incorporates the systematic part of the heteroscedasticity, at least to a good degree of approximation (with further safeguards possible by robust methods for the refined model).

For (iv), a problem about which relatively little is known is violation of the independence assumption, or robustness against unsuspected serial correlations. We are here roughly in a situation similar to the one before 1964 regarding outliers and other distortions of the distribution shape: before Huber's (1964) paper, only specific alternatives or narrow families of alternatives were investigated for their effect on customary estimators, and some new estimators were tried out in these situations in a rather ad hoc way. This highly important problem will be discussed separately in Section 8.1.

1.2c. The Frequency of Gross Errors

Gross errors, which are errors due to a source of deviations which acts only occasionally but is quite powerful, are the most frequent reasons for *outliers*, namely data which are far away from the bulk of the data, or more generally, from the pattern set by the majority of the data. Occasionally, distant outliers are due to a genuinely long-tailed distribution (this is even a frequent reason for moderate outliers); furthermore, a common reason for apparent outliers are model failures (choices of wrong models). It also happens not infrequently that only part of the data obeys a different model. A single outlier which is sufficiently far away can ruin, for example, a least-squares analysis completely; some sources for gross errors such as keypunch errors or wrong decimal points do indeed easily change values by orders of magnitude; and with the regrettable modern trend of putting masses of data unscreened into the computer, outliers can easily escape attention if no precautions are taken.

Some reasons for gross errors are copying errors, interchange of two values or groups of values in a structured design, inadvertent observation of a member of a different population, equipment failure, and also transient effects. With sufficient care, gross errors can often be prevented, and indeed there appear to exist—though probably rarely—high-quality data sets of thousands of data without gross errors; but the necessary care cannot always be afforded. Moreover, even with fully automatic data recording and properly working equipment, there may be large transient effects (cf. the example below). Distant gross errors are one of the most dangerous deviations from usual statistical assumptions; but they are also the deviation which can most easily be treated. The number of distant gross errors and other outliers which statisticians get to see is frequently decreased considerably below the original one because subject matter scientists often "clean" their data in some crude way before consulting the statistician. Even so, the frequency of gross errors varies considerably. Crudely speaking, one has to distinguish between high-quality data with no or virtually no gross errors, and routine data, with about 1–10% or more gross errors. Whenever a distant outlier occurs, *some* robust method is definitely needed, and be it just a subjective look at the data with subsequent special treatment of the outlier. On the other hand, for high-quality data, there is usually only a small increase in efficiency by the use of robust methods. This is not surprising and should not induce those who have the fortune of being able to work with high-quality data to deem robust methods altogether unnecessary. Moreover, even for high-quality data, the improvements possible by the use of good robust methods may still be important in certain cases.

Let us now consider some examples for the frequency of gross errors and other outliers in real data. As mentioned, there are many high-quality data sets where none can be found. The smallest positive fraction of outliers the writer came across was about 0.01%, or 1 in 10,000. One example occurred in the 1950 U.S. census data *after* careful screening and analysis of the data and was found by sheer luck; the fascinating detective story by Coale and Stephan (1962) shows how a tiny fraction of wrongly punched data could escape all controls since they were entirely possible, but produced some weird patterns in certain rare population subgroups. Another example (Huber, orally) was about a quarter million electroencephalographic data, fully automatically recorded by properly working equipment; the histogram looked beautifully normal, except for some jittering of the plotter way out in the tails, but the third and fourth moment were far too large. A search revealed that there was a huge spike of about two dozen data points when the equipment was switched on; these few transient points caused the high moments and the jitter in the plot. The point is that even in highest-quality data settings, there may be tiny fractions of outliers which sometimes may be hard to find, but which may upset parts of the statistical results if left untreated.

An example of 0.2% gross errors, again in a high-quality setting, are the waveguide data (Kleiner et al., 1979; Mallows, 1979): in a very smooth time series of about 1000 points, twice the measuring instrument duplicated the previous value, thereby completely distorting the low-power regions of the classical power spectrum.

An example where gross errors could be individually identified by comparing related data, are the seismogram readings in Freedman (1966), with 5–7% gross errors in routine data in science. Cochran (1947) cites an agricultural 4^2 with a large and impossible gross error (6%) that first went unnoticed. An agricultural analysis of covariance cited by Scheffé (1959, p. 217) where probably 7% (2 out of 30) values are interchanged, is discussed in Hampel (1976, 1985). The famous and frequently discussed "Bills of Mortality" contain 7–10% gross errors that apparently went unnoticed for centuries (Zabell, 1976). At the 1972 Oberwolfach meeting on medical statistics, A. J. Porth reported 8–12% gross errors as a result of a spot check on medical data in a clinic. (An unsubstantiated oral source claimed even 25% for U.S. clinics.) A very high fraction of gross errors (in this case copying errors and "corrections" by means of wrong theories) is contained in 50 ancient Babylonian Venus observations (Huber, 1974a)—certainly about 20%, probably 30%, and perhaps even 40%.

There are also some collections of different data sets which have been investigated for gross errors. Rosenthal (1978) cites 15 data sets from the behavioral sciences ranging from about 100 to about 10,000 observations and analyzed in final form, including any checking deemed necessary, by the respective researchers to their own satisfaction. Only one set (with about 700 data) contained no gross error after the "final" analysis—the maximum was 4% (in about 100 data) and the average about 1% gross errors.

A wealth of examples from industry, science, agriculture, and so on mostly taken from the literature (and *not* collected with the intention to provide examples for gross errors!) can be found in the books by Daniel and Wood (1971) and Daniel (1976). The frequency of outliers in the regression examples of the first book ranges from 0 to the 19% (4 in 21) transient values of the famous stack loss data, though 0 and 1% are more typical values. The second book considers about two dozen examples from the analysis of variance, about half of them in greater detail, and most of them published previously. About one-third shows no gross error (but often other peculiarities); almost one-half contains 3–6% gross errors, and, depending on one's selection criteria, about one-sixth to one-third has more than 10% (up to one-third) gross errors and deviating substructures (e.g., a simple model fitting only two-thirds of the data).

Such partial model failures should not be confused with gross errors, but both lead to outliers which should be detected; in the former case, these outliers are proper and often highly important observations. As an example,

Rayleigh noted that 7 out of 15 atomic weights of "nitrogen" (from air and from various chemical compounds) fell into a separate group and thus was led to the discovery of argon (cf. Tukey, 1977, p. 49). It is the statistician's task to try to fit a model as simple as possible, and if this can be done only for part of the data, he or she should discuss the remaining substructures separately, rather than fitting a complicated model to all data and thereby obscuring interesting features of the data. A simple model may fit to all data, or all but fractions of a percent, or less than half of the data. These problems of model fitting point in the direction of pattern recognition and cluster analysis and often need more than statistical routine techniques.

General statements about the frequency of gross errors seem to be rather uncommon; this is not suprising since so much depends on the specific circumstances. Paul Olmstead, cited by Tukey (1962, p. 14), maintains that engineering data typically involves about 10% "wild shots." Discussing various types of industrial data in his 1968 Berkeley courses, Cuthbert Daniel considered frequencies from less than 1% up to 10 and 20% as usual and cited, as rather exceptional, a set of about 3000 data points where he could not find anything wrong. For a high-quality sample survey, Freedman et al. (1978, p. 367) cite error rates of 0.5–1% interviewer errors as typical. There are cases (J. Kohlas, orally; A. W. Marshall, orally) where failure data are kept secret by companies, even from their own statisticians who are supposed to work on them. This shows how touchy an information the percentage of gross errors can be, and how there sometimes may be a tendency to downplay it or to shroud it into dubious disinformation. It is clear that the care with which data are obtained and processed is often one important factor for the frequency of gross errors, and it is not difficult to imagine that with moderate care, every 100th or even every 30th number comes out grossly wrong. With great care, the percentage of blunders can be pushed below 1%, even to 0%, though transient effects and other partial model failures are still a possible cause for outliers. Routine data treated with little individual care under great pressure of time can easily contain 10% or more stray values. We have to accept the whole range of data quality, as well as the different causes for outliers of which gross errors are but one, though the most conspicuous one; altogether, 1–10% gross errors in routine data seem to be more the rule rather than the exception.

1.2d. The Effects of Mild Deviations from a Parametric Model

Mild deviations from normality (or other parametric models), such as they occur in high-quality data, do not have the catastrophic effects of distant outliers on classical methods like mean, variance, or least squares; hence robust methods are not absolutely necessary in such a situation. However, the avoidable efficiency losses of classical methods can be much larger than is often naively assumed. Some typical values are not so much 90 or 99% efficiency loss; rather they may be more like 3 or 30%, but 30% efficiency loss is probably more than most believers in classical parametric procedures are willing to pay consciously.

Already, Fisher (1922) stressed the inefficiency of the arithmetic mean and variance in all but a very small region of Pearson curves around the normal. His argument was directed against K. Pearson's method of moments, but it is equally valid against the uncritical use of normal theory methods where there is no guarantee for normality (i.e., practically always). Fisher's method of maximum likelihood is equally unsafe in general unless the model is known exactly or long-tailed by itself; but we shall see that this method can be "robustified" so that it becomes reliable over a wide range of potential true models, with only a marginal loss of efficiency at the assumed ideal model (see Chapters 2 and 4).

While Fisher in this context makes no comparison with real data, the findings by Jeffreys and others mentioned above [Subsection 1.2b (iii)] suggest t-distributions with 3–9 degrees of freedom as the Pearson curves mimicking high-quality data. But the asymptotic efficiency of the arithmetic mean under t_ν is $1 - 6/\nu(\nu + 1)$ and that of the variance is $1 - 12/\nu(\nu - 1)$ (Fisher, 1922). Hence for t_9, t_5, and t_3 the efficiency of $\overline{X}$ is 93, 80, and 50%, respectively, and that of the variance and standard deviation is 83, 40, and 0%!

Other striking examples are due to Tukey (1960) who considers the location model $F(x - \theta)$ with $F(x) = (1 - \varepsilon)\Phi(x) + \varepsilon\Phi(x/3)$, Φ being the standard normal cumulative and ε ranging from 0–10%. This can be thought of as a mixture with gross errors of a realistic amount, but with unrealistic symmetry and closeness to the "good" data; or it can be thought of as a slight "fattening" of the "flanks" and close tails of a normal, as in high-quality data, still giving a distribution with quickly vanishing far tails and all moments. Tukey shows that the asymptotic efficiency of the mean decreases quickly from 1 (for $\varepsilon = 0\%$) to about 70% (for $\varepsilon = 10\%$), while that of the median increases from $2/\pi$ to about the same value and while other estimators (notably the "6%-trimmed mean", see Subsections 2.2b and 2.3b) have at least about 96% efficiency over the whole range.

The case for the variance is even worse. Tukey (1960) recalls the dispute between Fisher (1920) and Eddington about the use of standard deviation and mean deviation; but while the mean deviation is only 88% efficient at the strict normal, it suffices to take $\varepsilon = 0.18\%$ (!) in Tukey's contamination model to make the mean deviation more efficient than the standard deviation (cf. Huber, 1981, p. 3). This is a drastic example for the instability of relative efficiencies of two estimators, one of which is optimal at the normal, but both of which are nonrobust in a technical sense and neither of which is recommended.

We note in passing that this is also an example where concepts of robustness theory provide intuitive insight into the difference between two nonrobust estimators: the influence of outliers, as measured by the influence function (cf. Subsections 1.3d and 2.1b), increases quadratically for the standard deviation, but only linearly for the mean deviation, causing a noticeably higher "stability" of the latter, although both are qualitatively unstable. Basically the same phenomenon is also found in the quicker deterioration of the variance compared with the mean under t-distributions (cf. the numbers above) and in some aspects of the replacement of the F-test for equality of two variances by Levene's t-test applied to the absolute residuals (for which, however, the stability of the level is still more important).

Tukey (1960) remarks that Fisher was the only statistician queried by him who anticipated large effects of small contamination. He also shows that it is virtually impossible to distinguish between $\varepsilon = 0\%$ and $\varepsilon = 1\%$ even with a sample size of 1000, unless one or a few points arising from the contaminating distribution $\Phi(x/3)$ are rather extreme.

This brings us to an argument which is often presented roughly as follows: "Of course I would first throw away any outliers, but then I can use the classical methods." The overall procedure is then robust and not classical in the strict mathematical sense, even though the conditional procedure given that there are no outliers may be the classical one. Another question is how efficient such a combined procedure can be. It is precisely the doubtful outliers stemming from slightly elongated tails, not the distant outliers, which are hardest to deal with; and since even in the artificial model of "good," precisely normal data plus "bad" contamination there

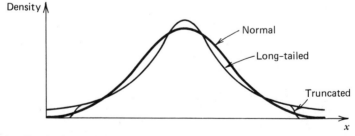

Figure 2.   Sketch of normal, long-tailed, and stochastically truncated long-tailed distribution.

will be errors of both kinds (wrong rejections and wrong detentions), no full efficiency can be expected even conditionally, after the (potential) rejection. In fact, a Monte Carlo study for sample size 20 showed that the better half of 14 classical rejection rules followed by the mean has about 15–20% avoidable efficiency loss under t_3, compared with better robust estimators which decrease the influence of doubtful outliers smoothly, not suddenly (Hampel, 1976, 1977; cf. also Hampel, 1978a, 1980 and see Section 1.4b below). Virtually the same results (about 15–20% efficiency loss) were found by Relles and Rogers (1977) in an interesting Monte Carlo study for subjective rejection. This shows that even if the data are "precleaned" by subjective or objective rejection methods, the remaining small deviations from normality (some of them only due to the rejection) can still easily cause avoidable efficiency losses of several tenths.

1.2e. How Necessary Are Robust Procedures?

A glance through the literature yields rather conflicting statements about the importance and necessity of robust procedures. Some authors, like Stigler (1977) and Hill and Dixon (1982), recommend only slightly robustified estimators, such as slightly trimmed means, and find them hardly superior to the arithmetic mean. Others, like Mallows (1979) and Rocke et al. (1982), give examples where very robust estimators are needed and are far superior to classical methods. Sometimes the same author finds examples for both situations (Spjøtvoll and Aastveit, 1980; Spjøtvoll and Aastveit, 1983).

Basically, there is no contradiction: for high-quality data or at least precleaned data without any outliers, robust methods are no more absolutely necessary; their main contribution lies hidden in the precleaning of the data from all outliers if there were any, or in the finding that the data are of high quality if there were no outliers. One might also take the risk of just hoping that there are no outliers and using classical methods without any check, dangerous and foolish as this may be. Even for high-quality data, good robust methods may still give a noticeable improvement over classical ones, as we have seen, but the size of this improvement is a second-order problem in practice and will differ from situation to situation. One would have to discuss details of the various studies.

For example, much can be said about Stigler's (1977) study, and much has already been said in the discussion following the paper; but, unfortunately, the comparisons given appear to be vacuous, since they compare essentially pseudorandom biases linearly, as was demonstrated by Pratt in the discussion and as was *not* refuted by the theorem in the reply. (The theorem was explicitly misinterpreted: "nondecreasing" is not the same as "increasing.") More revealing is Eisenhart's analysis in the discussion: not only are all robust estimators close together, and only the mean (and the "outmean") appears as an outlier (in full accordance with the example in Hampel, 1973a, and as in about half of Stigler's other data sets, if the median is ignored), but moreover the scientist Newcomb decided to choose a "robust" summary value and not one near the mean. More informative is also the figure in the paper which shows that if a t-distribution is fitted to the data sets, it should have less than 5 degrees of freedom, in full accordance with the results in Subsection 1.2b. It may be surmised that in a more meaningful analysis of these high-quality data the mean would still come out fairly good in absolute performance, though clearly worse than has been claimed.

In situations with outliers, the need for robust procedures is so obvious that it requires no further justification. A striking example is the waveguide time series discussed in Kleiner et al. (1979) where 0.2% tiny "outliers" destroy a large part of the power spectrum and reduce its sensitivity by several powers of ten. It should not be forgotten that outliers can occur even under ideal circumstances, and that they may be found and treated also by ad hoc subjective methods, if the necessary effort (and brain power!) can be afforded.

It seems the word is slowly spreading that the chi-square test and F-test for variances, as well as tests for random-effects models in the analysis of variance, are highly susceptible to slight nonnormality, in the sense that their level becomes very inaccurate; but many statisticians seem still to be unaware of these facts, more than half a century after E. S. Pearson's (1931) work and more than a quarter of a century after the papers by Box (1953) and Box and Andersen (1955) where they, incidentally, introduced the term "robust." Many more statisticians still believe in the robustness of the analysis of variance, at least for fixed-effect models. They are partly right in that the level of the tests in this case is fairly (though not very) robust; but the power (which is usually forgotten) is not. To be sure, the simplicity and elegance of least-squares methods is a strong temptation; and if distant outliers are treated by ad hoc methods, as often done by very good data analysts, the remaining losses of efficiency will be quite bearable. (They will be roughly in the 3–30% range rather than arbitrarily high.) It should also be remembered that robustness is only one aspect of good data analysis, and that other aspects like model choice, interpretation of outliers, interpretation of the main results, and so on are at least equally important.

One frequent "argument" in connection with least squares should still be mentioned, since it is no argument at all: namely that the Gauss–Markov theorem renders the least-squares estimators optimal among all linear

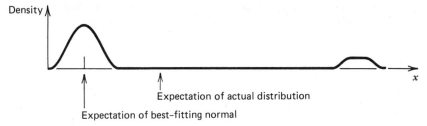

Figure 3.   Sketch of different aims of robust and nonparametric statistics.

estimators even if nothing is assumed about normality. The point is that *all* linear estimators are terribly bad except in a small neighborhood of the normal, as was shown already by Fisher (1922). Rejection of outliers is a nonlinear operation, and only if there are no outliers left are least-squares methods of moderate to good quality.

Probably one of the statements in the robustness literature which is most often misquoted is the statement that "the arithmetic mean, in its strict mathematical sense, is 'out'" (Andrews et al., 1972, p. 240). Those who deny it imply that they would use the mean in a parametric situation even if there are distant outliers present. (It may still be necessary to use the mean in a nonparametric situation when the true expectation of a completely unknown distribution is to be estimated, but such situations are rarer than it seems at first glance and have to be considered as ill-posed problems.) Those who cite the above sentence in isolation also seem to overlook the next sentences on p. 240 and the remarks on p. 243f. in Andrews et al. (1972) implying that the mean combined with any decent (subjective or objective) rejection rule (and this is the "mean" as used by good practicing statisticians) actually can survive and that for high-quality data (or after rejection) it is not even "mildly bad."

In summing up, we can answer the question of this subsection as follows:

Some robust methods (such as subjective or objective rejection of outliers, or rank-based methods) are necessary to prevent disasters due to distant outliers.

Good robust methods, as developed mainly in the last two decades, are necessary to prevent avoidable efficiency losses of, say, 3–30% or more.

Furthermore, the higher the dimension and complexity of the data set, the less suitable are subjective and simple ad hoc methods, and the more necessary are safe modern robust methods.

1.3. THE MAIN APPROACHES TOWARDS A THEORY OF ROBUSTNESS

1.3a. Some Historical Notes

Robust methods probably date back to the prehistory of statistics. Looking at the data and rechecking conspicuous observations is a step towards robustness; excluding highly deviant values is an informal robust procedure. The median is robust. Switching from the mean to the median in view of long-tailed data is a robust method. The mode (for discrete or grouped data) may be robust; it is so if the most probable value is clearly more probable than the second most probable value ("clearly" may have different meanings, depending on the formalization of the robustness problem chosen, in particular whether an asymptotic or a finite-sample one). Thus, the Greek besiegers of antiquity (cf. Rey, 1983) who counted the number of brick layers of the besieged city's wall and then took the mode of the counts in order to determine the necessary length of their ladders, may well have used a robust method.

   The discussion about the appropriateness of rejection of outliers goes back at least as far as Daniel Bernoulli (1777) and Bessel and Baeyer (1838), while Boscovich (1755) is known to have rejected outliers which course of action, according to Bernoulli (1777), was already a common practice among astronomers of their time. Also the trimmed mean has long been in use—see "Anonymous," 1821 (Gergonne, according to Stigler, 1976) or Mendeleev, 1895; it might be noted that outside sciences an asymmetrically trimmed mean (leaving out the worst value) also plays a role in the evaluation of certain sport performances with several judges some of whom might be biased. The first formal rejection rules apparently were given by Peirce (1852) and Chauvenet (1863), followed by Stone (1868), who uses a "modulus of carelessness," Wright (1884), Irwin (1925), Student (1927), Thompson (1935), Pearson and Chandra Sekar (1936), and many others.

   Student (1927) actually proposed repetition (additional observations) in the case of outliers, combined possibly with rejection, a refined technology which seems to have drawn little attention by statisticians, although related techniques seem to be rather common in parts of the sciences. One simple such technique (which the writer came across in consulting), which assumes prior information on the error, consists in taking two observations and, if they are too far apart, making a third one and taking the mean of the two closer ones. Such techniques need of course the possibility of making further observations, but if available, this class might be noticeably better for small samples than customary rejection rules. Sequential planning of this kind may be quite practical; other instances where it is called for are the breaking of confounding patterns in the analysis of variance, and the treatment of leverage points in multiple regression.

There are also early considerations of mixture models and of estimators (apart from trimmed means) which only partly downweight excessive observations (Glaisher, 1872–73; E. J. Stone, 1873; Edgeworth, 1883; Newcomb, 1886; Jeffreys, 1932, 1939). Newcomb posthumously (cf. Stigler, 1973) even "preinvented" a kind of one-step Huber-estimator. These attempts to "accommodate" the outliers, to render them harmless rather than to isolate and discard them, are very much in the spirit of modern robustness theory.

Some accounts of this early work are given in Barnett and Lewis (1978), Harter (1974–76), Huber (1972), and Stigler (1973).

After the normal distribution gained a central role as model distribution in the 19th century (cf. Gauss, 1823, for the motivation of its introduction), the system of Pearson curves by K. Pearson may be considered as a "supermodel" with two additional parameters, designed to accommodate a major part of the deviations from normality found in real data. Even though Pearson was probably thinking more of biological variation, and the distributions of data contaminated by gross errors may be quite different, good methods for Pearson curves still work in the presence of gross errors, basically since the Pearson system contains very long-tailed distributions. The same is true for many other "supermodels," including Bayesian ones of more recent origin.

The problem of unsuspected serial correlations (violation of the independence assumption) also found early attention (cf. Section 8.1).

With the success of Fisher's exact small-sample theory for the normal distribution, the latter regained a central position. However, soon after the first triumphs of the theory for exact parametric models, E. S. Pearson (1931) discovered the drastic nonrobustness of the tests for variances. His work (which included one of the early simulation studies—without computer!) can be seen as the beginning of systematic research into the robustness of tests. Techniques included moment calculations and use of Pearson curves and Edgeworth expansions, as well as sampling experiments. Some references are E. S. Pearson (1929, 1931), Bartlett (1935), Geary (1936, 1947), Gayen (1950), Box (1953), and Box and Andersen (1955). A useful discussion of this line of development can be found in Scheffé (1959, Chapter 10). During about the same period, "nonparametric" methods became popular, in particular rank tests (cf. Hodges and Lehmann, 1956, 1963), while permutation tests (randomization tests) were mainly used in theoretical work, since the feasibility of their application had to wait for the computer and the idea of randomly selected permutations. However, in most of the work on robustness of tests, the level of the test ("robustness of validity") was in the foreground, while the power ("robustness of efficiency") was all but ignored. The problems were difficult enough without considering the power. But it is easy to overlook that classical randomization tests "typically" have very little power when outliers are present (even though the level is being kept), and also some rank tests, such as the normal scores test, may have a fairly rapidly decreasing power if contamination is added. Another aspect which has to be kept in mind while reading robustness statements based on expansions and moments is that the neighborhoods in which the alternative distributions were lying were usually rather narrow; they may perhaps be suitable for high-quality data, but even a single outlier can cause arbitrarily high moments. Hence data with just one such outlier would be considered extremely "far away" from the model distribution. If it is argued that this should be so, for example, because the outlier is clearly visible, then one has to make precise that these expansions are only valid and useful *after* rejection of distant outliers, that is, after the first and most important step of robustification, or else under the condition of high-quality data.

The systematic attack on the problem of robust estimation started later than that on testing, be it because at that time testing was more popular with mathematical statisticians because of its elegant mathematical properties, be it because E. S. Pearson's findings caused much concern about testing with no corresponding obvious need in estimation, be it for other reasons. It was Tukey (1960) who, in summarizing earlier work of his group in the 1940s and 1950s, demonstrated the drastic nonrobustness of the mean and also investigated some useful robust alternatives (cf. also Subsection 1.2d). His work made robust estimation a general research area and broke the isolation of the early pioneers. Among a growing flood of papers which are too numerous to be reviewed here, were the first attempts at a manageable *and* rather realistic and comprehensive theory of robustness by Huber (1964, 1965, 1968) and Hampel (1968). Their main concepts and ideas will be discussed in the next subsections. These subsections may be read parallel with the more technical definitions and discussions given in Chapter 2.

1.3b. Huber's Minimax Approach for Robust Estimation

Huber's (1964) paper on "Robust estimation of a location parameter" formed the first basis for a theory of robust estimation. It is an important pioneer work, containing a wealth of material of which we can discuss only a part (cf. also the discussion in Section 2.7).

What are the main features of Huber's paper?

Huber introduced a flexible class of estimators, called "M-estimators," which became a very useful tool, and he derived properties like consistency and asymptotic normality. These estimators are just a slight generalization of maximum likelihood estimators: instead of solving $-\Sigma \log f(x_i - T) =$ min or $\Sigma f'/f(x_i - T) = 0$ for the location estimate T of θ, with the x_i distributed independently with density $f(x_i - \theta)$, Huber solves $\Sigma \rho(x_i - T)$ = min or $\Sigma \psi(x_i - T) = 0$ without assuming that ρ and ψ are of the form $-\log f$ or $-f'/f$ for any probability density f; and even if they are, f need not be the density of the x_i. In another paper, Huber (1967) calls them "maximum likelihood estimates under nonstandard conditions." The simple step of severing the tie between a maximum likelihood estimator and the model under which it was derived has far-reaching consequences. (It might be noted that in different contexts, and under different names, M-estimators were also proposed by other statisticians, such as Barnard and Godambe.)

Huber then introduced the "gross-error model:" instead of believing in a strict parametric model of the form $G(x - \theta)$ for known G (e.g., $G =$ the standard normal), he assumes that a (known) fraction ε ($0 \le \varepsilon < 1$) of the data may consist of gross errors with an arbitrary (unknown) distribution $H(x - \theta)$ (where θ is only introduced to preserve the form of a location model). The distribution underlying the observations is thus $F(x - \theta) =$ $(1 - \varepsilon)G(x - \theta) + \varepsilon H(x - \theta)$. This is the first time that a rather full kind of "neighborhood" of a strict parametric model is considered. Earlier work and also much later work on robustness either looks at a finite number of alternative models (e.g., normal, logistic, Cauchy) or enlarges the parametric model by introducing one or a few additional parameters (e.g., Tukey's 1960 model of the mixture of two normals, or the family of "Edgeworth distributions," using the first two terms of an Edgeworth expansion). Huber also considers much more general neighborhoods in an abstract setting, and he treats in detail the case of the Kolmogorov-distance neighborhood, consisting of all distributions whose Kolmogorov distance from some $G(x - \theta)$ (G assumed to be normal) is $\le \varepsilon$.

Huber's aim is to optimize the worst that can happen over the neighborhood of the model, as measured by the asymptotic variance of the estimator. He has to make some restrictions in order to be able to ignore or at least control the asymptotic bias (which in real life is unavoidable). But then he can use the formalism of a two-person zero-sum game: Nature chooses an F from the neighborhood of the model, the statistician chooses an M-estimator via its ψ, and the gain for Nature and loss for the statistician is the asymptotic variance $V(\psi, F)$ which under mild regularity conditions turns out to be $\int \psi^2 \, dF / (\int \psi' \, dF)^2$. Huber shows that under very general condi-

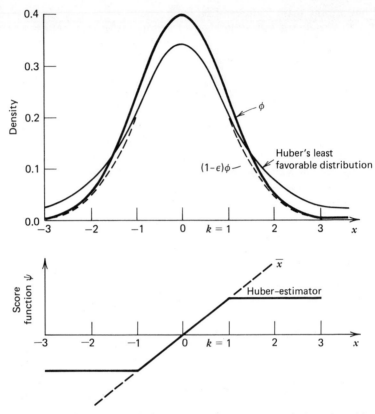

Figure 1. The saddlepoint in Huber's two-person zero-sum game: the least favorable distribution in the gross-error model and its maximum likelihood estimator.

tions there exists a saddlepoint of the game; in the case of the gross-error model, it consists of what has been called Huber's least favorable distribution, which is normal in the middle and exponential in the tails, and of the famous Huber-estimator, with $\psi(x) = \max(-k, \min(k, x))$, as the maximum likelihood estimator for the least favorable distribution and the minimax strategy of the statistician (cf. also Example 2 and Fig. 1 in Subsection 2.3a). More precisely, he obtains the class of Huber-estimators, with $k > 0$ depending on ε. Limiting cases, with $\varepsilon \to 0$ and 1 and $k \to \infty$ and 0, respectively, are arithmetic mean and median.

In the same paper, Huber (1964) also treats the problem of a scale parameter and of simultaneous estimation of location and scale; Huber (1973a) treats robust regression (though still without the problem of leverage points), and Huber (1977b) treats robust covariance matrices by means of the minimax approach, to mention its most important generalizations. Important surveys are Huber (1972, 1977a); see Huber (1981) for more details.

Some common objections against Huber's (1964) paper are discussed in Subsection 8.2a which can be read profitably at this point.

1.3c. Huber's Second Approach to Robust Statistics via Robustified Likelihood Ratio Tests

In the classical likelihood ratio test betwen a simple hypothesis and a simple alternative, a single observation (e.g., a gross error) can carry the test statistic to $\pm\infty$ if the likelihood ratio is unbounded. In his 1965 paper, Huber arrives at censored likelihood ratio tests which put a bound (possibly asymmetric) from above and below on the log likelihood ratio of each observation. He derives these tests by blowing up the hypotheses to various (nonoverlapping) neighborhoods (e.g., with respect to ε-contamination, total variation, Lévy distance, or Kolmogorov distance), and by looking for tests which maximize the minimum power over all alternatives, given a bound on the level for all members of the null hypothesis. Such a maximin test is often the ordinary likelihood ratio test between a least favorable pair of hypotheses which are "closest" to each other in the two neighborhoods and make the testing problem hardest; and the existence of such solutions is very closely tied to the question whether the neighborhoods can be described by "2-alternating Choquet capacities" (this is the famous main result of Huber and Strassen, 1973). These capacities are set functions which are more general than probabilities; in particular, the addition law of probabilities is replaced by inequalities. Capacities can be used to define sets of probability measures, namely by taking all probabilities which lie below (or above) a capacity (pointwise for all events). It turns out that the most common neighborhoods can all be described by 2-alternating capacities in this way (Huber and Strassen, 1973), making a mathematically elegant and deep theory nicely applicable.

The robust testing method can also be used to derive robust confidence intervals and point estimates of location (Huber, 1968): Given the length $2a > 0$ of the confidence interval, look for the estimate T which minimizes the maximum probability (over the neighborhoods of the parametric model distributions) of overshooting or undershooting the true θ by more than a. The estimate can be derived via a maximin test between $\theta = -a$ and $\theta = +a$. In the case of the normal location model, the optimal robust estimators in this sense are again the Huber-estimators (while for other location models the solution is different from the minimax solution). This approach yields exact (not asymptotic) solutions for every finite sample size, and there is no assumption of symmetry involved.

The difficult, elegant, and deep results using Choquet capacities had a strong impact on many mathematical statisticians; robustness theory became much more respectable in their eyes, and there is quite a bit of mathematical work continuing along these lines. However, the main drawback of this approach is its limited applicability; already the change from simple to composite parametric hypotheses and from location estimation with known scale to estimation with scale as a nuisance parameter, seems to present a barrier for further progress, not to mention more general estimation problems. Thus, from the point of view of practical relevance and applicability, this approach has been far less successful so far than the other two approaches discussed.

1.3d. The Approach Based on Influence Functions

This approach, which will be discussed in detail in this book, has also been called the "infinitesimal approach," although it also comprises an important global robustness aspect, namely the "breakdown point." It goes back to Hampel (1968), whose main results (about two-thirds of the thesis) were published in Hampel (1971, 1974). Survey articles are Hampel (1973a, 1978a), an elementary survey is Hampel (1980) (cf. also Hampel et al., 1982). In 1976 and 1977 the main optimality result was extended from the one-dimensional case to general parametric models, first in several unpublished talks and manuscripts by Hampel and by Krasker, later published in Hampel (1978a) in a preliminary form (with a misprint and a false claim, cf. Stahel, 1981a) and in Krasker (1980) with the first published proof in a special case (cf. Stahel, 1981a for an extensive discussion). Ronchetti (1979), Rousseeuw (1979), and Rousseeuw and Ronchetti (1981) have generalized the approach from estimators to tests. Since the technical details will be given later, we shall now try to give only a short, nontechnical overview.

The infinitesimal approach is based on three central concepts: qualitative robustness, influence function (with many derived concepts), and breakdown point. They correspond loosely to continuity and first derivative of a function, and to the distance of its nearest pole (or singularity). The approach is facilitated by the fact that many statistics depend only on the empirical cumulative distribution function of the data (or on the ecdf and n); in the language of functional analysis, they can therefore be considered as functionals on the space of probability distributions (or replaced by functionals for each n and usually also asymptotically), allowing the application of concepts like continuity and derivative.

Qualitative robustness is defined as equicontinuity of the distributions of the statistic as n changes; it is very closely related to continuity of the statistic viewed as functional in the weak topology. The precise relations between these two concepts and another very similar one, namely "Π-robustness," have been clarified by some theorems (Hampel, 1971; cf. also Huber, 1981; cf. also Subsection 2.2b). Qualitative robustness (or any of its variants) can be considered as a necessary but rather weak robustness condition; it eliminates already many classical procedures, but it tells us nothing about the differences between qualitatively robust procedures.

The richest quantitative robustness information is provided by the influence curve or influence function and derived quantities. Originally, it was called "influence curve" or IC in order to stress its geometric aspects as something that ought to be looked at and interpreted heuristically; later it was more often called "influence function" or IF, mainly in view of its generalization to higher-dimensional spaces. The influence function describes the (approximate and standardized) effect of an additional observation in any point x on a statistic T, given a (large) sample with distribution F (cf. the exact definition in Subsection 2.1b). Roughly speaking, the influence function $\mathrm{IF}(x; T, F)$ is the first derivative of a statistic T at an underlying distribution F, where the point x plays the role of the coordinate in the infinite-dimensional space of probability distributions. We thus can use a one-step Taylor expansion to replace our statistic T locally by a linear statistic (provided T is not "too nonlinear"). This gives us a simple and powerful tool. Since the true underlying distribution is assumed to lie in some neighborhood of the parametric model, and since for large n the empirical cumulative distribution will be close to the true underlying one, we can hope to get a lot of information on T in a full neighborhood by just studying the influence function and its properties at the exact parametric model. For larger deviations from the model, as they are more frequent for small n, we can either study the IF around some alternative away from the model, or integrate the IF with respect to F over a path of increasing contamination, or just make a bolder extrapolation from the model (which in fact will often suffice, as numerical examples indicate).

We need some guidance up to what distance from the model the local linearization provided by the IF can be used, and this guidance is provided by a simple quantitative global robustness measure, namely the breakdown point. Loosely speaking, it is the smallest fraction of gross errors which can carry the statistic over all bounds. Somewhat more precisely, it is the

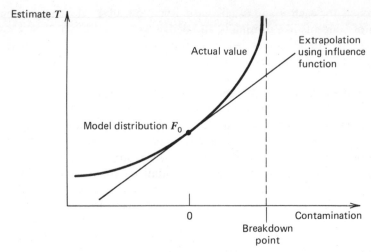

Figure 2. Extrapolation of a functional (estimator), using the infinitesimal approach. (Symbolic, using the analogue of an ordinary one-dimensional function.)

distance from the model distribution beyond which the statistic becomes totally unreliable and uninformative (cf. Subsection 2.2a for the formal definition). For example, the arithmetic mean and the median have breakdown points 0 and $\frac{1}{2}$, and the α-trimmed mean, which first takes away the αn smallest and the αn largest values before taking the mean (cf. Example 3 of Subsection 2.2b), has breakdown point α. In general, the breakdown point can lie between 0 and 1; a positive breakdown point is closely related to, though not identical with, qualitative robustness. An empirical rule of thumb, at least for M-estimators in simple location models, says that the linear extrapolation by means of the influence function tends to be quite accurate for distances up to one-quarter of the breakdown point, and it still seems to be quite usable for distances up to one-half of the breakdown point. Apart from these helpful indications, the breakdown point measures directly the global reliability of a statistic, one of its most important robustness aspects.

Huber (1972) likened these three robustness concepts to the stability aspects of, say, a bridge: (1) qualitative robustness—a small perturbation should have small effects; (2) the influence function measures the effects of infinitesimal perturbations; and (3) the breakdown point tells us how big the perturbation can be before the bridge breaks down.

There are a number of fine points, which have often led to misunderstandings, about the relation of the influence function, which is not a proper derivative in the sense of functional analysis, to such derivatives, as well as to qualitative robustness, which is not quite equivalent to boundedness of the influence function (cf. Hampel, 1968, 1971, 1974; Huber, 1981). There are also close and interesting relationships to and between the expansion of von Mises functionals, the expansion of Hoeffding's U-statistics, Hájek's projection method, and Tukey's jackknife, which were worked out by Mallows (1971). However, we shall now only present some of the concepts that can be derived from the IF. The two most important norms of the IF are the sup-norm over x, called the "gross-error sensitivity" γ^* as the central local robustness measure, measuring the maximum bias caused by infinitesimal contamination and thereby the stability of T under small changes of F, and the L_2-norm with respect to F, namely $\int IC^2\, dF$, yielding the well-known asymptotic variance of estimators (or the inverse efficacy of tests) as the basic efficiency measure. Both norms depend again on F, and can thus be considered as new functionals, and their infinitesimal stability (suitably standardized) can now in turn be measured by the "change-of-bias function" CBF (or "change-of-bias curve") and the "change-of-variance function" CVF (or "change-of-variance curve"). The sup-norms of these two functions, again providing simple summary measures, are "change-of-bias sensitivity" and "change-of-variance sensitivity" (cf. Hampel, 1973a, 1983a and Hampel et al., 1981). While the CBF has not yet been used much, except qualitatively for bounding the negative slope of redescending M-estimators, the CVF can be used to extrapolate the asymptotic variance at the model to a full neighborhood of it.

The most important robustness requirements, besides qualitative robustness, are a high breakdown point and a low gross-error sensitivity. (Some regularity conditions which imply qualitative robustness are needed for the sensitivity to be meaningful.) A breakdown point of $\frac{1}{2}$—the best possible

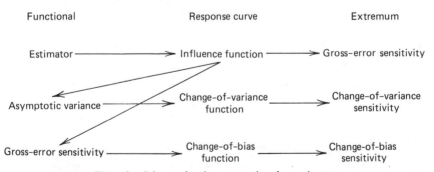

Figure 3. Scheme of various properties of an estimator.

one under some invariance conditions—is often easy to obtain (although there are problems with equivariant estimators in higher dimensions). A low gross-error sensitivity, however, is in conflict with the efficiency requirement of a low asymptotic variance under the parametric model. Both quantities have positive lower bounds, but in general these bounds cannot be reached simultaneously. There is an optimal class of compromise statistics, or "admissible robust statistics," given by Lemma 5 of Hampel (1968) and its generalizations, such that one bound has to be increased if the other one is decreased: the more robustness in this specific sense, the less efficiency, and vice versa. The optimal compromise estimators may also be called "robustified maximum likelihood estimators." The optimality property does not imply that these statistics have to be used, of course; there are practical considerations and other robustness concepts of varying importance (in addition to those already mentioned, also the "local-shift sensitivity" and the "rejection point"; cf. Subsection 2.1c), and a good practical procedure has to be a compromise between all relevant aspects, not just optimal with respect to one or two. On the other hand, the solutions to this particular optimality problem are not only very general, but also quite simple, intuitively interpretable, and without causing uniqueness or consistency problems, so they are actually good candidates for practical use. The solutions for the normal location model are again the Huber-estimators, which are thus optimal in at least three different senses.

The question which statistic to choose from the "admissible robust" class is answered as little by this optimality theory as it is for admissible procedures in decision theory. Practical considerations and numerical aspects come into play. One possibility for choice is to allow a fixed efficiency loss of, say, 5 or 10% under the ideal parametric model. Another possibility arises if the percentage of gross errors can be guessed, so that an approximate minimax strategy becomes feasible, as will be shown in the next subsection. A further path may be to minimize the estimated asymptotic

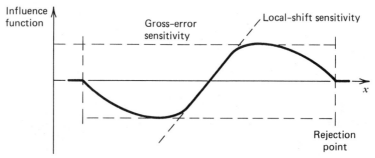

Figure 4.   Sketch of various properties of an influence function.

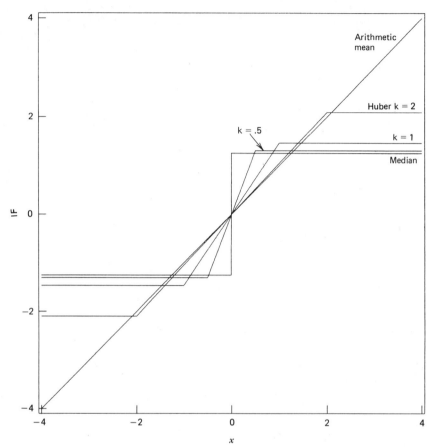

Figure 5. Influence functions of arithmetic mean, median, and some optimal compromises: Huber-estimators with $k = 2, 1, 0.5$.

variance within the optimal class, as in Huber's (1964) "Proposal 3." Contrary to the desire of many statisticians, there is no unique or at least distinguished optimal robust method, except for the extreme solutions, which in the normal location case are arithmetic mean (most efficient, but not robust) and median (lowest gross-error sensitivity, but not very efficient). Even after, for example, the efficiency loss has been fixed, there is yet another choice to be made between all asymptotically equivalent statistics which have the required local properties in infinitesimal neighborhoods of the parametric model, but which may differ considerably at some finite distance from the model. A practical solution is that the exact choice matters little in both cases if done within a reasonable range of possibilities,

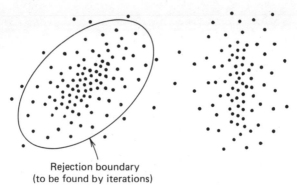

Rejection boundary
(to be found by iterations)

Figure 6.   Analysis of clusters, using redescending M-estimators.

and that in the second case the class of M-estimators is always a simple and reasonable choice.

In analogy to the condition of low gross-error sensitivity, which means a small asymptotic bias or a stable estimand, there is also the condition of a low change-of-variance sensitivity, which means a stable asymptotic variance of the estimate. Asymptotically, the variance tends to zero while the bias stays the same; but for each finite n (unless very large) we ought to combine the two quantities according to their relative importance. There is a similar conflict and a similar class of optimal compromises between the change-of-variance sensitivity and the asymptotic variance. In the normal location model this again yields the Huber-estimators (Rousseeuw, 1981a) —this is the fourth sense in which they are optimal.

The solution becomes especially interesting if the additional side condition is imposed that the influence function vanishes outside a certain region (meaning a low "rejection point" and leading to the "hyperbolic tangent estimators" in the location case—cf. Section 2.6 and Hampel, 1973a, 1974; Hampel et al., 1981). The hyperbolic tangent estimators will become especially important when the influence of nearby data should be made zero as quickly as possible. Some potential applications are the estimation of an approximate mode by means of redescending M-estimators (by the way, with the usual rate of convergence; by decreasing the rejection point with increasing n one may "asymptotically" estimate the strict mode); the estimation of change points in piecewise linear regression; the estimation of center and spread of clusters (or more general patterns) in the presence of other clusters; and perhaps the estimation of the structure of catastrophies and other complicated singularities.

Also the change-of-bias sensitivity leads to optimal robust compromises; in particular, the two- and three-part redescending M-estimators (cf.

Andrews et al., 1972) obviously have optimality properties of this kind (although they may not have been rigorously proven).

The concepts mentioned remain equally useful in general linear models; they only should be augmented somewhat. Suitable standardization of the concepts becomes more important. A new aspect in unbalanced designs is the influence of position in factor space (Hampel, 1973a) which together with the influence of residual forms the total influence. Also, in balanced designs with little redundance, as often in the analysis of variance, it may be useful to consider more refined variants of the breakdown point (Hampel, 1975).

The same concepts are of interest in general nonlinear models. The influence function remains applicable theoretically under some differentiability condition on the model, and practically as long as the curvature of the model is not too strong. Leverage points, namely points with high influence of position in factor space, are of increased interest. Especially important are breakdown aspects, such as necessary redundance in the presence of gross errors, and simple high-breakdown-point procedures which may be considered generalizations of the median, and which are useful as safe starting points for the iterations.

As to the finite-sample distributions for various statistics, the normal approximation is often sufficient for simple robust estimators if they are asymptotically normal at all. In fact, asymptotic normality will often be approached faster by more robust estimators (cf., e.g., the Cauchy in Andrews et al., 1972). Only robust estimators are safe starting points for one-step estimators, which then behave almost like the iterated ones. If (say, for $n < 10$) the normal approximation is not good enough, one can, in many cases, obtain very good "small sample asymptotic" approximations by means of the saddlepoint method and related techniques (Daniels, 1954; Hampel, 1973b; Field and Hampel, 1982; cf. also Section 8.5); these approximations are often excellent even in the extreme tails down to $n = 3$ or 4.

1.3e. The Relation between the Minimax Approach and the Approach Based on Influence Functions

Compare also the similar discussion in Section 2.7.

In the case of the normal location model, both the minimax approach and the infinitesimal approach using M-estimators yield exactly the same class of optimal robust estimators, namely the Huber-estimators, as we have seen. This seems to be a fortunate accident, but there are other examples which indicate that both approaches lead in general to different but numerically very similar classes of solutions. Two nice examples are the normal

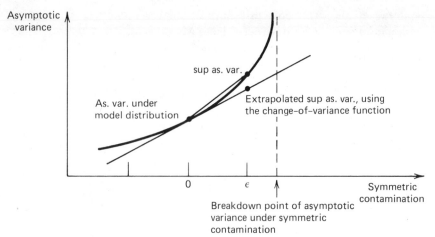

Figure 7. Asymptotic variance and its approximation, using the change-of-variance function. (Symbolic, using the analogue of an ordinary one-dimensional function.)

location model with R-estimators (estimators derived from rank tests) instead of M-estimators (cf., e.g., Hampel, 1983a), and the same model with a bound on the rejection point, leading to two slightly different classes of hyperbolic tangent estimators (cf., e.g., Hampel et al., 1981; also Subsection 2.6c and Section 2.7). This similarity should not be surprising. Both approaches try to do essentially the same, but one uses as "basis" the parametric model and the other one the least favorable distributions. The former approach is simpler and more general and corresponds to replacing a function by its tangent, while the latter approach is slightly more "realistic" in using the actual supremum of the function, but loosens the ties to the parametric model. (Since in a way two parametric models are used in parallel, the original one and the one consisting of the least favorable distributions for all parameter values, it is not so clear whether the estimator should be Fisher-consistent—estimating the right quantity—under the one or the other model.) See also Rousseeuw, 1981a for the relation between the two approaches.

Proponents of the minimax approach have maintained that the infinitesimal approach is totally unrealistic since it gives results only for infinitesimal neighborhoods, while their approach gives results for finite neighborhoods. This amounts to denying the practical power and usefulness of the differential calculus firmly established in mathematics and many fields of application for about three centuries. Two major reasons for using this calculus are just that it is simpler than consideration of the original function *and* that it gives very good results for *finite* neighborhoods. To show the accuracy and parsimony of such linear extrapolations in our statistical context, let us consider Huber's (1964) Table I with the upper bounds on the asymptotic variance of the Huber-estimator $H_1(k)$ with known scale for various values of k (corner point) and ε (symmetric contamination). The ten columns for varying ε can be replaced by just two columns and a simple formula, namely the asymptotic variance under the normal $V(H_1(k), \Phi)$, the change-of-variance sensitivity at the normal $\kappa^*(H_1(k), \Phi)$, and the extrapolation formula $\sup_G V(H_1(k), (1 - \varepsilon)\Phi + \varepsilon G) = V(H_1(k), \Phi)\exp(\varepsilon\kappa^*)$ (cf. Subsection 2.5a); see Table 1. This linear extrapolation (of the log variance) is exact up to three decimal places for $\varepsilon \leq 1\%$; the error is mostly a few permille for $\varepsilon = 5\%$ and mostly less than 1% for $\varepsilon = 10\%$, thus being negligible in the most important region. For $\varepsilon = 20\%$, it is mostly a few percent, and even for $\varepsilon = 50\%$ (the breakdown point if asymmetry were allowed) we obtain the order of magnitude quite reasonably.

Another, more involved example for the accuracy of the extrapolation from the parametric model can be found in Hampel (1983a); it implies that in some cases there were very simple quantitative relationships among the Monte Carlo variances in Andrews et al. (1972) which were overlooked by all analysts of the data, including this writer. A third example is found in Table 1 of Section 2.7. All examples show that there is a lot of almost redundance in the finite-neighborhood and finite-sample behavior of simple statistics like M-estimators.

The quality of the approximation suggests imitation of the minimax approach by means of the infinitesimal approach. In fact, it can be easily mimicked by replacing the exact asymptotic variance (itself only an approximation for finite n) by its extrapolated value. In the one-dimensional case, the approximative minimax problem for symmetric ε-contamination then reads: minimize $\log V(T, F_0) + \varepsilon\kappa^*(T, F_0)$ over T, where V is the asymptotic variance and κ^* the change-of-variance sensitivity of an estimator T under the model distribution F_0. Every solution T minimizes also $V(T, F_0)$ under a bound on κ^* and is thus a solution to a problem posed by the infinitesimal approach. In particular, it turns out that the Huber-estimators are also the *approximate* minimax solutions for the normal location model (this being their fifth optimality property, which is of course closely related to the fourth one).

We can go one step further and pose the approximate minimax problem for asymmetric contamination, by combining a bias term with the asymptotic variance to yield an "asymptotic" mean-squared error. This is particularly interesting because it automatically takes care of the interplay between

Table 1. Accuracy of Extrapolation of Asymptotic Variance to Finite Neighborhood[a]

k	$\kappa^*(H_1(k),\Phi)$	$V(H_1(k),\Phi)$	Maximum variance:	$\varepsilon = 0.01$	$\varepsilon = 0.05$	$\varepsilon = 0.1$	$\varepsilon = 0.2$	$\varepsilon = 0.5$
0.5	2.351	1.263	Exact[b]:	1.293	1.423	1.61	2.11	5.9
			Approx.[c]:	1.293	1.421	1.60	2.02	4.1
1.0	2.938	1.107	Exact[b]:	1.140	1.284	1.495	2.05	6.5
			Approx.[c]:	1.140	1.282	1.485	1.99	4.8
1.5	3.890	1.037	Exact[b]:	1.078	1.258	1.522	2.23	8.1
			Approx.[c]:	1.078	1.260	1.530	2.26	7.3
2.0	5.345	1.010	Exact[b]:	1.065	1.307	1.66	2.6	10.8
			Approx.[c]:	1.065	1.319	1.72	2.9	14.6

[a]Shown for the case of Huber's (1964) Table I.

[b]Exact: Maximum asymptotic variance of Huber-estimator $H_1(k)$ (fixed scale, corner point k) under fraction ε of symmetric contamination, as taken from Huber's (1964) Table I.

[c]Approx.: Extrapolation for maximum asymptotic variance by means of $V \cdot \exp(\varepsilon\kappa^*)$ using only the asymptotic variance $V = V(H_1(k),\Phi)$ and the change-of-variance sensitivity $\kappa^* = \kappa^*(H_1(k),\Phi)$ of the Huber-estimator under the standard normal Φ.

bias and variance as a function of the sample size n. A simple upper bound for the extrapolated asymptotic mean squared error is given by

$$\varepsilon^2 \gamma^{*2}(T, F_0) + V(T, F_0) \exp\big(\varepsilon\kappa^*(T, F_0)\big)/n,$$

where γ^* is the gross-error sensitivity. Consider the normal location model. Then (using results to be proved later) the Huber-estimator with the same asymptotic variance as the T which minimizes this expression (or else the median) has the lowest γ^* and the lowest κ^* among all M-estimators with this variance (or else among all M-estimators), hence it minimizes this expression among all M-estimators.

However, in general it may be that the maximum (approximate) mean-squared error is below this upper bound. (This paragraph is more technical and may be skipped by most readers.) We shall outline a possible proof that at least for M-estimators with fixed scale in the normal location case (and perhaps others) this does not seem to be the case. For this argument we have to use that for an M-estimator described by ψ, the IF $= \psi/\int\psi' \, d\Phi$ and the CVF $= 1 + \psi^2/\int\psi^2 \, d\Phi - 2\psi'/\int\psi' \, d\Phi$ (cf. Subsection 2.5a where the CVF is defined slightly differently without affecting the argument). The mean-squared error for a fraction ε of a contaminating distribution F becomes

$$\varepsilon^2\left(\int\psi \, dF \Big/ \int\psi' \, d\Phi\right)^2 + V(\psi, \Phi)\exp\left(\varepsilon\int\left(1 + \psi^2\Big/\int\psi^2 \, d\Phi - 2\psi'\Big/\int\psi' \, d\Phi\right) dF\right)\Big/n.$$

Assume the supremum over F is minimized by a ψ_0 which is not a Huber-estimator. Let ψ_H be the Huber-estimator with the same variance, so that

$$\int\psi_H^2 \, d\Phi \Big/ \left(\int\psi_H \, d\Phi\right)^2 = \int\psi_0^2 \, d\Phi \Big/ \left(\int\psi_0 \, d\Phi\right)^2.$$

Then ψ_H has a lower γ^* than ψ_0; hence, using the equality of the variances, sup $\psi_0^2/\int\psi_0^2 \, d\Phi >$ sup $\psi_H^2/\int\psi_H^2 \, d\Phi$, and since the remaining term in the CVF is at least arbitrarily close to zero if not zero or positive where $|\psi_0|$ reaches (or approaches) its sup, the F which is concentrated where $|\psi_0|$ reaches its sup pushes the mean-squared error already above that of the Huber-estimator which is a contradiction.

Thus, the Huber-estimators appear to be also the solutions for the linearized minimax problem with arbitrary *asymmetric* contamination. However, the parameter k of the Huber-estimator depends now not only on ε, but also on n. To give some crudely computed numbers as examples: for

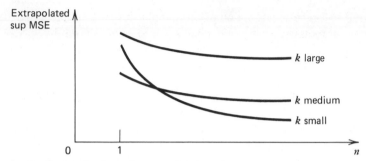

Figure 8. Qualitative sketch of the extrapolated maximum mean-squared error for various Huber-estimators (corners at $\pm k$) as a function of n, with the amount of contamination ε fixed.

$n = 5$ and $\varepsilon = 0.01$, $k \approx 2.0$; for $n = 5$ and $\varepsilon = 0.1$, $k \approx 1.2$, which are approximately the same values as for symmetric contamination since for such small n the variance is more important than the bias. On the other hand, for $n = 40$ and $\varepsilon = 0.01$, $k \approx 1.8$, and for $n = 40$ and $\varepsilon = 0.1$, $k \approx 0.7$ so that the bias caused by asymmetry becomes already noticeable, especially with the larger contamination, since its contribution grows quadratically with ε. For $n \to \infty$ we regain the conclusion already contained in Huber (1964), namely that the median is *the* choice for very large sample sizes. However, now we can also determine the minimax Huber-estimator for each ε and each n separately, without any artificial symmetry restriction.

1.3f. The Approach Based on Influence Functions as a Robustified Likelihood Approach, and Its Relation to Various Statistical Schools

Most discussions of the robustness theories by Huber and Hampel can give the impressions that they are part of decision theory, or at least the frequentist Neyman–Pearson theory. It is true that their historical development was greatly influenced by these theories, and Huber's minimax strategies certainly have strong ties to the theory of games. It is also true that the classical paradigms of decision theory have to be enlarged to include robustness aspects in addition to efficiency aspects in vector-valued loss functions (except where these vectors are reduced to single numbers as in the approach discussed at the end of Subsection 1.3e). But it would be narrow-minded to restrict the use of these robustness theories to the frequentist theories mentioned above.

For one thing, Bayesians are confronted with the same problem as frequentists, namely that their parametric models (of which they make strong use) are only approximately valid. Apart from the desideratum of stable inference under small changes of the parametric model, they have the additional problem of stability under small changes of the prior distributions. Even though the final form of inference may look different (e.g., may just consist of an a posteriori distribution of a parameter), concepts like influence function and breakdown point are equally valid and useful tools in the Bayesian as in the frequentist context. Only the extreme "omniscient" Bayesians who can put a quantitative prior probability distribution on everything will also put a prior on all possible deviations from the parametric model and then "integrate it out" without any formal need for stability considerations or model criticism. However, either the integration (and already the setting up of a prior over a sufficiently general neighborhood) poses huge problems both in theory and in practical applications, or the enlargements of the original parametric model are just too simple and restrictive to be realistic. That nevertheless the resulting procedures are sometimes robust and useful, can be seen by intuition or by means of the concepts of the classical robustness theory (which many Bayesians seem to be afraid to apply), but not within the Bayesian framework (apart from the "omniscient" attitude which closes itself off against any confrontation with reality). Because of its self-imposed isolation, much Bayesian robustness research lags behind its time by one to two decades. However, it is encouraging for the future development of the field that also more and more Bayesians seem to agree that robustness is a more basic concept than the division into Bayesian and frequentist methods.

A point which is often raised (cf., e.g., Dempster, 1975 and the following discussion) is that "frequentist" robustness methods do not condition on the particular data set at hand. This point is only partly justified. In fact, redescending M-estimators have the conditioning built into the estimator (the more outlying a point appears, the more it is downweighted), and while there are still quantitative refinements possible and desirable, this (approximate) conditioning is much stronger than the usual Bayesian (or likelihood) conditioning since it is valid not only for certain narrow enlargements of the original parametric model, but for a full broad neighborhood.

An aspect of the infinitesimal approach towards robustness which is very basic and equally important both for frequentists and Bayesians, is that it can be considered as a robustification of the likelihood function (and hence also of the likelihood approach). The likelihood in maximum likelihood estimation is replaced by a similar function whose derivative is required to be bounded and often continuous with bounded derivative itself. The similarity is measured in terms of the behavior at the parametric model. Correspondingly, likelihood ratios in testing are replaced by bounded approximations. As a consequence, the whole likelihood theory becomes much more elegant and satisfying. Regularity conditions simplify, and pathologies disappear. Let us mention some examples.

The renewed emphasis on functionals instead of arbitrary sequences of estimators has not only strengthened again the ties between statistics and functional analysis, it also gave more weight to Fisher's original definition of consistency ("Fisher-consistency") which is, in its realm of application, more natural and more satisfactory (and less probability dependent and closer to descriptive statistics) than the usual one with its highly arbitrary (including "superefficient") sequences of estimators. General sequences of estimators are not ruled out, but their properties should be judged for each n separately by an equivalent functional, not by some dubious "asymptotic" statement which may become meaningful only after 10^{10} observations. As to a specific functional, the expectation (including the arithmetic mean) is a functional which is nowhere continuous and not even defined in any full neighborhood with regard to the weak topology; it is replaced by continuous functionals defined everywhere. For a long time it was believed that Fréchet differentiability of estimators was a great exception, so that Reeds (1976) propagated the weaker concept of compact or Hadamard differentiability (cf. also Fernholz, 1983); but the belief arose because most common maximum likelihood estimators are not robust, while for robust M-estimators Boos and Serfling (1980) and Clarke (1983, 1984b) showed that Fréchet differentiability is indeed very often achieved. It can be surmised that theorems on the asymptotic behavior of maximum likelihood estimators do not require messier and stronger regularity conditions than those on Bayes estimators any more if the estimators are restricted to sufficiently regular robust replacements of the maximum likelihood estimators. A related example is the Chernoff–Savage theorem on the behavior of linear functions of the order statistics (L-estimators); the greatest technical problems are caused by the tails of the weight function and the underlying distributions, and these problems disappear completely if only robust L-estimators are considered, since their weight function has to be identically zero near both ends.

An interesting pathology which disappears under robustness considerations is the superefficiency of maximum likelihood estimators for models like the rectangular distribution. As mentioned in Subsection 1.2b (ii), it disappears already if only rounding effects are taken into account. But even if we believe temporarily in the fiction of infinite accuracy, superefficiency and robustness do not go along very well. As an example, let us consider n independent and identically distributed (i.i.d.) observations X_i distributed according to the rectangular distribution on $(0, \theta)$. The maximum likelihood estimate $X^{(n)}$ obviously is not robust. As a side remark, we note that within the ε-gross-error model there exist distributions like the uniform on $(0, \theta/(1 - \varepsilon))$ which are indistinguishable from a model distribution and hence lead to an unavoidable bias, like in other models. It is still possible to achieve superefficiency together with a high breakdown point (given infinite accuracy), for example, by $\hat{\theta} = \min(X^{(n)}, 3 \text{ med } X_i)$, since the breakdown point considers only the global behavior. However, if we require qualitative robustness or continuity, the estimators based on $X^{(n)}$ are unsuitable, as are estimators based on $X^{(n-k)}$ for fixed k as $n \to \infty$. We may, however, use a multiple of $X^{(\alpha n)}$ (for $\alpha < 1$) or of a trimmed mean or a variant of the Huber-estimator; these estimators are not only continuous, but even have an influence function and the usual asymptotic behavior, with the variance going to zero like n^{-1} and not like n^{-2}.

Not all problems of maximum likelihood estimators disappear: there may still be multiple solutions of the robustified likelihood equations. However, usually it will be possible to use a "monotone" estimator in order to identify the region of the proper solution; and since there are also results (Hampel, 1968, pp. 54 and 93) on when the proper solution is rather isolated from the others, it is normally not difficult to identify the proper solution (cf. also Clarke, 1984a).

A frequent misconception about robust statistics is that it replaces a strict parametric model by another parametric model which may perhaps be more complicated, perhaps more "robust," but which is equally strict. In fact, it replaces a parametric model by a full (and "nonshrinking") neighborhood of it. This important aspect has also consequences for the problem of deciding upon a parametric model.

Fisher (1922, p. 314) left the "question of specification" (of a parametric model), as he called it, entirely to the "practical statistician." Many statisticians then developed a kind of "logic," still found in many applied statistics books, which runs about as follows: Before applying the two-sample t-test, make a test of normality for each sample (e.g., a chi-squared goodness-of-fit test). If this result is not significant, you can believe in normality and can make an F-test for the equality of variances. If this one does not come out significant either, you can assume equality of variances and are allowed to make the t-test. In this argument, there is no word about possible lack of independence, no word about the fact that different goodness-of-fit tests may measure different and possibly unimportant aspects of the fit, no word about the catastrophic nonrobustness of the F-test for variances, no (or hardly any) word about what to do when one of the pretests comes out significant, no word about the fact that in sufficiently large samples practically all tests will come out significant, no word about the fact—perhaps stressed elsewhere in the same statistics book, and becoming popular in the philosophy of science—that a "sharp," nonisolated null hypothesis can never be statistically proven, only statistically disproven.

By contrast, it is possible to prove statistically that the true distribution lies in a certain *neighborhood* of a model distribution, and a suitable robust method is valid in this full neighborhood, not only at the strict parametric model. This dissolves the logical dilemma of classical statistics of having to work with a model which can at most be proven to be wrong and which in fact is known to be almost always wrong.

The fact that certain deviations from a model can be accommodated into valid and informative statistical procedures must not be taken to mean that in robust statistics one never changes the model. Model criticism and model selection are as important in robust statistics as they are in classical statistics, only the aims are slightly different: whereas in classical statistics the model has to fit all the data, in robust statistics it *may* be enough that it fits the majority of the data, the remainder being regarded as outliers. There are already formal methods of model choice based on robust versions of Mallows's C_p and Akaike's AIC (cf. Ronchetti, 1982b and Hampel, 1983b and Subsection 7.3d). Of course, they have to be used with the general reservation that a good model choice involves many additional informal aspects as well.

The "robustified likelihood" methods are equally general as classical likelihood methods, without suffering from all their defects. Since likelihood plays a central role in the Neyman–Pearson theory as well as in the Bayesian theory and of course in the pure likelihood approach, incorporation of robustness aspects into all these theories seems highly suitable. But the spirit of the infinitesimal approach is even closer to Fisher's approach to classical parametric statistics, and robust statistics, as the statistics of approximate parametric models, can be viewed as the natural extension of classical Fisherian statistics.

*1.4. REJECTION OF OUTLIERS AND ROBUST STATISTICS

1.4a. Why Rejection of Outliers?

This section treats a somewhat special topic and may be postponed until later, for example, until after Chapter 2, of which it makes some use. However, it should by no means be skipped if skipped means or any other objective or subjective rejection methods are of any interest to the reader.

Assume that nine measurements of the same quantity are "close together" and a tenth one is "far away." Then a common procedure in the application of statistics is to "reject" the "outlier," that is, to discard the distant value and proceed as if only the nine others were available. Sometimes the outlier is considered and interpreted separately, sometimes it is just "forgotten." The decision about what means "far away" can be done subjectively, that is, by just looking at the data and making a subjective decision, or by some formal, "objective" rule for the rejection of outliers, that is, some kind of test statistic. In practice, not infrequently the researcher finds subjectively that a value is probably an outlier, and then he looks for a rule which rejects it, in order to make his opinion more "objective." There are also several possibilities for the subsequent treatment of the outlier. It often can be checked, for example, by going to the original records, and not rarely it turns out to be a gross error (e.g., a copying error or an impossible value) which perhaps can even be corrected. (See Subsection 1.2c for the frequency of gross errors.) There is also the possibility of making one or several more observations if an outlier occurs, especially if there were very few observations to start with (cf. Subsection 1.3a).

Frequently (e.g., in breeding experiments or in the search for a new medicine, or in the search for a new elementary particle in physics) the outlier is a proper observation which is the most valuable one of all, and then even the rest of the sample may be forgotten. To give an example: According to C. Daniel (orally), his coauthor F. S. Wood of Daniel and Wood (1971) has two patents on outliers. In other cases the outlier (or a group of outliers) indicates a different model for all data (e.g., quadratic instead of linear dependence). In certain situations, the frequency of outliers is of direct interest as a measure of the quality and reliability of the data-producing system. A well-known class of examples are exams. It can happen also here that an outlier is the only correct value, and the majority of results are false. (Cf. also Wainer, 1983 on a case where an incorrectly scored item of the Preliminary Scholastic Aptitude Test was solved correctly only by a minority of the overall most intelligent students.) Another example where a sought-for result turned up only as a rare outlier is described by Hunt and Moore (1982, p. 33): Out of more than 200 persons claiming to be able to see the crescent of Venus with the naked eye, only two saw it—to their surprise—as opposite to the telescopic view shown to them before on the television screen. In some cases, a simple model holds only for part of the data, while the others need more parameters to describe them. One nice example (among others) is given in Daniel (1976, p. 162 ff.) where in a 33 × 31 analysis of variance out of seven columns with (crudely identified) large interactions, six (with 10 out of 12 large discrepancies) are chemically related and hence suggest a different interpretation from the rest, for which a main-effects model fits quite well.

We see that the detailed purposes in the rejection of outliers can be very different. However, two main aims emerge.

One is based on the observation that sometimes gross errors occur, and that a simple gross error which is also a clear outlier can be very detrimental to the statistical procedure used (such as arithmetic mean, standard deviation, or more generally the method of least squares). Surely, we would like to identify a gross error as such, and then remove it from the sample with good justification. However, this is not always possible; but the danger caused by the outlier remains and is much bigger than the threat of efficiency loss caused by rejection in case it was a "proper" observation. Moreover, it would often be extremely unlikely to have a "proper" observation from the assumed model distribution at such a distance. Therefore, a statistical argument is invoked to remain safe: If an outlier is too far away, it is deemed too unlikely under the parametric model used and hence rejected. It should be noted that the aim is safety of the main statistical analysis, and the "outlier test" is only a pretext to achieve safety if other methods (like checking the computations or the original records) fail. There would be no reason under this aspect to reject outliers if we used the model of the Cauchy distribution, because even very distant outliers have hardly any effect on the location estimate. (This becomes obvious if we look at the influence function of the maximum likelihood estimate; cf. Example 1 in Subsection 2.3a. There is, however, a limited but nonzero influence on the scale estimate which could cause limited concern; and the outliers could be interesting in their own right, as shall be discussed below.) We shall have to scrutinize how well the various methods for rejection of outliers, and also the overall approach of rejection as opposed to other robust techniques, achieve a satisfactory balance between safety and efficiency.

A complication arises since our parametric model may not be quite accurate, and a proper observation could easily occur at a distance which would be extremely unlikely under our model. Many statisticians believe an observation known to be "proper" should never be rejected. But a proper observation far away tells us that our model is wrong; if we change the model appropriately (to one with longer-tailed distributions or with an additional parameter), we find that the influence of our value on our previous parameter estimates will be greatly reduced; and if we are unable or unwilling to change the model (e.g., for reasons of simplicity), then we should at least render the observation harmless by rejecting it (or doing something similar).

A fundamentally different situation is the nonparametric one where we do not use any parametric model, mentioned repeatedly before (cf. mainly Subsections 1.1b and 1.2a). Clearly, if we have no redundancy, we cannot identify any outliers (in the sense of highly unlikely values) since everything is equally possible. We can still identify points with a high influence (in the sense of the influence function) on what we are estimating or testing, and have a sharp second look at them; but we do not have any probabilistic justification for rejecting them. Even worse, we do not even have the pragmatic justification for rejecting them as being potentially detrimental to what we are estimating, because what we are estimating is determined by *all* data, not only by some majority together with the redundance provided by some model. All we can do are special checks on highly influential points, and perhaps tentative analyses without these points in order to assess their potential dangers; but there is no justification within the data set, without additional knowledge from outside the data set, to reject them or to reduce their influence (cf. also Hampel, 1973a, p. 91; 1978a, p. 425f; 1980, p. 7).

The second main aim of rejection of outliers is to identify interesting values for special treatment. The values may be "good" outliers which indicate some unexpected or sought-for improvement; they may suggest an interesting new model or a new effect; or they may be gross errors which sometimes can be corrected or which can be studied in more detail. The conflict here is not between reliability and efficiency, rather it is between looking at too few special values and thereby missing important features of the data, and looking at too many values and thereby wasting too much time and energy. In practice (at least for a good data analysis) it is far more important not to miss any valuable features of the data than to raise a suspicion too often (especially since the suspicion will be abolished in the later steps of the analysis if it is unfounded).

If one considers the literature on the rejection of outliers (e.g., Hawkins, 1980; Beckman and Cook, 1983; and also large parts of Barnett and Lewis, 1978), one is led to believe that there is quite a different reason (or at least reasoning) for rejection. The outlier problem is isolated from the data-analytic context, it is even isolated from the overall formal statistical procedure (e.g., estimation of some parameter), and then it is considered as the problem of a statistical test of the null hypothesis of no outliers against some more or less special alternative, for example, three outliers on the right. The level of the test is usually chosen such that the null hypothesis is rarely rejected if it is correct, like in the classical Neyman–Pearson theory. It appears as if subsequent statistical analysis is to proceed as if the parametric model would hold perfectly for the remaining data: as if all gross errors and no "good" observations had been rejected. The question of consequences of statistical errors of both kinds in the outlier test on subsequent analysis is largely ignored. Apart from this, one can ask what the purpose of the outlier test is to be. Certainly, it is a useful side information to know how improbable an outlier would be under the strict parametric model; this could be measured by the P-value of an outlier test.

But by itself this information is of little use, except in extreme cases; and the intention of outlier tests does not seem to be to provide P-values, but statistical decisions using a fixed level. What can we learn from such a decision? We can learn (always with a small probability of erring, of course) that an observation is a clear outlier if the model is strictly correct. Neither do we learn that an observation is clearly not an outlier, nor can we infer that a "clear outlier" must belong to a different population, since its reason may just be that the model is slightly incorrect and that the true model is longer-tailed. More importantly, we learn nothing about the damage that undetected gross errors can still cause on the subsequent statistical analysis, and about the efficiency loss caused by wrongly rejecting "good" observations; and we learn almost nothing about which observations are worth a special look, except that some of them (the rejected ones) should definitely be included in a first round, although many interesting ones might still be missing. It thus seems that the paradigm of the Neyman–Pearson theory of testing has been implanted in a field where it is rather inappropriate.

It should be noted that there exist also attempts towards "Bayesian rejection of outliers." (Cf., e.g., the example and the bibliography in Dempster, 1975, as well as more recent work.) These attempts use a different framework of concepts, but otherwise their models are of similarly narrow scope as those of "orthodox" rejection.

Looking back, we find that there are good reasons for not treating all observations in the same way. It is the simplest though not necessarily best alternative to treat them in two mutually exclusive ways. We can try to see how well we can do quantitatively within this class of methods. But already now we notice some difficulties. Safety combined with good efficiency in the potential presence of gross errors is precisely one major aim of robust statistics; but good robust methods are more sophisticated than simple rejection of outliers, and the latter occurs only as a rather marginal and pathological example during the development of robustness theory [Huber, 1964, p. 79, (iv)] although this example will turn out to be superior to customary rejection rules. The criteria by which outlier rules are usually judged are not in concordance with the true aims. It will turn out that some outlier rules do not even fulfill their presumably most basic intuitive requirements. Or what shall one say about a rule for rejection of outliers which cannot even reject a single outlier out of, say, 10 or 20 observations at an arbitrarily large distance? The boundaries between rejection and nonrejection which yield good compromises between safety and efficiency will have to be found by methods other than levels of outlier tests (cf. Subsection 1.4b). These boundaries may not, and, in general, will not, coincide with the boundaries between "uninteresting" and "interesting" observa-

tions to be examined separately. The latter should include also all "doubtful" outliers which may be perfectly all right, but which in a broader context
(with additional information) may also turn out to be rather special.

This suggests at least three categories for the purpose of the individual
scrutiny of special values: "clear outliers" (perhaps even those rejected by a
good outlier rule, or those which are fully rejected by a redescending
M-estimator; cf. Section 2.6), "doubtful outliers" (all those at the fringes of
the sample about which no clear-cut a priori decision is possible; perhaps
those which are partly downweighted by a certain redescending M-estima-
tor), and the rest which is clearly uninteresting a priori. A more proper
reflection of reality is a continuously increasing curve of doubt whether an
observation is "proper" as it moves away from the center of the sample.
This reminds us of appropriate Bayesian a posteriori probabilities (although
most current Bayesian models are too narrow to be of general use), and it is
clearly reflected in the continuous downweighting by good redescending
robust estimators. Since such curves are more complicated to handle, the
classification into three or even two categories may be deemed sufficient for
crude practical purposes. One of the simplest reliable criteria for doing this

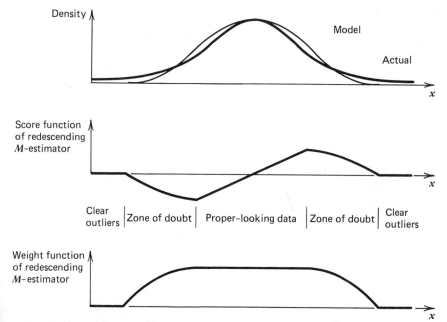

Figure 1. Trisection of data and approximate conditioning by redescending M-estimator in
full neighborhood of model.

is to use the residuals from a robust fit and to put the boundaries at certain multiples of the robustly estimated standard deviation, perhaps with a correction for small sample sizes. Obviously, such boundaries can be formulated in terms of (perhaps rather sophisticated) rejection rules.

There remains the question how well rejection rules can do for purposes of efficient and robust estimation, or in other words what the price is for declaring all observations either completely "good" or completely "bad" when the interest lies in a good parametric description of the majority of the data. This will be discussed in the next subsection.

1.4b. How Well Are Objective and Subjective Methods for the Rejection of Outliers Doing in the Context of Robust Estimation?

We shall now consider the robustness properties and efficiency losses of combined estimation procedures of the form: First reject outliers according to some method, then use a classical estimator, like the arithmetic mean, for the remaining observations. Our quantitative discussions shall be concentrated on the case of estimation of the location parameter of supposedly normal data, since this case is the simplest and by far the best known one, but the results carry over to much more general situations. We shall compare various objective rejection rules with each other, with subjective rejection and with other robust methods like Huber-estimators and redescending M-estimators. Two side results are the introduction of a fairly new simple rejection rule and a deeper understanding and explanation of the so-called "masking effect" in outlier rejection.

It is possible to obtain a very good qualitative and semiquantitative understanding for the behavior of rejection rules in the context of robust estimation by applying the theoretical robustness concepts to the combined rejection–estimation procedures. The influence function (cf. Subsections 1.3d and 2.6c and more generally Chapter 2) tells us a lot about the local behavior near the parametric model. Its most striking features are its high jumps at the "rejection points" which cause relatively large efficiency losses and whose height depends critically on the local density of the underlying distribution at the rejection points. The breakdown point (cf. Subsections 1.3d and 2.2a) of the combined rejection–estimation procedures tells us about their global reliability; upon computation of the breakdown points of various procedures we find striking differences between the qualities of different methods, including rather bad failures. These findings could be worked out in greater detail; for specific situations, they can also be made quantitative by means of Monte Carlo studies. Since according to experience Monte Carlo numbers, although rather restricted in their scope, are much more convincing for the great majority of statisticians than theoretical considerations of the above kind—this was also the reason for the Princeton Monte Carlo Study described in Andrews et al. (1972)—we shall now discuss selected parts of two Monte Carlo studies which were specifically done to obtain quantitative comparisons between rejection rules and other robust methods.

One Monte Carlo study was part of the unpublished continuation of the Princeton Monte Carlo Study mentioned above (cf. Gross and Tukey, 1973). Its general setup was the same as described in Andrews et al. (1972). It contained 32 variants of six different types of rejection rules (namely based on kurtosis, largest Studentized residual, Studentized range, Shapiro–Wilk statistic, Dixon's rule, and the new "Huber-type skipped mean"; cf. Grubbs, 1969 and Shapiro and Wilk, 1965 for the former five types). The variants include "one-step" and "iterated" versions and various significance levels and were applied to ten different distributions with sample size $n = 20$. Detailed analyses are contained in Hampel (1976, 1977, 1985), which should be read for more details (cf. also Huber, 1977a, p. 3; 1981, pp. 4–5; Donoho and Huber, 1983; Hampel, 1978a, 1980). We shall only discuss what are perhaps the most important and the most interesting results (cf. also the somewhat similar discussion in Hampel et al., 1982).

The other Monte Carlo study is by Relles and Rogers (1977) who plugged five statisticians into an investigation of the quantitative properties of subjective rejection by showing them many computer-generated samples from various t-distributions for sample sizes 10, 20, and 30 and asking them to give two subjective location estimates (one "seat-of-the-pants estimate" and one determined indirectly by the number of outliers rejected on each side). The efficiency losses for both estimators and all five statisticians were very similar and thus could be pooled. They range from about 15% to more than 20% for the most important distributions, compared with the best estimator for each distribution. Thus they may be considered as moderate, but not as small. We have to stress that the advantages of subjective rejection over objective rules lie elsewhere, namely in the chance for more immediate interpretation of the outliers, including a certain flexibility in dealing with unexpected situations. As a means for obtaining robust routine estimates, subjective rejection methods do prevent disasters (except with complicated and large data sets where they become extremely cumbersome), but they are only rather mediocre and can easily be beaten by good "objective" robust estimators. For our quantitative comparison given below, we can obtain the efficiency losses for Normal and t_3 directly from Relles and Rogers (1977), while those for the other two situations have to be extrapolated.

We shall consider only four distributions with 20 independent observations. One is the ideal model distribution, namely the Normal. Two mimic the effects of 5 and 10% "distant" outliers by taking (precisely) one and two observations respectively from $N(0, 100)$, a Normal with expectation zero and variance 100, and the others from $N(0, 1)$. We have seen before (cf. Subsection 1.2c) that a few percent gross errors are quite usual in data of average quality and that 10% gross errors are quite realistic for data with fairly low quality. It is to be expected that rejection rules are at their best when the outliers are far away (at least with high probability, as in our distributions), since then they are easy to separate from the "good" data. To mimic the slightly elongated tails of high-quality data in the absence of gross errors, we also include t_3, Student's t with 3 degrees of freedom [cf. Subsections 1.2b(iii) and 1.2d]. It turns out (cf. Hampel, 1976, 1985) that the estimators behave almost the same way under t_3 and under 15 $N(0, 1)$ & 5 $N(0, 9)$ which can be interpreted as having 25% gross errors rather close to the "good" data so that they often cannot be separated and thus cause bigger problems for the rejection rules.

The estimators considered here include the arithmetic mean in its strict mathematical sense, that is, without rejection of any outliers, as computed by the computer or as applied by a rather naive or else a rather "orthodox" statistician in the sense of being a firm believer in rigid and exact mathematical models and in a "monolithic, authoritarian" structure of statistics (cf. also Tukey, 1962, p. 13). Also included is the median which statisticians often use in view of plainly long-tailed distributions. The data-dependent switching between mean and median is already imitated in several proposals by Hogg in Andrews et al. (1972); they came out not too badly, but rather mediocre. As mentioned, we also include quantitative results on subjective rejection. The formal procedures of the type "rejection rule plus mean" include largest or maximum Studentized residual on two different levels (cf. Grubbs, 1969 or the other papers cited for technical details), which is probably one of the most popular rejection rules; it has been generalized to more complex designs (cf. Stefansky, 1971, 1972) and is also used informally and graphically in residual plots. Then there is the Studentized range (range divided by the standard deviation of the same sample) which has also been proposed but of which we can only hope that it is not used too often. The kurtosis is included with three different (approximate) levels, partly because of its relative reliability and partly to bring out a point concerning the levels.

Very simple rejection rules which are based on the robust median and median deviation rather than mean and standard deviation are the "Huber-type skipped means"; the rule with code name X84 says: Reject everything

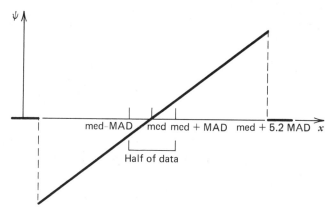

Figure 2.   Defining ψ-function of rejection rule X84, based on median and median deviation.

which is more than 5.2 median deviations (cf. Hampel, 1974 or Example 4 of Subsection 2.3a) away from the median and take the mean of the remainder. The median deviation, also called median absolute deviation (MAD), which should not be confused with the interquartile range, and which was already considered by Gauss (1816), is the median of the absolute residuals from the median. It is thus the natural scale counterpart to the median and a natural basis for rejection rules. These rules were investigated in more detail by Schweingruber (1980).

For comparison purposes we include two robust location estimators which treat outliers in different ways: the Huber-estimator H15 (with $k = 1.5$ and "Proposal 2 Scale"; cf. Subsection 4.2d, or Andrews et al., 1972) is the prototype of an estimator with monotone influence function which bounds the influence of outliers by "bringing them in" towards the majority, but still gives them maximum influence; the three-part descending M-estimator 25A (cf. Example 1 of Subsection 2.6a, with $a = 2.5$, $b = 4.5$, and $r = 9.5$ as multiples of the median deviation which is first estimated from the sample in order to make the estimator scale invariant, or cf. Andrews et al., 1972) is the prototype of a "smoothly rejecting" estimator which rejects distant outliers completely, but contrary to the "hard rejection rules" allows a zone of increasing doubt with continuously decreasing influence between the "clearly good" and "clearly bad" observations.

The entries in Table 1 are the percent efficiency losses compared with the best available estimator in the study of Andrews et al. (1972) and its continuation. The Monte Carlo variances of these estimators are also given in the table. Apparently, they are rather close to the asymptotically optimal

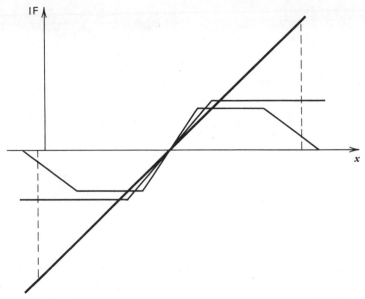

Figure 3. Four principal ways of treating an outlier: bounding its influence, smooth rejection, hard rejection, and no treatment at all. (Qualitative sketch.)

values, but we prefer values which can actually be reached for $n = 20$ and, besides, we need only relative comparisons.

Finally, the last column of Table 1 contains the breakdown points ε^* (cf. Subsection 2.2a) of the combined rejection-estimation procedures. These breakdown points tell us how many distant outliers can be safely rejected; and this information turns out to be the most important characteristic of the behavior of rejection rules, the second most important one being the efficiency loss under normality.

Let us now comment briefly on the numerical results in Table 1 and some additional aspects of the estimators.

The arithmetic mean without any rejection can be disastrously bad; even under t_3 it loses already about 50% efficiency.

The median is simple and robust, but moderately inefficient throughout. Its use can be recommended for moderately low efficiency requirements, but not for higher efficiency standards. Furthermore, its local behavior is very susceptible to changes of the density at the population median.

Subjective rejection loses about 10–20% efficiency; too much by some standards, but bearable by others.

Table 1. Monte Carlo Efficiency Losses (in %) and Breakdown Points ε^* for $n = 20$ for Rejection Rules with Mean and Other Estimators

			t_3	N	Outliers 1	Outliers 2	ε^*
$\bar{X}$ (arithmetic mean)			49	0	83	89	0%
Median			13	33	28	31	50%
Subjective rejection and mean			12	21	$\sim 16(?)$	$\sim 12(?)$	
Maximum Studentized residual,	$k = 2.56$,	$\alpha = 10\%$	19	4.4	1	3	12.7%
Maximum Studentized residual,	$k = 3.03$,	$\alpha = 1\%$	23	0.3	2	48	9.4%
Studentized range,	$k = 4.79$,	$\alpha = 1\%$	43	0.4	74	87	4.3%
Kurtosis (4th moment),	$k = 5$,	$\alpha = 1\%$	22	0.7	2	7	14.7%
Kurtosis (4th moment),	$k = 3.5$,	$\alpha = 10\%$	18	6	3	1	19%
Kurtosis (4th moment),	$k = 3$,	$\alpha = 50\%$	17	12	8	4	21%
"X84" (based on median and median deviation)			14	4.2	2	5	50%
"H15" (Huber-estimator, "Proposal 2 Scale", $k = 1.5$)			7	3.5	7	17	26%
"25A" (three-part redescending, corners at 2.5, 4.5,			5	4.4	1	1	50%
9.5 median deviations $\approx 1.7, 3.0,$							
6.4 standard deviations)							
Lowest available var (var$_{opt}$)			1.58	1.000	1.12	1.24	

The maximum Studentized residual with level $\alpha = 1\%$ cannot even safely reject two outliers (10%)! The Studentized range with $\alpha = 1\%$ cannot even reject a single distant outlier out of 20 observations (5%)! The explanation in both cases is given by the breakdown point. Typically, the test statistic for outliers tends to a limit as a contaminating point mass moves to infinity, and if this limit is below the critical value, the rule cannot reject and breaks down. (Cf. Hampel, 1976, 1985 for details.)

The breakdown point also explains the "masking effect" of outliers. It is said that an outlier masks a second one close by if the latter can be rejected alone, but not any more in "company." But in most cases this is nothing but the frequent situation that $1/n < \varepsilon^* < 2/n$ and a contaminating point mass is the "least favorable contamination" which makes rejection hardest. It should be clear that the ability to reject depends also on the proportion of outliers; after all, no invariant rule can reject a fraction of more than 50% outliers if these outliers are so "mean" that they form a beautiful mock sample elsewhere. As the table shows, most breakdown points are much lower than 50%. In abuse of the word, we might even say that a single outlier "masks itself" if $\varepsilon^* < 1/n$.

The result for the breakdown point of the maximum Studentized residual is perhaps the most important one for statistical practice, since this rule seems to be rather widely used. We note that this rule can be safely used only for rather good data, when 10% gross errors or more are out of the question. In addition, we need sample sizes of more than about 10 (with somewhat different critical values and hence breakdown points) in order to have a chance at all to reject one occasional single distant outlier. Ten replicates are a lot in, say, analysis-of-variance situations where there is usually rather little redundancy. Hence we are often forced to use other rejection rules. Of course, the positive level guarantees a positive percentage of wrong rejections under the ideal model, but these are rejections of nearby outliers, not of distant ones, as can be seen from the influence functions of mean, standard deviation, and their ratio. We might easily be trapped into distorting the level of the test under "ideal" circumstances if we use the formal rejection rule only for "doubtful" cases and deem its application superfluous for "obvious" outliers although these "obvious" outliers would not be formally rejected.

The rule based on fourth moment or kurtosis is the most reliable one among the three classical rules considered here. (We might mention here that the rule based on the Shapiro–Wilk test is still more reliable; cf. Hampel, 1976, 1985.) An interesting side point is that the rule with type-I-error probability of 50% still yields a fairly efficient and rather reliable estimator, showing that the level of the outlier test is rather meaningless if rejection is only a preliminary precaution before estimation.

The new simple rejection rule "X84" does about as well as the best other rejection rules in the situations considered, and it is much more reliable (in fact, "most reliable", with $\varepsilon^* = 50\%$) in other situations. Hence it pays to use residuals from a robust fit, not from a least-squares fit, both for detection and accommodation of outliers. For rather small samples, the inefficiency of median and median deviation might be thought to be a disadvantage; but both estimators can be replaced by more efficient and equally reliable estimators, and even the simple rule as given is at least reliable, while classical rules tend to break down completely with small samples, in the sense of not being able to reject a single distant outlier.

In a small Monte Carlo study, Schweingruber (1980) investigated the efficiency losses of Huber-type skipped means and related estimators for various distributions, sample sizes, and choices of the critical value k. Under the normal distribution, X84 (with $k \approx 5.2$ median deviations) loses about 2% efficiency (compared with the arithmetic mean) for $n = 40$, about 4–5% for $n = 20$, about 7–8% for $n = 10$, and between about 10 and 17% for n

between 8 and 4 (about 22% for $n = 3$). If $k \approx 4.5$ median deviations ("X42"), the corresponding losses are about 5, 8, 12, 13–20, and 25%. The losses stay below 10% down to $n = 4$ for $k \approx 8$ median deviations, while they are up to about 20% down to $n = 10$ for $k \approx 3.5$ median deviations. This shows that k has to be chosen fairly large, especially for small sample sizes, if the efficiency loss is to be kept small. Values of k around 5 (4.5–6) median deviations (corresponding to about 3 to 3.5 or even 4 standard deviations) appear to be frequently quite reasonable, according to present preliminary knowledge.

After discussing specific features of the rejection rules, we now come to some general properties. The most striking and most important feature is the jump in the efficiency loss as soon as the breakdown point is passed by the contamination. Another feature for whose full documentation we would need more numbers such as are given in Hampel (1976, 1985), is that different rules with the same efficiency losses under the Normal (but not necessarily the same level) show a rather similar behavior also in other situations, as long as they are not close to their respective breakdown points. Hence the breakdown point (global reliability) and the efficiency loss under the Normal (the "insurance premium" of Anscombe, 1960) are the most important numerical properties of the rejection rules.

Another general feature of all rejection rules is that they unnecessarily lose at least 10–15% efficiency (and sometimes much more) under t_3. As we have seen, t_3 is an example for outliers which are not so clear-cut; and if they become even less clear-cut—if they become concentrated near the "rejection points"—then all outlier rules can lose up to 100% efficiency. While some rejection rules cannot even cope with distant outliers, no rejection rule can cope well with gross errors and other contamination which is in the "region of doubt" close to the "good" data. This is the main, basic, and unavoidable weakness of all hard rejection rules, as is obvious from the shape of the influence function (see above).

We can contrast rejection rules with two other major ways of treating outliers. One is, in effect, to "move them in" close to the good data, as done in Huber-estimators, Hodges–Lehmann-estimator, trimmed and Winsorized means. Huber-estimators are asymptotically optimal for "smooth, nearby" contamination not much unlike t_3, and it is therefore not surprising that H15 is much better under t_3 than all rejection rules. It is also very safe against high concentrations of gross errors near the "good" data. On the other hand, clear outliers are not rejected with Huber-estimators, but still have a large though bounded influence; this results in avoidable efficiency losses of H15 and the like, compared with rejection rules, which for realistic amounts of outliers (say, 5–10%) are about 5–15%. These losses are thus about equally high as the avoidable losses of rejection rules under smooth, nearby contamination. In contemplating this limited weakness, one should keep in mind that robust estimators with monotone influence function (e.g., median and median deviation) are necessary at least as a first step in order to identify safely the majority of "good" data with the help of the unique solution of the M-estimator equation and to avoid getting stuck with a wrong solution out of several ones in the case of nonmonotone influence.

The other major way of treating outliers is to reject them "smoothly," with continuously decreasing weight and influence. It does not matter much whether the influence becomes strictly zero or only tends to zero as the outlier moves to infinity. In the case of 25A, it is strictly zero beyond about 6.4 robustly estimated standard deviations away from the location estimate (cf. Table 1). We see from the table that 25A deals excellently both with distant outliers and with smooth longer-tailed contamination; it has the highest possible breakdown point and, according to the way it is defined, is also very stable against localized contamination. It thus combines the good features of hard rejection rules and Huber-estimators without having their disadvantages. There is only one major danger with redescending M-estimators, namely that they redescend to zero too quickly. Furthermore, they need a preliminary robust estimate with monotone influence (as do hard rejection rules). Otherwise their detailed properties do not matter too much. We note that these "smoothly rejecting" estimators are a natural extension and a considerable improvement of classical "hard" rejection rules.

We can now summarize the behavior of rejection rules in the context of robust estimation.

1. Most importantly: Any way of treating outliers which is not totally inappropriate, prevents the worst.

Such methods include subjective rejection (except in complicated situations), objective rejection rules with high breakdown point, "objective" rules after visual inspection, or any decent robust estimate like median, H15, 25A, ... The efficiency losses might easily be 10 or even 50%, but hardly 90 or 99%.

Totally inappropriate are a nonrobust computer program (e.g., least squares!) *without* any built-in checks and without a careful follow-up residual analysis, and furthermore, some objective rejection rules, like Studentized range. The maximum Studentized residual, with its breakdown point around 10%, is a borderline case and often *not* safe enough!

2. For higher efficiency standards: Most methods still lose unnecessarily at least 5–20% efficiency in some realistic situations. These methods include subjective rejection, all objective rejection rules, Huber-estimators, and so on (but not smoothly redescending estimators such as 25A).

3. In a broader context: In general, one should not only detect and accommodate outliers, but also interpret (and perhaps correct) them. Moreover, one should incorporate the nonstatistical background knowledge about the plausibility of observations. This knowledge is at least as important as purely statistical arguments. There is also a general statistical background knowledge about frequent types of gross errors and other deviations from strict parametric models which should also be incorporated in the analysis of the data. When formal rejection methods are used, these considerations are the next step which should not be forgotten; when subjective methods are used, these considerations can be applied immediately, providing early feedback and thus often a quicker interpretation of the data.

On the other hand, formal identification of suspect observations (as the first step) may be much quicker and easier, and in complicated designs this may be the only possibility. Another disadvantage of subjective methods is that they do not provide proper confidence statements.

4. Identification of outliers and suspect outliers can be done much safer and better by looking at residuals from a robust fit, rather than a nonrobust fit. In particular "X84" is a very simple, relatively good rejection rule based on the "most robust" measures of location and scale.

5. Rejection rules with subsequent estimation are nothing but special robust estimators. Also estimation after subjective rejection can be viewed as a subjective robust procedure, about which even quantitative statements can be made. Both objective and subjective methods have a certain appealing conceptual simplicity in dividing all data into merely two classes, the "good" and the "bad" data, but their overall properties, compared with other robust estimators, are only mediocre to bad. Their properties can be lucidly described by the concepts of robustness theory. For example, the "masking effect" can be understood much better by means of the breakdown point.

It may be argued that general robust estimation (and robust statistics) is "merely" an offspring of preliminary rejection of outliers, developed in the search for better alternatives; but certainly by now rejection methods can and should be integrated into robust statistics.

EXERCISES AND PROBLEMS

Due to the informal character of this chapter, most exercises can only be solved *after* one has studied at least also Chapter 2. Moreover, several exercises require specialized knowledge from various other areas of statistics. Many exercises are fairly open-ended and flexible, which means that there is not one fixed answer, but rather the possibility of longer and shorter as well as better and less satisfying solutions.

The qualifications "short," "extensive," and "research" denote rather quick exercises, rather demanding exercises, and research problems, respectively.

Subsection 1.1b

1. Give an example for a two-sample randomization test (based on the difference between the means) which is clearly significant and which becomes nonsignificant by adding a single outlier to one sample.

2. What happens in Exercise 1 if medians are used instead of means? What are advantages and disadvantages of using medians?

3. Compute, for a small data set, the full distribution of the test statistic of a two-sample randomization test which uses the difference between two (proper) trimmed means.

4. What happens in Exercise 3 if one value is replaced by an outlier?

5. Do Exercises 3 and 4 for the usual randomization test based on the difference between means, and compare the results.

6. (Extensive) Show that the normal scores estimator soon has a bigger variance than the Hodges–Lehmann-estimator if one moves away from the normal distribution towards longer-tailed distributions. (You may use asymptotic formulas or Monte Carlo; mixtures of two normals, t-distributions, or other long-tailed distributions.)

7. (Research) Under what conditions are maximum likelihood estimators and Bayes estimators derived from a "supermodel" for location robust?

8. What types of deviations from a parametric model are taken into account in a neighborhood in Hellinger distance, total variation distance, and Kolmogorov distance?

9. What kinds of conditions (qualitatively) are needed on the prior distributions of Bayes estimators such that the estimators are insensitive against small changes of the prior?

Subsection 1.1c

10. Give an example for the normal location model of a Fisher-consistent estimator with breakdown point arbitrarily close to one.

11. Show how five regression diagnostics are related to influence functions.

12. Give a design in which a point is a leverage point under some models and no leverage point under others. Give a design in which the central point is a leverage point under some models.

Subsection 1.2b

13. When can a statistical null hypothesis be statistically proven?

14. (Research) Under what conditions on the underlying distribution and the influence function of an M-estimator is the estimator continuous or differentiable in suitable senses, if all data are rounded or grouped?

15. (Extensive) Work out the (approximate) variances of the maximum likelihood estimator for the exponential shift model with density $\exp(-(x - \theta))$, $x > \theta$, for different sample sizes, if all data are rounded to the nearest integer.

16. Show that the median is not continuous as a functional in any full Prohorov neighborhood of any distribution.

17. Show that the arithmetic mean has the lowest local-shift sensitivity among all location estimators.

18. Discuss the qualitative results of robustness theory for two- and three-parameter Gamma distributions.

19. Do Exercise 18 for Weibull distributions.

20. Do Exercise 18 for the negative binomial distribution.

Subsection 1.2c

21. Pick five real data sets and examine them for gross errors.

22. Pick a data set and let one value move to infinity. What are the effects on various statistics (e.g., estimates of location, slope in case of regression, scale, standardized residual)?

23. Pick a data set and add an observation somewhere. What are the effects? Compare with Exercise 22. When is the added observation harmless?

Subsection 1.2d

24. Plot some t-distributions on normal probability paper (e.g., $\nu = 9$, 5, and 3).

25. Do Exercise 24 for some of Tukey's mixtures of two normals (e.g., $\varepsilon = 1$, 5, and 10%).

26. Do Exercise 24 for some of Huber's least favorable distributions (e.g., $\varepsilon = 1$, 2, 5, and 10%).

27–29. In Exercises 24–26 respectively, for what sample sizes do the differences against the normal distribution become noticeable in the normal plots? Explain what you mean by "noticeable."

30–32. In Exercises 24–26 respectively, what is the approximate power of the test based on kurtosis against these alternatives to the normal for some sample sizes (e.g., $n = 10$, 20, 50, 100, 500, 1000, and 2000)?

33. Compute the relative asymptotic efficiency of the median deviation (MAD) compared with the mean deviation and with the standard deviation in Tukey's contamination model.

34. Do the comparisons of Exercise 33 (three ways) under t-distributions.

Subsection 1.3a

35. Describe more formally the robustness properties of the mode.

36. (Research) Investigate the robustness and efficiency properties of the rule based on taking a third observation, if necessary, which uses the mean of the two closest ones.

37. (Research; continued from Exercise 36) Compare with taking always two or always three observations.

38. (Research; continued from Exercise 36) Replace the mean of the closest two by other estimators.

39. (Research) Show that randomization tests using the mean have very little power for long-tailed distributions.

Subsection 1.3c

40. (Extensive) Using Huber (1968), find the solutions of Huber's second approach for double-exponential, logistic, and double-exponential extreme value distribution.

41. (Extensive; continued from Exercise 40) Compare them with the minimax and the infinitesimal solution.

Subsection 1.3d

42. (Research) Show that two-part redescending M-estimators like HMD (see Andrews et al., 1972) minimize the asymptotic variance under a bound on the change-of-bias sensitivity.

43. (Research) Do Exercise 42 for three-part redescending M-estimators and bounds on gross-error sensitivity and change-of-bias sensitivity.

Subsection 1.3e

44. (Extensive) Compute a short table for the approximate minimax mean-squared-error Huber-estimator as a function of ε and n.

Subsection 1.3f

45. (Research) Discuss a decision theoretic approach with the vector-valued risk function containing asymptotic variance and gross-error sensitivity.

46. (Research) Do Exercise 45 for asymptotic variance, gross-error sensitivity, and change-of-variance sensitivity.

47. (Extensive) Compute a downweighting scheme (and the corresponding M-estimator) based on a Bayesian outlier model for the normal location model and compare with other estimators.

48. (Research) Try to compute a downweighting scheme in the normal location model based on the likelihood that an observation is an outlier.

49. In the normal location model, by which functional would you replace a Bayes estimator for a given n? What robustness properties does it have?

50. (Short) Show that the expectation is nowhere continuous, and that the complement of its domain is dense in the weak topology.

51. Show that L-estimators with positive breakdown point have to have weight functions identically zero near zero and one. Compute the breakdown point.

52. Work out the details of the various estimators for the rectangular distribution on $(0, \theta)$.

53. Do Exercise 52 for the rectangular on $(\theta, \theta + 1)$.

54. (Extensive) For the Cauchy location model, discuss the possible solutions of the likelihood equations and the selection of the "proper" one.

55. Discuss the "logic" of testing a model before applying a method geared to the strict model. What are its hidden assumptions? What are its weaknesses?

56. (Research) Discuss the qualitative aspects of "robust" model selection. Show that the basic criterion for a good fit is vector valued, for example, fraction of outliers and scale of "good" data. What happens if a robust overall scale measure is used?

Subsection 1.4a

57. Give one example each of data where outliers stem from a long-tailed error distribution, a different model for part of the data, and a different model for all data.

58. Give two examples each for a parametric mean and a nonparametric mean.

59. Which observations are potentially dangerous for the model of a rather skew lognormal distribution? Which ones look like outliers?

60. What can be the effects of grouping for a lognormal distribution?

61. Where are the most dangerous "outliers" for Weibull distributions?

62. (Extensive) Work out an example of the consequences of statistical errors of both kinds in an outlier test on subsequent statistical analysis.

63. (Extensive) Analyze a Bayesian approach towards rejection of outliers using the concepts of robustness theory.

Subsection 1.4b

64. (Extensive) Carry out a small Monte Carlo study of subjective rejection.

65. (Research) Try to analyze a simple approximate model for data-dependent switching between mean and median.

66. Compute some asymptotic variances for the cases in Table 1 and compare them with the Monte Carlo variances.

67. Describe distributions where the sample median is "super-efficient"—that is, converges faster than at the usual rate; is qualitatively robust, but has infinite asymptotic variance; and is not even qualitatively robust.

68. Derive the breakdown points of some of the rejection procedures in Table 1.

69. Compute the breakdown points of three rejection rules with mean not discussed in Section 1.4.

70. Compute the breakdown point of the maximum Studentized residual for several sample sizes.

71. Compute the breakdown point of the maximum Studentized residual for some analysis-of-variance situations.

72. Show how with the Studentized range a single observation out of 20 can "mask itself." When is rejection most likely?

73. (Extensive) Show how with the Shapiro–Wilk test used as a rejection rule, two opposite outliers can "mask each other."

74. Show how under weak assumptions any estimator based on a rejection rule has infinite asymptotic variance under certain distributions.

75. (Short) Outline a model with outliers where an outlier rule is asymptotically "perfect."

76. Why is the trimmed mean not a rejection procedure?

77. What is the relation between trimmed means, Winsorized means, and Huber-estimators?

78. Why are Huber-estimators and similar estimators moderately inefficient in the presence of distant outliers?

79. Show that redescending M-estimators which redescend too quickly can have an infinite asymptotic variance.

80. Compare biweight, sine-estimator, three-part redescending estimator, and hyperbolic tangent estimator under the aspects of robustness theory.

81. Give examples for identification of outliers using general statistical experience.

82. Give examples for identification of outliers using subject matter knowledge.

CHAPTER 2

One-Dimensional Estimators

2.0 AN INTRODUCTORY EXAMPLE

To explore the various aspects of robust statistics, we start with a simple framework: the estimation of a one-dimensional parameter based on a collection of one-dimensional observations. The next chapter treats statistical tests in the same setup, and subsequent chapters deal with higher-dimensional problems such as multivariate location–covariance, regression analysis, and testing in linear models. However, most of the ideas and tools of robustness theory can already be explained in the one-dimensional case, as they were originally developed for that situation.

The most common one-dimensional problem is the estimation of a location parameter. Many studies of robustness have focused on this simple framework, as a first steppingstone to improve the understanding of goals and tools, and with the purpose of generalizing the insights obtained to more complex problems. (Unfortunately, this has left many statisticians with the impression that estimating one-dimensional location parameters is what robustness theory is all about, whereas this book intends to show the usefulness of robustness in a diversity of situations.) Since then, many tools constructed for one-dimensional location have proven their value in higher-order models.

As an example, let us look at the data by Cushny and Peebles (1905) on the prolongation of sleep by means of two drugs. The original data set was bivariate: for ten subjects, two different values were recorded (one for each drug). These data were used by Student (1908) as the first illustration of the paired t-test, to investigate whether a significant difference existed between the observed effects of both drugs. The data were then cited by Fisher (1925) and thereforth copied in numerous books as an example of a

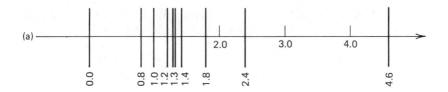

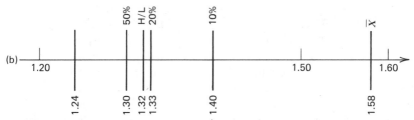

Figure 1. Cushny and Peebles data: (*a*) univariate sample of ten points and (*b*) six location estimates. $\bar{X}$ = arithmetic mean; 10% = 10%-trimmed mean; 20% = 20%-trimmed mean; H/L = Hodges–Lehmann estimator; 50% = median; and 1.24 = mean of data set without the observation 4.6.

normally distributed sample (see, e.g., Anderson, 1958). The ten pairwise differences (i.e., the set of differences between drug effects per subject) are the following:

$$0.0, 0.8, 1.0, 1.2, 1.3, 1.3, 1.4, 1.8, 2.4, 4.6.$$

Most authors have summarized this (supposedly normally distributed) uni-variate data set by its arithmetic mean, for instance when applying the *t*-test. However, a glance at these numbers (or at their plot in Figure 1*a*) reveals that the normality assumption is questionable, due to the occurrence of 4.6 which appears to be an outlier. Therefore, Pfanzagl (1968) and Tukey (1970/71) applied more robust procedures to this data.

Figure 1*b* contains some location estimates applied to this sample. The ordinary arithmetic mean $\bar{X}$ equals 1.58. The other estimates are robust, although not all to the same extent. The estimate denoted by 10% is the 10%-trimmed mean: one starts by removing the 10% largest and the 10% smallest observations, and computes the mean of the rest. In the present example, this amounts to deleting the largest observation (4.6) as well as the smallest (0.0) and computing the average of the remaining eight values, which yields 1.4. Analogously, the 20%-trimmed mean corresponds to the average of the middle six numbers, yielding 1.33. In the same spirit, the median (which equals 1.30 in this sample) can be denoted by 50%. The Hodges–Lehmann (1963) estimator H/L, which is defined as the median of all pairwise averages $(x_i + x_j)/2$ for $i = 1,\ldots,10$ and $j = 1,\ldots,10$ amounts to 1.32. For comparison, the average of the data set *without* the spurious 4.6 value equals 1.24. This number could be obtained by applying a *good* rejection rule (in the sense of Section 1.4) followed by the arithmetic mean, a combined procedure which may be considered a robust estimator.

Summarizing these results, we note that all the robust estimates range from 1.24 to 1.40, leaving a clear gap up to the arithmetic mean 1.58. Even estimators which are highly correlated with the mean in the case of strictly normal samples, like 10% and H/L (both being about 95% efficient) are now clearly separated from it. In this example, all the robust estimators give similar results. However, in general they may yield quite different answers, because some of the estimators are more robust than others (e.g., the median is much more robust than the 10%-trimmed mean). In multivariate situations, the differences between robust estimators become more important.

In order to understand why certain estimators behave the way they do, it is necessary to look at various measures of robustness and descriptive tools. A first question is to describe the effect of the outlier in the Cushny and Peebles data on, say, the arithmetic mean. For this purpose we can resort to the *influence function* (IF) of Section 2.1, which (roughly speaking) formalizes the bias caused by one outlier. (Subsection 2.1b contains the asymptotic definition of the IF, but Subsection 2.1e gives a finite-sample version which is illustrated on the Cushny and Peebles data.) A second question is How much contaminated data can an estimator stand before it becomes useless? In the above example, all robust estimators were able to deal with one outlier. However, if the data would contain two outliers the 10%-trimmed mean could be fooled, because it removes only one observation on either side: in case both outliers lie on the same side, one will remain. It takes three outliers (in a sample of ten points) to do the same thing to the 20%-trimmed mean, whereas the median can stand up to four outliers. This aspect is covered by the *breakdown point* of an estimator, discussed in Section 2.2. In Section 2.3 some useful types of estimators are introduced, mostly for location and scale parameters. By means of the influence function, optimal robust estimators are constructed in Section 2.4. The next tool for asserting robustness is the *change-of-variance function* (CVF) which reflects the influence of outliers on the variance of the estimator, and hence on the lengths of the confidence intervals (which hopefully remain relatively narrow). Section 2.6 deals with a class of refined robust estimators which can reject outliers completely, and studies their robustness by means of IF and CVF. In the last section, we compare our approach to Huber's (1964) minimax technique.

Robustness theory helps us to understand the behavior of statistical procedures in real-life situations. The various tools (like IF, CVF, and breakdown point) describe different aspects of this problem, and are therefore complementary to each other. When faced with new models or types of application other than the ones treated here, these principles offer guidance for choosing among several available statistical methods, or even for constructing new ones.

2.1. THE INFLUENCE FUNCTION

We start our investigation of robustness by means of the influence function (IF), which was introduced by Hampel (1968, 1974). The first subsection describes the necessary framework, followed by the definition of the IF in Subsection 2.1b. Then some useful robustness measures are derived from the IF, like the gross-error sensitivity, and a few examples are given. In the last subsection we look at some finite-sample modifications.

2.1a. Parametric Models, Estimators, and Functionals

Suppose we have one-dimensional observations $X_1, \ldots, X_n$ which are independent and identically distributed (i.i.d.). The observations belong to some sample space $\mathcal{X}$, which is a subset of the real line $\mathbb{R}$ (often $\mathcal{X}$ simply equals $\mathbb{R}$ itself, so the observations may take on any value). A *parametric model* consists of a family of probability distributions F_θ on the sample space, where the unknown parameter θ belongs to some parameter space Θ. In classical statistics, one then assumes that the observations X_i are distributed *exactly* like one of the F_θ, and undertakes to estimate θ based on the data at hand. Some well-known examples are:

1. $\mathcal{X} = \mathbb{R}$, $\Theta = \mathbb{R}$, and F_θ is the normal distribution with mean θ and standard deviation 1.
2. $\mathcal{X} = [0, \infty)$, $\Theta = (0, \infty)$, and F_θ is the exponential distribution with expectation θ.
3. $\mathcal{X} = \{0, 1, 2, \ldots\}$ (the set of nonnegative integers), $\Theta = (0, \infty)$, and F_θ is the Poisson distribution with risk parameter θ.
4. $\mathcal{X} = \{0, 1, \ldots, N\}$, $\Theta = (0, 1)$, and F_θ is the binomial distribution with probability θ of success.

Throughout this chapter, we will assume that Θ is an open convex subset of $\mathbb{R}$, and we will use the same notation F_θ for a probability distribution and its corresponding cumulative distribution function (cdf). Moreover, we assume that these F_θ have densities f_θ with respect to some σ-finite measure λ on $\mathscr{X}$ (in examples 1 and 2, λ would be the Lebesgue measure, and in examples 3 and 4 the counting measure). Indeed, many statistical procedures make use of the likelihood $f_\theta(x)$.

In the classical theory of statistical inference, one adheres strictly to these parametric models. However, in robustness theory we realize that the model $\{F_\theta; \ \theta \in \Theta\}$ is a mathematical abstraction which is only an *idealized approximation* of reality, and we wish to construct statistical procedures which still behave fairly well under deviations from this assumed model. Therefore, we do not only consider the distribution of estimators under the model, but also under other probability distributions. In this and the following chapters we shall treat deviations from the shape of F_θ, including gross errors (violations of the assumption of independence will be discussed in Section 8.1).

We identify the sample $(X_1, \ldots, X_n)$ with its empirical distribution G_n, ignoring the sequence of the observations (as is almost always done). Formally, G_n is given by $(1/n)\sum_{i=1}^n \Delta_{x_i}$, where Δ_x is the point mass 1 in X. As estimators of θ we consider real-valued statistics $T_n = T_n(X_1, \ldots, X_n) = T_n(G_n)$. In a broader sense, an estimator can be viewed as a sequence of statistics $\{T_n; \ n \geq 1\}$, one for each possible sample size n. Ideally, the observations are i.i.d. according to a member of the parametric model $\{F_\theta; \ \theta \in \Theta\}$, but the class $\mathscr{F}(\mathscr{X})$ of all possible probability distributions on $\mathscr{X}$ is much larger.

We consider estimators which are functionals [i.e., $T_n(G_n) = T(G_n)$ for all n and G_n] or can asymptotically be replaced by functionals. This means that we assume that there exists a functional $T:\text{domain}(T) \to \mathbb{R}$ [where the domain of T is the set of all distributions in $\mathscr{F}(\mathscr{X})$ for which T is defined] such that

$$T_n(X_1, \ldots, X_n) \underset{n \to \infty}{\to} T(G) \qquad (2.1.1)$$

in probability when the observations are i.i.d. according to the true distribution G in domain(T). We say that $T(G)$ is the *asymptotic value* of $\{T_n; \ n \geq 1\}$ at G. In addition to this property we often assume asymptotic normality, that is,

$$\mathscr{L}_G\big(\sqrt{n}\,[T_n - T(G)]\big) \xrightarrow[n \to \infty]{\text{weakly}} N(0, V(T, G)), \qquad (2.1.2)$$

where $\mathscr{L}_G$ means "the distribution of...under G'' and $V(T, G)$ is called the *asymptotic variance* of $\{T_n; n \geq 1\}$ at G. In practice, these conditions will often be met. Note that this true distribution G does not have to belong to the model. In fact, it will often deviate slightly from it.

In this chapter, we always assume that the functionals under study are Fisher consistent (Kallianpur and Rao, 1955):

$$T(F_\theta) = \theta \quad \text{for all } \theta \text{ in } \Theta, \tag{2.1.3}$$

which means that *at* the model the estimator $\{T_n; n \geq 1\}$ asymptotically measures the right quantity. The notion of Fisher consistency is more suitable and elegant for functionals than the usual consistency or asymptotic unbiasedness.

2.1b. Definition and Properties of the Influence Function

The cornerstone of the infinitesimal approach is the influence function (IF), which was invented by F. Hampel (1968, 1974) in order to investigate the infinitesimal behavior of real-valued functionals such as $T(G)$. The IF is mainly a *heuristic tool*, with an important intuitive interpretation. However, before using it as such, let us first reassure the mathematically oriented reader by discussing how it can be derived in a theoretical way, without going into details. We take a restricted point of view by choosing only one out of many formalizations, which is sufficient for our purposes. Suppose that domain (T) is a convex subset of the set of all finite signed measures on $\mathscr{X}$, containing more than one element. Following Huber (1977a) and Reeds (1976), we say that T is Gâteaux differentiable at the distribution F in domain (T), if there exists a real function a_1 such that for all G in domain (T) it holds that

$$\lim_{t \to 0} \frac{T((1-t)F + tG) - T(F)}{t} = \int a_1(x) \, dG(x)$$

which may also be written as

$$\frac{\partial}{\partial t} [T((1-t)F + tG)]_{t=0} = \int a_1(x) \, dG(x). \tag{2.1.4}$$

(Note that the left member is the directional derivative of T at F, in the direction of G.) There also exist stronger concepts, namely Fréchet differentiability and compact differentiability (Reeds, 1976, Section 2.2; Boos and

Serfling, 1980; Huber, 1981, Section 2.5; Clarke, 1983; Fernholz, 1983). The basic idea of differentiation of statistical functionals goes back to von Mises (1937, 1947) and Filippova (1962); one says in the above situation that T is a *von Mises functional*, with first kernel function a_1. By putting $G = F$ in (2.1.4) it is clear that

$$\int a_1(x)\, dF(x) = 0 \qquad (2.1.5)$$

so we may replace $dG(x)$ by $d(G - F)(x)$ in (2.1.4).

At this point the practical meaning of $a_1(x)$ is not yet evident; it is obscured by the fact that this kernel function occurs only implicitly in (2.1.4). An explicit expression may be obtained by inserting $G = \Delta_x$ (the probability measure which puts mass 1 at the point x) in (2.1.4), if Δ_x is in domain (T). Let us now take this expression as a definition:

Definition 1. The *influence function* (IF) of T at F is given by

$$\mathrm{IF}(x; T, F) = \lim_{t \downarrow 0} \frac{T((1 - t)F + t\Delta_x) - T(F)}{t} \qquad (2.1.6)$$

in those $x \in \mathcal{X}$ where this limit exists.

The existence of the influence function is an even weaker condition than Gâteaux differentiability, as is exemplified by Huber (1977a, p. 10). This makes the range of its applicability very large, as it can be calculated in all realistic situations without bothering about the (usually quite messy) regularity conditions. (It might even be attempted to verify such regularity conditions from the behavior of the IF in a full neighborhood of F.) If we replace F by $F_{n-1} \approx F$ and put $t = 1/n$, we realize that the IF measures approximately n times the change of T caused by an additional observation in x when T is applied to a large sample of size $n - 1$. (More details about such approximations can be found in Subsection 2.1e.)

The importance of the influence function lies in its heuristic interpretation: it describes the effect of an infinitesimal contamination at the point x on the estimate, standardized by the mass of the contamination. One could say it gives a picture of the infinitesimal behavior of the asymptotic value, so it measures the asymptotic bias caused by contamination in the observations. The IF was originally called the "influence curve" (Hampel, 1968, 1974); however, nowadays we prefer the more general name "influence function" in view of the generalizations to higher dimensions. [The notation

$\Omega(x; T, F)$ is also sometimes used in the literature.] If some distribution G is "near" F, then the first-order von Mises expansion of T at F (which is derived from a Taylor series) evaluated in G is given by

$$T(G) = T(F) + \int \mathrm{IF}(x; T, F)\, d(G - F)(x) + \text{remainder.} \qquad (2.1.7)$$

Mallows (1975) also studied terms of higher order.

Let us now look at the important relation between the IF and the asymptotic variance. When the observations X_i are i.i.d. according to F, then the empirical distribution F_n will tend to F by the Glivenko–Cantelli theorem. Therefore, in (2.1.7) we may replace G by F_n for sufficiently large n. Let us also assume that $T_n(X_1, \ldots, X_n) = T_n(F_n)$ may be approximated adequately by $T(F_n)$. Making use of (2.1.5), which we can rewrite as $\int \mathrm{IF}(x; T, F)\, dF(x) = 0$, we obtain

$$T_n(F_n) \simeq T(F) + \int \mathrm{IF}(x; T, F)\, dF_n(x) + \text{remainder.}$$

Evaluating the integral over F_n and rewriting yields

$$\sqrt{n}\,(T_n - T(F)) \simeq \frac{1}{\sqrt{n}} \sum_{i=1}^{n} \mathrm{IF}(X_i; T, F) + \text{remainder.}$$

The first term on the right-hand side is asymptotically normal by the central limit theorem. In most cases the remainder becomes negligible for $n \to \infty$, so T_n itself is asymptotically normal. That is, $\mathscr{L}_F(\sqrt{n}\,[T_n - T(F)])$ tends to $N(0, V(T, F))$, where the asymptotic variance equals

$$V(T, F) = \int \mathrm{IF}(x; T, F)^2\, dF(x). \qquad (2.1.8)$$

A rigorous mathematical treatment, using diverse sets of regularity conditions, can be found in Reeds (1976), Boos and Serfling (1980), and Fernholz (1983). But for us the IF remains chiefly a heuristic tool; in fact, when one has calculated the formal asymptotic variance of a certain estimator by means of (2.1.8), it is usually easier to verify the asymptotic normality in another way instead of trying to assess the necessary regularity conditions to make this approach rigorous. The important thing to remember is that (2.1.8) gives the right answer in all practical cases we know of. Furthermore, this formula can be used to calculate the asymptotic relative efficiency $\mathrm{ARE}_{T,S} = V(S, F)/V(T, F)$ of a pair of estimators $\{T_n; \ n \geq 1\}$ and $\{S_n; \ n \geq 1\}$.

It is also possible to recover the asymptotic Cramér–Rao inequality (or "asymptotic information inequality") for a sequence of estimators $\{T_n; \ n \geq 1\}$ for which the corresponding functional T is Fisher consistent. Denote the density of F_θ by f_θ, and put $F_* := F_{\theta_*}$ where θ_* is some fixed member of Θ. The Fisher information at F_* equals

$$J(F_*) = \int \left(\frac{\partial}{\partial \theta} [\ln f_\theta(x)]_{\theta_*} \right)^2 dF_*. \qquad (2.1.9)$$

Suppose that $0 < J(F_*) < \infty$. Assuming that the first-order von Mises approximation (2.1.7) behaves well, and making use of (2.1.5) and (2.1.3) we obtain

$$\frac{\partial}{\partial \theta} \left[\int \mathrm{IF}(x; T, F_*) \, dF_\theta \right]_{\theta_*} = \frac{\partial}{\partial \theta} [T(F_\theta)]_{\theta_*} = \left[\frac{\partial \theta}{\partial \theta} \right]_{\theta_*} = 1.$$

Changing the order of differentiation and integration, this yields

$$1 = \int \mathrm{IF}(x; T, F_*) \frac{\partial}{\partial \theta} [f_\theta(x)]_{\theta_*} d\lambda(x)$$

$$= \int \mathrm{IF}(x; T, F_*) \frac{\partial}{\partial \theta} [\ln f_\theta(x)]_{\theta_*} dF_*(x). \qquad (2.1.10)$$

Making use of Cauchy–Schwarz we obtain the Cramér–Rao inequality:

$$\int \mathrm{IF}(x; T, F_*)^2 \, dF_*(x) \geq \frac{1}{J(F_*)}, \qquad (2.1.11)$$

where equality holds if and only if

$$\mathrm{IF}(x; T, F_*) \text{ is proportional to } \frac{\partial}{\partial \theta} [\ln f_\theta(x)]_{\theta_*} \qquad (2.1.12)$$

(almost everywhere). This means that the estimator is asymptotically efficient if and only if $\mathrm{IF}(x; T, F_*) = J(F_*)^{-1}(\partial/\partial\theta)[\ln f_\theta(x)]_{\theta_*}$. (The factor $J(F_*)^{-1}$ comes from (2.1.10).) We shall see later on that this is indeed the IF of the maximum likelihood estimator. Making use of (2.1.8), one can easily calculate the (absolute) asymptotic efficiency of an estimator, which is given by $e := [V(T, F_*)J(F_*)]^{-1}$.

However, the IF did not become popular merely because of properties (2.1.8) and (2.1.11), useful as these may be, but mostly because it allows us to study several local robustness properties, to deepen our understanding of certain estimators, and to derive new estimators with prespecified characteristics.

2.1c. Robustness Measures Derived from the Influence Function

From the robustness point of view, there are at least three important summary values of the IF apart from its expected square. They were introduced by Hampel (1968, 1974). The first and most important one is the supremum of the absolute value. Indeed, we have seen that the IF describes the standardized effect of an (infinitesimal) contamination at the point x on (the asymptotic value of) the estimator. Therefore, one defines the *gross-error sensitivity* of T at F by

$$\gamma^* = \sup_x |\mathrm{IF}(x; T, F)|, \qquad (2.1.13)$$

the supremum being taken over all x where $\mathrm{IF}(x; T, F)$ exists. A more complete notation is $\gamma^*(T, F)$. The gross-error sensitivity measures the worst (approximate) influence which a small amount of contamination of fixed size can have on the value of the estimator. Therefore, it may be regarded as an upper bound on the (standardized) asymptotic bias of the estimator. It is a desirable feature that $\gamma^*(T, F)$ be finite, in which case we say that T is *B-robust* at F (Rousseeuw, 1981a). Here, the B comes from "bias." Typically, putting a bound on γ^* is the first step in robustifying an estimator, and this will often conflict with the aim of asymptotic efficiency. Therefore, we shall look for *optimal B-robust* estimators (Section 2.4) which cannot be improved simultaneously with respect to $V(T, F)$ and $\gamma^*(T, F)$. On the other hand, in most cases there exists a positive minimum of γ^* for Fisher-consistent estimators, belonging to the so-called *most B-robust* estimator (Subsections 2.5c and 2.6b).

The second summary value has to do with small fluctuations in the observations. When some values are changed slightly (as happens in rounding and grouping and due to some local inaccuracies), this has a certain measurable effect on the estimate. Intuitively, the effect of shifting an observation slightly from the point x to some neighboring point y can be measured by means of $\mathrm{IF}(y; T, F) - \mathrm{IF}(x; T, F)$, because an observation is added at y and another one is removed at x. Therefore, the standardized effect of "wiggling" around x is approximately described by a normalized difference or simply the slope of the IF in that point. A measure for the worst (approximate and standardized) effect of "wiggling" is therefore provided by the *local-shift sensitivity*

$$\lambda^* = \sup_{x \neq y} |\text{IF}(y; T, F) - \text{IF}(x; T, F)| / |y - x|, \qquad (2.1.14)$$

the smallest Lipschitz constant the IF obeys. However, note that even an infinite value of λ^* may refer to a very limited actual change, because of the standardization. Therefore, this notion is quite different from "the largest discontinuity jump of the IF."

It is an old robustness idea to reject extreme outliers entirely, and estimators with this property go back at least to Daniel Bernoulli in 1769 (Stigler, 1980). In the language of the influence function, this means that the IF vanishes outside a certain area. Indeed, if the IF is identically zero in some region, then contamination in those points does not have any influence at all. In case the underlying distribution F is symmetric (and putting the center of symmetry equal to zero), this allows us to define the third summary value, the *rejection point*

$$\rho^* = \inf\{r > 0; \text{IF}(x; T, F) = 0 \text{ when } |x| > r\}. \qquad (2.1.15)$$

(If there exists no such r, then $\rho^* = \infty$ by definition of the infimum.) All observations farther away than ρ^* are rejected completely. Therefore, it is a desirable feature if ρ^* is finite; such estimators will be studied in Section 2.6.

2.1d. Some Simple Examples

In order to clarify the notions introduced, we shall now treat some examples. Let us take the sample space $\mathcal{X} = \mathbb{R}$ and the parameter space $\Theta = \mathbb{R}$. One of the simplest statistical structures is the normal location model, given by $F_\theta(x) = \Phi(x - \theta)$, where Φ is the standard normal cdf with density $\phi(x) = (2\pi)^{-1/2}\exp(-\frac{1}{2}x^2)$. Let $\theta_0 = 0$, so $F_{\theta_0} = \Phi$.

Let us start by considering the arithmetic mean, probably the most widely known estimator of all times. Here $T_n = (1/n)\sum_{i=1}^n X_i$, and the corresponding functional $T(G) = \int u\, dG(u)$ is defined for all probability measures with existing first moment. Obviously T is Fisher consistent (2.1.3). From (2.1.6), it follows that

$$\text{IF}(x; T, \Phi) = \lim_{t \downarrow 0} \frac{\int u \, d\left[(1 - t)\Phi + t\Delta_x\right](u) - \int u \, d\Phi(u)}{t}$$

$$= \lim_{t \downarrow 0} \frac{(1 - t)\int u \, d\Phi(u) + t\int u \, d\Delta_x(u) - \int u \, d\Phi(u)}{t}$$

$$= \lim_{t \downarrow 0} \frac{tx}{t}$$

because $\int u \, d\Phi(u) = 0$; hence

$$\text{IF}(x; T, \Phi) = x. \tag{2.1.16}$$

Clearly, $\int \text{IF}(x; T, \Phi) \, d\Phi(x) = 0$ and $V(T, \Phi) = 1 = \int \text{IF}(x; T, \Phi)^2 \, d\Phi(x)$. Now $J(\Phi) = 1$, so $\int \text{IF}(x; T, \Phi)^2 \, d\Phi(x) = J(\Phi)^{-1}$, and indeed $\text{IF}(x; T, \Phi)$ is proportional to $(\partial/\partial\theta)[\ln f_\theta(x)]_{\theta_0}$. Therefore the arithmetic mean, which is the maximum likelihood estimator for this model, has asymptotic efficiency $e = 1$.

So far we have used the IF to show the good efficiency properties of the mean, but let us now turn to its robustness characteristics. First, the gross-error sensitivity γ^* equals infinity, so the mean is not B-robust. This reflects the fact that, for any sample size, even a single outlier can carry the estimate over all bounds (if it is far enough). It is of small comfort that the local-shift sensitivity $\lambda^* = 1$ is finite, which is much less important. [In fact, this is even the lowest value of λ^* possible for "well-behaved" Fisher-consistent functionals: from (2.1.10) it follows that $\int \text{IF}'(x; T, \Phi) \, d\Phi(x) = 1$ using $(\partial/\partial\theta)[f_\theta(x)]_0 = -(d/dx)\phi(x)$ and integrating by parts, hence it is impossible that $\text{IF}'(x; T, \Phi) < 1$ for all x.] Moreover, the rejection point ρ^* is infinite too. Concluding, we may say that the mean is least susceptible to rounding errors, but that it is dangerous in any situation where outliers might occur (which is to say, nearly always).

In the same framework one can also use the sample median. If n is odd, then T_n simply equals the middle order statistic; if n is even, then T_n is the average of the $(n/2)$th and the $(n/2 + 1)$th order statistics. The corresponding functional is given by $T(G) = G^{-1}(\frac{1}{2})$, where in the case of nonuniqueness one chooses the midpoint of the interval $\{t; G(t) = \frac{1}{2}\}$, and in the case that G has a discontinuous jump passing $\frac{1}{2}$, one chooses the point where this jump occurs. This functional is clearly Fisher consistent, and it will be

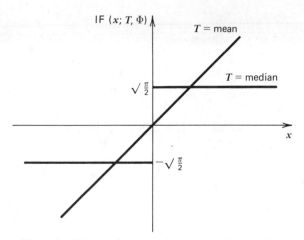

Figure 1. Influence functions of the mean and the median.

shown in Subsection 2.3a that

$$\text{IF}(x; T, \Phi) = \frac{\text{sign}(x)}{2\phi(0)}. \qquad (2.1.17)$$

(This influence function is drawn in Figure 1, together with that of the mean.) Again, $\int \text{IF}(x; T, \Phi) \, d\Phi(x) = 0$. The asymptotic variance is $V(T, \Phi) = \int \text{IF}(x; T, \Phi)^2 \, d\Phi(x) = (2\phi(0))^{-2} = \pi/2 \simeq 1.571$, so the asymptotic efficiency now equals $e = 2/\pi \simeq 0.637$.

The median possesses a finite gross-error sensitivity $\gamma^* = (2\phi(0))^{-1} = (\pi/2)^{1/2} \simeq 1.253$, so it is B-robust. Indeed, 1.253 is even the minimal value, as will be shown in Subsection 2.5c, so the median is most B-robust. Because the IF has a jump at zero we have $\lambda^* = \infty$, which means that the median is sensitive to "wiggling" near the center of symmetry. (However, this is not nearly as important as the fact that γ^* is finite.) And finally, again the rejection point ρ^* is infinite. This may seem surprising, but we stress that the median does *not* reject outliers, the latter having a certain (fixed) influence. Indeed, if one adds an outlier to the sample, then the "middle" order statistic will move in that direction.

Although the median is already much more robust than the mean, it can still be improved from our point of view. Indeed, it is possible to devise estimators such as the "three-part redescending M-estimators" and the "tanh-estimators" (see Section 2.6), which (for suitable choices of the constants involved):

Possess a γ^* which is still quite low.
Are much more efficient than the median.
Possess a low local-shift sensitivity λ^*.
Possess a finite rejection point ρ^*.

In order to show that robustness theory is not only for location parameters and symmetric distributions, let us now look at the Poisson model. Here the sample space $\mathscr{X}$ equals the set of nonnegative integers $\{0, 1, 2, \ldots\}$ and λ is the counting measure. The parametric model states that the observations $X_1, \ldots, X_n$ are i.i.d. with respect to some Poisson distribution F_θ, where the unknown risk parameter θ belongs to the sample space $\Theta = (0, \infty)$. The density f_θ of F_θ with respect to λ is given by

$$f_\theta(k) = \frac{\theta^k e^{-\theta}}{k!}.$$

This model has no symmetry or invariance. By maximizing $\Pi_{i=1}^n f_\theta(X_i)$ we immediately find that the maximum likelihood estimator for θ is $\overline{X} = (1/n)\Sigma_{i=1}^n X_i$. This estimator can also be written as the functional

$$T(F) = \int_{\mathscr{X}} u \, dF(u) = \int_{\mathscr{X}} uf(u) \, d\lambda(u) = \sum_{k=0}^{\infty} kf(k), \quad (2.1.18)$$

where F is any distribution on $\mathscr{X} = \{0, 1, 2, \ldots\}$ with density f. This functional is clearly Fisher consistent, because

$$T(F_\theta) = \sum_{k=0}^{\infty} k \frac{\theta^k e^{-\theta}}{k!} = \theta \sum_{k=1}^{\infty} \frac{\theta^{k-1} e^{-\theta}}{(k-1)!} = \theta$$

for all θ in $\Theta = (0, \infty)$. Let us now compute the influence function of T at some Poisson distribution $F_* = F_{\theta_*}$. Because of Definition 1, the contaminating point masses may only occur in points $x \in \mathscr{X}$, so the IF can only be calculated at integer values of x:

$$\mathrm{IF}(x; T, F_*) = \lim_{t \downarrow 0} \frac{\Sigma_{k=0}^{\infty} k \left[(1-t)f_*(k) + t1_{\{x\}}(k)\right] - \Sigma_{k=0}^{\infty} kf_*(k)}{t}$$

$$= \lim_{t \downarrow 0} \frac{t\Sigma_{k=0}^{\infty} k1_{\{x\}}(k) - t\Sigma_{k=0}^{\infty} kf_*(k)}{t}$$

$$= x - \theta_*. \quad (2.1.19)$$

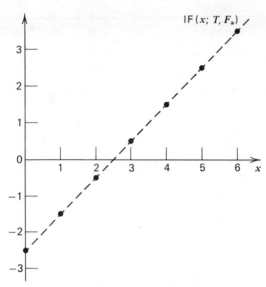

Figure 2. Influence function of the arithmetic mean at the Poisson distribution F_* with risk parameter $\theta_* = 2.5$. The influence function is only defined for integer arguments.

Figure 2 displays the influence function for a specific value of θ_*. Note that the IF is a discrete set of points! By means of (2.1.8) we can compute the asymptotic variance of our estimator:

$$V(T, F_*) = \int_{\mathscr{X}} \mathrm{IF}(x; T, F_*)^2 \, dF_*(x)$$

$$= \sum_{k=0}^{\infty} (k - \theta_*)^2 f_*(k) = \theta_*.$$

Because the Fisher information (2.1.9) equals $J(F_*) = \theta_*^{-1}$, it follows that the asymptotic efficiency is $e = [V(T, F_*)J(F_*)]^{-1} = 100\%$. However, the influence function is unbounded; hence $\gamma^* = \infty$ so the estimator is not B-robust.

2.1e. Finite-Sample Versions

The above definition of the influence function was entirely asymptotic, because it focused on functionals which coincide with the estimator's asymptotic value. However, there exist also some simple finite-sample versions, which are easily computed.

The simplest idea is the so-called *empirical influence function* (Hampel et al., 1982). The first definition is by means of an additional observation. Suppose we have an estimator $\{T_n; \ n \geq 1\}$ and a sample $(x_1, \ldots, x_{n-1})$ of $n - 1$ observations. Then the empirical IF of the estimator at that sample is a plot of

$$T_n(x_1, \ldots, x_{n-1}, x) \qquad (2.1.20)$$

as a function of x. Alternatively, one can define it by *replacing* an observation: when the original sample consists of n observations, one can replace one of them (say x_n) by an arbitrary x and again plot (2.1.20), which goes through the actual estimate $T_n(x_1, \ldots, x_n)$ for $x = x_n$. This second version is particularly useful when x_n is already an outlier. For example, for the Cushny and Peebles data we have $n = 10$ and $x_n = 4.6$. Figure 3 displays the empirical IF for the arithmetic mean $\bar{X}$ and the estimators 10% and 50%. It simply shows what the estimates might have been for other values of the last observation.

The curves for the arithmetic mean and the median look very much like the (asymptotic) influence functions in Figure 1 of this section. The most important aspect is that the empirical IF of the mean becomes unbounded for both $x \to +\infty$ and $x \to -\infty$, whereas the median remains constant. We will see later on that the asymptotic IF of 10% also looks very much like the empirical one in Figure 3.

The second tool is Tukey's (1970/71) *sensitivity curve*. Again, there are two versions, one with addition and one with replacement. In the case of an additional observation, one starts with a sample $(x_1, \ldots, x_{n-1})$ of $n - 1$ observations and defines the sensitivity curve as

$$SC_n(x) = n[T_n(x_1, \ldots, x_{n-1}, x) - T_{n-1}(x_1, \ldots, x_{n-1})]. \qquad (2.1.21)$$

This is simply a translated and rescaled version of the empirical IF. When the estimator is a functional [i.e., when $T_n(x_1, \ldots, x_n) = T(F_n)$ for any n, any sample $(x_1, \ldots, x_n)$, and corresponding empirical distribution F_n], then

$$SC_n(x) = \left[T\left(\left(1 - \frac{1}{n}\right)F_{n-1} + \frac{1}{n}\Delta_x \right) - T(F_{n-1}) \right] \Big/ \frac{1}{n}$$

where F_{n-1} is the empirical distribution of $(x_1, \ldots, x_{n-1})$. This last expression is a special case of (2.1.6), with F_{n-1} as an approximation for F and

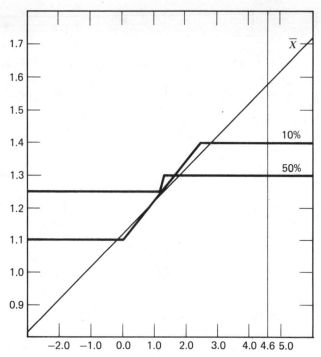

Figure 3. Empirical influence functions of the arithmetic mean, the 10%-trimmed mean, and the median, computed for the Cushny and Peebles data where the last observation 4.6 has been replaced by an arbitrary value x.

with contamination size $t = 1/n$. In many situations, $\text{SC}_n(x)$ will therefore converge to $\text{IF}(x; T, F)$ when $n \to \infty$.

Instead of basing $\text{SC}_n(x)$ on an actual sample $(x_1, \ldots, x_{n-1})$, one can also construct so-called *stylized sensitivity curves* (Andrews et al., 1972, p. 96) by starting with an artificial "sample" consisting of the $n - 1$ expected order statistics (from a random sample of $n - 1$). When F is the standard normal, these can easily be found. Another possibility is to put $x_i :=$ $\Phi^{-1}(i/n)$, where Φ is the standard normal cdf. In both situations the original "sample" is symmetric, so that often $T_{n-1}(x_1, \ldots, x_{n-1})$ will vanish, simplifying the computation. In Andrews et al. (1972, p. 97–101) many examples are plotted, including the arithmetic mean and the median which give rise to SC curves virtually identical to Figure 1.

A third important tool is the *jackknife* (Quenouille, 1956; Tukey, 1958). Consider again an estimator which is a functional, and compute $T_n(x_1, \ldots, x_n) = T(F_n)$ for a sample of size n. Then the *ith jackknifed pseudovalue* is defined as

$$T_{ni}^* = nT_n(x_1,\ldots,x_n) - (n-1)T_{n-1}(x_1,\ldots,x_{i-1},x_{i+1},\ldots,x_n).$$

$$(2.1.22)$$

For the arithmetic mean, we simply find $T_{ni}^* = x_i$. In general, a sample $(x_1,\ldots,x_n)$ and a functional T yield a pseudosample $(T_{n1}^*,\ldots,T_{nn}^*)$, which is used to compute a "corrected" estimate

$$T_n^* = \frac{1}{n} \sum_{i=1}^{n} T_{ni}^* \qquad (2.1.23)$$

which often has a smaller bias than the original T_n. The variance of the n pseudovalues equals

$$V_n = \frac{1}{n-1} \sum_{i=1}^{n} \left(T_{ni}^* - T_n^*\right)^2. \qquad (2.1.24)$$

Tukey (1958) found that $(1/n)V_n$ is often a good estimate of the variance of the original estimator T_n. In recent years, such variances have also been estimated by means of the *bootstrap*, a tool which has close connections to the jackknife (Efron, 1979). The jackknife is also related to the influence function. If we apply (2.1.6) to compute $\mathrm{IF}(x_i; T, F_n)$ and plug in $t = -1/(n-1)$ [negative values of t are allowed in Eq. (2.1.4)], we find

$$\left[T\!\left(\left(1 - \frac{-1}{n-1}\right)F_n + \left(\frac{-1}{n-1}\right)\Delta_{x_i}\right) - T(F_n)\right]\Big/ \frac{-1}{n-1}$$

$$= (n-1)\{T_n(x_1,\ldots,x_n) - T_{n-1}(x_1,\ldots,x_{i-1},x_{i+1},\ldots,x_n)\}$$

$$= T_{ni}^* - T_n(x_1,\ldots,x_n).$$

Formula (2.1.8) for the asymptotic variance gives rise to the following finite-sample approximation:

$$V(T,F) = \int \mathrm{IF}(x; T, F)^2 \, dF(x) = \mathrm{Var}_F[\mathrm{IF}(X; T, F)]$$

$$\approx \mathrm{Var}_{F_n}[\mathrm{IF}(X; T, F_n)] \approx \mathrm{Var}\{\mathrm{IF}(x_i; T, F_n); i = 1,\ldots,n\}$$

$$\approx \mathrm{Var}\{T_{ni}^* - T_n(x_1,\ldots,x_n); i = 1,\ldots,n\}$$

$$= \mathrm{Var}\{T_{ni}^*; i = 1,\ldots,n\} = V_n$$

which yields Eq. (2.1.24), corresponding to Tukey's estimate.

2.2. THE BREAKDOWN POINT AND QUALITATIVE ROBUSTNESS

2.2a. Global Reliability: The Breakdown Point

In the first section we introduced the influence function, which describes the infinitesimal stability of (the asymptotic value of) an estimator. The influence function is a collection of directional derivatives in the directions of the point masses Δ_x, and is usually evaluated at the model distribution F. It is a very useful tool, giving rise to important robustness measures such as the gross-error sensitivity γ^*, but there is one limitation—by construction, it is an entirely *local* concept. Therefore, it must be complemented by a measure of the *global* reliability of the estimator, which describes up to what distance from the model distribution the estimator still gives some relevant information.

First, let us start by giving a formal meaning to the word "distance." One distinguishes three main reasons why a parametric model does not hold exactly:

(i)   Rounding of the observations.

(ii)   Occurrence of gross errors.

(iii)   The model is only an idealized approximation of the underlying chance mechanism.

The sample space $\mathcal{X}$ is a subset of $\mathbb{R}$ as in Subsection 2.1a. The *Prohorov distance* (Prohorov, 1956) of two probability distributions F and G in $\mathcal{F}(\mathcal{X})$ is given by

$$\pi(F, G) := \inf\{\varepsilon;\ F(A) \leq G(A^\varepsilon) + \varepsilon \text{ for all events } A\}, \quad (2.2.1)$$

where A^ε is the set of all points whose distance from A is less than ε. When one looks closer at this definition, it becomes clear that it formalizes (i) and (ii) above. (Indeed, A^ε stands for rounding the observations of the event A, whereas the term $+\varepsilon$ indicates that a fraction ε of the data may be very differently distributed.) It also takes care of (iii) because this distance induces weak convergence on $\mathcal{F}(\mathcal{X})$. This is important because many approximations (e.g., by virtue of the central limit theorem) are in the sense of weak convergence.

Second, let us make more precise what is meant by saying that an estimator "still gives some relevant information." Let us recollect that the parameter space Θ is a subset of $\mathbb{R}$. For example, assume that the true value of the parameter θ at the underlying model distribution F is 3, say, and consider all distributions G at a Prohorov distance of, say, 0.25 or less from F. Clearly, our estimator becomes totally unreliable when the distributions in this set can yield estimates that are arbitrarily far away from 3. With this in mind, Hampel (1968, 1971) gave the following definition, generalizing an idea of Hodges (1967).

Definition 1. The *breakdown point* ε^* of the sequence of estimators $\{T_n; n \geq 1\}$ at F is defined by

$$\varepsilon^* := \sup\left\{ \varepsilon \leq 1; \text{ there is a compact set } K_\varepsilon \subsetneq \Theta \text{ such that} \right.$$

$$\left. \pi(F, G) < \varepsilon \text{ implies } G(\{T_n \in K_\varepsilon\}) \overset{n \to \infty}{\to} 1 \right\}. \qquad (2.2.2)$$

For instance, when $\Theta = \mathbb{R}$ we obtain $\varepsilon^* = \sup\{\varepsilon \leq 1;$ there exists r_ε such that $\pi(F, G) < \varepsilon$ implies $G(\{|T_n| \leq r_\varepsilon\}) \overset{n \to \infty}{\to} 1\}$. The breakdown point should formally be denoted as $\varepsilon^*(\{T_n; n \geq 1\}, F)$, but it usually does not depend on F. Compare also Subsection 1.3d for the intuitive interpretation of ε^*. Most importantly, it provides some guidance up to what distance from the model the local linear approximation provided by the IF can be used (see also Section 2.7).

Remark 1. There are some variants of the breakdown point, where in Definition 1 the Prohorov distance is replaced by the Lévy distance or the bounded Lipschitz metric, or even by the total variation distance or the Kolmogorov distance (for definitions and properties of these distances see Huber, 1981, Sections 2.3 and 2.4). One can also consider the *gross-error breakdown point* where $\pi(F, G) < \varepsilon$ is replaced by $G \in \{(1 - \varepsilon)F + \varepsilon H; H \in \mathscr{F}(\mathscr{X})\}$. Loosely speaking, this is the largest fraction of gross errors that never can carry the estimate over all bounds. This notion is much simpler and often leads to the same value of ε^*.

In recent years, people have become interested in finite-sample versions of the breakdown point. [Actually, already the definition of Hodges (1967) was written down for finite samples.] This yields much simpler notions, which do not contain probability distributions. Donoho and Huber (1983) list some possibilities; in our opinion, the most appealing and most general formalization corresponds to what they call "ε-replacement." A slightly different definition, which is more consistent with Definition 1, was given by Hampel et al. (1982, lecture 1). We shall now treat the latter.

Definition 2. The *finite-sample breakdown point* ε_n^* of the estimator T_n at the sample $(x_1, \ldots, x_n)$ is given by

$$\varepsilon_n^*(T_n; x_1, \ldots, x_n) := \frac{1}{n} \max\left\{ m; \max_{i_1, \ldots, i_m} \sup_{y_1, \ldots, y_m} |T_n(z_1, \ldots, z_n)| < \infty \right\},$$

$$(2.2.3)$$

where the sample $(z_1, \ldots, z_n)$ is obtained by replacing the m data points $x_{i_1}, \ldots, x_{i_m}$ by the arbitrary values $y_1, \ldots, y_m$.

Note that this breakdown point usually does not depend on $(x_1, \ldots, x_n)$, and depends only slightly on the sample size n. In many cases, taking the limit of ε_n^* for $n \to \infty$ yields the asymptotic breakdown point ε^* of Definition 1. Donoho and Huber (1983) take the smallest m for which the maximal supremum of $|T_n(z_1, \ldots, z_n)|$ is infinite, so their breakdown point equals $\varepsilon_n^* + 1/n$. For example, for the arithmetic mean we find $\varepsilon_n^* = 0$, whereas they obtain the value $1/n$.

In Subsection 1.4b, the breakdown point is used to investigate rejection rules for outliers in the one-dimensional location problem.

Finally, a word of caution. When computing some breakdown point, one must keep in mind that an estimator may sometimes "break down" in more ways than one. In certain situations, the possibility that the estimator becomes unbounded ("explosion") is not the only problem. For example, for scale estimators it is equally important to keep the estimate away from zero ("implosion"). This is covered by Definition 1, because the parameter space for scale is $(0, \infty)$ so any compact set has a positive distance from zero. Definition 2 is essentially for location estimators, and can be adapted to scale estimators by requiring that also $\min_{i_1, \ldots, i_m} \inf_{y_1, \ldots, y_m} T_n(z_1, \ldots, z_n) > 0$.

2.2b. Continuity and Qualitative Robustness

In the same context, Hampel (1971) also introduced some qualitative notions. The main idea is to complement the notion of differentiability (influence function) with continuity conditions, with respect to the Prohorov distance. Logically, continuity comes first; but since it yields only a simple dichotomy and not the rich quantitative structure of the IF, it has been deferred so far.

Definition 3. We say that a sequence of estimators $\{T_n; n \geq 1\}$ is *qualitatively robust at F* if for every $\varepsilon > 0$ there exists $\delta > 0$ such that for all

G in $\mathscr{F}(\mathscr{X})$ and for all n:

$$\pi(F, G) < \delta \Rightarrow \pi\big(\mathscr{L}_F(T_n), \mathscr{L}_G(T_n)\big) < \varepsilon.$$

Thus this definition describes equicontinuity of the distributions of T_n with respect to n.

Definition 4. A sequence $\{T_n; \, n \geq 1\}$ is called *continuous at F* when for every $\varepsilon > 0$ there exists $\delta > 0$ and n_0 such that for all $n, m \geq n_0$ and for all empirical cdf F_n, F_m:

$$\left.\begin{array}{l} \pi(F, F_n) < \delta \\ \pi(F, F_m) < \delta \end{array}\right\} \Rightarrow \big|T_n(F_n) - T_m(F_m)\big| < \varepsilon.$$

The notions of qualitative robustness and continuity are not identical, but they are closely linked by means of two theorems of Hampel (1971):

Theorem 1. A sequence $\{T_n; \, n \geq 1\}$ which is continuous at F and for which all T_n are continuous functions of the observations, is qualitatively robust.

Theorem 2. In case the estimators are generated by a functional T [i.e., $T_n(F_n) := T(F_n)$], then T is continuous (with respect to the Prohorov distance) at all F if and only if $\{T_n; \, n \geq 1\}$ is qualitatively robust at all F and satisfies $\pi\big(\mathscr{L}_F(T_n), \Delta_{T(F)}\big) \overset{n \to \infty}{\to} 0$ for all F.

Note that qualitative robustness is closely related to, but not identical with, a nonzero breakdown point.

Example 1. The arithmetic mean is nowhere qualitatively robust and nowhere continuous, with $\varepsilon^* = 0$. [This example shows that ordinary continuity of $T_n(x_1, \ldots, x_n)$ as a function of the observations $x_1, \ldots, x_n$ is not suitable for a robustness concept.]

Example 2. The median is qualitatively robust and continuous at F if $F^{-1}(\frac{1}{2})$ contains only one point, but note that it always satisfies $\varepsilon^* = \frac{1}{2}$.

Example 3. The α-trimmed mean $(0 < \alpha < \frac{1}{2})$, given by

$$T(F) = \int_{\alpha}^{1-\alpha} F^{-1}(t)\, dt / (1 - 2\alpha),$$

is qualitatively robust and continuous at all distributions, with $\varepsilon^* = \alpha$. This simple estimator is quite popular because it appeals to the intuition: one removes both the $[\alpha n]$ smallest and the $[\alpha n]$ largest observations, and calculates the mean of the remaining ones (cf. Subsection 2.3b).

Remark 2. Note that qualitative robustness is not equivalent to the B-robustness of Subsection 2.1b. The first condition has to do with continuity, whereas the second (requiring that the IF is bounded) is connected with differentiation. For example, at $F = \Phi$ the normal scores estimator (see Subsection 2.3c) is qualitatively robust but not B-robust, whereas the optimal L-estimator for the logistic distribution (see Subsection 2.3b) is B-robust but not qualitatively robust. However, for M-estimators there is a simple connection, as will be seen in Subsection 2.3a.

A slight variant of the concept of qualitative robustness, which has sometimes proven useful, is the notion of Π-robustness (Hampel, 1971, Section 5) which allows for some dependence within the n-tuples of observed outcomes, and also allows for slight changes of the distributions underlying the different observations. Thus one may avoid the strict assumption of independent identically distributed observations. (For a more thorough treatment of violations of the assumption of independence, see Section 8.1.)

More details about the breakdown point and qualitative robustness can be found in Hampel (1968, 1971, 1974, 1976) and Huber (1981, Sections 1.3 and 2.6; 1984).

2.3. CLASSES OF ESTIMATORS

2.3a. *M*-Estimators

General Case

Consider the estimation problem described in Subsection 2.1a. The well-known maximum likelihood estimator (MLE) is defined as the value $T_n = T_n(X_1, \ldots, X_n)$ which maximizes $\prod_{i=1}^{n} f_{T_n}(X_i)$, or equivalently by

$$\sum_{i=1}^{n} \left[-\ln f_{T_n}(X_i) \right] = \min_{T_n} ! \qquad (2.3.1)$$

where ln denotes the natural logarithm. Huber (1964) proposed to generalize this to

$$\sum_{i=1}^{n} \rho(X_i, T_n) = \min_{T_n} ! \qquad (2.3.2)$$

where ρ is some function on $\mathscr{X} \times \Theta$. Suppose that ρ has a derivative $\psi(x, \theta) = (\partial / \partial \theta)\rho(x, \theta)$, so the estimate T_n satisfies the implicit equation

$$\sum_{i=1}^{n} \psi(X_i, T_n) = 0. \qquad (2.3.3)$$

Definition 1. Any estimator defined by (2.3.2) or (2.3.3) is called an *M-estimator*.

The name "*M*-estimator" (Huber, 1964) comes from "generalized *maxi-mum* likelihood." Of course, (2.3.2) and (2.3.3) are not always equivalent, but usually (2.3.3) is very useful in the search for the solution of (2.3.2). The estimator is not altered when ψ is multiplied by any constant $r > 0$ [but a negative r would correspond to maximization in (2.3.2)]. For the sake of brevity, we shall often identify ψ with the *M*-estimator it defines.

If G_n is the empirical cdf generated by the sample, then the solution T_n of (2.3.3) can also be written as $T(G_n)$, where T is the functional given by

$$\int \psi(x, T(G)) \, dG(x) = 0 \qquad (2.3.4)$$

for all distributions G for which the integral is defined. Let us now replace G by $F_{t,x} = (1 - t)F + t\Delta_x$ and differentiate with respect to t, so

$$0 = \int \psi(y, T(F)) \, d(\Delta_x - F)$$

$$+ \int \frac{\partial}{\partial \theta} [\psi(y, \theta)]_{T(F)} \, dF(x) \cdot \frac{\partial}{\partial t} [T(F_{t,x})]_{t=0}$$

(if integration and differentiation may be interchanged). Making use of (2.1.6) and (2.3.4) we obtain

$$\text{IF } (x; \psi, F) = \frac{\psi(x, T(F))}{-\int (\partial / \partial \theta)[\psi(y, \theta)]_{T(F)} \, dF(y)} \qquad (2.3.5)$$

under the assumption that the denominator is nonzero. Therefore ψ is B-robust at F if and only if $\psi(\cdot, T(F))$ is bounded. By means of (2.1.8) the asymptotic variance may be calculated:

$$V(T, F) = \frac{\int \psi^2(x, T(F))\, dF(x)}{\left[\int (\partial/\partial\theta)[\psi(y, \theta)]_{T(F)}\, dF(y)\right]^2}.$$

Clarke (1983, 1984b) gave a mathematical justification of (2.3.5) by proving Fréchet differentiability of M-estimators in a general framework, even when ψ is not smooth. His results are very strong, because existence of the Fréchet derivative for statistical functionals implies existence of the weaker Hadamard derivative (also called compact derivative) of Reeds (1976) and Fernholz (1983), and the Gâteaux derivative discussed by Kallianpur (1963), a special case of which is the influence function.

The maximum likelihood estimator is also an M-estimator, corresponding to $\rho(x, \theta) = -\ln f_\theta(x)$, so $\mathrm{IF}(x; T, F_{\hat\theta}) = J(F_{\hat\theta})^{-1}(\partial/\partial\theta)[\ln f_\theta(x)]_{\hat\theta}$ and $V(T, F_{\hat\theta}) = J(F_{\hat\theta})^{-1}$, confirming (2.1.12).

When working with parametric models one keeps encountering the expression

$$s(x, \tilde\theta) := \frac{\partial}{\partial\theta}\left[\ln f_\theta(x)\right]_{\tilde\theta} = \frac{\partial}{\partial\theta}\left[f_\theta(x)\right]_{\tilde\theta}/f_{\tilde\theta}(x), \qquad (2.3.6)$$

which is generally called the *maximum likelihood scores function*, because the MLE corresponds to $\sum_{i=1}^{n} s(X_i, \mathrm{MLE}_n) = 0$.

We shall now use this function to find another expression for (2.3.5). Let T be Fisher consistent (2.1.3), that is,

$$\int \psi(x, \theta)\, dF_\theta(x) = 0 \quad \text{for all } \theta. \qquad (2.3.7)$$

By differentiation of (2.3.7) with respect to θ at some θ_*, we obtain another formula for the denominator of (2.3.5) at $F_* := F_{\theta_*}$. This yields

$$\mathrm{IF}(x; \psi, F_*) = \frac{\psi(x, \theta_*)}{\int \psi(y, \theta_*)(\partial/\partial\theta)[f_\theta(y)]_{\theta_*}\, d\lambda(y)}$$

$$= \frac{\psi(x, \theta_*)}{\int \psi(y, \theta_*)s(y, \theta_*)\, dF_*(y)} \qquad (2.3.8)$$

Note that this formula involves ψ only at θ_*. [Equation (2.3.8) could also be derived from (2.1.10).]

Location

When estimating location in the model $\mathscr{X} = \mathbb{R}$, $\Theta = \mathbb{R}$, $F_\theta(x) = F(x - \theta)$ it seems natural to use ψ-functions of the type

$$\psi(x, \theta) = \psi(x - \theta), \qquad (2.3.9)$$

where it is assumed that

$$\int \psi \, dF = 0 \qquad (2.3.10)$$

in order that T be Fisher consistent (2.1.3). Then

$$\mathrm{IF}(x; \psi, G) = \frac{\psi(x - T(G))}{\int \psi'(y - T(G)) \, dG(y)} \qquad (2.3.11)$$

and at the model distribution F we obtain

$$\mathrm{IF}(x; \psi, F) = \frac{\psi(x)}{\int \psi' \, dF} \qquad (2.3.12)$$

under the assumption that the denominator is nonzero. The IF is proportional to ψ, so it may be used itself as the defining ψ-function. (Hence, for every sufficiently regular location estimator there exists an M-estimator with the same IF at the model.) Huber (1964, p. 78) noted that ψ' may contain delta functions: If ψ is discontinuous in the point p with left and right limits $\psi(p-) \neq \psi(p+)$, then ψ' contains a delta function $[\psi(p+) - \psi(p-)]\delta_{(p)}$ which yields the term $[\psi(p+) - \psi(p-)]f(p)$ in the denominator of (2.3.12). Formula (2.3.8) avoids this difficulty.

The asymptotic variance at F can be calculated from (2.1.8), yielding

$$V(\psi, F) = \frac{\int \psi^2 \, dF}{\left(\int \psi' \, dF \right)^2}. \qquad (2.3.13)$$

Asymptotic normality with this variance may be shown rigorously under certain regularity conditions (Huber, 1964, 1967, 1981; Boos and Serfling, 1980; Clarke, 1983, 1984b). The Cramér–Rao inequality (2.1.11) holds with

Fisher information

$$J(F) = \int \left(\frac{f'}{f}\right)^2 dF. \tag{2.3.14}$$

In case the model distribution F is symmetric, it is natural to choose a skew-symmetric ψ-function [i.e., $\psi(-x) = -\psi(x)$]. In this situation all M-estimators with strictly monotone ψ satisfy the following properties: (1) if ψ is bounded then they are B-robust, qualitatively robust, and have breakdown point $\varepsilon^* = \frac{1}{2}$; and (2) if ψ is unbounded then they are neither B-robust nor qualitatively robust, and $\varepsilon^* = 0$ (Hampel, 1971; Huber, 1981, Section 3.2).

Remark 1. In case ψ is not strictly monotone, the solution of (2.3.3) may not be unique. For example, in the case of M-estimators with finite rejection point (which we shall deal with in Section 2.6), $\sum_{i=1}^{n}\psi(X_i - T_n)$ will vanish for all large absolute values of T_n. In that case there are three ways to define a sequence $\{T_n; \ n \geq 1\}$ of solutions which is unique and consistent: (1) take the global minimum of (2.3.2); (2) take the solution of (2.3.3) nearest to the sample median; or (3) use Newton's method, starting with the median as in Collins (1976) and Clarke (1984a). Methods (2) and (3) are quite easy to perform and inherit $\varepsilon^* = \frac{1}{2}$ from the median.

Example 1. The MLE corresponds to $\psi = -f'/f$ and possesses the smallest asymptotic variance $V(\psi, F)$ possible, namely the inverse of the Fisher information, by (2.1.12). At $F = \Phi$, we obtain the arithmetic mean with $\psi(x) = x$ and $V(\psi, \Phi) = J(\Phi)^{-1} = 1$. At the logistic distribution $F(x) = [1 + \exp(-x)]^{-1}$ we find $\psi(x) = 2F(x) - 1$ with $V(\psi, F) = J(F)^{-1} = 3$. At the double exponential or Laplace distribution ($f(x) = \frac{1}{2}\exp(-|x|)$) we obtain the median, given by $\psi_{\text{med}}(x) = \text{sign}(x)$ with $V(\psi_{\text{med}}, F) = J(F)^{-1} = 1$. At the Cauchy distribution, given by $f(x) = 1/[\pi(1 + x^2)]$, we find the nonmonotone $\psi(x) = 2x/(1 + x^2)$ with $V(\psi, F) = J(F)^{-1} = 2$.

Example 2. The *Huber estimator* was introduced (Huber, 1964) as the solution of the minimax problem sketched in Section 2.7, at $F = \Phi$. It is given by

$$\psi_b(x) = \min\{b, \max\{x, -b\}\} = x \cdot \min\left(1, \frac{b}{|x|}\right), \tag{2.3.15}$$

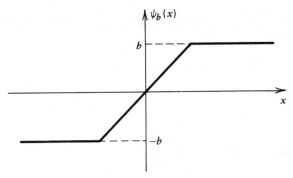

Figure 1. ψ-function defining the Huber-estimator with cutoff point b.

for $0 < b < \infty$ (see Fig. 1). It is the MLE for the distribution with density $f(x) = [\exp(-\int_0^x \psi_b(t)\, dt)]/\int [\exp(-\int_0^u \psi_b(t)\, dt)]\, du$, which is least favorable in the minimax sense (see Fig. 1 of Section 1.3). This estimator is also B-robust, qualitatively robust, and possesses $\varepsilon^* = \frac{1}{2}$. In the next section we shall even show it to be optimal B-robust.

Unfortunately, location M-estimators are usually not invariant with respect to scale, which is often a nuisance parameter. This problem can be solved by defining T_n through

$$\sum_{i=1}^{n} \psi\left(\frac{x_i - T_n}{S_n}\right) = 0,$$

where S_n is a robust estimator of scale. It is advisable to determine S_n first, by means of

$$S_n = 1.483 \,\mathrm{MAD}(x_i) = 1.483 \,\mathrm{med}_i\left\{\left|x_i - \mathrm{med}_j(x_j)\right|\right\}$$

which has maximal breakdown point $\varepsilon^* = 50\%$. On the other hand, one could also compute T_n and S_n simultaneously as M-estimators of location and scale, as in Huber's "Proposal 2" (Huber, 1964; 1981, Section 6.4), also discussed in Subsection 4.2d of the present book. However, simulation has clearly shown the superiority of M-estimators with initial scale estimate given by the MAD (Andrews et al., 1972, p. 239), which is also much easier to use than the simultaneous version. Therefore, we recommend the use of initial MAD scaling for M-estimators.

One-step M-estimators (sometimes called "*m*-estimators") are defined by

$$T_n = T_n^{(0)} + S_n^{(0)} \sum_{i=1}^{n} \psi \left(\frac{x_i - T_n^{(0)}}{S_n^{(0)}} \right) \bigg/ \sum_{i=1}^{n} \psi' \left(\frac{x_i - T_n^{(0)}}{S_n^{(0)}} \right)$$

(Bickel, 1975), where $T_n^{(0)}$ and $S_n^{(0)}$ are initial estimates of location and scale. The resulting T_n are the first step of an iterative algorithm for M-estimates (Huber, 1981, Section 6.7), but it turns out that they essentially have the same behavior (and also the same influence function, asymptotic variance, etc.) as their fully iterated versions, at least when ψ is odd and the underlying F is symmetric. In the case where ψ is not monotone, the one-step variant is even much safer to use, because it avoids the problem of uniqueness. However, one-step M-estimators should only be used when $T_n^{(0)}$ and $S_n^{(0)}$ are very robust: It is recommended to take

$$T_n^{(0)} = \text{med}(x_i),$$

$$S_n^{(0)} = 1.483 \text{MAD}(x_i).$$

Otherwise, the resulting estimate may be very nonrobust: It is shown (Andrews et al., 1972, p. 253) that starting from the arithmetic mean leads to disastrous results, and also that the interquartile range (having only a 25% breakdown point) is not nearly as good as the MAD initial scale.

Scale

The scale model is given by $F_\theta(x) = F(x/\theta)$, where $\mathcal{X} = \mathbb{R}$, $\Theta = (0, \infty)$, and $\theta_0 = 1$. An M-estimator for scale is defined by (2.3.3) with

$$\psi(x, \theta) = \psi(x/\theta). \tag{2.3.16}$$

We still need (2.3.10) for Fisher consistency. From (2.3.5) it follows that

$$\text{IF}(x; \psi, G) = \psi \left(\frac{x}{T(G)} \right) T(G) \bigg/ \int \psi' \left(\frac{y}{T(G)} \right) \frac{y}{T(G)} \, dG(y)$$

$$\tag{2.3.17}$$

under the assumption that the denominator is nonzero, so at the model distribution F we obtain

$$\text{IF}(x; \psi, F) = \frac{\psi(x)}{\int y \psi'(y) \, dF(y)}. \tag{2.3.18}$$

Again ψ' might contain delta functions, and the expected square of the IF equals the asymptotic variance. Cramér–Rao holds with Fisher information

$$J(F) = \int \left(x \frac{f'(x)}{f(x)} + 1 \right)^2 dF(x). \tag{2.3.19}$$

At the normal we find $J(\Phi) = 2$. When the model distribution F is symmetric, it is natural to use a symmetric ψ-function [i.e., $\psi(-x) = \psi(x)$]. When ψ is also strictly monotone for $x > 0$, then there are two possibilities: (1) if ψ is bounded then it is B-robust, qualitatively robust, and satisfies $\varepsilon^* = -\psi(0)/(\psi(\infty) - \psi(0)) \le \frac{1}{2}$; and (2) if ψ is unbounded then it is neither B-robust or qualitatively robust, with $\varepsilon^* = 0$ (see Huber, 1981, Section 5.2).

Example 3. The MLE is given by $\psi(x) = -x(f'(x)/f(x)) - 1$, and has the smallest asymptotic variance by (2.1.12), namely $1/J(F)$. At $F = \Phi$ we obtain $\psi(x) = x^2 - 1$, which is neither B-robust nor qualitatively robust, and has breakdown point $\varepsilon^* = 0$. A robustified version will be discussed in the next section.

Example 4. The M-estimator given by

$$\psi_{\mathrm{MAD}}(x) = \mathrm{sign}\big(|x| - \Phi^{-1}(\tfrac{3}{4})\big) \tag{2.3.20}$$

is the median of the absolute values of the observations, multiplied with $1/\Phi^{-1}(\tfrac{3}{4})$ to make it Fisher consistent for $F = \Phi$. In this situation there is no nuisance parameter. However, in the general case with unknown location parameter, one takes the *median of the absolute deviations from the median* (and then multiplies with the same constant $1/\Phi^{-1}(\tfrac{3}{4}) \simeq 1.483$). In this general case, we call this estimator the (standardized) *median (absolute) deviation (MAD)*. The influence function equals

$$\mathrm{IF}\,(x; \psi, \Phi) = \frac{1}{4\Phi^{-1}(\tfrac{3}{4})\phi\big(\Phi^{-1}(\tfrac{3}{4})\big)} \mathrm{sign}\big(|x| - \Phi^{-1}(\tfrac{3}{4})\big). \tag{2.3.21}$$

It holds that $\gamma^* = 1/[4\Phi^{-1}(\tfrac{3}{4})\phi(\Phi^{-1}(\tfrac{3}{4}))] \simeq 1.167$, so we have B-robustness. (This estimator is even most B-robust, as will be shown in Theorem 9 in Subsection 2.5e.) Moreover, $V(\psi, \Phi) = (\gamma^*)^2 \simeq 1.361$ by (2.1.8), and the asymptotic efficiency is $e = 1/[V(\psi, \Phi)J(\Phi)] = 1/[2(\gamma^*)^2] \simeq 0.367$. The MAD is also qualitatively robust and possesses maximal breakdown point $\varepsilon^* = -\psi(0)/(\psi(\infty) - \psi(0)) = \frac{1}{2}$.

2.3b. *L*-estimators

Location

Definition 2.   *L*-estimators are of the form

$$T_n(X_1, \ldots, X_n) = \sum_{i=1}^{n} a_i X_{i:n}, \qquad (2.3.22)$$

where $X_{1:n}, \ldots, X_{n:n}$ is the ordered sample and the a_i are some coefficients.

The name "*L*-estimators" comes from "*l*inear combinations of order statistics." A natural sequence of location estimators is obtained if the weights a_i are generated by

$$a_i = \int_{[(i-1)/n, i/n]} h \, d\lambda \Big/ \int_{[0,1]} h \, d\lambda, \qquad (2.3.23)$$

where $h: [0,1] \to \mathbb{R}$ satisfies $\int_{[0,1]} h \, d\lambda \neq 0$. Clearly, h is only defined up to a nonzero factor. (Therefore, one often works with the standardized version $h(u)/\int_{[0,1]} h \, d\lambda$.) Under diverse sets of regularity conditions (for a survey see Helmers, 1978, 1980), these estimators are asymptotically normal. The corresponding functional is

$$T(G) = \frac{\int x h(G(x)) \, dG(x)}{\int h(F(y)) \, dF(y)}. \qquad (2.3.24)$$

It is Fisher consistent because of the standardization, and its influence function is

$$\mathrm{IF}(x; T, F) = \frac{\int_{[0,x]} h(F(y)) \, d\lambda(y) - \int \left[\int_{[0,t]} h(F(y)) \, d\lambda(y) \right] dF(t)}{\int h(F(y)) \, dF(y)},$$

$$(2.3.25)$$

where the denominator is nonzero because it equals $\int_{[0,1]} h \, d\lambda$ (Huber, 1977a). It follows that

$$\frac{d}{dx} \mathrm{IF}(x; T, F) = \frac{h(F(x))}{\int h(F(y)) \, dF(y)}. \qquad (2.3.26)$$

Example 5.   The median corresponds to $h = \delta_{(1/2)}$, so $\int_{[0,1]} h \, d\lambda = 1$ and $T(G) = \int_{[0,1]} G^{-1}(y) h(y) \, d\lambda(y) = G^{-1}(\frac{1}{2})$. Its influence function thus equals

$$\text{IF } (x; T, F) = \frac{1}{2f\left(F^{-1}\left(\frac{1}{2}\right)\right)} \text{sign}\left(x - F^{-1}\left(\tfrac{1}{2}\right)\right). \qquad (2.3.27)$$

In case of a general combination of quantiles $T(G) = \sum_i \alpha_i G^{-1}(t_i)$ with $\sum_i \alpha_i = 1$, we have $h = \sum_i \alpha_i \delta_{(t_i)}$, $\int_{[0,1]} h \, d\lambda = 1$ and the IF has jumps of size $\alpha_i / f(F^{-1}(t_i))$ in the quantiles $F^{-1}(t_i)$. For example, Gastwirth (1966) proposed the "quick estimator" given by $\alpha_1 = 0.3$, $t_1 = \frac{1}{3}$, $\alpha_2 = 0.4$, $t_2 = \frac{1}{2}$, $\alpha_3 = 0.3$, $t_3 = \frac{2}{3}$.

Example 6. As in Example 1, let us now look for the estimators with maximal asymptotic efficiency. They are given by $h(F(x))$ proportional to $-[\ln f(x)]''$ because of (2.1.12). At $F = \Phi$ we obtain the arithmetic mean ($h \equiv 1$), and at the double exponential we find the median. At Huber's "least favorable distribution" (see Example 2) we obtain the α-*trimmed mean* (see Example 3 in Subsection 2.2b) given by $h = 1_{[\alpha, 1-\alpha]}$ for $\alpha = F(-b)$, which has the same IF at this model as the Huber estimator. However, note that the α-trimmed mean and its IF differ from the corresponding Huber-estimator outside this model; in particular, the α-trimmed mean achieves only $\varepsilon^* = \alpha$, whereas the Huber-estimator possesses $\varepsilon^* = \frac{1}{2}$! For the logistic distribution we obtain $h(t) = t(1 - t)$, and at the Cauchy distribution we find the peculiar weight function $h(t) = \cos(2\pi t)[\cos(2\pi t) - 1]$ (with $\int_{[0,1]} h \, d\lambda = \frac{1}{2}$) which is negative for $|t - \frac{1}{2}| > \frac{1}{4}$.

Remark 2. The logistic and the Cauchy L-estimators are important examples of estimators that are B-robust but not qualitatively robust, unlike the corresponding M-estimators with the same IF at the model. But why do L-estimators and M-estimators behave differently? For M-estimators the IF is always proportional to ψ at any distribution; but for L-estimators the shape of the IF depends on the distribution because of (2.3.26), so the IF itself is not very stable under small changes. For example, for the logistic L-estimator the derivative of the IF is proportional to $G(x)(1 - G(x))$, so a heavy-tailed G can make the IF unbounded.

Scale

L-estimators of scale are also defined by a T_n of the form (2.3.22), but now the weights are generated by

$$a_i = \int_{[(i-1)/n, \, i/n]} h \, d\lambda \bigg/ \int_{[0,1]} h(t) F^{-1}(t) \, d\lambda(t), \qquad (2.3.28)$$

where $h : [0, 1] \rightarrow \mathbb{R}$ is such that the denominator is nonzero. (At a symmetric model distribution, we usually choose a symmetric h in the location case and a skew-symmetric h in the scale case.) We obtain

$$T(G) = \frac{\int x h(G(x)) \, dG(x)}{\int y h(F(y)) \, dF(y)} \tag{2.3.29}$$

(some generalizations can be found in Huber, 1981, Section 5.3), and

$$\text{IF}(x; T, F) = \frac{\int_{[0, x]} h(F(y)) \, d\lambda(y) - \int \left[\int_{[0, t]} h(F(y)) \, d\lambda(y) \right] dF(t)}{\int y h(F(y)) \, dF(y)}.$$

$$\tag{2.3.30}$$

Example 7. The t-quantile range corresponds to $h = \delta_{(1-t)} - \delta_{(t)}$ for $0 < t < \frac{1}{2}$, and yields $T(G) = [G^{-1}(1 - t) - G^{-1}(t)] / [F^{-1}(1 - t) - F^{-1}(t)]$. When F is symmetric, then

$$\text{IF}(x; T, F) = \frac{1}{2 F^{-1}(1 - t) f(F^{-1}(1 - 2t))}$$

$$\times \left[(1 - 2t) 1_{\{ |x| > F^{-1}(1-t) \}} - 2t 1_{\{ |x| < F^{-1}(1-t) \}} \right]$$

$$\tag{2.3.31}$$

making use of $T(F) = \theta_0 = 1$. For example, for the *interquartile range* ($t = \frac{1}{4}$) at $F = \Phi$ we obtain (2.3.21), the IF of the MAD. Therefore, again $\gamma^* \simeq 1.167$ and $e \simeq 0.367$. The interquartile range shares the efficiency properties, B-robustness, and qualitative robustness of the MAD at the normal distribution, but there is one important difference—the MAD possesses $\varepsilon^* = \frac{1}{2}$ and the interquartile range only $\varepsilon^* = \frac{1}{4}$ (imagine all contamination on one side).

2.3c. *R*-Estimators

R-estimators go back to Hodges and Lehmann (1963); the name R-estimators comes from the fact that they are derived from *r*ank tests. In the one-sample case, they only exist for the location problem. Originally, these estimators were derived from one-sample tests, but we shall now discuss the more customary approach which uses two-samples tests (Huber, 1981, Section 3.4).

We first describe two-sample location rank tests. Let $X_1, \ldots, X_m$ and $Y_1, \ldots, Y_n$ be two samples with distributions $H(x)$ and $H(x + \Delta)$, where Δ is the unknown location shift. Let R_i be the rank of X_i in the pooled sample of size $N = m + n$. A rank test of $\Delta = 0$ against $\Delta > 0$ is based on a test statistic

$$S_N = \frac{1}{m} \sum_{i=1}^{m} a_N(R_i), \qquad (2.3.32)$$

with weights $a_N(i)$ that are only defined up to a positive affine transformation. Suppose the $a_N(i)$ are generated by a function $\phi : [0, 1] \to \mathbb{R}$ by means of

$$a_N(i) = N \int_{[(i-1)/N, i/N]} \phi(u)\, du, \qquad (2.3.33)$$

where ϕ is skew symmetric $[\phi(1 - u) = -\phi(u)]$, so $\int \phi(u)\, du = 0$.

Definition 3. An R-estimator of location can be defined as $T_n = T_n(X_1, \ldots, X_n)$, where T_n is chosen in order that (2.3.32) becomes as close to zero as possible when computed from the samples $X_1, \ldots, X_n$ and $2T_n - X_1, \ldots, 2T_n - X_n$.

The idea behind this definition is the following. From the original sample $X_1, \ldots, X_n$ we can construct a mirror image by replacing each X_i by $T_n - (X_i - T_n)$. We choose the T_n for which the test cannot detect any shift, which means that the test statistic S_N comes close to zero (although it often cannot become exactly zero, being a discontinuous function). Therefore, T_n corresponds to the functional T satisfying

$$\int \phi\left[\tfrac{1}{2}G(y) + \tfrac{1}{2}(1 - G(2T(G) - y))\right] dG(y) = 0. \qquad (2.3.34)$$

Let us suppose that F has a strictly positive and absolutely continuous density. If we replace G by $F_{t,x} = (1 - t)F + t\Delta_x$, differentiate with respect to t at zero, and perform some substitutions, we obtain

$$\mathrm{IF}(x; T, F) = \frac{U(x) - \int U\, dF}{\int U'\, dF}, \qquad (2.3.35)$$

where

$$U(x) = \int_{[0,x]} \phi'\left[\tfrac{1}{2}\{F(y) + 1 - F(2T(F) - y)\}\right] f(2T(F) - y)\, d\lambda(y).$$

$$(2.3.36)$$

When F is symmetric we have $T(F) = 0$, so $U(x) = \phi(F(x))$ and

$$\mathrm{IF}(x; T, F) = \frac{\phi(F(x))}{\int (\phi(F))'\, dF}. \qquad (2.3.37)$$

For monotone and integrable ϕ for which $T(F)$ is uniquely defined, T is continuous at F and the breakdown point ε^* is given by

$$\int_{[1/2,\, 1-\varepsilon^*/2]} \phi\, d\lambda = \int_{[1-\varepsilon^*/2,\, 1]} \phi\, d\lambda \qquad (2.3.38)$$

(see Hampel, 1971; Huber, 1981, Section 3.4).

Example 8. Again (2.1.12) yields the estimators with maximal efficiency; we obtain $\phi(F(x))$ proportional to $-f'(x)/f(x)$ for symmetric F. At $F = \Phi$ we find the *normal scores* estimator ($\phi(u) = \Phi^{-1}(u)$). It is qualitatively robust (Hampel, 1971, Section 7), but it is not B-robust because $\mathrm{IF}(x; T, \Phi) = x$. Its breakdown point is $\varepsilon^* = 2\Phi(-(\ln 4)^{1/2}) \approx 0.239$ by (2.3.38). At the logistic distribution we obtain the *Hodges–Lehmann* estimator $[\phi(u) = u - \tfrac{1}{2}]$ which is derived from the Wilcoxon test. It is given by the median of $\{(X_i + X_j)/2;\ i = 1,\ldots, n \text{ and } j = 1,\ldots, n\}$, and yields $\mathrm{IF}(x; T, F) = [F(x) - \tfrac{1}{2}]/\int f^2(z)\, dz$ and $\varepsilon^* = 1 - 2^{-1/2} \approx 0.293$. At the double exponential we again find the median, derived from the sign test (with $\phi(u) = -1$ for $u < \tfrac{1}{2}$ and $\phi(u) = 1$ for $u > \tfrac{1}{2}$). At the Cauchy distribution, we find $\phi(u) = -\sin(2\pi u)$.

Remark 3. Note that the normal scores estimator is qualitatively robust but not B-robust. The corresponding M-estimator, the arithmetic mean, is clearly not qualitatively robust. As in Remark 2, this different behavior lies in the fact that also for R-estimators the shape of the IF depends on the underlying distribution. In this case, the IF at a longer-tailed distribution will bend downwards. Also, note that R-estimators are awkward to handle theoretically in asymmetric situations. Concluding, we may say that M-

estimators are more straightforward to handle than either *L*- or *R*-estimators. Especially in multiparameter situations, they are much more flexible.

2.3d. Other Types of Estimators: *A*, *D*, *P*, *S*, *W*

At least five other types of estimators are being considered in robustness theory. Some of these techniques already play a prominent role, and some may do so in the future. We will just give a brief description of these estimators (in alphabetical order) with references for further study.

A-estimators (Lax, 1975) are based on formula (2.3.13) for the asymptotic variance of an *M*-estimator of location. The basic idea is to apply this formula to a finite sample, and to use the square root of the result as an estimator of scale. First, the data are standardized as follows:

$$u_i = \frac{x_i - \text{med}(x_j)}{1.483\text{MAD}(x_j)} \qquad (2.3.39)$$

and then the *A*-estimator S_n is defined by

$$S_n = 1.483\text{MAD}(x_j)\left[\frac{1}{n}\sum_{i=1}^{n}\psi^2(u_i)\right]^{1/2} \bigg/ \left|\frac{1}{n}\sum_{i=1}^{n}\psi'(u_i)\right| \qquad (2.3.40)$$

Several choices of ψ, together with some simulation results, can be found in Iglewicz (1983). Influence functions and breakdown points of these estimators were investigated by Shoemaker and Hettmansperger (1982) and Shoemaker (1984).

D-estimators (from "minimum *d*istance") go back to Wolfowitz (1957). Their basic principle is the following: When G_n is the empirical cdf, one chooses T_n as the value of θ for which $\pi(G_n, F_\theta)$ is minimal, where π is some measure of discrepancy between distributions. Often π corresponds to some badness-of-fit statistic (Kolmogorov, Cramèr–von Mises, etc.). Wolfowitz (1957) proved consistency of T_n. Knüsel (1969) examined *D*-estimators from the robustness point of view, and showed that many *D*-estimators may be considered as special cases of *M*-estimators. Beran (1977b) examined the situation where π is the Hellinger distance. Boos (1981) proved certain asymptotic results for *D*-estimators, and Parr and Schucany (1980) calculated their influence functions. Donoho (1981) investigated the breakdown behavior.

P-estimators (Johns, 1979) are generalizations of *P*itman estimators for location. They are defined by

$$T_n(x_1,\ldots,x_n) = \frac{\int \theta \prod_{i=1}^n \gamma(x_i - \theta)\, d\theta}{\int \prod_{i=1}^n \gamma(x_i - \theta)\, d\theta},\qquad (2.3.41)$$

where γ does not necessarily have to coincide with the unknown model density f. The advantage of *P*-estimators over *M*-estimators is that no iterative algorithm is necessary for their computation; but on the other hand numerical integration is required. Johns found that T_n corresponds to the functional T given by

$$\int \frac{\gamma'(x - T(G))}{\gamma(x - T(G))}\, dG(x) = 0,\qquad (2.3.42)$$

which leads to the following influence function:

$$\mathrm{IF}(x; T, F) = \frac{d}{dx}\ln\gamma(x)\Big/ \int \frac{d^2}{dx^2}\ln\gamma(x)\, dF(x)\qquad (2.3.43)$$

under certain regularity conditions. Note that for each such *P*-estimator there exists an *M*-estimator with the same IF (and hence the same asymptotic variance) by taking

$$\psi(x) = -\frac{d}{dx}\ln\gamma(x) = \frac{-\gamma'(x)}{\gamma(x)},\qquad (2.3.44)$$

and starting from an *M*-estimator one can find the corresponding *P*-estimator by putting

$$\gamma(x) = \exp(-\rho(x)).\qquad (2.3.45)$$

Therefore, Huber (1984) treats *M*-estimators and *P*-estimators in a parallel way when investigating their finite-sample breakdown points.

S-estimators (Rousseeuw and Yohai, 1984) are based on minimization of a *s*cale statistic. In particular, suppose we define the scale statistic $s(r_1,\ldots,r_n)$ by

$$\frac{1}{n}\sum_{i=1}^n \rho\left(\frac{r_i}{s}\right) = K,\qquad (2.3.46)$$

where K is taken to be $\int \rho(t)\, dF(t)$ at the model distribution F. The function ρ is assumed to be bounded. (Note that s is an M-estimator for scale.) Then the S-estimator of location $\hat{\theta}_n(x_1, \ldots, x_n)$ is defined by

$$\underset{\theta_n}{\text{minimize}}\; s(x_1 - \theta, \ldots, x_n - \theta), \qquad (2.3.47)$$

which at the same time yields a scale estimate

$$\hat{\sigma}_n = s(x_1 - \hat{\theta}_n, \ldots, x_n - \hat{\theta}_n). \qquad (2.3.48)$$

The influence function and asymptotic variance of $\hat{\theta}_n$ are equal to those of the location M-estimator constructed from the same function ρ, so they are given by (2.3.12) and (2.3.13) with $\psi = \rho'$. Actually, S-estimators were introduced because of their high breakdown point in multivariate situations such as regression (Rousseeuw and Yohai, 1984). They are generalizations of the least median of squares estimator (Hampel, 1975; Rousseeuw, 1984). In fact, S-estimators could be defined by means of any scale equivariant statistic [i.e., $s(\lambda r_1, \ldots, \lambda r_n) = |\lambda| s(r_1, \ldots, r_n)$ for all λ], in which case the class of S-estimators also encompasses least squares (arithmetic mean), least absolute deviations (median), least pth-power deviations (Gentleman, 1965), the method of Jaeckel (1972), and least trimmed squares (Rousseeuw, 1983b). In multiple regression, S-estimation is a projection pursuit technique (Donoho et al., 1985).

W-estimators (Tukey, 1970/71) are defined as a weighted mean of the observations:

$$T_n(x_1, \ldots, x_n) = \frac{\sum_{i=1}^n x_i w_i}{\sum_{i=1}^n w_i}, \qquad (2.3.49)$$

where the weights depend on the observations through

$$w_i = w(x_i - T_n). \qquad (2.3.50)$$

(When scale is a nuisance parameter, $x_i - T_n$ must be divided by a preliminary scale estimate S_n, preferably the MAD.) This means that T_n satisfies the equation

$$T_n = \frac{\sum_{i=1}^n x_i w(x_i - T_n)}{\sum_{i=1}^n w(x_i - T_n)}. \qquad (2.3.51)$$

In order to determine T_n, one usually resorts to an iterative algorithm: Starting from the median $T_n^{(0)}$ one computes

$$T_n^{(j+1)} = \frac{\sum_{i=1}^n x_i w\left(x_i - T_n^{(j)}\right)}{\sum_{i=1}^n w\left(x_i - T_n^{(j)}\right)} \qquad (2.3.52)$$

until the $T_n^{(j)}$ converge. (This is an example of *iteratively reweighted least squares*, a procedure which is well known in regression analysis.) The *one-step W-estimator* (sometimes called "*w*-estimator") $T_n^{(1)}$ has become popular because of its simple computation, while its performance is still very good. Note that fully iterated *W*-estimators are actually a variant of *M*-estimators, because (2.3.51) yields

$$\frac{\sum_{i=1}^n (x_i - T_n) w(x_i - T_n)}{\sum_{i=1}^n w(x_i - T_n)} = 0;$$

hence

$$\sum_{i=1}^n \psi(x_i - T_n) = 0$$

for $\psi(u) := u w(u)$. Therefore, *W*-estimators possess the same influence function and asymptotic variance as *M*-estimators.

2.4. OPTIMALLY BOUNDING THE GROSS-ERROR SENSITIVITY

2.4a. The General Optimality Result

The purpose of the present section is to construct estimators which are as efficient as possible, subject to an upper bound on their gross-error sensitivity γ^*. The latter is a robustness condition, because γ^* is the supremum of the absolute value of the influence function, so the (normalized) influence of any outlier cannot exceed γ^* in linear approximation. This first subsection contains a mathematical result, whereas the following subsections treat in detail the application of this principle to the classes of estimators introduced in the previous section.

Consider the situation of Subsection 2.1a. The parameter space Θ is an open convex subset of $\mathbb{R}$. The parametric model $\{F_\theta; \ \theta \in \Theta\}$ consists of distributions on $\mathcal{X}$ with strictly positive densities $f_\theta(x)$ with respect to some measure λ. Let θ_* belong to Θ, and put $F_* := F_{\theta_*}$ and $f_* := f_{\theta_*}$. Assume that $s(x, \theta_*)$ given by (2.3.6) exists for all x, that $\int s(x, \theta_*) \, dF_*(x) = 0$, and that the Fisher information

$$J(F_*) = \int s(x, \theta_*)^2 \, dF_*(x) \tag{2.4.1}$$

satisfies $0 < J(F_*) < \infty$. [The equality $\int s(x, \theta_*) \, dF_*(x) = 0$ is only a regularity condition, since it holds if integration with respect to x and differentiation with respect to θ may be interchanged.] For example, these conditions hold at the normal location model. Let us now state the optimality result.

Theorem 1 ("Lemma 5" of Hampel, 1968). Suppose the above conditions hold, and let $b > 0$ be some constant. Then there exists a real number a such that

$$\tilde{\psi}(x) := [s(x, \theta_*) - a]_{-b}^{b} \tag{2.4.2}$$

satisfies $\int \tilde{\psi} \, dF_* = 0$ and $d := \int \tilde{\psi}(y) s(y, \theta_*) \, dF_*(y) > 0$. Now $\tilde{\psi}$ minimizes

$$\int \psi^2 \, dF_* \Big/ \left[\int \psi(y) s(y, \theta_*) \, dF_*(y) \right]^2 \tag{2.4.3}$$

among all mappings ψ that satisfy

$$\int \psi \, dF_* = 0, \tag{2.4.4}$$

$$\int \psi(y) s(y, \theta_*) \, dF_*(y) \neq 0, \tag{2.4.5}$$

$$\sup_x \left| \psi(x) \Big/ \int \psi(y) s(y, \theta_*) \, dF_*(y) \right| < c := \frac{b}{d}. \tag{2.4.6}$$

Any other solution of this extremal problem coincides with a nonzero multiple of $\tilde{\psi}$, almost everywhere with respect to F_*.

The notation $[h(x)]^b_{-b}$ for an arbitrary function $h(x)$ means truncation at the levels b and $-b$:

$$[h(x)]^b_{-b} = -b \qquad \text{if } h(x) < -b$$

$$= h(x) \qquad \text{if } -b \leq h(x) \leq b$$

$$= b \qquad \text{if } b < h(x).$$

Proof. First, let us start by showing that such a value a exists. By the dominated convergence theorem, $\int [s(x, \theta_*) - \alpha]^b_{-b} \, dF_*(x)$ is a continuous function of α, and as $\alpha \to \pm \infty$ this integral tends to $\mp b$; hence there exists an α such that this integral becomes zero.

Second, let us show that $d > 0$. *Case (i)*: First assume that $|a| \leq b$. Then $d = \int (\tilde{\psi}(x) + a)s(x, \theta_*) \, dF_*(x)$ equals

$$\int_P \min\{s(x, \theta_*), a + b\} s(x, \theta_*) \, dF_*(x)$$

$$+ \int_N \max\{s(x, \theta_*), a - b\} s(x, \theta_*) \, dF_*(x),$$

where $P = \{x; s(x, \theta_*) > 0\}$ and $N = \{x; s(x, \theta_*) < 0\}$. Both integrands are nonnegative. Suppose that $d = 0$, hence both integrals are zero. In case $a \geq 0$ we consider the first one, because then $a + b > 0$. As the integrand is then strictly positive on P, it follows that $F_*(P) = 0$. But then $\int s(x, \theta_*) \, dF_*(x) = 0$ implies that also $F_*(N) = 0$. Finally, it follows that $J(F_*) = 0$ which is a contradiction. *Case (ii)*: If $a > b$, then $\int (\tilde{\psi} - (-b)) \, dF_* = b > 0$ and $s(x, \theta_*) > (a - b)$ on the set where the integrand does not vanish; hence $0 < \int (\tilde{\psi}(x) - (-b))s(x, \theta_*) \, dF_*(x) = d$ and similarly for $a < -b$.

Finally, let us now prove the optimality and the uniqueness of $\tilde{\psi}$. Take any ψ which is measurable and satisfies (2.4.4), (2.4.5), and (2.4.6). Without loss of generality we can assume that $\int \psi(x)s(x, \theta_*) \, dF_*(x) = d$, so we only have to minimize the numerator of (2.4.3). Using (2.4.4) we obtain

$$\int [(s(x, \theta_*) - a) - \psi(x)]^2 \, dF_*(x)$$

$$= \int (s(x, \theta_*) - a)^2 \, dF_*(x) - 2d + \int \psi^2 \, dF_*,$$

so it suffices to minimize the left member of this equation, which can be rewritten as

$$\int_{\{\bar{s} > b\}} (\bar{s} - \psi)^2 \, dF_* + \int_{\{|\bar{s}| \le b\}} (\bar{s} - \psi)^2 \, dF_* + \int_{\{\bar{s} < -b\}} (\bar{s} - \psi)^2 \, dF_*,$$

where $\bar{s}(x) = s(x, \theta_*) - a$. Because $|\psi| \le b$ by (2.4.6), this is minimized if and only if $\psi = \tilde{\psi}$ almost everywhere with respect to F_*. This ends the proof. □

2.4b. *M*-Estimators

Let us now apply this mathematical result to the *M*-estimators of Subsection 2.3a.

General Case

In the general estimation problem, an *M*-estimator is defined by (2.3.3) using a ψ-function of the type $\psi(x, \theta)$. Assume that the conditions of the preceding subsection hold, and that we have the Fisher-consistency condition (2.3.7), so the IF is given by (2.3.8). If we drop the argument θ_* from the notation $\psi(x, \theta_*)$ for a moment, then we note that the asymptotic variance $V(\psi, F_*)$ is given by (2.4.3) because of (2.1.8), and that the gross-error sensitivity $\gamma^*(\psi, F_*)$ is the left member of (2.4.6). Moreover, (2.4.4) is the Fisher-consistency condition (2.3.7) at θ_*, and condition (2.4.5) only states that the denominator of the IF is not zero. Therefore, the solution $\tilde{\psi}(\cdot, \theta_*)$ yields the smallest value of $V(\psi, F_*)$ among all functions $\psi(\cdot, \theta_*)$ satisfying $\gamma^*(\psi, F_*) \le c(\theta_*) = b(\theta_*)/d(\theta_*)$. [If the upper bound on γ^* were not imposed, then we would simply obtain the maximum likelihood estimator ($b = \infty$, $a = 0$). This classical solution can also be derived from (2.1.12), as we did in Example 1 in Subsection 2.3a.]

If we succeed in extending $\tilde{\psi}(\cdot, \theta_*)$ to a "nice" function $\tilde{\psi}(\cdot, \cdot)$ on $\mathcal{X} \times \Theta$, then we have constructed an *M*-estimator which is Fisher consistent at θ_* and which minimizes the asymptotic variance $V(\psi, F_*)$ subject to the upper bound $c(\theta_*)$ on the gross-error sensitivity $\gamma^*(\psi, F_*)$.

This result can be applied in the following way. One chooses a "nice" function $b(\theta)$ and carries out the construction of $\tilde{\psi}(\cdot, \theta)$ for each θ; then one checks whether the resulting function $\tilde{\psi}(\cdot, \cdot)$ determines an *M*-estimator (i.e., whether it is "smooth"; general conditions for this to be true are quite messy). In that case we say that $\tilde{\psi}(\cdot, \cdot)$ determines an *optimal B-robust M-estimator*.

Theoretically, it would be nicer to start with the upper bound $c(\theta)$ on γ^* and to determine $b(\theta)$ from it, but this is not feasible in the general case. (However, it can be done in certain special cases—see Subsection 2.5d.) Moreover, while every positive $b(\theta)$ can be chosen, not every positive c is possible, because typically there is a lower bound on γ^*. (An example of such a lower bound will be given in Theorem 3 in Subsection 2.5c.) Although in principle we have to choose a whole function $b(\theta)$, it is often possible to choose just the value of b for one θ and to determine the rest of the function by requiring equivariance, such as in the location and scale cases below.

Location

In the case of a location parameter (see Subsection 2.3a), the situation becomes much simpler. We have $\psi(x, \theta) = \psi(x - \theta)$, so the ψ-function at $\theta_0 = 0$ determines everything else. Put $F := F_{\theta_0}$. For the influence function (2.3.12) we obtain

$$\mathrm{IF}(x; \psi, F) = \frac{\psi(x)}{\int \psi(y) s(y, \theta_0)\, dF(y)} \qquad (2.4.7)$$

using Fisher consistency. One will choose b constant, so that a is also constant. The mapping $\tilde{\psi}$ determined by (2.4.2) is

$$\tilde{\psi}(x) = \left[-\frac{f'(x)}{f(x)} - a \right]_{-b}^{b}. \qquad (2.4.8)$$

It defines a location M-estimator which is Fisher-consistent (2.4.4) and for which the IF exists (2.4.5). Its asymptotic variance (2.4.3) is minimal for a given upper bound $c = b/d$ on $\gamma^*(\psi, F)$, so we conclude that $\tilde{\psi}$ is optimal B-robust.

When the distribution F is symmetric, we always have $a = 0$. At the standard normal ($F = \Phi$) we find

$$\tilde{\psi}(x) = [x]_{-b}^{b}, \qquad (2.4.9)$$

which is the Huber estimator (Example 2 in Subsection 2.3a). When $b \downarrow 0$, this estimator tends to the median. (See also Fig. 5 of Section 1.3.)

Remark. Let us give a geometric interpretation. In case F is symmetric, consider the vector space M of all measurable ψ which are square integrable

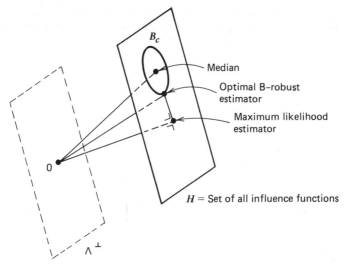

Figure 1. Geometrical representation of Theorem 1.

[necessary for $V(\psi, F)$], skew symmetric (implying Fisher consistency), and satisfy $\int \psi' \, dF = \int \psi \Lambda \, dF$ where $\Lambda = -f'/f$. It is endowed with the scalar product $\langle \psi_1, \psi_2 \rangle = \int \psi_1 \psi_2 \, dF$ which always exists by Cauchy–Schwarz, yielding the norm $\|\psi\|_2 = \langle \psi, \psi \rangle^{1/2}$. The orthogonal supplement $\Lambda^\perp$ of Λ is a linear subspace with co-dimension 1, and contains all ψ-functions which are not allowed because their IF does not exist (as $\int \psi' \, dF = \langle \psi, \Lambda \rangle = 0$.) The affine hyperplane H parallel to $\Lambda^\perp$ and going through $\Lambda / J(F)^{1/2}$ contains all ψ satisfying $\langle \psi, \Lambda \rangle = 1$, that is, all influence functions of elements of $M \setminus \Lambda^\perp$. For elements of H we have $\|\psi\|_2 = V(\psi, F)$; clearly, $\Lambda / J(F)^{1/2}$ (corresponding to the maximum likelihood estimator) minimizes $\|\psi\|_2$ in H. The intersection of $\{\|\psi\|_\infty \leq c\}$, a closed ball for the supremum-norm, and H is denoted by B_c; it contains all the influence functions with $\gamma^* \leq c$. (The median, which is most B-robust, minimizes $\|\psi\|_\infty$ in H.) Now Theorem 1 yields the element of B_c which minimizes $\|\psi\|_2$; in the proof it is determined as the element of B_c which is closest (in $\| \cdots \|_2$) to $\Lambda / J(F)^{1/2}$ (composition of projections).

Scale

M-estimation in the scale model was also treated in Subsection 2.3a. The ψ-function at $\theta_0 = 1$ determines everything, because $\psi(x, \theta) = \psi(x/\theta)$. The influence function (2.3.18) can be rewritten as (2.3.8) because of Fisher consistency [but note that the notation $s(x, \theta_0)$ has a different meaning than in the location case]. Therefore, the optimal B-robust scale M-estimator is also given by (2.4.2), yielding

$$\tilde{\psi}(x) = \left[-x \frac{f'(x)}{f(x)} - 1 - a \right]_{-b}^{b} \qquad (2.4.10)$$

which is a robustified version of the MLE. At the standard normal, we obtain

$$\tilde{\psi}(x) = [x^2 - 1 - a]_{-b}^{b}. \qquad (2.4.11)$$

When b is not too small, the scores function is only truncated above, and in this form it goes back to Huber (1964). When b becomes smaller there is truncation above and below, and for $b \downarrow 0$ we obtain the standardized median absolute deviation (Example 4 in 2.3a.)

2.4c. L-Estimators

Although M-estimators are very appropriate for robustness, the fact remains that many statisticians prefer L-estimators for their computational simplicity. Therefore, the construction of optimal B-robust estimators in this class is of interest. (This subsection and the following one contain results which appeared in Rousseeuw, 1979, 1982b.)

Location

Location L-estimators were described in Subsection 2.3b, and their IF equals (2.3.25). Performing the substitution

$$\chi(x) := \int_{[0, x]} h(F(y)) \, d\lambda(y) - \int \left[\int_{[0, t]} h(F(y)) \, d\lambda(y) \right] dF(t),$$

$$(2.4.12)$$

the influence function equals (2.3.12) with χ instead of ψ, which is a special case of (2.3.8). Its denominator equals $\int_{[0,1]} h \, d\lambda \neq 0$ so (2.4.5) is satisfied, and we obviously have (2.4.4). This enables us to apply Theorem 1, yielding a mapping $\tilde{\chi}$ given by (2.4.2). Usually there exists an admissible $\tilde{h}$ for which the corresponding χ is a multiple of $\tilde{\chi}$ (almost everywhere), and then the L-estimator in question possesses minimal asymptotic variance (2.4.3) among

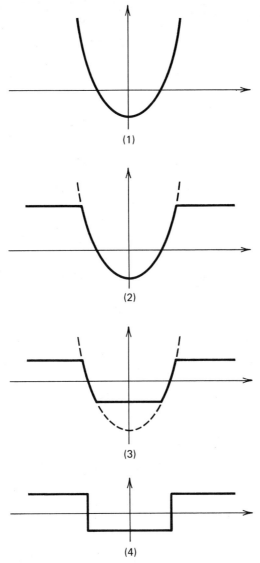

Figure 2. Influence function at $F = \Phi$ of the M-estimator for scale given by (2.4.11): (1) maximum likelihood estimator, corresponding to $b \uparrow \infty$; (2) solution with large b, only truncation from above; (3) solution with small b, truncation from above and below; and (4) standardized median absolute deviation, corresponding to $b \downarrow 0$.

all L-estimators with an upper bound (2.4.6) on $\gamma^*(T, F)$. We say that $\tilde{h}$ determines an optimal B-robust location L-estimator.

At the standard normal ($F = \Phi$) we obtain the α-trimmed mean

$$h = \tilde{\chi}' \circ \Phi^{-1} = 1_{[\alpha, 1-\alpha]} \tag{2.4.13}$$

with $\alpha = \Phi(-b)$, having the same IF as the Huber-estimator ψ_b. If we let b tend to zero, then $\gamma^* = c$ approaches its lower bound $\sqrt{\pi/2} \simeq 1.253$ and we obtain the median.

Scale

For scale L-estimators (see Subsection 2.3b) we apply the transformation (2.4.12) to (2.3.30), obtaining (2.3.18) which is a special case of (2.3.8). Therefore, our earlier result (2.4.11) can be translated. At the standard normal, the optimal B-robust scale L-estimators are given by

$$\begin{aligned} \tilde{h}(t) &= \Phi^{-1}(t) \quad \text{for } \left|\left(\Phi^{-1}(t)\right)^2 - 1 - a\right| \leq b \\ &= 0 \qquad\quad \text{elsewhere} \end{aligned} \tag{2.4.14}$$

for suitable a and b [i.e., the same values as in formula (2.4.11)]. If b tends to zero, then γ^* tends to its lower bound 1.167 and we obtain the interquartile range (Example 7 in Subsection 2.3b).

2.4d. R-estimators

For R-estimators (see Subsection 2.3c) we apply the transformation

$$\chi(x) = U(x) - \int U(t)\, dF(t) \tag{2.4.15}$$

to (2.3.35), again obtaining (2.3.12). Therefore, Theorem 1 can be applied: Usually there exists an admissible $\tilde{\phi}$ to which a χ corresponds which is a multiple of $\tilde{\chi}$ (almost everywhere), and then $\tilde{\phi}$ determines an optimal B-robust R-estimator. At the standard normal we obtain

$$\tilde{\phi}(u) = \left[\Phi^{-1}(u)\right]_{-b}^{b}, \tag{2.4.16}$$

which is a truncated normal scores function. For b tending to zero this again leads to the median. See Hampel (1983a) for some more details on R-estimators.

2.5. THE CHANGE-OF-VARIANCE FUNCTION

2.5a. Definitions

The influence function provides an intuitively appealing description of the local robustness of the *asymptotic value* of an estimator. However, it is also interesting to investigate the local robustness of the other very important asymptotic concept, namely the *asymptotic variance*, which is related to the length of confidence intervals. For this purpose one uses the *change-of-variance function* (*CVF*). It is defined in the framework of *M*-estimators of location, with precise mathematical conditions on the model distribution *F* and the ψ-function. This allows us to develop a rigorous treatment, investigating the relations between the various robustness notions which may be associated with the IF and the CVF. It is our opinion that a complete description of the robustness of a location *M*-estimator has to take into account both the IF and the CVF.

The results presented in this section were taken from Rousseeuw's (1981a, b, c; 1982a; 1983a) Ph.D. thesis and related papers: It is proved that the CVF leads to a more stringent robustness concept than the IF (Subsection 2.5b); that the median is the most robust estimator in both senses (Subsection 2.5c); and that the solutions to the problems of optimal robustness are the same, amounting at the standard normal to the Huber-estimator (Subsection 2.5d). The results are then generalized to scale estimation, and some further topics are discussed.

Our investigation takes place in the framework of *M*-estimation of a location parameter (Subsection 2.3a). The model is given by $\mathscr{X} = \mathbb{R}$, $\Theta = \mathbb{R}$ and the continuous distributions $F_\theta(x) = F(x - \theta)$. Here, the (fixed) model distribution *F* satisfies:

(F1)   *F* has a twice continuously differentiable density *f* (with respect to the Lebesgue measure λ) which is symmetric around zero and satisfies $f(x) > 0$ for all *x* in $\mathbb{R}$.

(F2)   The mapping $\Lambda = -f'/f = (-\ln f)'$ satisfies $\Lambda'(x) > 0$ for all *x* in $\mathbb{R}$, and $\int \Lambda' f \, d\lambda = -\int \Lambda f' \, d\lambda < \infty$.

The mapping Λ [which is clearly continuously differentiable from (F1)] is the ψ-function which corresponds to the maximum likelihood estimator. The condition $\Lambda'(x) > 0$ for all *x* in $\mathbb{R}$ (which is somewhat stronger than convexity of $-\ln f$) implies unimodality of *f*, because it follows that

$\Lambda(x) > 0$ for $x > 0$, so $f'(x) < 0$ for $x > 0$ and hence (using symmetry) $f(0) > f(x)$ for all $x \neq 0$. The condition $\int \Lambda' f \, d\lambda = - \int \Lambda f' \, d\lambda < \infty$ implies that the Fisher information

$$J(F) = \int \Lambda^2 \, dF > 0 \qquad (2.5.1)$$

satisfies $J(F) = \int \Lambda' \, dF < \infty$. It may be remarked that (F1) and (F2) imply that F satisfies Huber's conditions (1964, p. 80) on the model distribution, and thus his minimax asymptotic variance theorem (see Section 2.7 below) is applicable at F. Our favorite choice for F will be the standard normal distribution Φ with $\Lambda(x) = x$ (the MLE being the arithmetic mean) and $J(\Phi) = 1$. The logistic distribution provides another example, with $F(x) = 1/(1 + \exp(-x))$, $f(x) = F(x)(1 - F(x))$, $\Lambda(x) = \tanh(x/2) = 2F(x) - 1$, and $J(F) = \frac{1}{3}$. Note that in the first case Λ is unbounded and in the second it is bounded.

Let us recollect that an M-estimator of θ is given by

$$\sum_{i=1}^{n} \psi(X_i - T_n) = 0 \qquad (2.5.2)$$

and corresponds to the functional T defined by

$$\int \psi(x - T(G)) \, dG(x) = 0. \qquad (2.5.3)$$

Under certain regularity conditions (Huber, 1967), $\sqrt{n}\,(T_n - \theta)$ is asymptotically normal with asymptotic variance

$$V(\psi, G) = \int \psi^2 \, dG \bigg/ \left(\int \psi' \, dG \right)^2. \qquad (2.5.4)$$

We will examine the class Ψ consisting of all real functions ψ satisfying:

(i) ψ is well defined and continuous on $\mathbb{R} \setminus C(\psi)$, where $C(\psi)$ is finite. In each point of $C(\psi)$, there exist finite left and right limits of ψ which are different. Also $\psi(-x) = -\psi(x)$ if $\{x, -x\} \subset \mathbb{R} \setminus C(\psi)$ and $\psi(x) \geq 0$ for $x \geq 0$ not belonging to $C(\psi)$.

(ii) The set $D(\psi)$ of points in which ψ is continuous but in which ψ' is not defined or not continuous, is finite.

(iii) $\int \psi^2 \, dF < \infty$.

(iv) $0 < \int \psi' \, dF = - \int \psi(x) f'(x) \, dx = \int \Lambda \psi \, dF < \infty$.

To our knowledge, Ψ covers all ψ-functions ever used for this estimation problem. The condition that the left and right limits of ψ at a point of $C(\psi)$ be different, serves to eliminate unnecessary points of $C(\psi)$. Clearly, $C(\psi)$ and $D(\psi)$ are symmetric about zero, and have an empty intersection. From (i) and (iii) it follows that

$$\int \psi \, dF = 0 \tag{2.5.5}$$

which implies Fisher-consistency (2.3.10). From (i) and (iv) it follows that $0 < \int \psi^2 \, dF$. Therefore, if we define

$$A(\psi) := \int \psi^2 \, dF \quad \text{and} \quad B(\psi) := \int \psi' \, dF, \tag{2.5.6}$$

then $0 < A(\psi) < \infty$ and $0 < B(\psi) < \infty$ for all ψ in Ψ. Using (2.5.2) we say that the functions ψ_1 and ψ_2 in Ψ are *equivalent* if and only if $C(\psi_1) = C(\psi_2)$ and for all x not in this set we have $\psi_1(x) = r\psi_2(x)$ where $r > 0$.

Let us keep in mind that ψ' may contain delta functions (Huber, 1964, p. 78), so

$$\int \psi' \, dF = \int_{\mathbb{R}\setminus(C(\psi)\cup D(\psi))} \psi' \, dF + \sum_{i=1}^{m} \left[\psi(c_i +) - \psi(c_i -)\right] f(c_i), \tag{2.5.7}$$

where the first term is simply the classical integral of the piecewise continuous function ψ on $\mathbb{R}\setminus(C(\psi)\cup D(\psi))$, and $C(\psi) = \{c_1, \ldots, c_m\}$. This means that ψ' may be written as

$$\psi' 1_{\mathbb{R}\setminus(C(\psi)\cup D(\psi))} + \sum_{i=1}^{m} \left[\psi(c_i +) - \psi(c_i -)\right] \delta_{(c_i)}, \tag{2.5.8}$$

which is the sum of a "regular" part and a linear combination of delta functions $\delta_{(c_i)}$.

The local behavior of the asymptotic value of the M-estimator is described by its influence function (see Subsection 2.3a), given by

$$\text{IF}(x; \psi, F) = \frac{\psi(x)}{B(\psi)} \tag{2.5.9}$$

on $\mathbb{R} \setminus C(\psi)$. The gross-error sensitivity (see Subsection 2.1c) equals

$$\gamma^*(\psi, F) = \sup_{x \in \mathbb{R} \setminus C(\psi)} |IF(x; \psi, F)|. \qquad (2.5.10)$$

However, we feel that this only gives half of the picture, because it seems equally interesting to study the local behavior of the other very important asymptotic concept, namely the asymptotic variance (2.5.4). Consider any distribution G which has a symmetric density g and satisfies $0 < \int \psi^2 \, dG < \infty$ and $0 < \int \psi' \, dG < \infty$. Keeping in mind the interpretation of ψ', we can easily verify that

$$\frac{\partial}{\partial t} \left[V(\psi, (1 - t)F + tG) \right]_{t=0}$$

$$= \int \left[\frac{A(\psi)}{B(\psi)^2} \left(1 + \frac{\psi^2(x)}{A(\psi)} - 2\frac{\psi'(x)}{B(\psi)} \right) \right] dG(x). \qquad (2.5.11)$$

Therefore, in analogy with formula (2.1.4), we define:

Definition 1. The *change-of-variance function* $CVF(x; \psi, F)$ of ψ in Ψ at F is defined as the sum of the regular part

$$\frac{A(\psi)}{B(\psi)^2} \left(1 + \frac{\psi^2(x)}{A(\psi)} - 2\frac{\psi'(x)}{B(\psi)} \right) 1_{\mathbb{R} \setminus (C(\psi) \cup D(\psi))}(x)$$

which is continuous on $\mathbb{R} \setminus (C(\psi) \cup D(\psi))$, and

$$\frac{A(\psi)}{B(\psi)^2} \left(-\frac{2}{B(\psi)} \left[\sum_{i=1}^{m} (\psi(c_i +) - \psi(c_i -)) \delta_{(c_i)}(x) \right] \right).$$

It follows that $CVF(x; \psi, F)$ is continuous on $\mathbb{R} \setminus (C(\psi) \cup D(\psi))$ and that it is *symmetric*, whereas $IF(x; \psi, F)$ is skew symmetric. For continuous ψ no delta functions arise, and then we can simply write

$$CVF(x; \psi, F) = \frac{\partial}{\partial t} \left[V\left(\psi, (1 - t)F + t\left(\frac{1}{2}\Delta_x + \frac{1}{2}\Delta_{-x} \right) \right) \right]_{t=0},$$

where Δ_x is the probability measure putting mass 1 at x. Of course, (2.5.11) can also be generalized to other types of estimators.

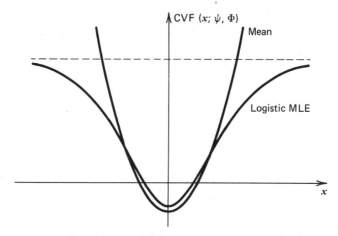

Figure 1. Change-of-variance functions of the arithmetic mean and the logistic maximum likelihood estimator, both evaluated at the standard normal distribution.

Example 1. For the MLE, we have $\mathrm{IF}(x; \Lambda, F) = \Lambda(x)/J(F)$ and $\mathrm{CVF}(x; \Lambda, F) = \{1 + [\Lambda^2(x) - 2\Lambda'(x)]/J(F)\}/J(F)$. For the arithmetic mean at $F = \Phi$ this yields $\mathrm{IF}(x; \Lambda, \Phi) = x$ and

$$\mathrm{CVF}(x; \Lambda, \Phi) = x^2 - 1. \tag{2.5.12}$$

Let us compare this to the maximum likelihood estimator for the logistic distribution, corresponding to $\psi(x) = [\exp(x) - 1]/[\exp(x) + 1]$. The resulting change-of-variance function at $F = \Phi$ is displayed in Figure 1, together with that of the arithmetic mean. Both curves are continuous, but the one is bounded while the other is not.

The median corresponds to $\psi_{\mathrm{med}}(x) = \mathrm{sign}(x)$, so

$$\mathrm{CVF}(x; \psi_{\mathrm{med}}, F) = \frac{2}{(2f(0))^2}\left[1_{\mathbf{R}\setminus\{0\}}(x) - \frac{1}{f(0)}\delta_{(0)}(x)\right]. \tag{2.5.13}$$

On the other hand, a Huber-type skipped mean, given by $\psi_{\mathrm{sk}(r)}(x) = x1_{[-r, r]}(x)$ where $0 < r < \infty$, yields a CVF containing two delta functions with *positive* factor, because of the downward jumps of $\psi_{\mathrm{sk}(r)}$ at r and $-r$.

The basic idea of the CVF was discovered first by F. Hampel in 1972, who used it in connection with the so-called "hyperbolic tangent estimators" which he conjectured to be optimal in some sense. He did not work out these ideas, although he referred to them briefly (Hampel, 1973a, p. 98;

1974, p. 393). The CVF was also used implicitly in a technical report by Boos and Serfling (1976). The CVF was rediscovered by Rousseeuw (1981a, b), who developed the present theory and proved the conjectured optimality result (see Subsection 2.6c below).

The CVF and the IF have many characteristics in common; for example, we have

$$\int \text{IF}(x; \psi, F) \, dF(x) = 0 \qquad (2.5.14)$$

and

$$\int \text{CVF}(x; \psi, F) \, dF(x) = 0 \qquad (2.5.15)$$

for all ψ in Ψ. However, these curves are not interpreted in the same way. Large positive and large negative values of the IF have qualitatively the same (unfavorable) meaning, namely a bias caused by contamination. (Therefore, γ^* is defined as the supremum of the *absolute value* of the IF.) On the other hand, one does not have to worry about large negative values of a CVF as much as about large positive values, since only the latter lead to wide confidence intervals. (A negative value of the CVF merely points to a decrease in V, indicating a higher accuracy and narrower confidence intervals, which is even beneficial.) This is in accordance with the reasoning behind Huber's minimax theory: There one is only concerned about the *large* values of $V(\psi, G)$ (where G belongs to some "neighborhood"; see Section 2.7), and not about small values. Therefore, we define:

Definition 2. The *change-of-variance sensitivity* $\kappa^*(\psi, F)$ is defined as $+\infty$ if a delta function with positive factor occurs in the CVF, and otherwise as

$$\kappa^*(\psi, F) := \sup\{\text{CVF}(x; \psi, F)/V(\psi, F); \, x \in \mathbb{R} \setminus (C(\psi) \cup D(\psi))\}.$$

Clearly, upward jumps of ψ (contributing only a negative delta function to the CVF) do not affect κ^*, but any downward jump makes it infinite. On a heuristic level, κ^* may be compared with the robustness measure $\sup V(\psi, G)$ (where G belongs to some "neighborhood") that was recently studied in detail by Collins (1977) and Collins and Portnoy (1981).

This change-of-variance sensitivity is identical to the κ^* occurring in Rousseeuw (1981a, b, c; 1982a) and Hampel, et al. (1981). However, in these papers the concept was derived in a slightly different way. The standardization by $V(\psi, F)$ was not done in the definition of κ^* (as in our Definition 2), but already in the function itself. Indeed, the change-of-variance *curve*

(CVC) introduced in these papers was defined by differentiation of $\ln V(\psi, G)$ and equals $CVF(x; \psi, F)/V(\psi, F)$, so κ^* was simply the supremum of this CVC. However, in higher dimensions $V(\psi, G)$ becomes a matrix, and different standardizations are possible. Therefore, we nowadays prefer to work with the CVF, which can be easily generalized.

Example 2. For the arithmetic mean we have $\kappa^* = \infty$, and for the median $\kappa^* = 2$. Any Huber-type skipped mean has $\kappa^* = \infty$, which corresponds to Lemma 3.2 of Collins (1976), stating that $\sup V(\psi_{\text{sk}(r)}, G) = \infty$ for all gross-error neighborhoods of $F = \Phi$. (In the proof, Collins constructs contaminating distributions with concentrated mass around the points where the CVF becomes unbounded.)

2.5b. B-robustness versus V-robustness

For any M-estimator we now have the IF with γ^* on the one hand, and the CVF with κ^* on the other hand. We recollect that an M-estimator is called B-*robust* (from "bias") when γ^* is finite, and we say it is V-*robust* (from "variance") when κ^* is finite. For example, in Figure 1 we see that the logistic MLE is V-robust, whereas the arithmetic mean is not. We shall now show that the concept of V-robustness is stronger than the concept of B-robustness. (Note that the conditions on F and ψ of Subsection 2.5a are assumed to hold throughout Section 2.5.)

Theorem 1. For all ψ in Ψ, V-robustness implies B-robustness. In fact, $\gamma^*(\psi, F) \leq [(\kappa^*(\psi, F) - 1)V(\psi, F)]^{1/2}$.

Proof. Suppose that ψ belongs to Ψ, $\kappa^*(\psi, F)$ is finite, and that there exists a point x_0 for which $|IF(x_0; \psi, F)| > [(\kappa^* - 1)V(\psi, F)]^{1/2}$. Without loss of generality, $x_0 \notin D(\psi)$ and $x_0 > 0$. If we had $\psi'(x_0) \leq 0$, then $\kappa^* \geq 1 + \psi^2(x_0)/A(\psi) - 2\psi'(x_0)/B(\psi) \geq 1 + IF(x_0; \psi, F)^2/V(\psi, F) > 1 + (\kappa^* - 1) = \kappa^*$, a contradiction. Therefore $\psi'(x_0) > 0$; hence there exists $\varepsilon > 0$ such that $\psi'(t) > 0$ for all t in $[x_0, x_0 + \varepsilon)$, and thus $\psi(x) > \psi(x_0)$ for all x in $(x_0, x_0 + \varepsilon]$. We now show that $\psi(x) > \psi(x_0)$ for all $x > x_0$, $x \notin C(\psi)$. Suppose the opposite were true; then $x_0 + \varepsilon \leq x' := \inf\{x > x_0; x \notin C(\psi)$ and $\psi(x) \leq \psi(x_0)\} < \infty$. As in points of $C(\psi)$ only upward jumps are allowed (otherwise $\kappa^* = \infty$), we have $x' \notin C(\psi)$. [Take some $c \in C(\psi)$, $c > x_0$. If $\psi(c -) < \psi(x_0)$ then $x' < c$, and if $\psi(c -) \geq \psi(x_0)$ then $\psi(c +) > \psi(x_0)$, so $c \neq x'$.] Hence ψ is continuous at x', so $\psi(x') = \psi(x_0)$. Clearly, $\psi(x) > \psi(x_0)$ for all x in $(x_0, x') \backslash C(\psi)$. Now there exists a point x'' in $(x_0, x') \backslash (C(\psi) \cup D(\psi))$ such that $\psi'(x'') \leq 0$, because otherwise we could show that $\psi(x') > \psi(x_0)$ by starting in x_0 and going to the right, using the upward jumps in points of $C(\psi)$. Since we must also have $\psi^2(x'') > \psi^2(x_0)$, it would hold that $1 + \psi^2(x'')/A(\psi) - 2\psi'(x'')/B(\psi) \geq 1 + \psi^2(x'')/A(\psi) > 1 + (\kappa^* - 1) = \kappa^*$, a contradiction. We conclude that $\psi(x) > \psi(x_0)$ for all $x > x_0$, $x \notin C(\psi)$.

We now proceed to the final contradiction. Because $C(\psi) \cup D(\psi)$ is finite, we can assume from now on that $[x_0, +\infty) \cap (C(\psi) \cup D(\psi))$ is empty. Then $\psi^2(x) - 2\psi'(x)A(\psi)/B(\psi) \leq b^2$ for $x > x_0$, where $b = [(\kappa^* - 1)A(\psi)]^{1/2}$. Because $\psi^2(x) > \psi^2(x_0) > b^2$, this gives $\psi'(x)(2A(\psi)/B(\psi)) \geq \psi^2(x) - b^2 > 0$. Therefore, $\psi'(x)/[\psi^2(x) - b^2] \geq d := (B(\psi)/2A(\psi)) > 0$. For all $x \geq x_0$, we define

$$R(x) := -\frac{1}{b} \coth^{-1}\left(\frac{\psi(x)}{b}\right).$$

It follows that R is well defined and differentiable with derivative $\psi'(x)/[\psi^2(x) - b^2]$ on $[x_0, \infty)$. Therefore, $R(x) - R(x_0) \geq d(x - x_0)$ for all $x \geq x_0$, which implies that $\coth^{-1}(\psi(x)/b) \leq b(dx_0 - R(x_0)) - dx)$. The left-hand side of the latter inequality is strictly positive because $\psi(x)/b > 1$, but the right-hand side tends to $-\infty$ when $x \to \infty$, which is clearly impossible. This ends the proof. $\qquad\square$

Remark 1.   Clearly ψ is B-robust if and only if $|\psi|$ is bounded. From Theorem 1 it follows that ψ is V-robust if and only if $|\psi|$ is bounded and ψ' is bounded from below. (The first bound follows from Theorem 1, and in its turn implies the second by the definition of the CVF; the reverse is obvious.) Therefore, both B-robustness and V-robustness are independent of F (as far as the regularity conditions on ψ and F hold.)

Theorem 2.   For nondecreasing ψ in Ψ, V-robustness and B-robustness are equivalent. In fact, $\kappa^*(\psi, F) = 1 + \gamma^*(\psi, F)^2/V(\psi, F)$.

Proof.   The one implication follows from Theorem 1. For the other, assume that ψ is B-robust. Because ψ is monotone, the CVF can only contain negative delta functions, which do not contribute to κ^*. For all $x \notin (C(\psi) \cup D(\psi))$ it holds that $\psi'(x) \geq 0$, so $1 + \psi^2(x)/A(\psi) - 2\psi'(x)/B(\psi) \leq 1 + \psi^2(x)/A(\psi) \leq 1 + \gamma^{*2}/V(\psi, F)$; hence ψ is also V-robust. On the other hand, it follows from Theorem 1 that also $\kappa^* \geq 1 + \gamma^{*2}/V(\psi, F)$. This ends the proof. $\qquad\square$

Remark 2.   In general we do not have equivalence, as is exemplified by the Huber-type skipped mean (Example 1), which is B-robust but not

V-robust. Even counterexamples with continuous ψ may be constructed (Rousseeuw, 1981a).

Example 3. The median, defined by the nondecreasing function $\psi_{\mathrm{med}}(x) = \mathrm{sign}(x)$, is both B-robust and V-robust. The same holds for the MLE if Λ is bounded, as is the case when F is the logistic distribution, where $\gamma^*(\Lambda, F) = 3$ and $\kappa^*(\Lambda, F) = 4$. If Λ is unbounded (e.g., at $F = \Phi$), then the MLE is neither B-robust nor V-robust.

2.5c. The Most Robust Estimator

Let us now determine those estimators which do not only have finite sensitivities, but which even possess the smallest sensitivities possible. An estimator minimizing γ^* we call *most B-robust*, and when it minimizes κ^* we say it is *most V-robust*.

Theorem 3. The median is the most B-robust estimator in Ψ. For all ψ in Ψ we have $\gamma^*(\psi, F) \geq 1/(2f(0))$, and equality holds if and only if ψ is equivalent to ψ_{med}.

Proof. Assume that ψ is bounded, otherwise $\gamma^*(\psi, F) = \infty$ and there is nothing left to prove. Clearly, $\sup\{|\psi(x)|; \; x \in \mathbb{R} \setminus C(\psi)\} > 0$, or else we would have $A(\psi) = 0$. Then

$$B(\psi) = \int |\Lambda| \, |\psi| \, dF \leq \sup_{x \in \mathbb{R} \setminus C(\psi)} |\psi(x)| \int |\Lambda| \, dF$$

because $\Lambda(x)$ and $\psi(x)$ are positive for $x > 0$ and negative for $x < 0$. It holds that $\int |\Lambda| \, dF = 2\int_0^\infty - f'(x) \, dx = 2f(0)$; hence $\gamma^*(\psi, F) = \sup|\psi(x)|/B(\psi) \geq 1/(2f(0)) = \gamma^*(\psi_{\mathrm{med}}, F)$. For the uniqueness part, suppose that some ψ in Ψ satisfies $\gamma^*(\psi, F) = 1/(2f(0))$. This implies $\int |\Lambda| \, |\psi| \, dF = \int |\psi| \sup_{x \in \mathbb{R} \setminus C(\psi)} |\psi(x)| \, dF$, where $f(y) > 0$ for all y and $|\Lambda(y)| > 0$ for all $y \neq 0$ by (F2). By means of some elementary analysis, it follows for all y in $\mathbb{R} \setminus C(\psi)$ that $|\psi(y)| = \sup_{x \in \mathbb{R} \setminus C(\psi)} |\psi(x)|$, which is a strictly positive finite constant. Suppose w.l.o.g. that this constant equals 1. But from $\psi(x) \geq 0$ for positive x in $\mathbb{R} \setminus C(\psi)$ and skew symmetry of ψ, this implies $\psi(x) = \mathrm{sign}(x)$ for all $x \neq 0$ not belonging to $C(\psi)$. Therefore $C(\psi) = \{0\}$ and $\psi = \psi_{\mathrm{med}}$, which ends the proof. (For more details, see Rousseeuw, 1981b.)   □

Example 4. The minimal value of γ^* equals $\gamma^*(\psi_{\mathrm{med}}, F) = \sqrt{\pi/2} \approx 1.2533$ at the normal and $\gamma^*(\psi_{\mathrm{med}}, F) = 2$ at the logistic distribution.

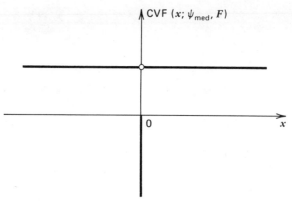

Figure 2. Change-of-variance function of the median. The downward "spike" in the center indicates a negative delta function.

Remark 3. P. Huber (1964, p. 83) already showed in a different setting that the median gives the smallest asymptotic bias. His reasoning already contained the value $1/(2\phi(0)) = \sqrt{\pi/2} \simeq 1.2533$ of $\gamma^*(\psi_{med}, \Phi)$, although the IF had not yet been formally defined at that time.

Theorem 4. The median is also the most V-robust estimator in Ψ. For all ψ in Ψ we have $\kappa^*(\psi, F) \geq 2$, and equality holds if and only if ψ is equivalent to ψ_{med}.

Proof. It holds that $V(\psi, F) = A(\psi)/B^2(\psi) = \int \mathrm{IF}^2(x; \psi, F) \, dF(x) \leq (\gamma^*)^2$. Making use of Theorem 1, it follows that $\kappa^*(\psi, F) \geq 1 + \gamma^*(\psi, F)^2/V(\psi, F) \geq 2$. If equality holds, then $V(\psi, F) = \gamma^*(\psi, F)^2$ which implies that ψ is equivalent to ψ_{med}.   □

Figure 2 displays the CVF of the median, given by (2.5.13). It shows that the variance of the median can be drastically reduced by adding "extra mass" exactly *at* the true median of the underlying distribution. If placed anywhere else, contamination will only have a constant effect (in linear approximation).

2.5d. Optimal Robust Estimators

In Subsection 2.4b we determined optimal B-robust M-estimators by means of Hampel's optimality result. We showed that certain M-estimators had minimal asymptotic variance for a given upper bound c on γ^*. However, some questions remained unanswered:

What is the range of c yielding these solutions?
Are there other solutions for other ranges of c?

In the present framework it is possible to give detailed answers to these questions, for both B-robustness and V-robustness.

The M-estimators corresponding to

$$\psi_b(x) = [\Lambda(x)]^b_{-b} \qquad (2.5.16)$$

(where $0 < b < \|\Lambda\| := \sup_x |\Lambda(x)|$) were introduced by Huber (1964, p. 80). In the special case $F = \Phi$ they correspond to $\psi_b(x) = [x]^b_{-b}$ and are called Huber-estimators (Example 2 of Subsection 2.3a.)

Lemma 1. The mapping $b \to \gamma^*(\psi_b, F)$ is an increasing continuous bijection from $(0, \|\Lambda\|)$ onto $(\gamma^*(\psi_{\mathrm{med}}, F), \gamma^*(\Lambda, F))$.

Proof. The mapping $b \to \gamma^*(\psi_b, F) = b/B(\psi_b)$ is clearly continuous. We first prove that it is strictly increasing. We know that $B(\psi_b) = 2\int_{[0, \Lambda^{-1}(b)]} \Lambda'(x)f(x)\,d\lambda(x)$; performing the substitution $u = \Lambda(x)$, we obtain $B(\psi_b) = 2\int_0^b f(\Lambda^{-1}(u))\,du$. Now define $S(z) = 2\int_0^z f(\Lambda^{-1}(u))\,du$ for all z in $[0, \|\Lambda\|)$; clearly, $S(0) = 0$. On the other hand, $S'(z) = 2f(\Lambda^{-1}(z))$ which is strictly decreasing; hence S is strictly concave. If $0 < b_1 < b_2 < \|\Lambda\|$ then $[S(b_1) - S(0)]/[b_1 - 0] > [S(b_2) - S(0)]/[b_2 - 0]$, so $\gamma^*(\psi_{b_1}, F) = b_1/S(b_1) < b_2/S(b_2) = \gamma^*(\psi_{b_2}, F)$. Moreover, $\lim_{b \to \|\Lambda\|} \gamma^*(\psi_b, F) = \gamma^*(\Lambda, F)$. By L'Hôpital's rule, $\lim_{0 < b \to 0} \gamma^*(\psi_b, F) = \lim_{0 < b \to 0} (1/2f(\Lambda^{-1}(b))) = 1/(2f(0)) = \gamma^*(\psi_{\mathrm{med}}, F)$. □

Rousseeuw (1981b) also proved that $b \to V(\psi_b, F)$ is a decreasing continuous bijection from $(0, \|\Lambda\|)$ onto $(V(\Lambda, F), V(\psi_{\mathrm{med}}, F))$, but we do not need that result here.

Theorem 5. The only optimal B-robust estimators in Ψ are (up to equivalence) given by $\{\psi_{\mathrm{med}}, \psi_b \ (0 < b < \infty)\}$ if Λ is unbounded, and by $\{\psi_{\mathrm{med}}, \psi_b \ (0 < b < \|\Lambda\|), \Lambda\}$ otherwise.

Proof. From Theorem 1 of Section 2.4 and Lemma 1 of this section it follows that for each constant c in $(1/(2f(0)), \gamma^*(\Lambda, F))$ there exists a unique b in $(0, \|\Lambda\|)$ such that $\gamma^*(\psi_b, F) = c$, and that this ψ_b minimizes $V(\psi, F)$ among all ψ in Ψ which satisfy $\gamma^*(\psi, F) \leq c$; moreover, any other solution is equivalent to ψ_b. Theorem 3 implies that no solution can exist for any $c < 1/(2f(0)) = \gamma^*(\psi_{\text{med}}, F)$. If one puts $c = 1/(2f(0))$, then only mappings equivalent to ψ_{med} can satisfy $\gamma^*(\psi, F) \leq c$, and therefore the median itself is automatically optimal B-robust. Let us now consider $\gamma^*(\Lambda, F)$. If $\gamma^*(\Lambda, F) = \infty$, then the MLE is not B-robust. If $\gamma^*(\Lambda, F) < \infty$, the consideration of any $c \geq \gamma^*(\Lambda, F)$ always yields Λ itself because Λ minimizes $V(\psi, F)$ in Ψ. [Indeed, $B^2(\psi) = (\int \Lambda \psi \, dF)^2 \leq J(F) A(\psi)$ by the Cauchy–Schwarz inequality, so $V(\psi, F) = A(\psi)/B^2(\psi) \geq 1/J(F) = V(\Lambda, F)$.] $\Box$

We say that an M-estimator is *optimal V-robust* when it minimizes $V(\psi, F)$ for a given upper bound k on $\kappa^*(\psi, F)$.

Lemma 2. The mapping $b \to \kappa^*(\psi_b, F)$ is an increasing continuous bijection from $(0, \|\Lambda\|)$ onto $(\kappa^*(\psi_{\text{med}}, F), \kappa^*(\Lambda, F))$.

Proof. The mapping $b \to \kappa^*(\psi_b, F) = 1 + b^2/A(\psi_b)$ is strictly increasing, because its derivative equals $4b \int_0^{\Lambda^{-1}(b)} \Lambda^2(x) f(x) \, dx / A^2(\psi_b) > 0$. By L'Hôpital's rule, $\lim_{0 < b \to 0} \kappa^*(\psi_b, F) = 1 + \lim_{0 < b \to 0} 2b/[4b(1 - F(\Lambda^{-1}(b)))] = 2$. On the other hand, $\lim_{b \to \|\Lambda\|} \kappa^*(\psi_b, F) = 1 + \|\Lambda\|^2/A(\Lambda)$, which equals $\kappa^*(\Lambda, F)$ by Theorem 2. $\Box$

Figure 3 shows the change-of-variance function of a Huber-estimator. On the interval $(-b, b)$ it is parabolic, like the CVF of the arithmetic mean in Figure 1. On (b, ∞) and $(-\infty, -b)$ the CVF is a constant, and in b and $-b$ it is not defined. We shall see in Theorem 6 that this estimator has maximal asymptotic efficiency subject to an upper bound on κ^*. (In order to avoid the discontinuity in the CVF one might use the logistic MLE instead, which has nearly the same behavior, and the CVF of which in Figure 1 looks like a smoothed version of that of the Huber-estimator.)

Theorem 6. The optimal V-robust M-estimators coincide with the estimators listed in Theorem 5.

Proof. From Lemma 2 it follows that for each k in $(2, \kappa^*(\Lambda, F))$ there is a unique b in $(0, \|\Lambda\|)$ for which $\kappa^*(\psi_b, F) = k$. Let us now show that ψ_b is optimal V-robust. Take any ψ in Ψ satisfying $V(\psi, F) < V(\psi_b, F)$. Then $\gamma^*(\psi, F) > \gamma^*(\psi_b, F)$ because ψ_b is optimal B-robust. From Theorem 1 it follows that $\kappa^*(\psi, F) \geq 1 + \gamma^*(\psi, F)^2/V(\psi, F) > 1 + \gamma^*(\psi_b, F)^2/V(\psi_b, F)$. The latter expression equals $\kappa^*(\psi_b, F)$ by Theorem 2; hence $\kappa^*(\psi, F) > \kappa^*(\psi_b, F)$. This proves that ψ_b is optimal V-robust.

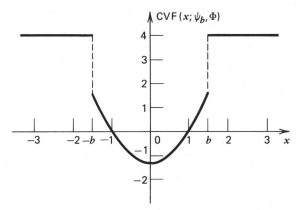

Figure 3. Change-of-variance function of the Huber-estimator with $b = 1.5$ evaluated at the standard normal distribution.

Now take another function χ minimizing $V(\psi, F)$ subject to $\kappa^*(\psi, F) \le k$. But then $V(\chi, F) = V(\psi_b, F)$ and $\gamma^*(\chi, F) \le [(\kappa^*(\chi, F) - 1)V(\chi, F)]^{1/2} \le [(k - 1)V(\psi_b, F)]^{1/2} = c$, so χ is equivalent to ψ_b by Theorem 5. Theorem 4 implies that no solution can exist for any $k < 2$, and Theorem 2 entails that $\kappa^*(\psi, F)$ is finite if and only if ψ is bounded. The rest of the proof is analogous to that of Theorem 5.   □

When the Huber-estimators $\psi_b(x) = [x]_{-b}^{b}$ are considered at the standard normal distribution, one obtains the following expressions for the quantities described above:

$$A = A(\psi_b) = 2\Phi(b) - 1 - 2b\phi(b) + 2b^2(1 - \Phi(b)),$$

$$B = B(\psi_b) = 2\Phi(b) - 1,$$

$$V = V(\psi_b, \Phi) = A/B^2,$$

$$e = e(\psi_b, \Phi) = B^2/A,$$

$$\gamma^* = \gamma^*(\psi_b, \Phi) = b/B,$$

$$\kappa^* = \kappa^*(\psi_b, \Phi) = 1 + b^2/A,$$

$$\lambda^* = \lambda^*(\psi_b, \Phi) = 1/B,$$

Table 1. Huber-Estimators at $F = \Phi$

b	A	B	V	e	γ^*	κ^*	λ^*
0.0	0.00000000	0.00000000	1.5708	0.6366	1.2533	2.0000	∞
0.1	0.00946861	0.07965567	1.4923	0.6701	1.2554	2.0561	12.5540
0.2	0.03576156	0.15851942	1.4232	0.7027	1.2617	2.1185	6.3084
0.3	0.07576610	0.23582284	1.3624	0.7340	1.2721	2.1879	4.2405
0.4	0.12649241	0.31084348	1.3091	0.7639	1.2868	2.2649	3.2171
0.5	0.18512837	0.38292492	1.2625	0.7921	1.3057	2.3504	2.6115
0.6	0.24908649	0.45149376	1.2219	0.8184	1.3289	2.4453	2.2149
0.7	0.31604157	0.51607270	1.1866	0.8427	1.3564	2.5504	1.9377
0.8	0.38395763	0.57628920	1.1561	0.8650	1.3882	2.6669	1.7352
0.9	0.45110370	0.63187975	1.1298	0.8851	1.4243	2.7956	1.5826
1.0	0.51605855	0.68268949	1.1073	0.9031	1.4948	2.9378	1.4648
1.1	0.57770496	0.72866788	1.0880	0.9191	1.5096	3.0945	1.3724
1.2	0.63521478	0.76986066	1.0718	0.9330	1.5587	3.2669	1.2989
1.3	0.68802633	0.80639903	1.0580	0.9451	1.6121	3.4563	1.2401
1.4	0.73581588	0.83848668	1.0466	0.9555	1.6697	3.6637	1.1926
1.5	0.77846522	0.86638560	1.0371	0.9642	1.7313	3.8903	1.1542
1.6	0.81602712	0.89040142	1.0293	0.9716	1.7969	4.1372	1.1231
1.7	0.84869059	0.91086907	1.0229	0.9776	1.8663	4.4052	1.0979
1.8	0.87674726	0.92813936	1.0178	0.9825	1.9394	4.6955	1.0774
1.9	0.90056035	0.94256688	1.0137	0.9865	2.0158	5.0086	1.0609
2.0	0.92053693	0.95449974	1.0104	0.9897	2.0953	5.3453	1.0477
2.1	0.93710425	0.96427116	1.0078	0.9922	2.1778	5.7060	1.0371
2.2	0.95069027	0.97219310	1.0059	0.9942	2.2629	6.0910	1.0286
2.3	0.96170849	0.97855178	1.0043	0.9957	2.3504	6.5006	1.0219
2.4	0.97054680	0.98360493	1.0032	0.9968	2.4400	6.9348	1.0167
2.5	0.97755998	0.98758067	1.0023	0.9977	2.5314	7.3935	1.0126
2.6	0.98306545	0.99067762	1.0017	0.9983	2.6245	7.8764	1.0094
2.7	0.98734148	0.99306605	1.0012	0.9988	2.7189	8.3835	1.0070
2.8	0.99062765	0.99488974	1.0008	0.9992	2.8144	8.9142	1.0051
2.9	0.99312667	0.99626837	1.0006	0.9994	2.9109	9.4682	1.0037
3.0	0.99500728	0.99730020	1.0004	0.9996	3.0081	10.0452	1.0027
4.0	0.99987950	0.99993666	1.0000	1.0000	4.0003	17.0019	1.0001
5.0	0.99999889	0.99999943	1.0000	1.0000	5.0000	26.0000	1.0000
∞	1.00000000	1.00000000	1.0000	1.0000	∞	∞	1.0000

where λ^* is the local-shift sensitivity as defined in Subsection 2.1c. These numbers are listed in Table 1 for b ranging from 0.1 to 5.0. Also the limits for $b \downarrow 0$ (median) and for $b \uparrow \infty$ (arithmetic mean) are included. (See also Fig. 5 of Section 1.3.)

2.5e. M-Estimators for Scale

Let us now generalize the CVF to M-estimators of scale (see Subsection 2.3a). The model is given by $\mathcal{X} = \mathbb{R}$, $\Theta = \{\sigma > 0\}$, and $F_\sigma(x) = F(x/\sigma)$. The (fixed) model distribution F satisfies (F1) and (F2) as in Subsection 2.5a. An M-estimator of σ is given by

$$\sum_{i=1}^{n} \chi(X_i/S_n) = 0 \qquad (2.5.17)$$

and corresponds to the functional S given by

$$\int \chi(x/S(G))\, dG(x) = 0. \qquad (2.5.18)$$

Under certain regularity conditions, $\sqrt{n}\,(S_n - \sigma)$ is asymptotically normal with asymptotic variance

$$V_1(\chi, G) = \int \chi^2\, dG \Big/ \left(\int x\chi'(x)\, dG(x) \right)^2. \qquad (2.5.19)$$

The class Ψ^1 consists of all functions χ satisfying:

(i) χ is well-defined and continuous on $\mathbb{R} \setminus C(\chi)$, where $C(\chi)$ is finite. In each point of $C(\chi)$ there exist finite left and right limits of χ which are different. Also $\chi(-x) = \chi(x)$ if $\{-x, x\} \subset \mathbb{R} \setminus C(\chi)$, and there exists $d > 0$ such that $\chi(x) \leq 0$ on $(0, d)$ and $\chi(x) \geq 0$ on (d, ∞).

(ii) The set $D(\chi)$ of points in which χ is continuous but in which χ' is not defined or not continuous is finite.

(iii) $\int \chi\, dF = 0$ and $\int \chi^2\, dF < \infty$.

(iv) $0 < \int x\chi'(x)\, dF(x) = \int (x\Lambda(x) - 1)\chi(x)\, dF(x) < \infty$.

The condition $\int \chi\, dF = 0$ reflects Fisher-consistency (2.1.3). Let

$$A_1(\chi) = \int \chi^2\, dF \quad \text{and} \quad B_1(\chi) = \int x\chi'(x)\, dF(x), \qquad (2.5.20)$$

so $IF(x; \chi, F) = \chi(x)/B_1(\chi)$. Applying Cauchy–Schwarz to $(B_1(\chi))^2$ and making use of (2.3.19) we see that the asymptotic variance satisfies $V_1(\chi, F)$ $= A_1(\chi)/(B_1(\chi))^2 \geq 1/J(F)$ (Cramér–Rao). Two functions χ_1 and χ_2 are said to be equivalent when $C(\chi_1) = C(\chi_2)$ and $\chi_1(x) = r\chi_2(x)$ for all x not in this set, where $r > 0$. Consider any distribution G which has a symmetric density g and satisfies $\int \chi\, dG = 0$, $0 < \int \chi^2\, dG < \infty$, and $0 < \int x\chi'(x)\, dG(x) < \infty$. Then (2.5.11) is replaced by

$$\frac{\partial}{\partial t}\left[V_1(\chi, (1 - t)F + tG)\right]_{t=0}$$

$$= \int \frac{A_1(\chi)}{B_1(\chi)^2}\left(1 + \frac{\chi^2(x)}{A_1(x)} - 2\frac{x\chi'(x)}{B_1(x)}\right)dG(x) \quad (2.5.21)$$

so we define the change-of-variance function by

$$CVF(x; \chi, F) = \frac{A_1(\chi)}{B_1(\chi)^2}\left(1 + \frac{\chi^2(x)}{A_1(x)} - 2\frac{x\chi'(x)}{B_1(x)}\right)$$

$$(2.5.22)$$

which is symmetric and may contain delta functions as in Definition 1. Again

$$\int CVF(x; \chi, F)\, dF(x) = 0$$

for all χ in Ψ^1. The change-of-variance sensitivity $\kappa^*(\chi, F)$ is defined as in Definition 2.

Example 5. The maximum likelihood estimator is given by $\chi_{MLE}(x) = x\Lambda(x) - 1$. At $F = \Phi$ this yields $\chi_{MLE}(x) = x^2 - 1$, so $A_1(\chi_{MLE}) = B_1(\chi_{MLE}) = J(\Phi) = 2$; hence

$$CVF(x; \chi_{MLE}, \Phi) = \tfrac{1}{4}(x^4 - 6x^2 + 3). \quad (2.5.23)$$

This function is displayed in Figure 4. The variance of the estimator decreases somewhat when contamination is placed in the regions where the CVF is negative, that is, when the contaminating x's have absolute values roughly between 1 and 2. Contamination elsewhere will increase the variance of the estimator, although the effect of "inliers" (points near zero) is

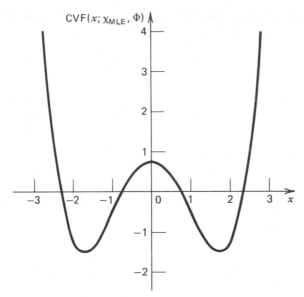

Figure 4. Change-of-variance function of the standard deviation at $F = \Phi$.

relatively harmless when compared to the effect of "gross outliers" (points with large $|x|$) where the CVF is of the order of x^4. Therefore, $\kappa^*(\chi_{MLE}, \Phi) = \infty$.

On the other hand, for the (standardized) MAD (Example 4 in Subsection 2.3a) we obtain $A_1(\chi_{MAD}) = 1$, $B_1(\chi_{MAD}) = 4q\phi(q)$, and

$$CVF(x; \chi_{MAD}, \Phi) = \frac{2}{(4q\phi(q))^2}\left(1 - \frac{1}{2q\phi(q)}\left[\delta_{(q)}(x) + \delta_{(-q)}(x)\right]\right)$$

(2.5.24)

where $q = \Phi^{-1}(\tfrac{3}{4})$; hence $\kappa^* = 2$.

Theorem 7. For all χ in Ψ^1, V-robustness implies B-robustness. In fact, $\gamma^*(\chi, F) \le [(\kappa^*(\chi, F) - 1)V_1(\chi, F)]^{1/2}$.

Proof. Suppose that $\kappa^*(\chi, F)$ is finite and that there exists some x_0 for which $|IF(x_0, \chi, F)| > [(\kappa^* - 1)V_1(\chi, F)]^{1/2}$. W.l.o.g. put $x_0 \notin D(\chi)$ and $x_0 > 0$. It follows that $|\chi(x_0)| > b$, where this time $b := [(\kappa^* - 1)A_1(\chi)]^{1/2}$. If $\chi'(x_0) \le 0$, then $1 + \chi^2(x_0)/A_1(\chi) - 2\chi'(x_0)/B_1(\chi) > 1 + b^2/A_1(\chi)$

$= \kappa^*$, a contradiction. Therefore, $\chi'(x_0) > 0$. *Case A*: Assume that $\chi(x_0)$
> 0. There exists $\varepsilon > 0$ such that $\chi'(t) > 0$ for all t in $[x_0, x_0 + \varepsilon)$, so
$\chi(x) > \chi(x_0)$ for all x in $(x_0, x_0 + \varepsilon]$. As in Theorem 1, it follows that
$\chi(x) > \chi(x_0) > b$ for all $x > x_0$, $x \notin C(\chi)$ because only upward jumps of
χ are allowed for positive x. As $C(\chi) \cup D(\chi)$ is finite, we may assume that
$[x_0, +\infty) \cap (C(\chi) \cup D(\chi))$ is empty. It holds that $1 + \chi^2(x)/A_1(\chi) -$
$2x\chi'(x)/B_1(\chi) \le \kappa^*$; hence $\chi^2(x) - 2x\chi'(x)A_1(\chi)/B_1(\chi) \le b^2$ for all x
$\ge x_0$. Therefore $\chi'(x)/[\chi^2(x) - b^2] \ge d/x$, where this time $d :=$
$B_1(\chi)/2A_1(\chi)$. Putting

$$R(x) := -\frac{1}{b} \coth^{-1}\left(\frac{\chi(x)}{b}\right) \quad \text{and} \quad P(x) := d\ln(x)$$

it follows that $R'(x) \ge P'(x)$ for all $x \ge x_0$; hence $R(x) - R(x_0) \ge P(x)$
$- P(x_0)$, and thus $\coth^{-1}(\chi(x)/b) \le b[P(x_0) - R(x_0) - d\ln(x)]$. However, the left member is positive because $\chi(x) > b$ and the right member
tends to $-\infty$ for $x \to \infty$, a contradiction. *Case B*: Assume $\chi(x_0) < 0$.
There exists $\varepsilon > 0$ such that $\chi'(t) > 0$ for all t in $(x_0 - \varepsilon, x_0]$, so $\chi(x) <$
$\chi(x_0)$ for all x in $[x_0 - \varepsilon, x_0)$. As in case A we show that $\chi(x) < \chi(x_0) <$
$(-b)$ for all $0 < x < x_0$, $x \notin C(\chi)$ because only upward jumps of χ are
allowed for positive x. Because $C(\chi) \cup D(\chi)$ is finite, we may assume that
$(0, x_0] \cap (C(\chi) \cup D(\chi))$ is empty. As in case A, $R'(x) \ge P'(x)$ for all x in
$(0, x_0]$; hence $R(x_0) - R(x) \ge P(x_0) - P(x)$, and thus $\coth^{-1}(\chi(x)/b) \ge$
$b[P(x_0) - R(x_0) - d\ln(x)]$. However, the left member is negative because
$\chi(x) < (-b)$, and the right member tends to $+\infty$ for $x \downarrow 0$, which is a
contradiction. Therefore, the desired inequality is proved in both cases. $\quad\square$

Theorem 8. If $\chi \in \Psi^1$ is nondecreasing for $x \ge 0$, then V-robustness
and B-robustness are equivalent. In fact, $\kappa^*(\chi, F) = 1 + \gamma^*(\chi, F)^2/$
$V_1(\chi, F)$.

Proof. Analogous to Theorem 2. $\quad\square$

Theorem 9. The (standardized) median deviation is the most B-robust
estimator in Ψ^1. For all χ in Ψ^1 we have $\gamma^*(\chi, F) \ge 1/[4F^{-1}(\frac{3}{4})f(F^{-1}(\frac{3}{4}))]$,
and equality holds if and only if χ is equivalent to χ_{MAD}.

Proof. Take some χ in Ψ^1. From $\int \chi\, dF = 0$ it follows that

$$B_1(\chi) = \int (x\Lambda(x) - t\Lambda(t))\chi(x)\, dF(x)$$

$$\le \sup_{x \in \mathbb{R}\backslash C(\chi)} |\chi(x)| \int |x\Lambda(x) - t\Lambda(t)|\, dF(x)$$

for all $t > 0$. The last factor may be calculated by replacing $\int$ by $2\int_0^t + 2\int_t^\infty$, yielding

$$\int |x\Lambda(x) - t\Lambda(t)|\, dF(x) = (t\Lambda(t) - 1)(4F(t) - 3) + 4tf(t).$$

In order to find its minimum we differentiate with respect to t, yielding $(t\Lambda(t))'(4F(t) - 3)$ which is zero if and only if $t = F^{-1}(\frac{3}{4})$, negative for smaller t, and positive for larger t. The minimal value therefore equals $4F^{-1}(\frac{3}{4})f(F^{-1}(\frac{3}{4}))$; hence $\gamma^*(\chi, F) = \sup_{x \in \mathbb{R} \setminus C(\chi)} |\chi(x)|/B_1(\chi) \geq 1/[4F^{-1}(\frac{3}{4})f(F^{-1}(\frac{3}{4}))] = \gamma^*(\chi_{\text{MAD}}, F)$, and as in Theorem 3 one shows that equality holds if and only if χ is equivalent to χ_{MAD}.   □

Theorem 10. The (standardized) median deviation is also the most V-robust estimator in Ψ^1. For all χ in Ψ^1 we have $\kappa^*(\chi, F) \geq 2$, and equality holds if and only if χ is equivalent to χ_{MAD}.

Proof. Analogous to Theorem 4.   □

Let us now turn to the problem of optimal robustness. For each $b > 0$, there exists a unique constant a such that

$$\chi_b(x) := [x\Lambda(x) - 1 - a]_{-b}^{b} \qquad (2.5.25)$$

satisfies $\int \chi_b\, dF = 0$ and therefore belongs to Ψ^1. The mapping $b \to \gamma^*(\chi_b, F)$ is an increasing bijection from $(0, \infty)$ onto $(\gamma^*(\chi_{\text{MAD}}, F), \infty)$, and $b \to \kappa^*(\chi_b, F)$ is an increasing bijection from $(0, \infty)$ onto $(2, \infty)$.

Theorem 11. The only optimal B-robust estimators in Ψ^1 are (up to equivalence) given by $\{\chi_{\text{MAD}}, \chi_b \ (0 < b < \infty)\}$. They also coincide with the optimal V-robust estimators.

Proof. Analogous to Theorems 5 and 6, making use of the fact that the MLE is never B-robust because $x\Lambda(x) - 1$ is unbounded as $\Lambda'(x) > 0$ for all x by condition (F2).   □

The mappings χ_b are shown in Figure 2 of Section 2.4 for different values of b. Note that the truncation from above and below in (2.5.25) has been caused by bounding the CVF only from above!

*2.5f.   Further Topics

An M-Estimator Is Characterized by Its CVF

It is obvious from (2.5.9) that a location M-estimator is characterized by its influence function, up to equivalence (in the sense of Subsection 2.5a):

Lemma 3.   Let ψ_1 and ψ_2 belong to Ψ. If $IF(x; \psi_1, F)$ and $IF(x; \psi_2, F)$ are identical, then ψ_1 and ψ_2 are equivalent.

Rousseeuw (1983a) showed that an M-estimator is also characterized by the local behavior of its asymptotic variance at F:

Theorem 12.   Let ψ_1 and ψ_2 belong to Ψ. If $CVF(x; \psi_1, F)$ and $CVF(x; \psi_2, F)$ are identical, then ψ_1 and ψ_2 are equivalent.

Proof.   Put $C := C(\psi_1) = C(\psi_2)$ and $D := D(\psi_1) = D(\psi_2)$. It holds that $CVF(x; \psi_1, F) = CVF(x; \psi_2, F)$ for all x in $\mathbb{R} \setminus (C \cup D)$ and that $\psi_1(c_i +) - \psi_1(c_i -) = \psi_2(c_i +) - \psi_2(c_i -)$ for all c_i in C. We may suppose w.l.o.g. that

$$B(\psi_1) = B(\psi_2),$$

so it remains to prove $\psi_1 \equiv \psi_2$. Suppose w.l.o.g. that

$$A(\psi_2) \le A(\psi_1).$$

Now assume that $\psi_1 \not\equiv \psi_2$. Then there exists x_0 in $[0, \infty) \setminus C$ such that $\psi_2(x_0) > \psi_1(x_0)$. (Otherwise, $\psi_2 \le \psi_1$ on $[0, \infty) \setminus C$, and in some point we must have *strict* inequality as $\psi_1 \not\equiv \psi_2$; but then $B(\psi_2) = 2\int_{[0, \infty)} \Lambda \psi_2 \, dF < B(\psi_1)$, a contradiction.) Suppose w.l.o.g. that $x_0 \notin D$. It is impossible that $(\psi_2)'(x_0) \le (\psi_1)'(x_0)$, because then $CVF(x_0; \psi_2, F) > CVF(x_0; \psi_1, F)$. Thus there exists $\varepsilon > 0$ such that $\psi_2 > \psi_1$ on $[x_0, x_0 + \varepsilon)$. By means of a reasoning analogous to part of the proof of Theorem 1, it follows that $\psi_2(x) > \psi_1(x)$ for all $x \in [x_0, \infty) \setminus C$.

There exists a finite $K \ge x_0$ such that $(C \cup D) \subset (-K, K)$, so ψ_1 and ψ_2 are continuously differentiable and satisfy $\psi_2(x) > \psi_1(x)$ on $[K, \infty)$, where it also holds that $(\psi_2)^2(x)/A(\psi_1) - 2(\psi_2)'(x)/B(\psi_1) \le (\psi_2)^2(x)/A(\psi_2) - 2(\psi_2)'(x)/B(\psi_2) = (\psi_1)^2(x)/A(\psi_1) - 2(\psi_1)'(x)/B(\psi_1)$. Putting $a := B(\psi_1)/(2A(\psi_1)) > 0$ and observing that $[(\psi_2)^2(x) - (\psi_1)^2(x)] > 0$ we find

$$\frac{(\psi_2)'(x) - (\psi_1)'(x)}{(\psi_2)^2(x) - (\psi_1)^2(x)} \ge a > 0.$$

Making use of $[(\psi_2)^2(x) - (\psi_1)^2(x)] = [\psi_2(x) - \psi_1(x)][\psi_2(x) + \psi_1(x)]$ $\geq [\psi_2(x) - \psi_1(x)]^2$, this becomes

$$\frac{(\psi_2)'(x) - (\psi_1)'(x)}{(\psi_2(x) - \psi_1(x))^2} \geq a.$$

For all $y \geq K$, we define

$$R(y) := -1/(\psi_2(y) - \psi_1(y)).$$

On $[K, \infty)$ the mapping R is well defined, strictly negative, and continuously differentiable with derivative $[(\psi_2)'(x) - (\psi_1)'(x)]/(\psi_2(x) - \psi_1(x))^2$ $\geq a$. Hence, $R(x) - R(K) \geq a(x - K)$, so $ax \leq aK - R(K) < \infty$ for all $x \geq K$, which is clearly impossible. $\square$

Remark 4. It follows that if $0 < \mathrm{ARE}_{\psi_1, \psi_2}(F) < \infty$ and $(\partial/\partial t)[\mathrm{ARE}_{\psi_1, \psi_2}((1 - t)F + tG)]_{t=0} = 0$ for all "nice" G as in (2.5.11), then the CVF's are identical and ψ_1 and ψ_2 must be equivalent. This result can be applied when the ARE of two location M-estimators equals a constant $0 < k < \infty$ on some (infinitesimal or other) "neighborhood" of F, in which case equivalence follows, so $k = 1$.

Some Generalizations

In connection with the change-of-variance function, it is also possible to deal with *asymmetric* contamination. The asymptotic variance of a location M-estimator at some asymmetric distribution is still given by the expected square of its influence function. We put $\tilde{F}_t = (1 - t)F + tG$, where G is not necessarily symmetric, and obtain

$$\frac{\partial}{\partial t}\left[\frac{\int \psi^2(x - T(\tilde{F}_t)) \, d\tilde{F}_t(x)}{\left(\int \psi'(x - T(\tilde{F}_t)) \, d\tilde{F}_t(x)\right)^2}\right]_{t=0}$$

$$= \int\left[\frac{A(\psi)}{B(\psi)^2}\left(1 + \frac{\psi^2(x)}{A(\psi)} - 2\frac{\psi'(x)}{B(\psi)}\right)\right] dG(x)$$

$$-2\frac{A(\psi)}{B(\psi)^2}\frac{\partial}{\partial t}\left[T(\tilde{F}_t)\right]_{t=0}\int\left(\frac{\psi(x)\psi'(x)}{A(\psi)} - \frac{\psi''(x)}{B(\psi)}\right) dF(x).$$

$$(2.5.26)$$

Clearly, $\psi\psi'$ and ψ'' are skew symmetric because ψ is symmetric, so the last term vanishes and (2.5.26) reduces to (2.5.11), yielding the same CVF and change-of-variance sensitivity as before. This means that it is allowed to replace the equation

$$\mathrm{CVF}(x;\psi,F) = \frac{\partial}{\partial t}\left[V\left(\psi,(1-t)F + t\left(\tfrac{1}{2}\Delta_x + \tfrac{1}{2}\Delta_{-x}\right)\right)\right]_{t=0}$$

$$(2.5.27)$$

(which followed Definition 1 of Subsection 2.5a) by the equation

$$\mathrm{CVF}(x;\psi,F) = \frac{\partial}{\partial t}\left[V\left(\psi,(1-t)F + t\Delta_x\right)\right]_{t=0} \qquad (2.5.28)$$

In this form the CVF will be generalized to regression (see Subsection 6.3c).

It is also possible to define the CVF for other types of estimators. For this purpose one extends (2.5.11), where the asymptotic variance is given by the expected square of the influence function at the contaminated distribution. Unfortunately, the resulting expressions are often rather complicated. For instance, Ronchetti (1979) computed a variant of the CVF for R-estimators. Moreover, one cannot simply translate the results for M-estimators to other cases. Indeed, one might hope that an L- or R-estimator which has the same IF at F as some M-estimator, would also possess the same CVF at F. This is not the case because for L- and R-estimators the shape of the IF depends on the underlying distribution (see Remark 2 in Subsection 2.3b and Remark 3 in Subsection 2.3c), which clearly modifies the asymptotic variance at contaminated distributions. Of course, the median (which is an M-, L-, and R-estimator) always yields (2.5.13), and the mean (which is also an L-estimator) still gives (2.5.12), as does the normal scores estimator (Example 8 in Subsection 2.3c). However, for the Hodges–Lehmann estimator we obtain

$$\mathrm{CVF}(x;T,\Phi)/V(T,\Phi) = 4\left(1 - \frac{\phi(x)}{\int\phi^2(u)\,du}\right)$$

$$= 4 - 8\sqrt{\pi}\,\phi(x) \qquad (2.5.29)$$

so its change-of-variance sensitivity equals $\kappa^* = 4$. If we now construct the location M-estimator with the same influence function at Φ as the

Hodges–Lehmann estimator, then we obtain $\psi(x) = \Phi(x) - \frac{1}{2}$, which yields a CVF different from (2.5.29). Ronchetti and Yen (1985) extended Theorem 4 and a weaker form of Theorem 6 to R-estimators, yielding the median and the truncated normal scores estimator given by (2.4.16).

Alternative Definitions

We shall now discuss two notions that are comparable to the change-of-variance sensitivity κ^*. The first is Hampel's original suggestion, going back to 1972. He proposed to calculate the supremum of the CVF as in Definition 2, but without the standardization by $1/V(\psi, F)$. Therefore, this sensitivity equals

$$\chi^*(\psi, F) := V(\psi, F)\kappa^*(\psi, F). \qquad (2.5.30)$$

Because χ^* is directly proportional to κ^*, it is clear that Theorems 1 and 2 can be translated immediately. Unfortunately, the same is not true for Theorems 4 and 6. Indeed, in Figure 5 the sensitivity χ^* of the Huber-estimators ψ_b is plotted against b. It is clear that the median (corresponding to $b = 0$) is no longer most robust in the sense of χ^* (although it was most robust in the sense of κ^*), because its sensitivity χ^* is not even minimal in this small family of estimators. Moreover, if the Huber-estimators were optimal robust in this sense, then the graph of χ^* would be strictly increasing because $V(\psi_b, \Phi)$ is strictly decreasing.

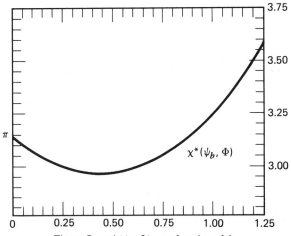

Figure 5. $\chi^*(\psi_b, \Phi)$ as a function of b.

A similar graph can be drawn for the most V-robust and the optimal V-robust estimators of Subsections 2.6b and 2.6c. Therefore, the "median-type hyperbolic tangent estimator" is no longer most robust in the sense of χ^*, and the "tanh-estimators" are no longer optimal robust [one only has to plot $\chi^*(\chi_{r,k})$ as a function of k, for a fixed value of r].

The second notion is more promising. It has often turned out to be more convenient to work with the inverse of $V(T, F)$ than with $V(T, F)$ itself. For example, when developing his minimax asymptotic variance theory, Huber (1964, Section 8) preferred the utility function $V(\psi, F)^{-1}$ over the loss function $V(\psi, F)$ because of its useful convexity properties. Moreover, $V(T, F)^{-1}$ is nicely bounded between 0 and $J(F)$ by Cramér–Rao. The *asymptotic efficacy* of T at F is defined as

$$E(T, F) := V(T, F)^{-1}. \qquad (2.5.31)$$

This name comes from the one-dimensional testing problem (see Subsection 3.2b); the asymptotic efficacy of tests goes back to Pitman (Noether, 1955) and generalizes (2.5.31). Now we can introduce the *efficacy influence function* (EIF) of estimators by differentiation of $E(T, F)$:

$$\frac{\partial}{\partial t}\left[E(T, (1 - t)F + tG)\right]_{t=0} = \int EIF(x; T, F)\, dG(x) \qquad (2.5.32)$$

so the EIF is simply proportional to the CVF:

$$EIF(x; T, F) = -E(T, F)^2 CVF(x; T, F). \qquad (2.5.33)$$

One defines the corresponding sensitivity η^* as the supremum of $\{-EIF(x; T, F)\}$ in the sense of Definition 2; hence

$$\eta^*(T, F) = E(T, F)\kappa^*(T, F). \qquad (2.5.34)$$

The EIF and η^* were discovered in 1978 by Rousseeuw and Ronchetti in connection with their work on robust testing. The advantage of this approach lies in its broad applicability, covering estimators and tests at the same time.

Let us now consider location M-estimators with $\psi \in \Psi$, and investigate what happens to our previous results. Theorems 1 and 2 can be translated immediately, as ψ is EIF-robust if and only if it is V-robust (2.5.34).

Now $b \to \eta^*(\psi_b, F)$ is a strictly increasing continuous bijection from $(0, \|\Lambda\|)$ onto $(\eta^*(\psi_{med}, F), \eta^*(\Lambda, F))$ because both $E(\psi_b, F)$ and $\kappa^*(\psi_b, F)$ are increasing continuous bijections (Rousseeuw, 1981b). Together with much other empirical evidence, this seems to indicate that

$$\eta^*(\psi_{\text{med}}, F) = 8f^2(0) \qquad (2.5.35)$$

is the minimal value of η^* in Ψ, which would mean that the median is the most EIF-robust *M*-estimator in Ψ. Unfortunately, no proof of this conjecture is available. (However, a simple partial result holds. In the class $\Psi' = \{\psi \in \Psi; E(\psi, F) \geq E(\psi_{\text{med}}, F)\}$ of *M*-estimators that are at least as efficient as the median, which in practice is a reasonable restriction, we have $\eta^*(\psi, F) \geq 2E(\psi_{\text{med}}, F) = 8f^2(0)$ by Theorem 4, and equality holds if and only if ψ is equivalent to ψ_{med}.) It is also conjectured that the most *V*-robust estimators of the next section (Theorem 3 in Subsection 2.6b) are at the same time most EIF-robust.

On the other hand, it is easy to prove that the Huber-estimators ψ_b, which are optimal *V*-robust by Theorem 6, are also optimal EIF-robust. Indeed, ψ_b maximizes $E(\psi, F)$ among all ψ satisfying $\eta^*(\psi, F) \leq kE(\psi_b, F)$. (To see this, take any ψ in Ψ which satisfies this side condition. It follows that $E(\psi, F) \leq E(\psi_b, F)$, because otherwise $\kappa^*(\psi, F) > k$ by Theorem 6, and then $\eta^*(\psi, F) = \kappa^*(\psi, F)E(\psi, F) > kE(\psi_b, F)$.) Analogously, one shows that the optimal *V*-robust estimators of the next section (Theorem 6 in Subsection 2.6c) are at the same time optimal EIF-robust.

The disadvantages of the notions χ^* and η^* discussed in this subsection justify our choice of the change-of-variance sensitivity κ^* as the most convenient measure for describing the local robustness of the asymptotic variance of estimators. This notion κ^* can also easily be generalized to tests (see Section 3.5).

2.6. REDESCENDING *M*-ESTIMATORS

2.6a. Introduction

In the previous section, we have investigated the local behavior of location *M*-estimators by means of the influence function (*B*-robustness) and the change-of-variance function (*V*-robustness). In the present section we remain in the same framework, but we restrict consideration to location *M*-estimators which are able to reject extreme outliers entirely, which implies that their ψ-function vanishes outside some central region. This means that we shall focus our attention to the subclass

$$\Psi_r := \{\psi \in \Psi; \psi(x) = 0 \text{ for all } |x| \geq r\} \qquad (2.6.1)$$

of the class Ψ introduced in Subsection 2.5a, where $0 < r < \infty$ is a fixed constant. This is equivalent to putting an upper bound $\rho^*(\psi) \le r$ on the *rejection point* ρ^* (2.1.15), which means that all observations lying farther away than r are discarded (or "rejected"). Such M-estimators are said to be *redescending*, but one has to be careful with this name because some authors also use it when the ψ-function merely tends to zero for $|x| \to \infty$ (see, e.g., Holland and Welsch, 1977). Most M-estimators that have been mentioned in earlier sections are not redescending, except for the Huber-type skipped mean (Example 1 in Subsection 2.5a). Let us now give some examples with continuous ψ-functions:

Example 1. The M-estimators corresponding to

$$
\begin{aligned}
\psi_{a,b,r}(x) &= x & 0 \le |x| \le a \\
&= a\,\mathrm{sign}(x) & a \le |x| \le b \\
&= a\frac{r - |x|}{r - b}\,\mathrm{sign}(x) & b \le |x| \le r \\
&= 0 & r \le |x|,
\end{aligned}
\qquad (2.6.2)
$$

where $0 < a \le b < r < \infty$ are sometimes called "hampels" because they were introduced by F. Hampel in the Princeton Robustness Study (Andrews et al., 1972, Sections 2C3, 7A1, and 7C2), where they were very successful. The so-called *two-part redescending M*-estimators correspond to $\psi_{a,a,r}$ for $0 < a < r < \infty$, whereas we speak of *three-part redescending M*-estimators (see Fig. 1) when $0 < a < b < r < \infty$.

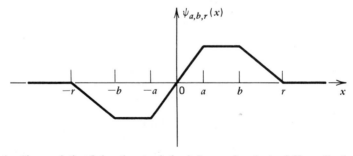

Figure 1. Shape of the Ψ-function (and the influence function) of Hampel's three-part redescending M-estimator.

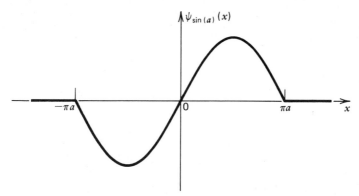

Figure 2. Andrew's sine function.

Example 2. The use of the *sine function* (see Fig. 2) was advocated by Andrews (Andrews et al., 1972):

$$\psi_{\sin(a)}(x) = \sin\left(\frac{x}{a}\right)1_{[-\pi a, \pi a]}(x). \tag{2.6.3}$$

Example 3. Some very smooth ψ-function, the *biweight* (or "bisquare") shown in Figure 3, was proposed by J. Tukey and has become increasingly popular (Beaton and Tukey, 1974):

$$\psi_{\text{bi}(r)}(x) = x(r^2 - x^2)^2 1_{[-r,r]}(x). \tag{2.6.4}$$

Surprisingly enough, certain redescending *M*-estimators are relatively old. The estimator of Smith (going back to 1888) turns out to be an

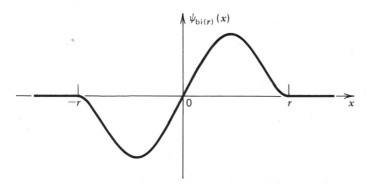

Figure 3. Tukey's biweight.

M-estimator, with

$$\psi_{\mathrm{Sm}(r)}(x) = x(r^2 - x^2)1_{[-r,r]}(x)$$

(Stigler, 1980, p. 218). In the same paper, Stigler even mentions the suggestion of Daniel Bernoulli (in 1769) to use

$$\psi_{\mathrm{Ber}(r)}(x) = x(r^2 - x^2)^{1/2}1_{[-r,r]}(x).$$

Remark 1. Although redescending M-estimators were very successful in the Princeton Robustness Study, some people still avoid them because of computational problems. Indeed, the solution of $\sum_{i=1}^{n}\psi(x_i - T_n) = 0$ is not unique because the left member will vanish for all far-away values of T_n. However, there are several ways out of this: One can either (1) take the global minimum of $\sum_{i=1}^{n}\rho(x_i - T_n)$; (2) select the solution of $\sum_{i=1}^{n}\psi(x_i - T_n) = 0$ nearest to the sample median; or (3) use Newton's method, starting with the median. As the "uniqueness problem" is really caused by the bad behavior of iterative algorithms, the simplest solution is to compute only a *one-step M-estimator* as described in Subsection 2.3a, starting from the median. This estimator has breakdown point 50% and the same IF and CVF as the fully iterated M-estimator (at least when ψ is odd and F is symmetric), while at the same time the uniqueness problem does not occur and the finite-sample behavior is very good (Andrews et al., 1972).

A second problem often cited in connection with redescending M-estimators is their sensitivity to a wrong scale. In most practical situations the scale is unknown, so T_n is defined through

$$\sum_{i=1}^{n} \psi\left(\frac{x_i - T_n}{S_n}\right) = 0,$$

and for one-step M-estimators by

$$T_n = T_n^{(0)} + \frac{S_n \sum_{i=1}^{n}\psi\left((x_i - T_n^{(0)})/S_n\right)}{\sum_{i=1}^{n}\psi'\left((x_i - T_n^{(0)})/S_n\right)},$$

where the initial estimate of location $T_n^{(0)}$ is the median of the observations $x_1, \ldots, x_n$. When the scale estimate S_n is way off, this might lead to wrong results because then some outliers might no longer be rejected. However, this can easily be avoided by using

$$S_n = 1.483\mathrm{MAD}(x_i) = 1.483\,\mathrm{med}_i\{|x_i - \mathrm{med}_j(x_j)|\},$$

which can be calculated immediately from the data, and does not have to be computed simultaneously with T_n. This S_n has a 50% breakdown point, and is very easy to use. Moreover, M-estimators with this auxiliary scale estimate give very good finite-sample results (Andrews et al., 1972, p. 239).

One other problem can occur with one-step M-estimators: for certain functions ψ and certain samples it may happen that the denominator $\sum_{i=1}^{n}\psi'([x_i - T_n^{(0)}]/S_n)$ becomes zero. However, this can easily be avoided by replacing this denominator by $n\int\psi'(z)\,d\Phi(z)$. Another solution is to switch to a one-step W-estimator (see Subsection 2.3d):

$$W_n = \frac{\sum_{i=1}^{n}x_i w\big([x_i - \mathrm{med}(x_i)]/[1.483\mathrm{MAD}(x_i)]\big)}{\sum_{i=1}^{n}w\big([x_i - \mathrm{med}(x_i)]/[1.483\mathrm{MAD}(x_i)]\big)}$$

where the nonnegative weight function w is related to ψ through $w(u) = \psi(u)/u$. This computation is very simple, and does not lead to problems because the denominator cannot become zero. (It suffices to make sure that $w(u)$ is strictly positive for $0 \le |u| \le 0.68$ because then at least half of the data give a strictly positive contribution to the denominator.) This approach avoids all the problems that were mentioned for redescenders.

In this section, we shall investigate what happens to the robustness concepts of the previous section if we restrict ourselves to the class Ψ_r of redescending ψ-functions with rejection point at most equal to r. Theorem 1 in Subsection 2.5b becomes useless since each ψ in Ψ_r is already B-robust by the Weierstrass theorem, as the piecewise continuous mapping ψ must be bounded on $[-r, r]$. Theorem 2 in Subsection 2.5b becomes meaningless since Ψ_r does not contain any monotone mappings. On the other hand, there exist elements of Ψ_r which are not V-robust, such as the Huber-type skipped mean, the CVF of which contains delta functions with positive sign.

The lower bounds for γ^* and κ^* as given by Theorems 3 and 4 in Subsection 2.5c remain valid, but they can no longer be reached because the median is not redescending. Indeed, in Subsection 2.6b exact bounds will be given for this case, and it will turn out that the most B-robust and the most V-robust estimators are no longer the same.

Moreover, the problems of optimal robustness will have to be tackled anew, because the mappings ψ_b of Subsection 2.5d do not belong to Ψ_r either. These investigations will be carried out in Subsection 2.6c, where it will be seen that the optimal B-robust and the optimal V-robust estimators in Ψ_r are also different.

The results of this section were taken from Rousseeuw's (1981b) Ph.D. thesis. Some parts already appeared in Rousseeuw (1981c; 1982a) and Hampel et al. (1981).

2.6b. Most Robust Estimators

We first look for the most B-robust estimator in Ψ_r. Inspired by Theorem 3 in Subsection 2.5c, we introduce

$$\psi_{\mathrm{med}(r)}(x) := \mathrm{sign}(x)1_{[-r,r]}(x) \tag{2.6.5}$$

(see Fig. 4). We call this estimator a *skipped median*, because observations farther away than r are skipped. [Here "skipped" refers to the Huber-type skipped mean, and not to Tukey's skipping procedures (Andrews et al., 1972).] We have $A(\psi_{\mathrm{med}(r)}) = 2F(r) - 1$ and $B(\psi_{\mathrm{med}(r)}) = 2[f(0) - f(r)]$. For not too small values of r, it appears that $\psi_{\mathrm{med}(r)}$ becomes an acceptable alternative to the median.

Theorem 1. The skipped median is the most B-robust estimator in Ψ_r. For all ψ in Ψ_r we have $\gamma^*(\psi, F) \geq 1/\{2[f(0) - f(r)]\}$, and equality holds if and only if ψ is equivalent to $\psi_{\mathrm{med}(r)}$.

Proof. The proof is an adaptation of Theorem 3 in Subsection 2.5c, where the real line is replaced by $[-r, r]$.   □

Unfortunately, the downward jumps of $\psi_{\mathrm{med}(r)}$ at r and $-r$ imply that $\kappa^*(\psi_{\mathrm{med}(r)}, F) = \infty$. Therefore, the most V-robust and the most B-robust estimators can no longer be the same.

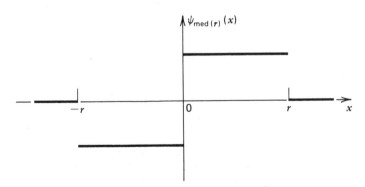

Figure 4. Skipped median.

For brevity, we introduce

$$K(r) := \int_{[-r,r]} \Lambda^2 \, dF \qquad (2.6.6)$$

which is an increasing continuous bijection of $(0, \infty)$ onto $(0, J(F))$. At $F = \Phi$ we obtain

$$K(r) = 2\Phi(r) - 1 - 2r\phi(r).$$

Theorem 2. There exist constants κ_r and B_r such that

$$\chi_r(x) := (\kappa_r - 1)^{1/2} \tanh\left[\tfrac{1}{2} B_r(\kappa_r - 1)^{1/2}(r - |x|)\right] \operatorname{sign}(x) 1_{[-r,r]}(x)$$

belongs to Ψ_r and satisfies $A(\chi_r) = 1$ and $B(\chi_r) = B_r$. Moreover, $\kappa^*(\chi_r, F) = \kappa_r > 2F(r)/(2F(r) - 1) > 2$ and $0 < 1/V(\chi_r, F) = B_r^2 < K(r) < J(F)$.

Proof. The constant $0 < r < \infty$ is fixed. For each pair of positive constants (κ, B) we introduce the notation

$$\eta_{\kappa,B}(x) := (\kappa - 1)^{1/2} \tanh\left[\tfrac{1}{2} B(\kappa - 1)^{1/2}(r - |x|)\right]$$
$$\times \operatorname{sign}(x) 1_{[-r,r]}(x).$$

We must prove that there exists a pair of values κ and B such that $A(\eta_{\kappa,B}) = 1$ and $B(\eta_{\kappa,B}) = B$. These equations are equivalent to

$$2(\kappa - 1) \int_{[0,r]} \tanh^2\left[\tfrac{1}{2} B(\kappa - 1)^{1/2}(r - x)\right] dF(x) = 1 \qquad (A)$$

and

$$B\left[1 + (\kappa - 1)(2F(r) - 1)\right] /$$
$$\left\{(\kappa - 1)^{1/2} \tanh\left[\tfrac{1}{2} B(\kappa - 1)^{1/2} r\right]\right\} = 4f(0). \qquad (B)$$

1. We start by "solving" (A): we prove that for each $\kappa > 2F(r)/(2F(r) - 1)$ there exists a value $B \in (0, \infty)$ such that $A(\eta_{\kappa,B}) = 1$. Let $\kappa > 2F(r)/(2F(r) - 1)$ be fixed. Now the left member of (A) is strictly increasing and continuous in B. For $B \downarrow 0$ it tends to zero, and for $B \uparrow \infty$ it tends to $(\kappa - 1)(2F(r) - 1) > 1$. The mean value theorem yields a unique solution B of (A), which we denote by $B(\kappa)$. By elementary but involved arguments, we can verify that $B(\kappa)$ is strictly decreasing and continuous in κ. When $\kappa \downarrow 2F(r)(2F(r) - 1)$ it holds that $B(\kappa) \uparrow \infty$, and $B(\kappa)(\kappa - 1)^{1/2}$ tends to zero when $\kappa \downarrow 0$.

2. If we now substitute $B = B(\kappa)$ in equation (B), we only have to "solve" this equation in the single variable κ. We first investigate what happens if $\kappa \uparrow \infty$. Using the first part of the proof, the left member of (B) then tends to $2(2F(r) - 1)/r$ using L'Hôpital's rule. Now we know that $(F(r) - \frac{1}{2})/r = [F(r) - F(0)]/[r - 0] = f(\xi) < f(0)$ where $0 < \xi < r$, so $2(2F(r) - 1)/r < 4f(0)$. We conclude that for $\kappa \uparrow \infty$ the left member of (B) becomes strictly smaller than the right one. We now investigate what happens for $\kappa \downarrow 2F(r)/(2F(r) - 1)$. Because then $B(\kappa) \uparrow \infty$ (part 1), clearly the left member of (B) tends to infinity, thereby becoming strictly larger than the right member. We conclude the existence of a solution κ, which we denote by κ_r, and we also put $B_r := B(\kappa_r)$.

3. This solution χ_r belongs to Ψ_r with $C(\chi_r) = \{0\}$ and $D(\chi_r) = \{-r, r\}$. By construction, $\kappa_r > 2F(r)/(2F(r) - 1) > 2$. To prove

Table 1.   Values of κ_r and B_r at $F = \Phi$

r	κ_r	B_r	e	γ^*
2.0	4.457305	0.509855	0.2600	2.6946
2.5	3.330328	0.604034	0.3649	2.0688
3.0	2.796040	0.668619	0.4471	1.7491
3.5	2.505102	0.711310	0.5060	1.5694
4.0	2.331507	0.739426	0.5468	1.4610
4.5	2.221654	0.758161	0.5748	1.3922
5.0	2.149604	0.770809	0.5941	1.3471
5.5	2.101379	0.779423	0.6075	1.3168
6.0	2.068765	0.785313	0.6167	1.2964
7.0	2.031553	0.792091	0.6274	1.2731
8.0	2.014392	0.795236	0.6324	1.2623
10.0	2.002953	0.797340	$\cdot$ 0.6358	1.2552
∞	2.000000	0.797885	0.6366	1.2533

$\kappa^*(\chi_r, F) = \kappa_r$, note that $\mathrm{CVF}(x; \chi_r, F)/V(\chi_r, F) = \kappa_r$ for $0 < |x|$ $< r$. For more details of this proof, see Rousseeuw (1981b).   □

In Table 1 some values of κ_r and B_r can be found for $F = \Phi$. Note that the asymptotic efficiency e equals B_r^2 because $A(\chi_r) = 1 = J(\Phi)$. Uniqueness of κ_r and B_r will follow from Theorem 3. Looking at Table 1, it appears that the estimator corresponding to χ_r yields an acceptable alternative to the median (corresponding to $r = \infty$) provided r is not too small. Apart from a finite γ^* and ρ^* it also possesses a finite κ^*, so χ_r is "more robust" than $\chi_{\mathrm{med}(r)}$. In fact, Theorem 3 states that χ_r is most V-robust in Ψ_r, a property shared by the median in Ψ. Moreover, $C(\chi_r) = \{0\} = C(\psi_{\mathrm{med}})$ and in both cases the jump is upward, so both estimators have the same behavior at the center. Because of all this, we call the estimator determined by χ_r a *median-type hyperbolic tangent estimator* (see Fig. 5).

Remark 2. The constants κ_r and B_r of Table 1 were computed in the following way. Consider

$$\chi_r(x) = \alpha_r \tanh\left[\beta_r(r - |x|)\right]\mathrm{sign}(x)\mathbf{1}_{[-r,r]}(x),$$

where $\alpha_r := (\kappa_r - 1)^{1/2}$ and $\beta_r := \frac{1}{2}B_r(\kappa_r - 1)^{1/2}$. The constants κ_r and B_r are defined implicitly through the equations $A(\chi_r) = 1$ and $B(\chi_r) = B_r$, which are equivalent to $\int \chi_r^2 d\Phi = 1$ and $B_r(1 + \alpha_r^2(2\Phi(r) - 1)) = 4\phi(0)(\alpha_r \tanh(r\beta_r))$. From the given value of r, the starting values $2\Phi(r)/(2\Phi(r) - 1)$ for κ_r and $K(r)^{1/2}$ for B_r are computed. At each step we obtain new values by

$$B_{r\,\mathrm{new}} := \left[4\phi(0)\alpha_r \tanh(r\beta_r)/\left(1 + \alpha_r^2(2\Phi(r) - 1)\right)\right]_{\mathrm{old}},$$

$$\kappa_{r\,\mathrm{new}} := \kappa_{\mathrm{old}} + \Delta\kappa_r,$$

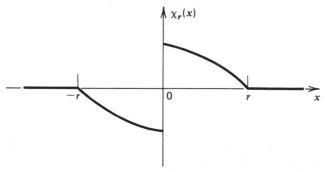

Figure 5.   Median-type tanh-estimator.

where $\Delta\kappa_r = 1 - \int\chi^2_{r\,\text{old}}\,d\Phi$ is computed by means of Romberg integration. The iteration continues till both B_r and κ_r have increments smaller than 10^{-10}, converging quite rapidly. Then γ^* is calculated by $\gamma^* = \chi_r(0 +)/B_r = (\alpha_r/\beta_r)\tanh(r\beta_r)$, whereas always $\lambda^* = \infty$.

Theorem 3. The median-type tanh-estimator is the most V-robust estimator in Ψ_r. For all ψ in Ψ_r we have $\kappa^*(\psi, F) \geq \kappa_r$, and equality holds if and only if ψ is equivalent to χ_r.

The proof of this result is rather technical and can be found in Rousseeuw (1981b, 1982a).

2.6c. Optimal Robust Estimators

In connection with optimal B-robustness in Ψ we worked with the mappings ψ_b (2.5.16), which unfortunately do not belong to Ψ_r. To repair this, we define

$$\psi_{r,b} := [\Lambda]^b_{-b}1_{[-r,r]} \tag{2.6.7}$$

for all $0 < r < \infty$ and $0 < b < \Lambda(r)$. This means that $\psi_{r,b} = \psi_b 1_{[-r,r]}$, so at $F = \Phi$ we could say that $\psi_{r,b}$ determines a *skipped Huber estimator* (see Fig. 6a). If we let b tend to $\Lambda(r)$, then the horizontal parts disappear and we obtain

$$\tilde{\psi}_r := \Lambda 1_{[-r,r]}, \tag{2.6.8}$$

which corresponds to the (*Huber-type*) *skipped mean* [given by $\psi_{\text{sk}(r)}(x) = x1_{[-r,r]}(x)$] at $F = \Phi$ (see Fig. 6b). We easily verify that $A(\tilde{\psi}_r) = B(\tilde{\psi}_r) = K(r)$ by (2.6.6), so $V(\tilde{\psi}_r, F) = 1/K(r)$ and $\psi^*(\tilde{\psi}_r, F) = \Lambda(r)/K(r) < \infty$. Lemma 1 of subsection 2.5d can be translated immediately:

Lemma 1. The mapping $b \to \gamma^*(\psi_{r,b}, F)$ is an increasing continuous bijection from $(0, \Lambda(r))$ onto $(\gamma^*(\psi_{\text{med}(r)}, F), \gamma^*(\tilde{\psi}_r, F))$.

Analogously, $b \to V(\psi_{r,b}, F)$ is a decreasing continuous bijection from $(0, \Lambda(r))$ onto $(V(\tilde{\psi}_r, F), V(\psi_{\text{med}(r)}, F))$ (Rousseeuw, 1981b). Therefore, the $\psi_{r,b}$ bridge the gap between $\psi_{\text{med}(r)}$ (which in Ψ_r plays the role of the median) and $\tilde{\psi}_r$ [which possesses the smallest $V(\psi, F)$ in Ψ_r, as $B^2(\psi) \leq K(r)A(\psi)$ for all ψ in Ψ_r by Cauchy–Schwarz].

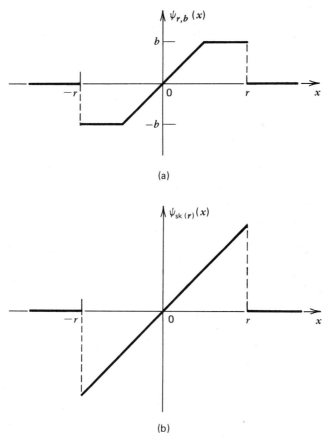

Figure 6. Optimal *B*-robust redescending *M*-estimators: (*a*) skipped Huber-estimator and (*b*) skipped mean.

Theorem 4. The only optimal *B*-robust estimators in Ψ_r are (up to equivalence) given by $\{\psi_{\text{med}(r)}, \psi_{r,b}(0 < b < \Lambda(r)), \tilde{\psi}_r\}$.

Proof. The proof mimics that of Theorem 5 of Subsection 2.5d, making use of Theorem 1 and keeping in mind that $\gamma^*(\tilde{\psi}_r, F)$ is always finite. □

Remark 3. There are some variants of this result when also the local-shift sensitivity λ^* (2.1.14) is considered. In the sense of λ^*, the two-part redescending *M*-estimators (Example 1) with $a = r/2$ are most robust (Collins, 1976, Theorem 3.2). When there are simultaneous upper bounds on

γ^* and λ^*, then the problem of optimal robustness at $F = \Phi$ yields the three-part redescending M-estimators (Example 1), the redescending parts of which are not steeper than the ascending part (Collins, 1976, Remark 3.7).

Unfortunately, all ψ-functions of Theorem 4 possess downward jumps at r and $-r$, hence they all have infinite change-of-variance sensitivities. Therefore, the optimal V-robust and the optimal B-robust estimators can no longer be the same.

Theorem 5 (Existence of tanh-estimators).   For each $k > \kappa_r$ there exist A, B, and p such that

$$\chi_{r,k}(x) := \Lambda(x) \qquad\qquad\qquad\qquad 0 \le |x| \le p$$

$$:= \left(A(k-1)\right)^{1/2}\tanh\!\left[\tfrac{1}{2}\!\left((k-1)B^2/A\right)^{1/2}(r - |x|)\right]\mathrm{sign}(x)$$

$$p \le |x| \le r$$

$$:= 0 \qquad\qquad\qquad\qquad\qquad\qquad r \le |x|, \qquad (2.6.9)$$

where $0 < p < r$ satisfies

$$\Lambda(p) = \left(A(k-1)\right)^{1/2}\tanh\!\left[\tfrac{1}{2}\!\left((k-1)B^2/A\right)^{1/2}(r - p)\right] \qquad (2.6.10)$$

belongs to Ψ_r and satisfies $A(\chi_{r,k}) = A$, $B(\chi_{r,k}) = B$, and $\kappa^*(\chi_{r,k}, F) = k$. Moreover, $0 < A < B < K(r) < J(F) < \infty$ and $V(\tilde{\psi}_r, F) < V(\chi_{r,k}, F) < V(\chi_r, F)$.

Proof.   Let $k > \kappa_r$ be fixed. The problem is equivalent to the existence of constants $s > B_r$ and $p \in (0, r)$ such that the mapping $\xi_{s,p}$ defined by

$$\xi_{s,p}(x) = \Lambda(x)(k-1)^{1/2}\tanh\!\left[\tfrac{1}{2}(k-1)^{1/2}s(r - p)\right]/\Lambda(p)$$

$$0 \le |x| \le p$$

$$= (k-1)^{1/2}\tanh\!\left[\tfrac{1}{2}(k-1)^{1/2}s(r - |x|)\right]\mathrm{sign}(x)$$

$$p < |x| \le r$$

$$= 0 \qquad\qquad\qquad\qquad\qquad\qquad |x| > r$$

satisfies the equations

$$\int \xi_{s,p}^2 \, dF = 1 \tag{A'}$$

and

$$\int \Lambda \xi_{s,p} \, dF = s. \tag{B'}$$

It is clear that such a solution $\xi_{s,p}$ would be continuous, and continuously differentiable on $\mathbb{R} \setminus \{-r, -p, p, r\}$. (In order to return to the original problem, we calculate $A := (\Lambda(p)/\{(k - 1)^{1/2} \tanh[\frac{1}{2}(k - 1)^{1/2} s(r - p)]\})^2$, $B := s\sqrt{A}$ and $\chi_{r,k} := \xi_{s,p}\sqrt{A}$.)

1. We start by "solving" (A') by showing that for each $s \geq B_r$ there exists some $p \in (0, r)$ satisfying (A'). Let $s \geq B_r$ be fixed. Now $\int \xi_{s,p}^2 \, dF$ is continuous and strictly decreasing in p, tending to zero for $p \uparrow r$. Moreover, $\lim_{p \downarrow 0} \int \xi_{s,p}^2 \, dF > \int \chi_r^2 \, dF = 1$. Therefore there exists a unique value of $p \in (0, r)$ satisfying (A'), which we denote by $p(s)$. We can verify that $p(s)$ is a strictly increasing continuous function of s. For $s = B_r$ we define $p_0 := p(B_r) \in (0, r)$.

2. We now substitute $p = p(s)$ in (B'), so we are left with the task of solving the resulting equation in s. Now $\int \Lambda \xi_{s,p(s)} \, dF$ is continuous in s; we shall first investigate what happens for $s = B_r$, so $p = p_0$. For brevity we denote $(k - 1)^{1/2} \tanh[\frac{1}{2}(k - 1)^{1/2} B_r(r - p_0)]/\Lambda(p_0)$ by α. We see that $\int_{[-r,r]}(\chi_r - \alpha\Lambda)^2 \, dF = 1 + \alpha^2 K(r) - 2\alpha B(\chi_r)$ and that $\int_{[-r,r]}(\xi_{B_r,p_0} - \alpha\Lambda)^2 \, dF = 1 + \alpha^2 K(r) - 2\alpha B(\xi_{B_r,p_0})$. We know that $\xi_{B_r,p_0}(x) = \alpha\Lambda(x)$ on $[0, p_0]$, and because $k > \kappa_r$ we can verify that $\chi_r(x) < \xi_{B_r,p_0}(x) < \alpha\Lambda(x)$ on (p_0, r). It follows that $\int_{[-r,r]}(\chi_r - \alpha\Lambda)^2 \, dF > \int_{[-r,r]}(\xi_{B_r,p_0} - \alpha\Lambda)^2 \, dF$, and therefore $B(\chi_r) < B(\xi_{B_r,p_0})$. Using $B(\chi_r) = B_r$ (see Theorem 2), we see that for $s = B_r$ the right member of (B') is strictly *smaller* than the left one. When $s \uparrow \infty$, the left member of (B') remains bounded from above by $(k - 1)^{1/2} K(r)/\Lambda(p_0) < \infty$, so the right member of (B') becomes strictly *larger* than the left one. Therefore there must exist a solution $s \geq B_r$ of $\int \Lambda \xi_{s,p(s)} \, dF = s$, hence $\xi_{s,p(s)}$ satisfies both (A') and (B'). Returning to the original problem, $\chi_{r,k} := \xi_{s,p(s)}\sqrt{A}$ satisfies $A(\chi_{r,k}) = A$ and $B(\chi_{r,k}) = B$.

3. The solution $\chi_{r,k}$ belongs to Ψ_r with $C(\chi_{r,k}) = \phi$ and $D(\chi_{r,k}) = \{-r, -p, p, r\}$. We note that $|\chi_{r,k}(x)| = |\Lambda(x)|$ for $0 \leq |x| \leq p$ and that $|\chi_{r,k}(x)| < |\Lambda(x)|$ for $p < |x| \leq r$, so $A = \int_{[-r,r]}|\chi_{r,k}|^2 \, dF < \int_{[-r,r]}|\Lambda||\chi_{r,k}| \, dF = B$. Again using these inequalities we further obtain $B = \int_{[-r,r]}|\Lambda||\chi_{r,k}| \, dF < \int_{[-r,r]}|\Lambda|^2 \, dF = K(r) < J(F)$. By means of the

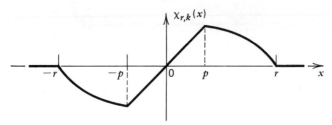

Figure 7.   Shape of the ψ-function (and the influence function) of a tanh-estimator, for $F = \Phi$.

Cauchy–Schwarz inequality we obtain $B^2 = (\int_{[-r,r]}|\Lambda| \, |\chi_{r,k}| \, dF)^2 < (\int_{[-r,r]}|\Lambda|^2 \, dF) \times (\int_{[-r,r]}|\chi_{r,k}|^2 \, dF) = K(r)A$, hence $V(\tilde{\psi}_r, F) = K(r)^{-1} < A/B^2 = V(\chi_{r,k}, F)$. On the other hand we have $s > B_r$ where s must equal $B/\sqrt{A}$, hence $V(\chi_{r,k}, F) = A/B^2 < 1/B_r^2 = V(\chi_r, F)$. (More details can be found in Rousseeuw, 1981b.)   □

The estimators determined by these $\chi_{r,k}$ are called *tanh-estimators*. They go back to Hampel in 1972, who conjectured them to be optimal in some sense. Indeed, in Theorem 6 it will be shown that they are optimal V-robust, also entailing uniqueness of A, B, and p. Now, why is Theorem 5 necessary? When looking for the optimal V-robust estimators, we first remark that the central part of $\chi_{r,k}$ has to be proportional to Λ in order to achieve a high asymptotic efficiency at the model. It also becomes clear that CVF $(x; \chi_{r,k}, F)/V(\chi_{r,k}, F) \equiv k$ has to be satisfied on the redescending parts of $\chi_{r,k}$. (Therefore $\chi_{r,k}$ may not redescend too steeply, or else $-\chi'_{r,k}$ would become too large, which is not allowed because $-\chi'_{r,k}$ occurs in the CVF.) From this differential equation we obtain (2.6.9) with the hyperbolic tangent, where (2.6.10) makes this function continuous. However, the constants A and B in these expressions also have to equal $\int \chi_{r,k}^2 \, dF$ and $\int \chi'_{r,k} \, dF$, and it becomes necessary to show that such values A, B, and p do exist.

At $F = \Phi$ we can replace $\Lambda(x)$ by x in (2.6.9) and $\Lambda(p)$ by p in (2.6.10), obtaining Figure 7. The corresponding change-of-variance function is given in Figure 8. For this case, Table 2 gives a list of constants A, B, and p corresponding to some values of r and k. We selected the entries of the table on the basis of r, k, γ^*, λ^* (2.1.13), and the asymptotic efficiency $e = 1/V(\chi_{r,k}, \Phi)$ at the model. Indeed, our values of r lie in the range commonly used, and we want $k = \kappa^*$ not too large (at most five, say), γ^* and λ^* not too large (say, not much larger than two), and at the same time a high value of e. Of course, there is some arbitrariness in the choice of reasonable bounds of γ^*, κ^*, and λ^*, depending on the amount of

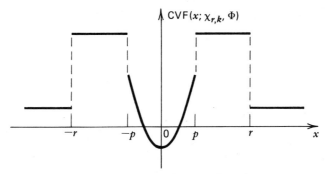

Figure 8. Change-of-variance function of a tanh-estimator.

Table 2. Hyperbolic Tangent Estimators for $F = \Phi$

r	k	A	B	p	e	γ^*	λ^*
3.0	4.0	0.493810	0.628945	1.096215	0.8011	1.7429	1.5900
	4.5	0.604251	0.713572	1.304307	0.8427	1.8279	1.7500
	5.0	0.680593	0.769313	1.470089	0.8696	1.9109	2.0000
4.0	4.0	0.725616	0.824330	1.435830	0.9365	1.7418	1.5000
	4.5	0.804598	0.877210	1.634416	0.9564	1.8632	1.7500
	5.0	0.857044	0.911135	1.803134	0.9686	1.9790	2.0000
5.0	4.0	0.782111	0.867443	1.523457	0.9621	1.7563	1.5000
	4.5	0.849105	0.910228	1.715952	0.9758	1.8852	1.7500
	5.0	0.893243	0.937508	1.882458	0.9840	2.0079	2.0000
6.0	4.0	0.793552	0.875862	1.541383	0.9667	1.7598	1.5000
	4.5	0.857058	0.915911	1.730683	0.9788	1.8896	1.7500
	5.0	0.899024	0.941556	1.895246	0.9861	2.0129	2.0000

contamination one might expect. It seems that $r = 4.0$, $k = 4.5$ are fair default values.

Remark 4. The constants A, B, and p of Table 2 were computed iteratively. The starting values for A, B, and p were $K(r) = 2\Phi(r) - 1 - 2r\phi(r)$, $2\Phi(r) - 1$, and $r/2$, respectively. A new value of p is found by applying Newton–Raphson to (2.6.10), and new values of A and B are found by $A_{\text{new}} = \int (\chi_{r,k}^2)_{\text{old}} d\Phi$ and $B_{\text{new}} = \int (\chi'_{r,k})_{\text{old}} d\Phi$ using Romberg integration. The iteration continued until a precision of 10^{-10} was reached, converging rapidly. Finally, also $e = B^2/A$, $\gamma^* = p/B$, and $\lambda^* = \max\{1/B, (k-1)/2\}$ are listed.

Theorem 6 (Optimality of tanh-estimators). The only optimal V-robust estimators in Ψ_r are (up to equivalence) given by $\{\chi_r, \chi_{r,k}(k > \kappa_r)\}$.

Proof. Let $k > \kappa_r$ be fixed. We now have to show that the mapping $\chi_{r,k}$ minimizes $V(\psi, F)$ among all $\psi \in \Psi_r$ which satisfy $\kappa^*(\psi) \leq k$.

1. It is assumed that $\kappa^*(\psi) \leq k < \infty$, hence for all c_i in $C(\psi)$ the "jump" $\psi(c_i +) - \psi(c_i -)$ has to be positive. This implies that ψ is continuous at r and $-r$, where $\psi(r) = \psi(-r) = 0$. Because multiplication of ψ with any positive constant changes nothing, it suffices to prove that $\chi_{r,k}$ minimizes $A(\psi)$ among all $\psi \in \Psi_r$ which are subject to

$$\mathrm{CVF}(x; \psi, F)/V(\psi, F) \leq k \quad \text{for all } x \in \mathbb{R} \setminus [C(\psi) \cup D(\psi)] \quad (C)$$

and

$$B(\psi) = B(\chi_{r,k}). \quad (D)$$

For all $\psi \in \Psi_r$ satisfying (D) one verifies that $\int_{[-r,r]}(\Lambda - \psi)^2 dF = \int(\tilde{\psi}_r - \psi)^2 dF = K(r) - 2B(\chi_{r,k}) + A(\psi)$, so we are left with the task of minimizing $\int_{[-r,r]}(\Lambda - \psi)^2 dF$.

2. Suppose $\chi_{r,k}$ is not optimal. Then there exists a mapping $\psi^* \in \Psi_r$ satisfying (C), (D), and

$$A(\psi^*) < A(\chi_{r,k}).$$

Therefore $\int_{[-r,r]}(\Lambda - \psi^*)^2 dF < \int_{[-r,r]}(\Lambda - \chi_{r,k})^2 dF$, so there is a point $x_0 \in [-r, r] \setminus C(\psi^*)$ such that $(\Lambda(x_0) - \psi^*(x_0))^2 < (\Lambda(x_0) - \chi_{r,k}(x_0))^2$. We may assume $x_0 \notin D(\psi^*)$. Also $x_0 \neq r$, because $\psi^*(r) = 0 = \chi_{r,k}(r)$. Let $x_0 \geq 0$ w.l.o.g. It follows that $x_0 \in (p, r)$, and because $\chi_{r,k}$ is strictly decreasing on (p, r) we know that $\Lambda(x_0) > \chi_{r,k}(x_0)$, hence $\psi^*(x_0) > \chi_{r,k}(x_0) > 0$. Define $x_1 := \sup\{x; x_0 < x \text{ and } \psi^*(y) > \chi_{r,k}(y)$ for all $y \in (x_0, x_1) \setminus C(\psi^*)\}$. Because ψ^* and $\chi_{r,k}$ are continuous at x_0 we have $x_0 < x_1$; on the other hand $x_1 \leq r$ by definition of Ψ_r. Because at points of $C(\psi^*)$ only upward "jumps" are allowed, we have $x_1 \notin C(\psi^*)$ and therefore $\psi^*(x_1) = \chi_{r,k}(x_1)$. Also, for all $y \in (x_0, x_1) \setminus C(\psi^*)$ we see that $\psi^*(y) > \chi_{r,k}(y) > 0$, and thus $(\psi^*)^2(y) > (\chi_{r,k})^2(y)$.

3. Now suppose that for all $x \in (x_0, x_1) \setminus [C(\psi^*) \cup D(\psi^*)]$ we would have $(\chi_{r,k})'(x) \le (\psi^*)'(x)$. Denote by $\{y_1, \ldots, y_m\}$ the intersection $(x_0, x_1) \cap [C(\psi^*) \cup D(\psi^*)]$ where $x_0 < y_1 < \cdots < y_m < x_1$. In each point y_i we have $(\psi^* - \chi_{r,k})(y_i -) = \psi^*(y_i -) - \chi_{r,k}(y_i) \le \psi^*(y_i +) - \chi_{r,k}(y_i) = (\psi^* - \chi_{r,k})(y_i +)$. The mapping $(\psi^* - \chi_{r,k})$ is continuously differentiable in between these points, hence

$$\left(\psi^* - \chi_{r,k}\right)(x_0) \le \left(\psi^* - \chi_{r,k}\right)(y_i -) \le \left(\psi^* - \chi_{r,k}\right)(y_1 +) \le \cdots$$

$$\le \left(\psi^* - \chi_{r,k}\right)(y_m -) \le \left(\psi^* - \chi_{r,k}\right)(y_m +) \le \left(\psi^* - \chi_{r,k}\right)(x_1).$$

But then $(\psi^* - \chi_{r,k})(x_1) \ge (\psi^* - \chi_{r,k})(x_0) > 0$, in contradiction to $\psi^*(x_1) = \chi_{r,k}(x_1)$. This proves the existence of a point $x_2 \in (x_0, x_1) \setminus [C(\psi^*) \cup D(\psi^*)]$ satisfying $(\psi^*)'(x_2) < (\chi_{r,k})'(x_2)$ and also (by part 2 of the proof) $(\psi^*)^2(x_2) > (\chi_{r,k})^2(x_2)$. Combining everything, it follows that

$$\mathrm{CVF}\left(x_2, \psi^*, F\right)/V(\psi^*, F) > \mathrm{CVF}\left(x_2, \chi_{r,k}, F\right)/V(\chi_{r,k}, F) = k$$

contradicting (C). Therefore $\chi_{r,k}$ is optimal.

4. Any other solution is equivalent to $\chi_{r,k}$. Suppose another solution $\bar{\psi}$ exists. W.l.o.g. we may put $B(\bar{\psi}) = B(\chi_{r,k})$ and therefore also $A(\bar{\psi}) = A(\chi_{r,k})$ because both are optimal. As in parts 2 and 3, no point $x_0 \in (p, r) \setminus C(\bar{\psi})$ can exist such that $\bar{\psi}(x_0) > \chi_{r,k}(x_0)$, hence

$$\left(\Lambda(x) - \bar{\psi}(x)\right)^2 \ge \left(\Lambda(x) - \chi_{r,k}(x)\right)^2 \quad \text{for } p < |x| < r, x \notin C(\bar{\psi}).$$

By part 1, $\int_{[-r,r]} (\Lambda - \chi_{r,k})^2 \, dF = \int_{[-r,r]} (\Lambda - \bar{\psi})^2 \, dF$, hence $\int_{(p,r)} (\Lambda - \chi_{r,k})^2 \, dF = \int_{(0,r)} (\Lambda - \bar{\psi})^2 \, dF + \int_{(p,r)} (\Lambda - \bar{\psi})^2 \, dF$. This implies that $\bar{\psi} = \chi_{r,k}$ on $(p, r) \setminus C(\bar{\psi})$, hence $\int_{(0,r)} (\Lambda - \bar{\psi})^2 \, dF = 0$ from which it follows that also $\bar{\psi} = \chi_{r,k}$ on $(0, r) \setminus C(\bar{\psi})$. Finally, $C(\bar{\psi}) = \phi$ and $\bar{\psi} = \chi_{r,k}$ everywhere. (More details can be found in Rousseeuw, 1981b.) □

Remark 5. This last situation completes the picture. Theorems 5 and 6 of Subsection 2.5d (optimal robustness in Ψ) depended on the upper extremes $\gamma^*(\Lambda, F)$ and $\kappa^*(\Lambda, F)$ being finite or infinite. In Theorem 4 we saw that $\gamma^*(\tilde{\psi}_r, F)$ was always finite regardless of F, so $\tilde{\psi}_r$ is itself optimal *B*-robust. In Theorem 6, $\kappa^*(\tilde{\psi}_r, F)$ is always infinite, so $\tilde{\psi}_r$ is never *V*-robust.

Remark 6. We further note that optimal *B*- and *V*-robustness in Ψ_r lead to markedly different estimators, which was not the case in Ψ. Indeed, the

ψ-functions of Theorem 1 (most B-robust) and Theorem 4 (optimal B-robust) have abrupt downward jumps at r and $-r$, indicating "hard" outlier rejection, whereas those of Theorem 3 (most V-robust) and Theorem 6 (optimal V-robust) all redescend in a smooth way.

Not only do tanh-estimators possess a finite rejection point ρ^* and a finite change-of-variance sensitivity κ^*, they also have other characteristics that indicate a high degree of robustness. In the first place, they are qualitatively robust (Section 2.2), and their breakdown point ε^* equals the maximal value $\frac{1}{2}$. Furthermore, they have a low gross-error sensitivity γ^* and a finite local-shift sensitivity λ^*, contrary to $\hat{\psi}_r$ which is a limiting case of our estimators. [This can be seen as follows: If $k \to \infty$, then p tends to r and the tanh-part in (2.6.9) becomes very steep and causes in the limit a discontinuity at r, and thus also an infinite λ^*. This indicates that the resulting estimator is very sensitive to the local behavior of the underlying distribution in the neighborhood of r and $-r$, whereas tanh-estimators reject "smoothly".] In addition to all these advantages, tanh-estimators with well-chosen constants are quite efficient, as indicated by their asymptotic efficiency e in Table 2.

Tanh-estimators are refinements of Hampel's three-part redescending M-estimators (Example 1). Like the latter, they should be combined with a robust scale estimate (preferably the median absolute deviation MAD) to ensure equivariance in the frequently occurring case where scale is a nuisance parameter.

Let us now compare tanh-estimators with some of the well-known redescending location M-estimators. We "standardize" the competitors as follows: They have to belong to Ψ_r with $r = 4$, and their gross-error sensitivities at Φ must be equal. (Let us put $\gamma^* = 1.6749$, which is the value for the biweight for $r = 4$.) The estimators under study are the following: sine (Example 2 with $a = 1.142$), biweight (Example 3 with $r = 4$), three-part redescending (Example 1 with $a = 1.31$, $b = 2.039$, $r = 4$), Huber–Collins [formula (2.7.5) with $p = 1.277$, $x_1 = 1.344$, $r = 4$], and tanh-estimator ($r = 4$, $k = 3.732$, $p = 1.312$, $A = 0.667$, $B = 0.783$). Table 3 lists the asymptotic efficiencies of these estimators at the normal model, as well as their asymptotic variances under different distributions. (The expression 5%3N denotes the distribution $0.95\Phi(x) + 0.05\Phi(x/3)$, and t_3 is the Student distribution with 3 degrees of freedom.)

In turns out that the five redescending estimators in the table have a similar behavior. The first four estimators possess asymptotic efficiencies close to 91%, whereas one gains about 1% with the tanh-estimator, and the comparison of the asymptotic variances also seems slightly favorable for the

Table 3. Comparison of Some Redescending *M*-Estimators[a]

Estimator	e	5%3N	10%10N	t_3	25%3N	Cauchy
Sine (Andrews)	0.9093	1.1991	1.2691	1.5769	1.7687	2.2688
Biweight (Tukey)	0.9100	1.1978	1.2683	1.5708	1.7645	2.2593
Huber–Collins	0.9107	1.1966	1.2689	1.5581	1.7583	2.2591
Three-part (Hampel)	0.9119	1.1954	1.2662	1.5783	1.7603	2.3306
Tanh-estimator	0.9205	1.1866	1.2590	1.5625	1.7579	2.2977
Huber-estimator	0.9563	1.1649	1.4385	1.5663	1.7877	2.7890
Scaled logistic MLE	0.9344	1.1872	1.4624	1.5380	1.7989	2.6390

The table header "Asymptotic Variances" spans the columns 5%3N, 10%10N, t_3, 25%3N, and Cauchy.

[a] The estimators under study are the following: sine, $\psi(x) = \sin(x/a)$ for $|x| < \pi a$ and zero otherwise, with $a = 1.142$; biweight, $\psi(x) = x(r^2 - x^2)^2$ for $|x| < r$ and zero otherwise, with $r = 4$; Huber–Collins, $p = 1.277$, $x_1 = 1.344$, $r = 4$; three-part redescending, ψ bends at 1.31, 2.039, 4; tanh-estimator, $r = 4$, $k = 3.732$, $p = 1.312$, $A = 0.667$, $B = 0.783$; Huber-estimator; ψ bends at $b = 1.4088$; and scaled logistic MLE, $\psi(x) = [\exp(x/a) - 1]/[\exp(x/a) + 1]$ with $a = 1.036$. All estimators satisfy $\gamma^* = 1.6749$ at the standard normal distribution, where also the asymptotic efficiency e is evaluated. The abbreviation 5%3N stands for the distribution $0.95\Phi(x) + 0.05\Phi(x/3)$ and t_3 is the Student distribution with 3 degrees of freedom.

latter estimator. For smaller values of r these differences will become larger, and vice versa.

For comparison, Table 3 also contains two *M*-estimators with monotone ψ. In order to make them comparable, their tuning constants are also chosen to attain $\gamma^* = 1.6749$ exactly. The first is the Huber-estimator with bending constant $b = 1.4088$. The second is the so-called "scaled logistic MLE" corresponding to $\psi(x) = [\exp(x/a) - 1]/\exp(x/a) + 1]$, which is the maximum likelihood estimator of location at the logistic distribution $F(x) = 1/\{1 + \exp(-x/a)\}$, where the constant a is put equal to 1.036 in order to attain the same value of γ^*. Because no finite rejection point was imposed, these estimators can achieve 1 to 3% more efficiency at the normal. They also do rather well at relatively short-tailed distributions like 5%3N, t_3, and 25%3N. However, their variance goes up considerably at distributions which produce larger outliers: At 10%10N they lose 15% efficiency with respect to the redescenders, and at the Cauchy distribution even 20%. At such distributions they suffer from the defect that they can never reject an outlier, no matter how far away.

Table 4.   A Schematic Summary of B-Robustness and V-Robustness Results in the Framework of M-Estimation of Location

Section 2.5

Ψ	Robust [finite sensitivity]	Most Robust [minimal sensitivity]	Optimal Robust [sensitivity $\leq k$, minimize $V(\psi, F)$]
B (bias) IF, γ^*	V-robust implies B-robust (Theorem 1),	Median (Theorem 3)	$\psi_{\text{med}}, \psi_b$, sometimes MLE (Theorem 5) at Φ: Huber-estimator
V (variance) CVF, κ^*	equivalent for monotone ψ (Theorem 2)	Median (Theorem 4)	$\psi_{\text{med}}, \psi_b$, sometimes MLE (Theorem 6) at Φ: Huber-estimator

Section 2.6

Ψ_r	Robust [finite sensitivity]	Most Robust [minimal sensitivity]	Optimal Robust [sensitivity $\leq k$, minimize $V(\psi, F)$]
B (bias) IF, γ^*	All ψ in Ψ_r are B-robust (Weierstrass)	Skipped median (Theorem 1)	$\psi_{\text{med}(r)}, \psi_{r,b}, \tilde{\psi}_r$ (Theorem 4); at Φ: skipped Huber-estimator
V (variance) CVF, κ^*	Huber-type skipped mean: not V-robust	Median-type tanh-estimator (Theorems 2 and 3)	$\chi_r, \chi_{r,k}$ tanh-estimator (Theorems 5 and 6)

2.6d.   Schematic Summary of Sections 2.5 and 2.6

Table 4 shows a schematic summary of the results of Sections 2.5 and 2.6.

*2.6e.   Redescending M-Estimators for Scale

M-estimators for scale were studied in Subsection 2.3a, and their CVF was introduced in Subsection 2.5e. Let us now restrict consideration to redescending M-estimators for scale, that is, to members of the class

$$\Psi_r^1 = \left\{ \chi \in \Psi^1; \chi(x) = 0 \quad \text{for all} \quad |x| > r \right\},$$

where $0 < r < \infty$ is fixed and Ψ^1 is as in Subsection 2.5e. We shall construct optimal V-robust estimators in Ψ_r^1.

Suppose that for some positive k there exist constants A_1, B_1, p, and a such that the mapping

$$\tilde{\chi}_{r,k}(x) := x\Lambda(x) - 1 + a \qquad\qquad 0 \le |x| \le p$$

$$:= \left(A_1(k-1)\right)^{1/2}\tanh\!\left[\tfrac{1}{2}\!\left((k-1)B_1^2/A_1\right)^{1/2}\!\left(\ln(r) - \ln(|x|)\right)\right]$$

$$p \le |x| \le r$$

$$:= 0 \qquad\qquad r \le |x| \qquad (2.6.11)$$

with

$$p\Lambda(p) - 1 + a = \left(A_1(k-1)\right)^{1/2}\tanh\!\left[\tfrac{1}{2}\!\left((k-1)B_1^2/A_1\right)^{1/2}\!\ln(r/p)\right]$$

$$(2.6.12)$$

satisfies $\int \tilde{\chi}_{r,k}\, dF = 0$, $A_1(\tilde{\chi}_{r,k}) = A_1$, $B_1(\tilde{\chi}_{r,k}) = B_1 > 0$, and $1 + (a - 1)^2/A_1 \le k$. It follows that $0 < p < r$ and that $0 < a < 1$ ($a > 0$ because otherwise $\int \tilde{\chi}_{r,k}\, dF < \int(x\Lambda(x) - 1)\, dF = 0$, and $a < 1$ because otherwise $\tilde{\chi}_{r,k}$ would be nonnegative; hence $\int \tilde{\chi}_{r,k}\, dF > 0$). The function $\tilde{\chi}_{r,k}$ is continuous (see Fig. 9), and belongs to Ψ_r^1.

Let us calculate the change-of-variance sensitivity of $\tilde{\chi}_{r,k}$. The CVF of this estimator is displayed in Figure 10. On $(-r, -p)$ and (p, r) it holds that $\mathrm{CVF}(x; \tilde{\chi}_{r,k}, F)/V(\tilde{\chi}_{r,k}, F) \equiv k$; actually, by means of this equation we found the above expression. For $|x| > r$ it holds that $\mathrm{CVF}(x; \tilde{\chi}_{r,k}, F)/V(\tilde{\chi}_{r,k}, F) \equiv 1 < k$ because $1 + (a - 1)^2/A_1 \le k$. Using the fact that $x\Lambda(x) - 1 + a$ is monotone for $x \ge 0$, positive in p, and negative in 0, we can define the number d in $(0, p)$ by means of $d\Lambda(d) - 1 + a = 0$. On $[-d, d]$ it holds that $\mathrm{CVF}(x; \tilde{\chi}_{r,k}, F)/V(\tilde{\chi}_{r,k}, F) \le 1 +$

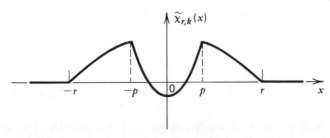

Figure 9. Optimal redescending *M*-estimator of scale, for $F = \Phi$.

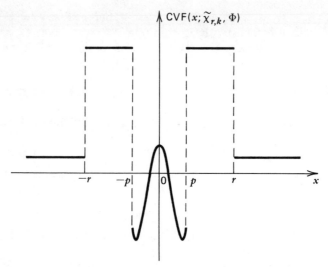

Figure 10. Change-of-variance function of an optimal redescending M-estimator of scale.

$(x\Lambda(x) - 1 + a)^2/A_1 \leq 1 + (a - 1)^2/A_1 \leq k$. [The condition $1 + (a - 1)^2/A_1 \leq k$ is also necessary, because otherwise $\kappa^*(\tilde{\chi}_{r,k}, F) \geq$ CVF$(0; \tilde{\chi}_{r,k}, F)/V(\tilde{\chi}_{r,k}, F) = 1 + (a - 1)^2/A_1 > k$.] On $[d, p)$ and $(-p, -d]$ we verify that CVF$(x; \tilde{\chi}_{r,k}, F)/V(\tilde{\chi}_{r,k}, F) \leq 1 + (p\Lambda(p) - 1 + a)^2/A_1 \leq 1 + A_1(k - 1)/A_1 = k$, making use of the fact that the square of a hyperbolic tangent is always smaller than 1. We conclude that $\kappa^*(\tilde{\chi}_{r,k}, F) = k$.

Theorem 7. The mapping $\tilde{\chi}_{r,k}$, satisfying the above conditions, is optimal V-robust in Ψ_r^1. [It minimizes $V_1(\chi, F)$ among all χ in Ψ_r^1 which satisfy $\kappa^*(\chi, F) \leq k$, and every other solution is equivalent to it.]

Proof. We show that $\tilde{\chi}_{r,k}$ minimizes $A_1(\chi)$ subject to $\chi \in \Psi_r^1$, $B_1(\chi) = B_1(\tilde{\chi}_{r,k})$, and $\kappa^*(\chi, F) \leq k$. It holds that

$$\int [x\Lambda(x) - 1 + a - \chi(x)]^2 dF(x)$$

$$= \int [x\Lambda(x) - 1 + a]^2 dF(x) - 2B_1(\tilde{\chi}_{r,k}) + A_1(\chi),$$

so it suffices to minimize the left member of this equality. The rest of the proof is analogous to Theorem 6.   □

Table 5.   Optimal *V*-Robust Redescending *M*-Estimators of Scale for $F = \Phi$

r	*k*	A_1	B_1	*p*	*a*	*e*	γ^*	λ^*
3.0	6.0	0.167554	0.319908	1.079721	0.486660	0.3054	2.0395	6.7502
	7.0	0.281931	0.465252	1.249943	0.393619	0.3839	2.0547	5.3732
	8.0	0.391789	0.590050	1.381465	0.329294	0.4443	2.0977	4.6825
4.0	6.0	0.453990	0.709875	1.405979	0.293372	0.5550	1.7893	3.9612
	7.0	0.620910	0.885645	1.560177	0.227110	0.6316	1.8758	3.5233
	8.0	0.765560	1.026474	1.684394	0.181935	0.6882	1.9670	3.2819
5.0	6.0	0.637109	0.926499	1.549797	0.220007	0.6737	1.7505	3.3455
	7.0	0.812012	1.095522	1.694232	0.167696	0.7390	1.8604	3.0930
	8.0	0.957789	1.227069	1.812263	0.132274	0.7860	1.9694	2.9538
6.0	6.0	0.739162	1.038695	1.620538	0.187503	0.7298	1.7461	3.1203
	7.0	0.909846	1.196082	1.757414	0.142829	0.7862	1.8655	2.9386
	8.0	1.049618	1.317061	1.870063	0.112557	0.8263	1.9815	2.8398

At $F = \Phi$, (2.6.11) and (2.6.12) are simplified because $\Lambda(x) = x$. Table 5 gives some values of the defining constants (which are unique by Theorem 7), as well as the asymptotic efficiency $e = 1/(2V_1(\tilde{\chi}_{r,k}, \Phi)) = B_1^2/2A_1$ [because $J(\Phi) = 2$], $\gamma^* = (p^2 - 1 + a)/B_1$, and λ^*. Note that the values of e are usually smaller than those of Table 2, which is a typical phenomenon—for example, compare $e \simeq 0.637$ of the median with $e \simeq 0.367$ of the median deviation. The algorithm we used was a modification of the one described in Remark 4, but at each step also the value of a was adjusted by means of $a_{\text{new}} := [a - \int \tilde{\chi}_{r,k} \, d\Phi]_{\text{old}}$.

Remark 7.   Note that the expression of $\tilde{\chi}_{r,k}(x)$ for $p \leq |x| \leq r$ is very similar to that of our tanh-estimator of location (2.6.9). (This is not so surprising, because one can transform a scale problem into a location problem by means of the transformation $y = \ln(|x|)$. However, since the direct approach is straightforward, we shall not pursue this line of thought any further.) A simpler expression for these parts of $\tilde{\chi}_{r,k}$ is

$$\tilde{\chi}_{r,k}(x) = \left(A_1(k-1) \right)^{1/2} \frac{(r^2)^\beta - (x^2)^\beta}{(r^2)^\beta + (x^2)^\beta} \qquad p \leq |x| \leq r,$$

where β equals $\frac{1}{2}((k-1)B_1^2/A_1)^{1/2}$.

2.7. RELATION WITH HUBER'S MINIMAX APPROACH

The foundations of modern robustness theory were laid by P. Huber in his 1964 paper. He introduced M-estimators, providing a flexible framework for the study of robust estimation of location (see Subsection 2.3a). Furthermore, he determined the M-estimators that are optimal in a minimax sense. For this purpose he considered the gross-error model, a kind of "neighborhood" of the symmetric model distribution F. It is defined by

$$\mathscr{P}_{\varepsilon} := \{(1 - \varepsilon)F + \varepsilon H; \ H \text{ is a } symmetric \text{ distribution}\}, \quad (2.7.1)$$

where $0 < \varepsilon < 1$ is fixed. (Later he also considered other neighborhoods, such as those based on total variation distance.) The asymptotic variance of an M-estimator defined by some function ψ at a distribution G of $\mathscr{P}_{\varepsilon}$ is given by $V(\psi, G)$ as in (2.3.13). Huber's idea was to minimize the maximal asymptotic variance over $\mathscr{P}_{\varepsilon}$, that is, to find the M-estimator ψ_0 satisfying

$$\sup_{G \in \mathscr{P}_{\varepsilon}} V(\psi_0, G) = \min_{\psi} \ \sup_{G \in \mathscr{P}_{\varepsilon}} V(\psi, G). \quad (2.7.2)$$

This is achieved by finding the least favorable distribution F_0, that is, the distribution minimizing the Fisher information $J(G)$ over all $G \in \mathscr{P}_{\varepsilon}$. Then $\psi_0 = -F_0''/F_0'$ is the maximum likelihood estimator for this least favorable distribution.

At $F = \Phi$, this minimax problem yields the Huber-estimator

$$\psi_0(x) = [x]_{-b}^{b}$$

(see Example 2 of Subsection 2.3a). The value b corresponds to the amount ε of gross-error contamination by means of the equation

$$2\Phi(b) - 1 + 2\phi(b)/b = 1/(1 - \varepsilon).$$

The least favorable distribution F_0 has the density $f_0(x) = (1 - \varepsilon)(2\pi)^{-1/2}\exp(-\rho(x))$, where $\rho(x) = \int_0^x \psi_0(t)\, dt$.

Huber's work for M-estimators was extended to L-estimators and R-estimators (see Subsections 2.3b and 2.3c) by Jaeckel (1971). The problem of calculating $\sup V(\psi, G)$ for various types of ψ-functions and neighborhoods was studied by Collins (1976, 1977) and Collins and Portnoy (1981). We shall not describe the minimax approach any further here, but refer the interested reader to Huber's (1981) book instead.

The minimax variance approach and the influence function approach have existed next to each other for quite some time. Because the Huber-estimator was the initial outcome of both methodologies (Huber, 1964; Hampel, 1968), it was tacitly assumed by many statisticians that minimax variance and (in our terminology) optimal B-robustness were related in some way. However, this is not the right link. The actual connections (which were clarified by Hampel and Rousseeuw) will be described in this section. Indeed, Sections 2.5 and 2.6 contain many hints that it is really the change-of-variance sensitivity $\kappa^*(\psi, F)$ which is related to $\sup\{V(\psi, G); G \in \mathscr{P}_\varepsilon\}$ as indicated by its intuitive interpretation, by similar results like Remark 2 of Subsection 2.5a, and by the fact that for redescenders the Huber–Collins minimax variance estimators correspond to optimal V-robust estimators, while being very different from the optimal B-robust estimators which involve "hard" rejection.

Let us make this argumentation more precise. It turns out that it is possible to calculate a good approximation of $\sup\{V(\psi, G); G \in \mathscr{P}_\varepsilon\}$ by means of the change-of-variance sensitivity κ^*:

$$\sup_{G \in \mathscr{P}_\varepsilon} V(\psi, G) = \exp\left(\sup_H \left[\ln V(\psi, (1 - \varepsilon)F + \varepsilon H) \right] \right)$$

$$\simeq \exp\left(\sup_H \left[\ln V(\psi, F) + \varepsilon \int \left[\mathrm{CVF}(x; \psi, F)/V(\psi, F) \right] dH(x) \right] \right)$$

$$= \exp\left(\ln V(\psi, F) + \varepsilon \cdot \sup_x \left[\mathrm{CVF}(x; \psi, F)/V(\psi, F) \right] \right)$$

$$= V(\psi, F)\exp\left(\varepsilon \cdot \kappa^*(\psi, F) \right). \tag{2.7.3}$$

Table 1 illustrates the quality of this approximation at the standard normal distribution $F = \Phi$. The estimators under study are the following. For the

Table 1. Approximation of $\sup V(T, G)$

Estimator	$\varepsilon = 1\%$		$\varepsilon = 5\%$		$\varepsilon = 10\%$	
	sup	(2.7.3)	sup	(2.7.3)	sup	(2.7.3)
Median	1.603	1.603	1.741	1.736	1.939	1.919
Hodges–Lehmann	1.090	1.090	1.286	1.279	1.596	1.562
Huber ($b = 1.4$)	1.086	1.086	1.256	1.257	1.507	1.510

median we have $V(\psi, \Phi) \simeq 1.571$ and $\kappa^* = 2$. The Hodges–Lehmann estimator (Example 8 of Subsection 2.3c) is not an M-estimator, but it can be verified that $V(T, \Phi) \simeq 1.047$ and $\kappa^* = 4$. The latter estimator is compared to the Huber-estimator with $b = 1.4$ which has the same asymptotic variance at $F = \Phi$, but possesses $\kappa^* = 3.6637$ (see Table 1 of Subsection 2.5d). Some entries of the present table come from Huber (1964, Table 1; 1981, exhibit 6.6.2).

Especially for small ε, the approximation (2.7.3) is excellent. Therefore, we can apply it to Huber's minimization problem

$$\underset{\psi}{\text{minimize}} \ \underset{G \in \mathscr{P}_\varepsilon}{\sup} \ V(\psi, G)$$

yielding

$$\underset{\psi}{\text{minimize}} \ V(\psi, F) \exp\left(\varepsilon \cdot \kappa^*(\psi, F) \right). \tag{2.7.4}$$

But any solution $\tilde{\psi}$ of the latter problem has to minimize $V(\psi, F)$ subject to an upper bound $\kappa^*(\psi, F) \leq k := \kappa^*(\tilde{\psi}, F)$! This means that $\tilde{\psi}$ is *optimal V-robust* at F (see Subsection 2.5d). Therefore, it is no coincidence that the Huber-estimator, introduced as the solution to the minimax problem (2.7.2), also turns out to be optimal V-robust (Rousseeuw, 1981a). [Of course, the Huber-estimator is also optimal B-robust, but this is due to the strong relation (Theorem 2 of Section 2.5) between γ^* and κ^* for monotone ψ-functions.]

Let us now investigate this connection for redescending M-estimators (see Section 2.6). In 1972, Huber considered the minimax problem for ψ-functions vanishing outside of $[-r, r]$ (see Huber, 1977c). The latter idea was rediscovered and treated extensively by Collins (1976). The mapping ψ_{HC} which minimizes $\sup\{V(\psi, G); G \in \mathscr{P}_\varepsilon\}$ in this class can be written as

$$\psi_{\text{HC}}(x) = \Lambda(x) \qquad\qquad\qquad 0 \leq |x| \leq p$$

$$= x_1 \tanh[\tfrac{1}{2} x_1 (r - |x|)] \text{sign}(x) \qquad p \leq |x| \leq r$$

$$= 0 \qquad\qquad\qquad\qquad\qquad r \leq |x|, \tag{2.7.5}$$

where p and x_1 satisfy $\Lambda(p) = x_1 \tanh[\tfrac{1}{2} x_1 (r - p)]$. For a table of values of p and x_1 at $F = \Phi$, see Collins (1976, p. 79). These estimators are very similar to our tanh-estimators $\chi_{r,k}$ (2.6.9) which we found as solutions to the problem of optimal V-robustness. (It is true that their constants are not identical, but in practice the difference between these estimators is very small.) On the other hand, the Huber–Collins estimator (2.7.5) differs a lot from the optimal B-robust estimator (2.6.7), for which $\kappa^* = \infty$ and $\sup\{V(\psi, G); G \in \mathscr{P}_\varepsilon\} = \infty$ (because its ψ-function has downward discontinuities corresponding to "hard" rejection). This shows that *the minimax variance criterion corresponds to optimal V-robustness and not to optimal B-robustness.*

Huber also considered the *minimax bias* problem (1981, Section 4.2). Here, the problem is to

$$\underset{\psi}{\text{minimize}} \ \underset{G \in \mathscr{A}_\varepsilon}{\sup} \ |T(G) - T(F)|, \qquad (2.7.6)$$

where T is the functional underlying the estimator and $\mathscr{A}_\varepsilon$ is the *asymmetric ε-contamination model* $\{(1 - \varepsilon)F + \varepsilon H; \ H$ is any distribution$\}$. Making use of

$$\underset{G \in \mathscr{A}_\varepsilon}{\sup} \ |T(G) - T(F)| = \underset{H}{\sup} |T((1 - \varepsilon)F + \varepsilon H) - T(F)|$$

$$\simeq \underset{H}{\sup} \left| \varepsilon \int \mathrm{IF}(x; \psi, F) \, dH(x) \right|$$

$$= \varepsilon \underset{x}{\sup} |\mathrm{IF}(x; \psi, F)|$$

$$= \varepsilon \cdot \gamma^*(\psi, F), \qquad (2.7.7)$$

this corresponds to

$$\underset{\psi}{\text{minimize}} \ \varepsilon \cdot \gamma^*(\psi, F) \qquad (2.7.8)$$

which determines the *most B-robust* estimator (see Subsection 2.5c). Indeed, (2.7.6) and (2.7.8) yield the same estimator, namely the median.

Table 2 shows the corresponding notions of the full neighborhood approach on the left and those of our approach on the right.

Let us now look at another way, independently discovered by Huber, to visualize the connections between the various robustness notions. For this purpose, make a plot of the maximal asymptotic bias $\sup\{|T(F) -$

Table 2.   Related Concepts

A. Bias

$\sup_{G\in\mathscr{A}_\varepsilon} \lvert T(G) - T(F)\rvert$ is finite	$\gamma^*(\tilde{\psi}, F)$ is finite
	B-robust
$\tilde{\psi}$ minimizes $\sup_{G\in\mathscr{A}_\varepsilon} \lvert T(G) - T(F)\rvert$ *minimax bias*	$\tilde{\psi}$ minimizes $\gamma^*(\psi, F)$ *most B-robust*
$\tilde{\psi}$ minimizes $V(\psi, F)$ subject to $\sup_{G\in\mathscr{A}_\varepsilon} \lvert T(G) - T(F)\rvert \leq$ constant	$\tilde{\psi}$ minimizes $V(\psi, F)$ subject to $\gamma^*(\psi, F) \leq c$ *optimal B-robust*

B. Variance

$\sup_{G\in\mathscr{P}_\varepsilon} V(\tilde{\psi}, G)$ is finite $\Leftrightarrow \sup_{G\in\mathscr{P}_\varepsilon} [V(\psi, G) - V(\psi, F)]$ is finite	$\kappa^*(\psi, F)$ is finite *V-robust*
$\tilde{\psi}$ minimizes $\sup_{G\in\mathscr{P}_\varepsilon} [V(\psi, G) - V(\psi, F)]$	$\tilde{\psi}$ minimizes $\kappa^*(\psi, F)$ *most V-robust*
$\tilde{\psi}$ minimizes $\sup_{G\in\mathscr{P}_\varepsilon} V(\psi, G)$ *minimax variance* $\Rightarrow \tilde{\psi}$ minimizes $V(\psi, F)$ subject to $\sup_{G\in\mathscr{P}_\varepsilon} [V(\psi, G) - V(\psi, F)] \leq$ constant	$\tilde{\psi}$ minimizes $V(\psi, F)$ subject to $\kappa^*(\psi, F) \leq k$ *optimal V-robust*

$T(G)\rvert\,; G \in \mathscr{A}_\varepsilon\}$ as a function of the fraction ε of contamination. The value of ε for which this function becomes infinite (i.e., its *vertical asymptote*) is the (gross-error) breakdown point ε^*. (When the estimator is so bad that $\varepsilon^* = 0$, as in the case of the arithmetic mean, the whole function collapses, so we assume that $\varepsilon^* > 0$ as in Fig. 1.) If the function is *continuous* at $\varepsilon = 0$ the estimator is "qualitatively robust" in the sense that the maximum possible asymptotic bias goes to zero as the fraction of contamination goes to zero. For well-behaved estimators, the gross-error sensitivity γ^* of the estimator equals the *slope* of this function at $\varepsilon = 0$ by (2.7.7).

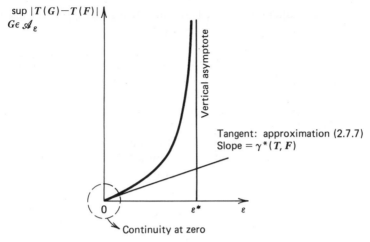

Figure 1. Plot of $\sup\{|T(G) - T(F)|;\ G \in \mathcal{A}_\varepsilon\}$ as a function of ε.

This representation describes the robustness notions as elementary calculus properties of a function of one argument, namely its continuity, differentiability, and vertical asymptote. It also visualizes very neatly why the breakdown point ε^* is needed to complement the influence function: The breakdown point tells us up to which distance the "linear approximation" provided by the influence function is likely to be of value. Indeed, for

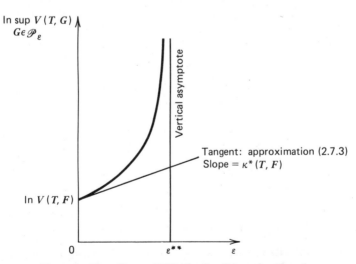

Figure 2. Plot of $\ln \sup\{V(T, G);\ G \in \mathcal{P}_\varepsilon\}$ as a function of ε.

$\varepsilon \geq \varepsilon^*$ the approximation (2.7.7) clearly fails. A reasonable rule of thumb is to use (2.7.7) only up to half the breakdown point, that is, for amounts of contamination $\varepsilon \leq \varepsilon^*/2$.

The same kind of thing can also be done for the asymptotic variance, by plotting $\sup\{V(T, G); G \in \mathscr{P}_\varepsilon\}$ on a logarithmic scale (as in Fig. 2) versus the amount of contamination ε. The value of ε for which this function becomes infinite is called the (*asymptotic*) *variance breakdown point* (Huber, 1981, Sections 1.4 and 6.6) and denoted by ε^{**}. The slope of the function at $\varepsilon = 0$ can be recognized as the change-of-variance sensitivity κ^*, because of (2.7.3). Again, the variance breakdown point tells us up to which amount of contamination the approximation (2.7.3) can be used.

In Subsection 1.3e, γ^* and κ^* are combined to study the mean-squared error of an estimator.

EXERCISES AND PROBLEMS

Subsection 2.1a

1. The α-trimmed mean $(0 < \alpha < \frac{1}{2})$ of a sample $x_1, \ldots, x_n$ is defined as follows. Let $[\alpha n]$ denote the integer part of αn. One removes the $[\alpha n]$ smallest observations and the $[\alpha n]$ largest observations, and takes the mean of the remaining ones. Show that such estimators are asymptotically functionals, that is, find a functional T such that (2.1.1) holds. What is $T(F_n)$, when F_n denotes the empirical distribution of the sample $x_1, \ldots, x_n$?

2. What is the relation between consistency and Fisher consistency (implications, counterexamples)?

Subsection 2.1b

3. (Short) Show that in all "reasonable" cases the asymptotic variance of an estimator can be obtained from (2.1.8).

Subsection 2.1d

4. Derive the influence function of the standard deviation and the variance. What is the relationship between both? What are γ^*, λ^*, and ρ^*?

5. (Extensive) Derive—heuristically—the IF for a maximum likelihood estimator of location. Consider the special cases of the normal distri-

bution, logistic, and Cauchy. Compare the shapes of the three curves: Are they bounded or not? Monotone? What do they tend to for $x \to \infty$? What are γ^*, λ^*, and ρ^*?

6. Compute the IF for the α-trimmed mean, defined in Exercise 1. What are γ^*, λ^*, and ρ^*? What happens if we let α tend to $\frac{1}{2}$?

7. The α-Winsorized mean $(0 < \alpha < \frac{1}{2})$ of a sample $x_1, \ldots, x_n$ is defined as follows. First, order the sample, obtaining $x_{1:n} \leq x_{2:n} \leq \cdots \leq x_{n:n}$. Then replace the $[\alpha n] = h$ smallest observations by $x_{h+1:n}$, hence counting this value $(h + 1)$ times. Analogously, replace the h largest observations by $x_{n-h:n}$. Then compute the arithmetic mean of this "modified" sample. Find the IF of this estimator, and compare it to that of the α-trimmed mean (Exercise 6). Also compare γ^* and λ^*.

8. Compute the (discrete) influence function of the maximum likelihood estimator of the "success" parameter p in the binomial model $\{ B(N, p); p \in (0, 1) \}$.

Subsection 2.1e

9. Construct the empirical IF for the estimators 20% and H/L at the Cushny and Peebles data, as in Figure 3. Are these functions bounded or unbounded? Also compare the empirical IF of 20% with the asymptotic IF of Exercise 6.

10. The "shorth" is defined as the mean of the shortest half of the sample. To be precise, let $h = [n/2] + 1$. Then consider all "half-samples" $\{ x_{j:n}, \ldots, x_{j+h-1:n} \}$ consisting of h subsequent ordered observations, and take the arithmetic mean of the half-sample with shortest length $x_{j+h-1:n} - x_{j:n}$. Construct the empirical IF and some stylized sensitivity curves for this estimator. Try to derive its asymptotic IF. What happens? Relate this to the behavior of the stylized sensitivity curves for $n \to \infty$.

11. Construct the empirical IF also for the Bickel–Hodges estimator, defined as $\text{median}_i \, (x_{i:n} + x_{n+1-i:n})/2$. Plot some stylized sensitivity curves for increasing n. Can the asymptotic IF be computed?

Subsection 2.2a

12. Consider the Cushny and Peebles data listed in the beginning of the chapter, but without the last value 4.6. Compute the estimates $\overline{X}$, 10%,

20%, and 50% (and perhaps also H/L) for this sample of nine points. Then add one outlier with value 10.0 to the sample, and recompute the estimates for this data set with ten points. Repeat this with two outliers at 10.0, with three, and so on, until there are nine outliers. Then make a plot of the results, with the number of outliers (from 0 to 9) on the horizontal axis, and the estimate on the vertical axis. This shows the evolution of the estimators with increasing amounts of contamination. Which estimator is most resistant?

13. Compute the asymptotic (gross-error) breakdown point ε^* and the finite-sample breakdown point ε_n^* for the median and the α-trimmed mean.

14. Compute (heuristically) ε^* and ε_n^* for the Hodges–Lehmann estimator and the Bickel–Hodges estimator.

15. Compute ε^* and ε_n^* for the standard deviation, the interquartile range, and the median absolute deviation from the median, defined as MAD $= \text{med}_i\{|x_i - \text{med}_j(x_j)|\}$. At what kind of distributions is the interquartile range equal to twice the MAD?

16. Show that for location-invariant estimators always $\varepsilon^* \leq \frac{1}{2}$. Find the corresponding upper bound on the finite-sample breakdown point ε_n^*, both in the case where n is odd or n is even.

Subsection 2.2b

17. Why is ordinary continuity of $T_n(x_1, \ldots, x_n)$ as a function of the observations $x_1, \ldots, x_n$ not suitable for a robustness concept?

18. Give an example of a distribution at which the median is qualitatively robust, but where its IF does not exist.

19. Give an example of a distribution at which the median is not even qualitatively robust.

20. Consider again the maximum likelihood estimator at the Poisson model, treated in Subsection 2.1d. Is it qualitatively robust? What is its breakdown point?

Subsection 2.3a

21. Compare the behavior of the MLE of location for the normal, logistic, and Cauchy distribution.

22. Derive the IF of the MLE of location for Student t-distributions. What happens when the number of degrees of freedom tends to infinity?

23. For which distributions is the mean deviation $(1/n)\sum_{i=1}^{n}|x_i|$ the MLE for scale?

24. (Extensive) For which distributions is the median deviation $\text{med}_i|x_i|$ the MLE for scale? (Solution in Hampel, 1974, p. 389.)

25. The gamma distributions are defined on the positive half-line with densities $f(x) = [\sigma^p\Gamma(p)]^{-1}x^{p-1}\exp(-x/\sigma)$ with $\sigma > 0$ and $p > 0$, where σ is a scale parameter and p characterizes the shape of the distribution. Derive the IF of the MLE for the scale parameter, keeping p fixed.

26. Do the same for the shape parameter of this distribution, keeping σ fixed.

27. Give an example of a maximum likelihood estimator of location which cannot be written in the form (2.3.3).

28. Show that M-estimators of location with finite rejection point cannot be written as MLE for any model distribution.

29. Compute the Huber estimator with $b = 1.5$ for the Cushny and Peebles data by trial and error, starting from the median and using 1.483MAD as initial scale estimate. Also compute the corresponding one-step M-estimator.

30. Which functional corresponds to one-step M-estimators? Derive its influence function when ψ is odd and F is symmetric. (Assume that the scale parameter is known in advance, and put it equal to 1.)

Subsection 2.3b

31. Derive the IF of location L-estimators from (2.3.24).

32. Make a plot of the IF of Gastwirth's "quick estimator" at the normal, and compare it to that of the median.

33. Show that location L-estimators with finite rejection point ρ^* at the standard normal distribution will contain negative weights. What happens to ρ^* when the same estimator is used at another distribution?

34. Show that the highest breakdown point of scale L-estimators is 25%, taking into account both explosion ($S \to \infty$) and implosion ($S \to 0$) of the estimators.

Subsection 2.3c

35. Derive the IF of R-estimators from (2.3.34).

36. Derive the asymptotic variance of the Hodges–Lehmann estimator at the logistic distribution by means of its IF.

Subsection 2.3d

37. (Short) Which A-estimator corresponds to the ψ-function of the arithmetic mean?

38. (Short) Derive the IF of a P-estimator from (2.3.42).

39. (Short) Use (2.1.12) to find the most efficient P-estimator at a given model distribution F.

40. (Short) Which scale statistic is necessary to obtain the arithmetic mean as an S-estimator? Which one for the median?

41. Describe the S-estimator defined by $s(r_1, \ldots, r_n) = \text{med}_i(|r_i|)$ when n is odd (related to Exercise 10). What is its finite-sample breakdown point?

42. Apply both the w-estimator and the W-estimator corresponding to the Huber-estimator with $b = 1.5$ to the Cushny and Peebles data. (Start from the median, and use 1.483MAD as your preliminary scale estimate.)

Subsection 2.4b

43. The logistic model with unknown scale parameter σ is given by $F_\sigma(x) = 1/\{1 + \exp[-x/\sigma]\}$. Compute the MLE and its IF, and compare it to that of the standard deviation. Apply Theorem 1 to obtain an optimal B-robust estimator.

44. The Cauchy model with unknown scale parameter σ is given by $f_\sigma(x) = \sigma^{-1}\{\pi[1 + (x/\sigma)^2]\}^{-1}$. Compute the IF of the MLE, and the corresponding γ^*. Is it still necessary to apply Theorem 1 in order to convert the MLE into a B-robust estimator?

45. Apply Theorem 1 to find a robust variant of the MLE for the scale parameter σ of gamma distributions (Exercise 25), keeping the shape parameter fixed.

46. (Extensive) Apply Theorem 1 to the Poisson example of Subsection 2.1d. Sketch the IF of the resulting optimal B-robust estimator when b is large and when b is small.

47. (Extensive) Apply Theorem 1 to the binomial distributions with unknown success parameter.

Subsection 2.5a

48. Compute the CVF of the location M-estimator corresponding to $\psi(x) = \Phi(x) - \frac{1}{2}$.

49. Consider the "scaled logistic MLE" corresponding to $\psi(x) = [\exp(x/a) - 1]/[\exp(x/a) + 1]$ where a is some positive constant. Draw the CVF of this location M-estimator for some value of a. What happens if $a \uparrow \infty$? What happens if $a \downarrow 0$?

50. Compute the CVF of the Huber-type skipped mean, and make a plot. Is it possible to see the delta functions as limits of something more natural? (Construct a sequence of continuous ψ-functions converging to $\psi_{\text{sk}(r)}$, and investigate their CVF.)

Subsection 2.5b

51. Construct an M-estimator with continuous ψ-function which is B-robust but not V-robust.

Subsection 2.5c

52. Give the formula of the asymptotic variance of the median at a contaminated distribution $(1 - \varepsilon)F + \varepsilon G$, where $0 < \varepsilon < \frac{1}{2}$ and G is symmetric around zero. What happens if $g(0) = f(0)$, so at zero the contaminated density looks like f? How large is the effect when g is more spread out than f, so $g(0) < f(0)$? How large is the effect when g is more concentrated than f, so $g(0) > f(0)$? Try to interpret the CVF of the median from this behavior.

Subsection 2.5d

53. Plot the CVF of a Huber-estimator with rather large bending constant b. What happens if $b \uparrow \infty$? Conversely, investigate what happens if $b \downarrow 0$.

Subsection 2.5e

54. Plot the CVF of the (standardized) MAD, and compare it to that of the median (Fig. 2).

55. Consider the (standardized) mean deviation $\sqrt{\pi/2}\,(1/n)\sum_{i=1}^{n}|x_i|$, corresponding to $\chi(x) = |x| - \sqrt{2/\pi}$. (Here, the constants $\sqrt{\pi/2} \simeq 1.2533$ and $\sqrt{2/\pi} \simeq 0.79788$ enter only to ensure Fisher consistency.) Plot its CVF at $F = \Phi$, and compare its general shape to that of the standard deviation (Fig. 4). What happens now for $|x| \to \infty$? Is the mean deviation therefore more robust or less robust than the standard deviation, from the point of view of the CVF?

Subsection 2.5f

56. Derive Eq. (2.5.29), the CVF (at $F = \Phi$) of the Hodges–Lehmann (H/L) estimator. Compare it to the CVF (at $F = \Phi$) of the MLE for the logistic distribution, which has the same IF *at the logistic* as the H/L estimator (see Fig. 1). Are they identical? Also compare it to the CVF at $F = \Phi$ of $\psi(x) = \Phi(x) - \frac{1}{2}$ (Exercise 48), which has the same IF *at the normal* as H/L.

57. What does the plot of $\kappa^*(\psi_b, \Phi)$ look like, as compared to Figure 5? (You can make use of theoretical results and/or Table 1.) Do the same for the plot of $\eta^*(\psi_b, \Phi)$.

58. (Research) Is the median most EIF-robust?

Subsection 2.6a

59. Plot the CVF of a three-part redescending M-estimator and a biweight, both with the same constant r. Compare them from the point of view of shape and continuity.

60. Why does a one-step M-estimator have the same CVF as its fully iterated version? Use the fact that ψ is odd and that the CVF is constructed from symmetric contaminated distributions in (2.5.11).

61. Compute the empirical IF of a one-step "sine" estimator with $a = 1.5$ and MAD scaling for the Cushny and Peebles data, as in Figure 3 of Section 2.1. Compare it with the influence function (Fig. 2 of Section 2.6).

62. Is the maximum likelihood estimator of location corresponding to the Cauchy distribution also a redescending M-estimator by our defini-

tion? Compute its CVF at $F = \Phi$ and compare its shape to that of either the biweight or the sine.

Subsection 2.6b

63. Is it possible to compute one-step M-estimators when the ψ-function consists of constant parts, as in the case of the median and the skipped median? Could this be repaired by replacing the expression $(1/n)\sum_{i=1}^{n}\psi'((x_i - T_n^{(0)})/S_n)$ by a fixed value? Which value would you choose in order to have the same IF at normal distributions as the original M-estimator?

64. Compute the CVF of the skipped median at $F = \Phi$. What is κ^*?

65. Compute the CVF of the median-type tanh-estimator at $F = \Phi$. Does the delta function cause us trouble? What happens (roughly) to the asymptotic variance if "contamination" occurs right there?

Subsection 2.6c

66. Compute the empirical IF of the one-step tanh-estimator with $r = 5$, $k = 4$ and MAD scaling for the Cushny and Peebles data, as in Figure 3 of Section 2.1. Compare it with the influence function (Fig. 7 of Section 2.6).

67. Derive the expression of λ^* for tanh-estimators, and verify it with the first two entries of λ^* in Table 2.

68. (Short) Table 3 contains M-estimators with redescending ψ and with bounded monotone ψ, but as yet none with unbounded ψ. Compute the asymptotic variance of the arithmetic mean, corresponding to $\psi(x) = x$, at the distributions of Table 3. (This can easily be done, even by hand, because the asymptotic variance of $\overline{X}$ simply equals the variance of the underlying distribution.) Add this line to the table, and compare it to the previous lines.

69. Compute (approximately) the change-of-variance sensitivities of the estimators in Table 3, and compare them to the entries of the table.

Subsection 2.6e

70. Find the expression of γ^* and λ^* for the optimal V-robust redescending M-estimator of scale, and use them to verify some entries of Table 5.

Section 2.7

71. Approximate the maximal asymptotic variance of the Huber-estimator with $b = 1.0$ in gross-error neighborhoods around $F = \Phi$ with $\varepsilon = 0.01, 0.05,$ and 0.1. The exact values are $1.140, 1.284,$ and 1.495 (Huber, 1964). Use Table 1 of Subsection 2.5d.

72. What is the maximal asymptotic variance of the arithmetic mean in a gross-error neighborhood around $F = \Phi$? What do you obtain with the approximation (2.7.3)?

73. Compute the approximate maximal asymptotic variance over gross-error neighborhoods around $F = \Phi$ with $\varepsilon = 5\%, 10\%,$ and 25%, for the tanh-estimator in Table 3 of Subsection 2.6c. [*Hint*: the numbers necessary for applying (2.7.3) can be computed from A, B, and k.] Are the entries of that table for 5%3N, 10%10N, and 25%3N indeed below their approximate upper bounds?

74. Show that the Huber–Collins estimator is never completely identical to the tanh-estimator. (*Hint*: the redescending part of ψ_{HC} satisfies a differential equation $\psi_{\mathrm{HC}}^2(x) - 2\psi_{\mathrm{HC}}'(x) \equiv$ constant.) Which differential equation does the redescending part of $\psi_{r,k}$ satisfy? Can the coefficients of both equations be the same? (Use the fact that $A < B$ by Theorem 5 of Subsection 2.6c).

75. Construct the complete diagram of Figure 1 (linear approximation included) for the median at $F = \Phi$, making use of the following values of $\sup\{|T(G) - T(\Phi)|; G \in \mathscr{A}_\varepsilon\}$ taken from (Huber, 1981, p. 104):

ε	.01	.02	.05	.10	.15	.20	.25	.30	.40	.50
sup	.0126	.0256	.066	.139	.222	.318	.430	.567	.968	∞

76. Do the same for Figure 2, making use of the following values of $\sup\{V(T,G); G \in \mathscr{P}_\varepsilon\}$ taken from (Huber, 1981, p. 105):

ε	0.01	0.02	0.05	0.10	0.15	0.20	0.25	0.30	0.40	0.50
sup	1.60	1.64	1.74	1.94	2.17	2.45	2.79	3.21	4.36	6.28

Note that $\varepsilon^{**} > \varepsilon^*$ for this estimator, as well as for many other estimators (at least when variance is only considered for neighborhoods constructed with *symmetric* contamination, otherwise ε^{**} decreases).

CHAPTER 3

One-Dimensional Tests

3.1. INTRODUCTION

Although the word "robustness" was first used in statistics by G.E.P. Box (1953) in connection with tests, there is far less literature on robust testing than on robust estimation. Until recently, most theoretical work centered around the case of a simple alternative. In 1965, P. Huber invented the censored likelihood ratio test by robustifying the Neyman–Pearson lemma (Huber, 1965, 1968; Huber and Strassen, 1973). This work led to an approach using shrinking neighborhoods (Huber-Carol, 1970; Rieder, 1978, 1980, 1981). Rieder (1982) also proposed an extension of the notion of qualitative robustness to rank tests.

This chapter contains the work of Rousseeuw and Ronchetti (1979, 1981) and some related material. In Section 3.2, the influence function of Chapter 2 is adapted to non-Fisher-consistent functionals in order to investigate the local robustness of test statistics. This extension inherits many useful properties, including some on asymptotic efficiency. Functionals in two variables, arising from two-sample tests, are also treated. A relation with the stability of level and power and a connection with Hodges–Lehmann-type shift estimators are given. In Section 3.3, the theory is illustrated by one- and two-sample rank tests. In Section 3.4, the optimal B-robust tests are determined (Rousseeuw, 1979, 1982b; Ronchetti, 1979). The change-of-variance function of Section 2.5 is generalized to tests in Section 3.5. Recently, two related approaches have appeared (Lambert, 1981; Eplett, 1980), which are compared to our work in Section 3.6. Finally, it is shown in Section 3.7 that the case of a simple alternative can also be treated by means of the influence function.

187

One way to investigate the robustness of a test is to study the stability of its level and power under small changes of the model distribution. As an example, consider tests for comparing variances, the framework in which Box (1953) coined the term "robustness."

First, suppose there are only two samples, consisting of the observations $X_1, X_2, \ldots, X_m$ and $Y_1, Y_2, \ldots, Y_n$. The classical assumption is that the X_i and the Y_i are normally distributed. Denote their unknown variances by σ_x^2 and σ_y^2. The null hypothesis states that $\sigma_x^2 = \sigma_y^2$, and the alternative hypothesis is simply $\sigma_x^2 \neq \sigma_y^2$. The classical F-test proceeds as follows: One computes the sample variances s_x^2 and s_y^2, and rejects the null hypothesis if the test statistic $T = s_x^2/s_y^2$ satisfies either $T > c$ or $T < 1/c$, where $c > 1$ is some critical value determined by means of the F-distribution and the nominal level.

Now suppose there are k samples, with $k > 2$, and that one wants to test the null hypothesis that all variances are equal ($\sigma_1^2 = \sigma_2^2 = \cdots = \sigma_k^2$) against the alternative that not all variances are equal. Then one uses the Bartlett test, which generalizes the F-test, in the following way. One computes all sample variances $s_1^2, s_2^2, \ldots, s_k^2$, and calculates the test statistic $\tilde{T}$ by

$$\tilde{T} = \frac{\text{arithmetic mean of } s_1^2, \ldots, s_k^2}{\text{geometric mean of } s_1^2, \ldots, s_k^2}.$$

The null hypothesis is then rejected if $\tilde{T} > \tilde{c}$, where $\tilde{c}$ is some critical value determined under the assumption that all samples follow a normal distribution.

Let us now look at the *actual level* of this test under different distributions (assuming that the samples are large). That is, suppose that all

Table 1. **Actual Level (in Large Samples) of the Bartlett Test When the Observations Come from a Slightly Nonnormal Distribution**[a]

Distribution	Actual Level		
	$k = 2$	$k = 5$	$k = 10$
Normal	5.0%	5.0%	5.0%
t_{10}	11.0%	17.6%	25.7%
t_7	16.6%	31.5%.	48.9%

[a] From Box, 1953.

observations come from t_7, the Student distribution with 7 degrees of freedom, whereas the critical value $\tilde{c}$ was derived under the assumption of normality. The actual level is then the probability that $\tilde{T} > \tilde{c}$ when in reality all variances are equal. As t_7 is very similar to the normal distribution, one would expect that the actual level would be close to the nominal one (say, 5%). However, Table 1 shows that the differences may be dramatic: A statistician who believes to be using the 5% level might actually be working with a level of 48.9%! For small samples, this example was investigated by Rivest (1984). In the present chapter, also the *power* of tests will be studied.

3.2. THE INFLUENCE FUNCTION FOR TESTS

We start our robustness investigation by introducing the influence function for tests as defined by Rousseeuw and Ronchetti (1979, 1981). The first subsection describes the background and gives the basic definition, both in the one-sample and the two-sample case. In Subsection 3.2b it is shown how the IF may be used for computing asymptotic efficiencies of tests. Subsection 3.2c contains the important relation between the IF and the stability of level and power, and subsection 3.2d gives a connection with shift estimators.

3.2a   Definition of the Influence Function

The One-Sample Case

For one-sample tests, we work in the same framework as for one-dimensional estimators. That is, we suppose we have one-dimensional observations $X_1, \ldots, X_n$ which are independent and identically distributed (i.i.d.). The observations belong to some sample space $\mathscr{X}$, which is a subset of the real line $\mathbb{R}$ (often $\mathscr{X}$ simply equals $\mathbb{R}$ itself, so the observations may take on any value). The (fixed) parametric model is a family of probability distributions F_θ on the sample space, where the unknown parameter θ belongs to some parameter space Θ which is an open convex subset of $\mathbb{R}$. One then wants to test the null hypothesis

$$H_0: \theta = \theta_0$$

by means of a test statistic $T_n(X_1, \ldots, X_n)$. When the alternative hypothesis is one-sided, for example,

$$H_1: \theta > \theta_0$$

(or H_1: $\theta < \theta_0$), one compares T_n with a critical value $h_n(\alpha)$ depending on the level α, and applies a rule like

$$\text{reject } H_0 \text{ if and only if } T_n > h_n(\alpha).$$

In the two-sided case

$$H_1: \theta \neq \theta_0,$$

the rule becomes

$$\text{reject } H_0 \text{ if and only if } T_n < h'_n(\alpha) \text{ or } T_n > h''_n(\alpha).$$

We call the set $\{T_n < h'_n(\alpha) \text{ or } T_n > h''_n(\alpha)\}$ the *critical region* of the test.

We identify the sample $X_1, \ldots, X_n$ with its empirical distribution G_n, ignoring the sequence of the observations, so $T_n(X_1, \ldots, X_n) = T_n(G_n)$. As in Chapter 2, we consider statistics which are functionals or can asymptotically be replaced by functionals. This means that we assume that there exists a functional T: domain $(T) \rightarrow \mathbb{R}$ (where the domain of T is the collection of all distributions on $\mathscr{X}$ for which T is defined) such that

$$T_n(X_1, \ldots, X_n) \underset{n \to \infty}{\rightarrow} T(G)$$

in probability when the observations are i.i.d. according to the true distribution G in domain (T). Note that G does not have to belong to the parametric model $\{F_\theta; \ \theta \in \Theta\}$, and in most applications will deviate slightly from it. (In practice, one often does not work with T_n but with $n^{1/2}T_n$ or nT_n, as is the case for rank statistics. This factor is not relevant, however.)

The ordinary influence function for functionals T was introduced in Chapter 2. It is most useful in connection with Fisher-consistent estimators, that is, when it is required that

$$T(F_\theta) = \theta \quad \text{for all } \theta \text{ in } \Theta.$$

Since test statistics are usually *not* Fisher consistent, we modify the definition of the influence function to become more useful in this context. Let the mapping ξ_n: $\Theta \rightarrow \mathbb{R}$ be defined by $\xi_n(\theta) := E_\theta[T_n]$, and put $\xi(\theta) := T(F_\theta)$. We assume that:

(i) $\xi_n(\theta)$ converges to $\xi(\theta)$ for all θ.
(ii) ξ is strictly monotone with a nonvanishing derivative, so that ξ^{-1} exists.

Define $U(G)$ as $\xi^{-1}(T(G))$; this functional gives the parameter value which the true underlying distribution G would have if it belonged to the model. This U is clearly Fisher consistent, since $U(F_\theta) = \xi^{-1}(T(F_\theta)) = \theta$ for all θ. We now consider Hampel's influence function of this new functional U. Denote by Δ_x the probability measure which puts mass 1 in the point x.

Definition 1. The test influence function of T at F is defined as

$$\mathrm{IF}_{\mathrm{test}}(x; T, F_\theta) = \mathrm{IF}(x; U, F_\theta)$$

$$= \lim_{t \downarrow 0} \frac{U((1 - t)F_\theta + t\Delta_x) - U(F_\theta)}{t} \qquad (3.2.1)$$

in those x where it exists.

It may seem strange that the IF is defined on the test statistic, and not on things like level and power. However, we shall see in Subsection 3.2c that these influences are proportional to $\mathrm{IF}_{\mathrm{test}}$.

Remark 1. The influence function represents the influence of an outlier in the sample on the value of the (standardized) test statistic, and hence on the decision (acceptance or rejection of H_0) which is based on this value. Its interpretation is therefore quite analogous to that of Hampel's influence function (and it is also subject to the same type of regularity conditions). It seems justified to transfer the robustness measures γ^*, λ^*, and ρ^* of Subsection 2.1c to this case; for example, we put

$$\gamma^*_{\mathrm{test}}(T, F_\theta) := \sup_x |\mathrm{IF}_{\mathrm{test}}(x; T, F_\theta)|. \qquad (3.2.2)$$

A bounded influence function thus indicates a finite gross-error sensitivity.

Remark 2. Generally, U is hard to write down explicitly. However, the test influence function can be constructed easily, since

$$\mathrm{IF}_{\mathrm{test}}(x; T, F_\theta) = \frac{\partial}{\partial t} \left[T((1 - t)F_\theta + t\Delta_x) \right]_{t=0} / \xi'(\theta)$$

$$= \mathrm{IF}(x; T, F_\theta) / \xi'(\theta) \qquad (3.2.3)$$

by (3.2.1) and (ii).

Remark 3. We say that two functionals T^1 and T^2 are equivalent if $T^1(G) = h(T^2(G))$ for all G, where h is a strictly monotone mapping with a nonvanishing derivative. In this case, T^1 and T^2 determine the same statistical test; but then they also yield the same influence function, because $U^1 = U^2$.

For Fisher-consistent estimators, Definition 1 coincides with Definition 1 of Section 2.1 since in this case $\xi(\theta) = T(F_\theta) = \theta$ for all θ, so $U = T$. Therefore, the new definition generalizes the old one.

From now on, we shall always calculate the $\mathrm{IF}_{\mathrm{test}}$ at the null hypothesis (put $\theta = \theta_0$ and denote $F := F_{\theta_0}$) for simplicity.

The Two-Sample Case

Suppose we have two samples $X_1, \ldots, X_m$ and $Y_1, \ldots, Y_n$. Often, one assumes that the distributions G and F underlying these samples satisfy a relation of the type $G(x) = F(x - \theta)$ for all x ("location" or "shift" model). The null hypothesis states

$$H_0: \theta = \theta_0,$$

where θ_0 is usually taken to be zero, and the alternative hypothesis may be one-sided or two-sided. For testing H_0 we can use a statistic $T_{m,n}(X_1, \ldots, X_m; Y_1, \ldots, Y_n)$ which has to be compared to some critical value(s).

In order to do asymptotics, we suppose that m and n depend on the total sample size $N = m + n$, and that both tend to infinity as $N \to \infty$, with $\lim_{N \to \infty} m/N = \lambda$ where $0 < \lambda < 1$. Now assume the existence of a functional T such that $T_{m,n}(X_1, \ldots, X_m; Y_1, \ldots, Y_n)$ tends to $T(H^1, H^2)$ in probability when $N \to \infty$ and the observations are i.i.d. according to H^1 and H^2.

Suppose $T_{m,n}$ is invariant with respect to an identical shift of both samples, that is,

$$T_{m,n}(X_1 - a, \ldots, X_m - a; Y_1 - a, \ldots, Y_n - a)$$

$$= T_{m,n}(X_1, \ldots, X_m; Y_1, \ldots, Y_n) \quad \text{for all } a.$$

We want to calculate $\mathrm{IF}_{\mathrm{test}}$ in a pair (G, F) with $G(x) = F(x - \theta)$. In this case, the distributions underlying both samples have the same (known) shape, so the expected value of $T_{m,n}$ only depends on the shift parameter θ;

we may therefore denote it by

$$\xi_{m,n}(\theta) := E_\theta[T_{m,n}].$$

We assume that $\xi_{m,n}(\theta)$ tends to $\xi(\theta)$, defined in the same way as the value of T. We say that T is Fisher consistent when $\xi(\theta) = \theta$ for all θ. Again we suppose (ii) holds, and we define $U(H^1, H^2) := \xi^{-1}(T(H^1, H^2))$ for all H^1 and H^2.

A fundamental difference to one-sample statistics arises: Outliers may occur in the first sample, in the second, and in both. Thus there are a priori also three different influence functions (which are, however, usually strongly linked, as we shall see in the next subsection). Let $G_{t,x} = (1 - t)G + t\Delta_x$ and $F_{t,y} = (1 - t)F + t\Delta_y$.

Definition 2.   Under the above assumptions, we define

$$\text{IF}_{\text{test},1}(x; T, G, F) = \lim_{t \downarrow 0} \frac{U(G_{t,x}, F) - U(G, F)}{t},$$

$$\text{IF}_{\text{test},2}(y; T, G, F) = \lim_{t \downarrow 0} \frac{U(G, F_{t,y}) - U(G, F)}{t},$$

$$\text{IF}_{\text{test}}(x, y; T, G, F) = \lim_{t \downarrow 0} \frac{U(G_{t,x}, F_{t,y}) - U(G, F)}{t}$$

in all points where the limits exist.

Note that $\text{IF}_{\text{test}}(x, y; T, G, F)$ is a surface in a three-dimensional space. The interpretation and remarks of the one-sample case remain valid; for instance, for the calculation we use

$$\text{IF}_{\text{test},1}(x; T, G, F) = \frac{\partial}{\partial t} [T(G_{t,x}, F)]_{t=0} / \xi'(\theta), \qquad (3.2.4)$$

and so on. We now also have a definition for the influence function of a Fisher-consistent two-sample estimator, for which $\xi(\theta) = \theta$, so the denominator of (3.2.4) becomes 1.

We usually calculate the influence function at the null hypothesis (put $\tilde{\theta} = \theta_0$ so $F = G$) and use the notations $\text{IF}_{\text{test},1}(x; T, F)$, $\text{IF}_{\text{test},2}(y; T, F)$, and $\text{IF}_{\text{test}}(x, y; T, F)$.

The case of a parameter of relative scale $[G(x) = F(x/\theta)$ with $\theta_0 = 1]$ can be treated analogously.

3.2b. Properties of the Influence Function

The One-Sample Case

Property 1.   Under the assumptions of 3.2a, we have

$$\int \mathrm{IF}_{\text{test}}(x; T, F)\, dF(x) = 0.$$

Proof.   This follows from the same property for estimators and (3.2.3).
$\square$

With estimators, an important role is played by (2.1.8):

$$\int \mathrm{IF}(x; T, F)^2\, dF(x) = V(T, F), \qquad (3.2.5)$$

where $V(T, F)$ is the asymptotic variance of the sequence T_n. This equality also holds for non-Fisher-consistent sequences, but it has little relevance in this case since one needs Fisher consistency for the Cramér–Rao inequality (2.1.11). Therefore, $V(T, F)$ gives insufficient information for efficiency considerations in the case of tests. With this in mind, we shall extend (3.2.5) to test statistics in terms of the test influence function.

We need Pitman's theorem as it can be found in Noether (1955) with the $m_i = 1$, $\delta_i = \frac{1}{2}$. Put $F := F_{\theta_0}$. The *asymptotic (Pitman) efficacy* E of T at F is defined as

$$E(T, F) = \lim_{n \to \infty} \left[\xi_n'(\theta_0)\right]^2 / (n \mathrm{Var}_F(T_n)). \qquad (3.2.6)$$

This quantity is very useful. For instance, Pitman's theorem states that the asymptotic relative efficiency of two tests can be computed as the ratio of their efficacies:

$$\mathrm{ARE}_{1,2} = E(T_1, F) / E(T_2, F). \qquad (3.2.7)$$

Also the asymptotic power can be expressed by means of the efficacy. The asymptotic power is defined as the limit (for $n \to \infty$) of the power of the test at the alternative $\theta_n = \theta_0 + \Delta n^{-1/2}$, and can be computed as

$$\beta = 1 - \Phi\left(\lambda_{1-\alpha} - \Delta\sqrt{E}\right) \qquad (3.2.8)$$

where E is the asymptotic efficacy $E(T, F)$ and Δ is some positive constant.

Property 2. Under the assumptions of Subsection 3.2a and (3.2.5) we have $\int \mathrm{IF}_{\mathrm{test}}(x; T, F)^2 \, dF(x) = E(T, F)^{-1}$.

Proof. We have $\mathrm{IF}_{\mathrm{test}}(x; T, F) = (\partial/\partial t)[T(F_{t,x})]_{t=0}/\xi'(\theta_0)$ by (3.2.3). Combining $\int\{(\partial/\partial t)[T(F_{t,x})]_{t=0}\}^2 \, dF(x) = V(T, F)$ with $V(T, F) = \lim_{n \to \infty} n \, \mathrm{Var}_F(T_n)$, we obtain $\int \mathrm{IF}_{\mathrm{test}}(x; T, F)^2 \, dF(x) = V(T, F)/[\xi'(\theta_0)]^2 = \lim_{n \to \infty} n \, \mathrm{Var}_F(T_n)/[\xi'_n(\theta_0)]^2 = E(T, F)^{-1}$. $\qquad\Box$

Now one can also reformulate (3.2.7) and (3.2.8) in terms of the $\mathrm{IF}_{\mathrm{test}}$.

Remark 4. Applying (3.2.6) to a Fisher-consistent sequence of estimators, we see that $E(T, F)$ equals $V(T, F)^{-1}$, so (3.2.7) still holds. Since Definition 1 applies to such sequences and yields the estimator influence function, Property 2 is true in general.

The Fisher information $J(F)$ is given by (2.1.9), and we denote the density of F_θ by f_θ.

Property 3 (Asymptotic Cramér–Rao Inequality). Under the assumptions of Subsection 3.2a and (3.2.5), we have $\int \mathrm{IF}_{\mathrm{test}}(x; T, F)^2 \, dF(x) \geq J(F)^{-1}$. The statistics T_n are asymptotically efficient (meaning that equality holds) if and only if $\mathrm{IF}_{\mathrm{test}}(x; T, F)$ is proportional to $(\partial/\partial\theta)[\ln f_\theta(x)]_{\theta_0}$.

Proof. This follows from (2.1.11) and (2.1.12) because U is Fisher consistent. $\qquad\Box$

One defines the (absolute) asymptotic efficiency e of T at F as

$$e = E(T, F)/J(F). \tag{3.2.9}$$

By means of Property 3 one can find the classically optimal procedures, which have $e = 1$. In Section 3.4, optimal B-robust tests will be developed.

The Two-Sample Case

Property 4. Under the assumptions of Subsection 3.2a, it holds that

$$\int \mathrm{IF}_{\mathrm{test},1}(x; T, F) \, dF(x) = \int \mathrm{IF}_{\mathrm{test},2}(y; T, F) \, dF(y)$$

$$= \iint \mathrm{IF}_{\mathrm{test}}(x, y; T, F) \, dF(x) \, dF(y) = 0.$$

We now investigate how $\mathrm{IF}_{\mathrm{test},1}$, $\mathrm{IF}_{\mathrm{test},2}$, and $\mathrm{IF}_{\mathrm{test}}$ are related.

Property 5. Under the assumptions of Subsection 3.2a, it holds that $\mathrm{IF}_{\mathrm{test}}(x, y; T, F) = \mathrm{IF}_{\mathrm{test},1}(x; T, F) + \mathrm{IF}_{\mathrm{test},2}(y; T, F)$.

Proof. Apply the chain rule to Definition 2.   □

From now on, we shall always assume condition (S)

$$\mathrm{IF}_{\mathrm{test},1}(x; T, F) = -\mathrm{IF}_{\mathrm{test},2}(x; T, F) \qquad (S)$$

holds. It reflects a certain symmetry of T, namely "treating both samples in the same way," so $\mathrm{IF}_{\mathrm{test}}(x, x; T, F) = 0$ for all x. [For a counterexample, rewrite (3.3.7) with two estimators S_m and S_n' which have a different influence function.]

When we calculate the first-order terms in the von Mises expansion of T, we obtain the following approximations (where F_m^1 and F_n^2 are the empirical distributions of the samples):

$$T\left(F_m^1, F_n^2\right) \simeq T(F, F) + \int \frac{\partial}{\partial t}\left[T(F_{t,x}, F)\right]_{t=0} dF_m^1(x)$$

$$+ \int \frac{\partial}{\partial t}\left[T(F, F_{t,y})\right]_{t=0} dF_n^2(y)$$

$$\simeq T(F, F) + \frac{1}{m}\sum_{i=1}^{m} \frac{\partial}{\partial t}\left[T(F_{t,x_i}, F)\right]_{t=0}$$

$$+ \frac{1}{n}\sum_{j=1}^{n} \frac{\partial}{\partial t}\left[T(F, F_{t,y_j})\right]_{t=0}$$

so when $T_N(F_m^1, F_n^2)$ is approximated adequately by $T(F_m^1, F_n^2)$ we obtain

$$\mathrm{Var}_{F,F}(T_N) \simeq \frac{1}{m}\mathrm{Var}_F\left\{\frac{\partial}{\partial t}\left[T(F_{t,x}, F)\right]_{t=0}\right\}$$

$$+ \frac{1}{n}\mathrm{Var}_F\left\{\frac{\partial}{\partial t}\left[T(F, F_{t,y})\right]_{t=0}\right\}.$$

Let us now impose condition (A), which states that asymptotically this approximation becomes exact:

$$\lim_{N \to \infty} N \operatorname{Var}_{F,F}(T_N) = \lim_{N \to \infty} N \left[\frac{1}{m} \operatorname{Var}_F \left\{ \frac{\partial}{\partial t} \left[T(F_{t,x}, F) \right]_{t=0} \right\} \right.$$

$$\left. + \frac{1}{n} \operatorname{Var}_F \left\{ \frac{\partial}{\partial t} \left[T(F, F_{t,y}) \right]_{t=0} \right\} \right]. \quad (A)$$

Property 6. Under the assumptions of Subsection 3.2a, (A), and (S) we obtain

$$\int \mathrm{IF}_{\text{test},1}(x; T, F)^2 \, dF(x) = \lambda(1 - \lambda) E^{-1} = \int \mathrm{IF}_{\text{test},2}(y; T, F)^2 \, dF(y)$$

and

$$\iint \mathrm{IF}_{\text{test}}(x, y; T, F)^2 \, dF(x) \, dF(y) = 2\lambda(1 - \lambda) E^{-1}.$$

Proof. Because of Property 5 and (S) we only have to prove the first equality. Let us calculate the asymptotic variance $V(T, F)$ of $T_N(X_1, \ldots, X_m; Y_1, \ldots, Y_n)$ by means of (A) and (S):

$$V(T, F) = \lim_{N \to \infty} N \operatorname{Var}_{F,F}(T_N)$$

$$= \lim_{N \to \infty} N \left(\frac{1}{m} + \frac{1}{n} \right) \int \left\{ \frac{\partial}{\partial t} \left[T(F_{t,x}, F) \right]_{t=0} \right\}^2 dF(x)$$

so $\int \{ (\partial/\partial t)[T(F_{t,x}, F)]_{t=0} \}^2 \, dF(x) = \lambda(1 - \lambda)V(T, F)$, and using (3.2.4) the rest of the proof follows as in Property 2. □

Now e can be defined as

$$e = E/[\lambda(1 - \lambda)J(F)]. \quad (3.2.10)$$

The equation $e = [J(F) \int \mathrm{IF}_{\text{test},1}(x; T, F)^2 \, dF(x)]^{-1}$ thus holds for one- and two-sample estimators and tests. (In the one-sample case, $\mathrm{IF}_{\text{test},1}$ coincides with $\mathrm{IF}_{\text{test}}$.) This property will be used when determining the optimal B-robust tests in Section 3.4.

3.2c.   Relation with Level and Power

We now examine the influence of contamination on the level and on the power of a test. Let S be a one-sample test of level α for $\theta = \theta_0$ against $\theta > \theta_0$, defined by a sequence of test statistics T_n which converges to the functional T. We consider the standardized statistics $U_n = \xi_n^{-1}(T_n)$ which converge to $U = \xi^{-1}(T)$. The critical values $k_n(\alpha)$ for U_n are given by $F_{\theta_0}\{U_n \geq k_n(\alpha)\} = \alpha$.

In order to define the asymptotic power of such a test, one usually constructs a sequence of alternatives $\theta_n = \theta_0 + \Delta n^{-1/2}$ where $\Delta > 0$ (Noether, 1955). Let us now assume that these alternatives are contaminated. When θ_n tends to θ_0, the contamination must tend to zero equally fast, or its effect will soon dominate everything else and give divergence. [The same idea was used by Huber-Carol, 1970 and Rieder, 1978.] Therefore we put $t_n = tn^{-1/2}$ and define the contaminated distribution

$$F_{n,t,x}^P = (1 - t_n)F_{\theta_n} + t_n\Delta_x$$

for the power, and

$$F_{n,t,x}^L = (1 - t_n)F_{\theta_0} + t_n\Delta_x$$

for the level. By means of the quantities

$$P_{n,t,x} = F_{n,t,x}^P\{U_n \geq k_n(\alpha)\}$$

and

$$L_{n,t,x} = F_{n,t,x}^L\{U_n \geq k_n(\alpha)\},$$

we can introduce a *level influence function* (LIF) by

$$\text{LIF}(x; S, F) = \lim_{n \to \infty} \frac{\partial}{\partial t}[L_{n,t,x}]_{t=0} \qquad (3.2.11)$$

and a *power influence function* (PIF) by

$$\text{PIF}(x; S, F, \Delta) = \lim_{n \to \infty} \frac{\partial}{\partial t}[P_{n,t,x}]_{t=0}. \qquad (3.2.12)$$

The interpretation is clear: If the observations are contaminated by outliers, this influences the probability that the test rejects the null hypothesis. For instance, the true distribution can become part of the alternative instead of the null hypothesis, because of this contamination. (This can happen also for rank tests, notwithstanding the fact that such tests have a constant level in their original framework. Actually, one-sample rank tests maintain their level among *symmetric* distributions, whereas $F_{n,t,x}^P$ and $F_{n,t,x}^L$ are asymmetric. And two-sample rank tests only maintain their level when the distributions underlying both samples have exactly the same shape, which is no longer true in the presence of contamination.)

By means of the asymptotic normality

$$\mathscr{L}_G\left(\sqrt{n}\left(U_n - E_G[U_n]\right)\middle/\left[\int \mathrm{IF}_{\text{test}}(x; T, G)^2\, dG\right]^{1/2}\right) \to \mathscr{N}(0, 1)$$

we can evaluate these influence functions, obtaining

$$\mathrm{LIF}(x; S, F) = \sqrt{E}\,\phi(\lambda_{1-\alpha})\mathrm{IF}_{\text{test}}(x; T, F) \qquad (3.2.13)$$

and

$$\mathrm{PIF}(x; S, F, \Delta) = \sqrt{E}\,\phi\left(\lambda_{1-\alpha} - \Delta\sqrt{E}\right)\mathrm{IF}_{\text{test}}(x; T, F), \quad (3.2.14)$$

where ϕ is the standard normal density, $\lambda_{1-\alpha}$ is the $(1-\alpha)$ quantile of the standard normal, and $E = E(T, F)$. Therefore, both the LIF and the PIF are directly proportional to the IF.

With two-sample tests, we use $(F_{n,t,x}^P, F_{n,t,y}^L)$ for the power, and $(F_{n,t,x}^L, F_{n,t,y}^L)$ for the level, obtaining

$$\mathrm{LIF}(x, y; S, F) = \sqrt{E}\,\phi(\lambda_{1-\alpha})\mathrm{IF}_{\text{test}}(x, y; T, F) \qquad (3.2.15)$$

and

$$\mathrm{PIF}(x, y; S, F, \Delta) = \sqrt{E}\,\phi\left(\lambda_{1-\alpha} - \Delta\sqrt{E}\right)\mathrm{IF}_{\text{test}}(x, y; T, F).$$

$$(3.2.16)$$

For one-sample rank tests, we can apply Corollary 5.1 of Rieder (1981) to compute the maximal asymptotic level and the minimal asymptotic power in the case of a fraction ε of *asymmetric* gross errors, yielding (in our notation)

$$\alpha_{\max} = 1 - \Phi\left(\lambda_{1-\alpha} - \varepsilon\sqrt{E}\,\gamma_{\text{test}}^*\right),$$

$$\beta_{\min} = 1 - \Phi\left(\lambda_{1-\alpha} - \Delta\sqrt{E} + \varepsilon\sqrt{E}\,\gamma_{\text{test}}^*\right). \qquad (3.2.17)$$

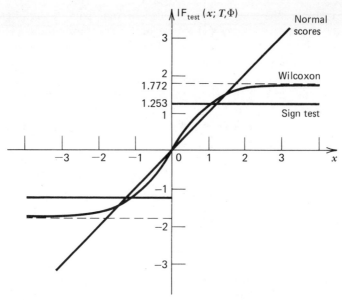

Figure 1. Influence function of the sign test, the Wilcoxon test, and the normal scores test, at $F = \Phi$.

Also in this case, one can construct an influence function version. Formula (2.7.7) told us how to use the influence function of an estimator to approximate the maximal bias over an asymmetric gross-error neighborhood $\mathscr{A}_\varepsilon$. In exactly the same way, we can approximate $\alpha_{\max}$ and $\beta_{\min}$ by extrapolating (3.2.13) and (3.2.14), yielding

$$\alpha_{\max} \simeq \alpha + \varepsilon\sqrt{E}\,\phi(\lambda_{1-\alpha})\gamma^*_{\text{test}},$$

$$\beta_{\min} \simeq \beta - \varepsilon\sqrt{E}\,\phi(\lambda_{1-\alpha} - \Delta\sqrt{E})\gamma^*_{\text{test}}, \qquad (3.2.18)$$

where β is the asymptotic power (3.2.8) at the model. Note that these approximations are linear in ε.

Let us compare (3.2.17) and (3.2.18) at the standard normal $F = \Phi$, with nominal level $\alpha = 5\%$. We shall show in the next section that the sign test satisfies

$$\text{IF}_{\text{test}}(x; T, \Phi) = \frac{\text{sign}(x)}{2\phi(0)} \qquad (3.2.19)$$

so $\gamma^*_{\text{test}}(T, \Phi) = 1/[2\phi(0)] = \sqrt{\pi/2} \simeq 1.2533$ by (3.2.2) and $E(T, \Phi) = [\int \text{IF}_{\text{test}}(x; T, \Phi)^2\, d\Phi(x)]^{-1} = [2\phi(0)]^2 = 2/\pi \simeq 0.6366$ by Property 2.

Table 1. **Maximal Asymptotic Level and Minimal Asymptotic Power of the Sign Test**[a]

			Power			
	Level		$\Delta = 0.5$		$\Delta = 3.0$	
ε	α_{max}	(3.2.18)	β_{min}	(3.2.18)	β_{min}	(3.2.18)
0.00	0.0500	0.0500	0.1067	0.1067	0.7731	0.7731
0.01	0.0510	0.0510	0.1048	0.1049	0.7707	0.7701
0.05	0.0553	0.0551	0.0975	0.0975	0.7577	0.7580
0.10	0.0612	0.0603	0.0914	0.0883	0.7419	0.7430

[a] Comparison between asymptotic minimax and influence function extrapolation.

Table 1 contains α_{max} and β_{min} of this test, together with the extrapolation (3.2.18) which appears to be quite good.

For the one-sample Wilcoxon test it holds that

$$\mathrm{IF}_{test}(x; T, \Phi) = \frac{\Phi(x) - \frac{1}{2}}{\int \phi^2(u) \, du} \qquad (3.2.20)$$

which yields $\gamma^*_{test}(T, \Phi) \simeq 1.772$, $E(T, \Phi) \simeq 0.955$, and the resulting maximal level and minimal power in Table 2.

Note that the level of the sign test is more stable than that of the Wilcoxon test. The latter test has a better power in the uncontaminated

Table 2. **Maximal Asymptotic Level and Minimal Asymptotic Power of the Wilcoxon Test**

			Power			
	Level		$\Delta = 0.5$		$\Delta = 3.0$	
ε	α_{max}	(3.2.18)	β_{min}	(3.2.18)	β_{min}	(3.2.18)
0.00	0.0500	0.0500	0.1238	0.1238	0.9009	0.9009
0.01	0.0518	0.0518	0.1203	0.1203	0.8979	0.8979
0.05	0.0595	0.0589	0.1069	0.1061	0.8849	0.8858
0.10	0.0705	0.0679	0.0918	0.0884	0.8673	0.8707

case, but loses power more rapidly for increasing ε. (Indeed, at $\Delta = 0.5$ the power of the Wilcoxon and the sign test become the same for $\varepsilon \simeq 0.1$, and for larger ε the Wilcoxon power falls below that of the sign test.) Therefore, the sign test is to be preferred over the Wilcoxon test when the amount of contamination is moderately large. This resembles the choice between the median and the Hodges–Lehmann (H/L) estimator: H/L has a lower asymptotic variance at the normal and when the amount of contamination ε is small, but the median is to be preferred for larger ε and is more efficient than H/L at long-tailed distributions.

On the other hand, tests with $\gamma_{\text{test}}^* = \infty$ (like the normal scores test) are always dangerous, because (3.2.17) yields $\alpha_{\max} = 1$ and $\beta_{\min} = 0$ for *any* fraction $\varepsilon > 0$ of contamination! This resembles the position of the arithmetic mean in estimation theory, because its maximal asymptotic variance is infinite in any gross-error neighborhood.

Remark. By means of so-called *small-sample asymptotic techniques* (Daniels, 1954; Hampel, 1973b; Field and Hampel, 1983), one can find good approximations to the distribution of U_n and use them to compute another level and power influence function. These can be viewed as finite-sample improvements of LIF and PIF and can be used to approximate the actual level and power of a test over a gross-error model. Numerical calculations show the great accuracy of these approximations down to very small sample sizes. Details are provided by Field and Ronchetti (1985). More on small sample asymptotics can be found in Section 8.5.

3.2d. Connection with Shift Estimators

Suppose the model is $F_\theta(x) = F(x - \theta)$ with $\theta_0 = 0$; we make the same assumptions concerning T_n and T as for Definition 1. Denote the translation of a distribution G by b as $G_b(x) = G(x - b)$. Assume that $T(G_b)$ is strictly increasing and continuous in b (for all G in a certain domain depending on the form of T) and that a (necessarily unique) solution of the equation $T(G_b) = T(F)$ exists.

Definition 3. We define the "shift estimate of location" $D(G)$ as the unique solution of $T(G_{(-D)}) = T(F)$.

Assume that D is the limit of its finite-sample versions

$$D_n(\{X_i\}) = \tfrac{1}{2}\inf\{d; T_n(\{X_i - d\}) < h_n(\alpha)\}$$
$$+ \tfrac{1}{2}\sup\{d; T_n(\{X_i - d\}) > h_n(\alpha)\}.$$

[The scale case with $F_\theta(x) = F(x/\theta)$ and $\theta_0 = 1$ can be treated analogously, as well as the two-sample problem.]

For example, from one-sample location rank statistics T_n we find some R-estimators, as they were first discovered by Hodges and Lehmann (1963). [In Subsection 2.4c we constructed them from two-sample tests, whereas they are formulated from a one-sample rank test approach by Hettmansperger and Utts (1977)]. The idea is to construct estimators from tests with good properties. This approach is not limited to location—for instance, one may obtain estimators of the ratio of scale parameters in the two-sample problem by inverting scale rank statistics.

Restricting ourselves for simplicity to the one-sample location case, we prove the following property:

Property 7. Under the assumptions of Definitions 1 and 3, $\mathrm{IF}_{\mathrm{test}}(x; T, F) = \mathrm{IF}(x; D, F)$ for all x.

Proof. From Definition 3 it follows that for all x and t

$$T(F) = T\left[\left(F_{t,x}\right)_{(-D_{t,x})}\right]$$

with $F_{0,x} = F$ and $D_{0,x} = 0$. Hence

$$\frac{\partial}{\partial t}\left[U\{(F_{t,x})_{(-D_{t,x})}\}\right]_{t=0} = \frac{\partial}{\partial t}\left[\xi^{-1}\left(T\{(F_{t,x})_{(-D_{t,x})}\}\right)\right]_{t=0} = 0.$$

Using the chain rule, we find

$$\frac{\partial}{\partial t}\left[U(F_{t,x})\right]_{t=0} + \frac{\partial}{\partial t}\left[U\left(F_{-D_{t,x}}\right)\right]_{t=0} = 0.$$

Now $U(F_{(-D_{t,x})}) = -D_{t,x}$ because U is Fisher consistent, and therefore this equation becomes $\mathrm{IF}_{\mathrm{test}}(x; T, F) - \mathrm{IF}(x; D, F) = 0$, which ends the proof. $\square$

Properties 2 and 7, together with Pitman's theorem, entail Theorem 6 of Hodges and Lehmann (1963) which derives the asymptotic efficiency of the Hodges–Lehmann estimator from that of the Wilcoxon rank test. This is particularly important because the asymptotic efficiency of the Wilcoxon rank test is rather high (Hodges and Lehmann 1956).

3.3. CLASSES OF TESTS

3.3a. The One-Sample Case

The estimators T_n of the types M, L, and R (Section 2.4) correspond to Fisher-consistent functionals. However, these statistics can be used just as well for testing $\theta = \theta_0$ against some one-sided or two-sided alternative; in practice one would be inclined to work with

$$S_n = \sqrt{n}\,(T_n - \theta_0) \qquad \text{(type I)} \tag{3.3.1}$$

because of the asymptotic normality $\mathscr{L}(S_n) \overset{\text{weak}}{\to} \mathscr{N}(0, V(T, F_{\theta_0}))$ at the null hypothesis. The advantage of (3.3.1) lies in the fact that it suffices to compile tables for the exact critical values up to a certain N, after which the asymptotics take over. In this way we construct tests of the types M, L, and R, having the same IF as the estimators from which they are derived. Some examples of classical M-tests include the z-test for location ($\psi(x, \theta) = x - \theta$) and the χ^2-test for scale ($\psi(x, \theta) = (x/\theta)^2 - 1$) which will be robustified in the next section. M-tests of type I have the additional advantage that even the (very robust) redescending ψ-functions of Section 2.6 can be used.

It is also possible to construct another type of M-test (Ronchetti, 1979; Sen, 1982) by means of

$$S_n = \frac{1}{\sqrt{n}} \sum_{i=1}^{n} \psi(X_i) \qquad \text{(type II)}, \tag{3.3.2}$$

where ψ satisfies $\int \psi\, dF_{\theta_0} = 0$ and $\int \psi^2\, dF_{\theta_0} < \infty$. From $\int \psi\, dF_{\theta_0} = 0$ it follows that $T_n = (1/\sqrt{n})S_n \to 0$ (a.e.) at the null hypothesis, by means of the Kolmogorov theorem. However, in general, T_n is not Fisher consistent. Also making use of $\int \psi^2\, dF_{\theta_0} < \infty$, we see that $\mathscr{L}(S_n) \overset{\text{weak}}{\to} \mathscr{N}(0, \int \psi^2\, dF_{\theta_0})$ at the null hypothesis by the central limit theorem. Now $\xi(\theta) = T(F_\theta) = \int \psi\, dF_\theta$, so $\xi'(\theta_0) = (\partial/\partial\theta)[\int \psi\, dF_\theta]_{\theta_0}$. When $\xi'(\theta_0) \neq 0$, we obtain

$$\text{IF}_{\text{test}}(x; T, F_{\theta_0}) = \frac{\psi(x)}{\int \psi(y)(\partial/\partial\theta)[f_\theta(y)]_{\theta_0}\, d\lambda(y)} \tag{3.3.3}$$

which is equal to (2.3.8) for the corresponding M-estimator. Therefore, M-tests of the types I and II have the same asymptotic behavior at the null hypothesis. [In the special case of testing against a simple alternative, tests given by (3.3.2) were already treated by H. Rieder, 1978, who denoted ψ by "IC" without actually defining an influence function. Note that this function is determined only up to a factor.] The disadvantage of (3.3.2) lies in the fact that redescending ψ cannot be used in this framework, since the null hypothesis would be accepted when *all* observations belong to a distant alternative.

One-sample rank tests are known only for the location problem (Hajek and Sidak, 1967). Put $F_\theta(x) = F(x - \theta)$, where $\theta_0 = 0$ and F is symmetric with a positive absolutely continuous density f. Starting from the sample $X_1, \ldots, X_n$, we denote by R_i^+ the rank of $|X_i|$. The test statistics are

$$T_n = \frac{1}{n} \sum_{i=1}^{n} a_n^+(R_i^+)\operatorname{sign}(X_i) \qquad (3.3.4)$$

(in practice one often uses nT_n), where the scores $a_n^+(i)$ are nondecreasing and only defined up to a positive factor. Suppose there is a score-generating function $\phi^+: [0,1] \to \mathbb{R}$ which is square integrable and nondecreasing; score-generating means that $\lim_{n \to \infty} \int_0^1 \{a_n^+(1 + [un]) - \phi^+(u)\}^2 du = 0$, where $[un]$ is the largest integer which is not larger than un. The ϕ^+ of the most efficient rank tests are

$$\phi^+(u, f) = -\frac{f'[F^{-1}(\frac{1}{2} + \frac{1}{2}u)]}{f[F^{-1}(\frac{1}{2} + \frac{1}{2}u)]}.$$

In general, the T_n of (3.3.4) converge to the functional

$$T(G) = \int \phi^+[G(|x|) - G(-|x|)]\operatorname{sign}(x)\, dG(x) \qquad (3.3.5)$$

which is not Fisher consistent. Let us now compute its influence function. From Hajek and Sidak (1967, p. 220), we find that $\xi_n(\theta) \to \theta \int_0^1 \phi^+(u)\phi^+(u, f)\, du$, so $\xi'(0) = \int_0^1 \phi^+(u)\phi^+(u, f)\, du$. As for the numerator of (3.2.3) we break the integration at zero, perform some partial integrations, and obtain

$$\mathrm{IF}_{\text{test}}(x; T, F) = \frac{\phi^+[2F(|x|) - 1]\operatorname{sign}(x)}{\int_0^1 \phi^+(u)\phi^+(u, f)\, du}. \qquad (3.3.6)$$

One could also use Property 7; $D(G)$ is determined by

$$\int \phi^+[G(|x - D| + D) - G(-|x - D| + D)]\operatorname{sign}(x - D)\, dG(x) = 0.$$

Now replace all G by $F_{t,x}$, all D by $D_{t,x}$, and put the derivative with respect to t at 0 of the entire expression equal to zero; then solve the resulting equation in $(\partial/\partial t)[D_{t,x}]_{t=0}$, which again yields (3.3.6).

Examples. The *sign* test has $\phi^+(u) = 1$, so $IF_{\text{test}}(x; T, F) =$ $\text{sign}(x)/2f(0)$, the IF of the median. The *one-sample Wilcoxon* test has $\phi^+(u) = u$, so its IF is that of the Hodges–Lehmann estimator, which is derived from it. The *Van der Waerden–Van Eeden* and the *normal scores* tests both have $\phi^+(u) = \Phi^{-1}(\frac{1}{2} + \frac{1}{2}u)$, so their influence function equals that of the normal scores estimator. At Φ the influence function is $IF_{\text{test}}(x; T, \Phi) = x$, so (Property 3) the latter tests are asymptotically efficient there. The sign and the Wilcoxon tests have a bounded influence function; their efficiencies at the normal distribution, given by Property 2, are $2/\pi \cong 0.64$ and $3/\pi \cong 0.95$.

Note that γ_{test}^* is finite when the scores function ϕ^+ is bounded, as is the case for the sign and the Wilcoxon tests. Tests with unbounded ϕ^+ (like the normal scores test) satisfy $\gamma_{\text{test}}^* = \infty$ which implies that $\alpha_{\max} = 1$ and $\beta_{\min} = 0$ by (3.2.17).

3.3b. The Two-Sample Case

We are now in the situation of Definition 2 of Section 3.2, where the model is $G(x) = F(x - \theta)$ with $\theta_0 = 0$ or $G(x) = F(x/\theta)$ with $\theta_0 = 1$. A Fisher consistent statistic can be constructed out of a one-sample M-, L-, or R-estimator in a simple way: Apply it to both samples, obtaining $S_m(X_1, \ldots, X_m)$ and $S_n(Y_1, \ldots, Y_n)$ and let

$$T_N(X_1, \ldots, X_m; Y_1, \ldots, Y_n) = S_m(X_1, \ldots, X_m) - S_n(Y_1, \ldots, Y_n)$$

$$(3.3.7)$$

in the location case. (Replace subtraction by division in the scale problem.) This T_N can serve as an estimator or a test statistic. Clearly,

$$IF_{\text{test}}(x, y; T, F) = IF(x; S, F) - IF(y; S, F). \qquad (3.3.8)$$

The classical test for location is the z-test, where S is the M-estimator with $\psi(x, \theta) = x - \theta$, and hence $IF_{\text{test},1}(x; T, \Phi) = x$. In the scale model, we obtain the F-test with $\psi(x, \theta) = (x/\theta)^2 - 1$; hence $IF_{\text{test},1}(x; T, \Phi)$ $= \frac{1}{2}(x^2 - 1)$. This influence function describes the extremely high sensitivity of the F-test to outliers.

Two-sample rank statistics provide interesting examples of functionals that fail to be Fisher consistent. Let F be symmetric with a positive absolutely continuous density f. Location rank tests are based on the ranks R_i of X_i in the pooled sample; in Hajek and Sidak (1967) the following test statistics are studied:

$$T_N = \frac{1}{m} \sum_{i=1}^{m} a_N(R_i) \qquad (3.3.9)$$

with nondecreasing $a_N(i)$, only defined up to a positive affine transformation. Suppose a score-generating function ϕ exists which is odd [meaning $\phi(1 - u) = -\phi(u)$], so $\int \phi[F(x)]\, dF(x) = 0$. The ϕ of the most efficient rank test is given by

$$\phi(u, f) = -\frac{f'[F^{-1}(u)]}{f[F^{-1}(u)]}.$$

Now T_N corresponds to the functional

$$T(H^1, H^2) = \int \phi[\lambda H^1(x) + (1 - \lambda)H^2(x)]\, dH^1(x) \quad (3.3.10)$$

which is not Fisher consistent. Applying Definition 2 of Section 3.2 one obtains, independently of λ,

$$IF_{\text{test},1}(x; T, F) = \phi[F(x)] \Big/ \int (\phi(F(y)))'\, dF(y)$$

$$= \phi[F(x)] \Big/ \int_0^1 \phi(u)\phi(u, f)\, du. \qquad (3.3.11)$$

When using Property 7, one solves

$$\frac{\partial}{\partial t}\left[\int \phi[\lambda F_{t,x}(y + D_{t,x}) + (1 - \lambda)F_{t,x}(y)]\, dF_{t,x}(y + D_{t,x}) \right]_{t=0} = 0$$

in $(\partial/\partial t)[D_{t,x}]_{t=0}$, again yielding (3.3.11).

Since property (S) is satisfied, only $IF_{\text{test},1}$ must be written down. One-sample and two-sample rank tests are said to be *"similar"* when $\phi^+(u)$

$= \phi(\frac{1}{2} + \frac{1}{2}u)$. In this case, the following three functions are identical:

1. The $\text{IF}_{\text{test},1}$ of the two-sample rank test.
2. The IF_{test} of the "similar" one-sample rank test.
3. The IF of the R-estimator constructed with the same function ϕ (see Subsection 2.3c).

Examples. The *median* test has $\phi(u) = -1$ for $u < \frac{1}{2}$, $\phi(u) = 1$ for $u > \frac{1}{2}$, and is similar to the one-sample sign test of Subsection 3.3a. The *two-sample Wilcoxon* test corresponds to $\phi(u) = u - \frac{1}{2}$ (up to a factor) and is similar to its one-sample counterpart. The *Van der Waerden* and the *Fisher–Yates–Terry–Hoeffding-normal scores* tests have $\phi(u) = \Phi^{-1}(u)$ and thus they are also similar to their one-sample versions. The results of Subsection 3.3a (e.g., the values of e) thus remain valid.

Two-sample rank tests for *scale* are based on (3.3.9) with different scores $a_1(i)$ corresponding to a score-generating function ϕ_1, this time supposed to be even ($\phi_1(1 - u) = \phi_1(u)$), and nondecreasing for $u \geq \frac{1}{2}$. Because ϕ_1 is defined up to a positive affine transformation, we may still assume $\int \phi_1[F(y)] \, dF(y) = 0$. The most efficient ϕ_1 is now

$$\phi_1(u, f) = -1 - F^{-1}(u)\{f'[F^{-1}(u)]\}/\{f[F^{-1}(u)]\}.$$

From Definition 2 of Subsection 3.2 it follows that

$$\text{IF}_{\text{test},1}(x; T, F) = \phi_1[F(x)] \Big/ \int y(\phi_1(F(y)))' \, dF(y)$$

$$= \phi_1[F(x)] \Big/ \int_0^1 \phi_1(u)\phi_1(u, f) \, du. \qquad (3.3.12)$$

Examples. The *quartile* test (see Hajek and Sidak, 1967 for this and other tests) satisfies

$$\phi_1(u) = 1 \qquad \text{for } u < \tfrac{1}{4} \text{ or } u > \tfrac{3}{4}$$

$$= -1 \qquad \text{for } \tfrac{1}{4} < u < \tfrac{3}{4}.$$

Therefore, $\text{IF}_{\text{test},1}$ equals the IF of the interquartile range estimator (Example 7 in Subsection 2.3b). The *Ansari–Bradley* and the *Mood* tests correspond to $\phi_1(u) = |u - \frac{1}{2}| - \frac{1}{4}$ and $\phi_1(u) = (u - \frac{1}{2})^2 - \frac{1}{12}$; both have a bounded influence function too. The *Klotz* as well as the *Capon-normal scores* tests correspond to $\phi_1(u) = [\Phi^{-1}(u)]^2 - 1$ so their influence function at Φ equals that of the F-test, which implies that there are asymptotically efficient at Φ. The former tests have (at Φ) the asymptotic efficiencies 0.37, 0.61, and 0.76, in accordance with Property 6. By means of Subsection 3.2d we can define two-sample estimators of scale out of these tests, which then have the same influence functions.

3.4. OPTIMALLY BOUNDING THE GROSS-ERROR SENSITIVITY

Let us now repeat what we did in Section 2.4, where we selected the estimators (called optimal B-robust) which minimize the asymptotic variance subject to an upper bound on the gross-error sensitivity. The performance of a *test* is measured by its asymptotic (Pitman) efficacy $E(T, F)$, which is inversely proportional to $\int \mathrm{IF}_{\text{test}}(x; T, F)^2 \, dF(x)$ in the one-sample case (Property 2 of Subsection 3.2b), and inversely proportional to $\int \mathrm{IF}_{\text{test},1}(x; T, F)^2 \, dF(x)$ in the two-sample case (Property 6 of Subsection 3.2b). In the one-sample case, the gross-error sensitivity can simply be defined as

$$\gamma^*_{\text{test}}(T, F) = \sup_x |\mathrm{IF}_{\text{test}}(x; T, F)|. \tag{3.4.1}$$

In the two-sample case, we see from Property 5 and (S) of Subsection 3.2b that $\mathrm{IF}_{\text{test},1}$ determines everything, so we can put

$$\gamma^*_{\text{test}}(T, F) = \sup_x |\mathrm{IF}_{\text{test},1}(x; T, F)|. \tag{3.4.2}$$

Concluding, our aim in both cases is to minimize the expected square of the influence function subject to a bound c on the test gross-error sensitivity (Rousseeuw, 1979, 1982b; Ronchetti, 1979). By (3.2.8) this is equivalent to the principle:

Find a test which maximizes the asymptotic power, subject to a bound on the influence function at the null hypothesis.

For this purpose we make use of Theorem 1 of Subsection 2.4a, with $\theta_* = \theta_0$ and $F := F_{\theta_0}$. If the bound $\gamma^* \leq c$ were not imposed, then we could simply use the Cramér–Rao inequality (Property 3 of Subsection 3.2b) which gives us the classically optimal tests (often with $\gamma^* = \infty$).

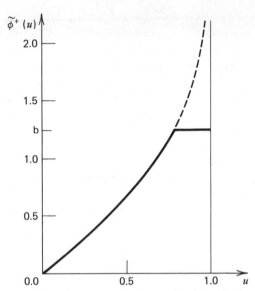

Figure 1. Truncated normal scores function (3.4.4).

For the one-sample tests based on estimators of the type M, L, or R by means of (3.3.1) or (3.3.2) there is no problem at all, because they inherit the influence function of the original estimator. Therefore, we only have to insert the optimal robust solutions of Section 2.4.

For one-sample location rank tests, the IF is given by (3.3.6). Putting

$$\chi(x) := \phi^+(2F(|x|) - 1)\mathrm{sign}(x) \qquad (3.4.3)$$

we recognize formula (2.3.12) with χ instead of ψ. This enables us to apply Theorem 1 of Subsection 2.4a, yielding $\tilde{\chi}$ given by (2.4.8). At the standard normal ($F = \Phi$), the optimal B-robust solution is given by

$$\tilde{\phi}^+(u) = \min\left\{ \Phi^{-1}\left(\tfrac{1}{2} + \tfrac{1}{2}u\right), b \right\} \qquad (3.4.4)$$

which is a *truncated version of the normal scores function*. For $b \downarrow 0$ we obtain the sign test, which is most B-robust (meaning that it minimizes γ^*_{test}) by Theorem 3 in Subsection 2.5c.

For the two-sample statistics constructed from M-, L-, or R-estimators by means of (3.3.7), which can be used as "shift" estimators or test statistics, there is again no problem because $\mathrm{IF}_{\mathrm{test},1}$ equals the IF of the

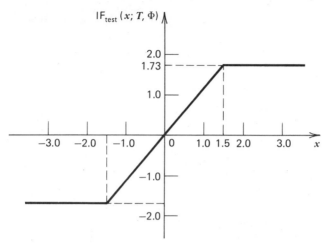

Figure 2. Influence function of the truncated normal scores test with $b = 1.5$, at $F = \Phi$.

original estimator, so our previous optimality results can be translated directly.

The IF of two-sample rank statistics for location is given by (3.3.11), which can be reduced to (2.3.12) by means of

$$\chi(x) := \phi(F(x)). \tag{3.4.5}$$

When the skew-symmetric mapping $\tilde{\chi}$ of (2.4.8) is nondecreasing, the mapping $\tilde{\phi}(u) = \tilde{\chi}(F^{-1}(u))$ is an acceptable scores function which determines an optimal B-robust test. At $F = \Phi$, we obtain

$$\tilde{\phi}(u) = \left[\Phi^{-1}(u)\right]_{-b}^{b}, \tag{3.4.6}$$

a truncation of the Van der Waerden scores. For $b \downarrow 0$ we find the median test, which is most B-robust (Theorem 3 in Subsection 2.5c).

In the case of two-sample rank statistics for scale (3.3.12) we put $\chi(x) = \phi_1(F(x))$, obtaining (2.3.18). If $\tilde{\chi}$ given by (2.4.10) is nondecreasing for positive arguments, then $\tilde{\phi}_1(u) = \tilde{\chi}(F^{-1}(u))$ determines an optimal B-robust solution. At $F = \Phi$, this amounts to

$$\tilde{\phi}_1(u) = \left[\left(\Phi^{-1}(u)\right)^2 - 1 - a\right]_{-b}^{b}, \tag{3.4.7}$$

which is a robustification of the Klotz scores function. For $b \downarrow 0$ we find the quartile test, which is most B-robust by Theorem 9 in Subsection 2.5e.

When two-sample estimators of location shift or relative scale are derived from two-sample rank statistics as in Subsection 3.2d, one only has to apply the optimal robust scores determined above, because the resulting estimators possess the same influence functions.

3.5. EXTENDING THE CHANGE-OF-VARIANCE FUNCTION TO TESTS

In Subsection 3.2b we saw that the performance of a test is measured by the asymptotic efficacy $E(T, F)$ given by (3.2.6). This quantity can be calculated using Property 2 of Subsection 3.2b in the one-sample case, and by Property 6 of the same section in the two-sample case. By means of the asymptotic efficacy one can evaluate the asymptotic power (3.2.8), the relative asymptotic efficiency (3.2.7), and the absolute asymptotic efficiency (3.2.9) or (3.2.10).

Let us now investigate the local robustness of the functional $E(T, \cdot)$ at F, where T is a functional corresponding to a sequence of test statistics and F is a symmetric cdf for which $E(T, F)$ exists. The *change-of-efficacy function* (CVF) of T at F is defined by

$$\frac{\partial}{\partial t}\left[E^{-1}(T,(1-t)F + tG)\right]_{t=0} = \int \mathrm{CVF}_{\mathrm{test}}(x; T, F)\, dG(x)$$

$$(3.5.1)$$

for all *symmetric G* for which this makes sense. In its original form, this function was introduced by Rousseeuw and Ronchetti in 1978. Like the change-of-variance function (2.5.11) it may contain delta functions, and it is also symmetric in x. In the case of a Fisher-consistent estimator it holds that $E(T, F) = V(T, F)^{-1}$, so the $\mathrm{CVF}_{\mathrm{test}}$ coincides with the CVF in that case. Therefore, the $\mathrm{CVF}_{\mathrm{test}}$ generalizes the CVF of estimators. Note that even in the case of two-sample statistics, the $\mathrm{CVF}_{\mathrm{test}}$ is a function of one variable.

We only worry about *large* values of the $\mathrm{CVF}_{\mathrm{test}}$, since they point to a decrease of E. Therefore, we define the *change-of-efficacy sensitivity* $\kappa^*(T, F)$ as $+\infty$ if a delta function with positive factor occurs in the $\mathrm{CVF}_{\mathrm{test}}$, and otherwise as

$$\kappa^*(T, F) := \sup_{x}\left\{\mathrm{CVF}_{\mathrm{test}}(x; T, F)E(T, F)\right\} \qquad (3.5.2)$$

where the supremum is taken over those x where $\text{CVF}_{\text{test}}(x; T, F)$ is continuous. [We use the same notation as for the change-of-variance sensitivity (Definition 2 of Subsection 2.5a) because they coincide in the case of estimators.] When $\kappa^*(T, F)$ is finite we say that T is *V-robust* at F.

For one-sample or two-sample tests based on M-estimators of location or scale by means of (3.3.1), (3.3.2), or (3.3.7) it holds that $E(T, F)$ is proportional to $V(T, F)^{-1}$ (there is a factor depending on λ in the two-sample case), so

$$\text{CVF}_{\text{test}}(x; T, F)E(T, F) = \text{CVF}(x; \psi, F)/V(\psi, F). \quad (3.5.3)$$

Therefore, all the examples and results of Sections 2.5 and 2.6 remain valid: *V*-robustness implies *B*-robustness but not conversely, and the most *V*-robust and optimal *V*-robust ψ-functions have already been constructed.

One-sample rank tests (3.3.4) and two-sample rank tests (3.3.9) for location are called "similar" when $\phi^+(u) = \phi(\frac{1}{2} + \frac{1}{2}u)$. The following functions are then identical, at any F:

1. The $\text{IF}_{\text{test},1}$ of the two-sample rank test.
2. The IF_{test} of the similar one-sample rank test.
3. The IF_1 of the corresponding two-sample shift estimator (as in Subsection 3.2d).
4. The IF of the R-estimator constructed with the same ϕ (see Subsection 2.3c).

Therefore, these four procedures have proportional asymptotic efficacies by Properties 2 and 6 of Subsection 3.2b, and hence they possess the same κ^*. For example, the Hodges–Lehmann estimator, the one-sample Wilcoxon test, the two-sample Wilcoxon test and the corresponding two-sample shift estimator all satisfy

$$\text{CVF}_{\text{test}}(x; T, \Phi)E(T, \Phi) = 4 - 8\sqrt{\pi}\,\phi(x) \quad (3.5.4)$$

[where $\phi(x)$ is the standard normal density], so $\kappa^* = 4$. For the median, the one-sample sign test, and the two-sample median test it holds that

$$\text{CVF}_{\text{test}}(x; T, \Phi)E(T, \Phi) = 2\left(1 - \frac{1}{\phi(0)}\delta_{(0)}(x)\right), \quad (3.5.5)$$

so $\kappa^* = 2$, which is the minimal value by Theorem 4 of Subsection 2.5c. For the normal scores estimator, the Van der Waerden–Van Eeden, and the Fisher–Yates–Terry–Hoeffding tests we find

$$\mathrm{CVF}_{\text{test}}(x; T, \Phi) = x^2 - 1, \tag{3.5.6}$$

which we recognize as the CVF of the arithmetic mean, hence $\kappa^* = \infty$.

The change-of-efficacy sensitivity can be used to approximate the minimal asymptotic efficacy when the underlying distribution belongs to a *symmetric* gross-error neighborhood $\mathcal{P}_\varepsilon$ of F. Following (2.7.3), we find

$$\inf_{G \in \mathcal{P}_\varepsilon} E(T, G) \approx E(T, F)\exp\left[-\varepsilon\kappa^*(T, F)\right]. \tag{3.5.7}$$

Because the asymptotic power (3.2.8) is monotone in E, this yields an approximation to the minimal asymptotic power over a symmetric gross-error neighborhood:

$$\inf_{G \in \mathcal{P}_\varepsilon} \beta(T, G) = 1 - \Phi\left(\lambda_{1-\alpha} - \Delta\left[\inf_{G \in \mathcal{P}_\varepsilon} E(T, G)\right]^{1/2}\right)$$

$$\approx 1 - \Phi\left(\lambda_{1-\alpha} - \Delta[E(T, F)]^{1/2}\exp\left[-\tfrac{1}{2}\varepsilon\kappa^*(T, F)\right]\right).$$

$$\tag{3.5.8}$$

There are two essential differences between this approximation and (3.2.17). First, (3.5.8) deals with symmetric contamination, and (3.2.17) with asymmetric contamination. Second, (3.5.8) tells us what happens for a change in the distribution underlying the observations, affecting both the null hypothesis and the alternatives, whereas (3.2.17) applies to outliers in the sample. For instance, for two-sample location tests, this aspect becomes clearly visible: The $\mathrm{CVF}_{\text{test}}$ approximation gives us the minimal power when the shifted distributions become longer-tailed, so both samples change in the same way (in such a situation, the level of rank tests will remain constant), whereas the $\mathrm{IF}_{\text{test},1}$ describes what happens if there is an outlier in the first sample, which may turn acceptance into rejection or vice versa (therefore, such contamination influences both level and power).

*3.6. RELATED APPROACHES

3.6a. Lambert's Approach

In this subsection we present the connections between our influence function for tests (Rousseeuw and Ronchetti, 1979, 1981) and that of Lambert (1981). In order to compare them, we first summarize the basic concepts of Bahadur and Pitman efficiency and the notion of P-value.

Bahadur and Pitman Efficiency

Bahadur (1960) calls the sequence of tests statistics $\{T_n;\ n \geq 1\}$ a *standard sequence* if the following conditions are satisfied:

(B1)   There exists a continuous cdf F such that for all x:

$$\lim_{n \to \infty} P_{\theta_0}\{T_n < x\} = F(x).$$

(B2)   There exists a constant $0 < a < \infty$ such that for $x \to \infty$:

$$\ln(1 - F(x)) = -\tfrac{1}{2}ax^2(1 + o(1)).$$

(B3)   There exists a function $b(\theta)$ on $\Theta \setminus \{\theta_0\}$ with $0 < b(\theta) < \infty$, such that for each $\theta \neq \theta_0$: $\lim_{n \to \infty} P_\theta\{|n^{-1/2}T_n - b(\theta)| > x\} = 0$ for every $x > 0$.

For any such standard sequence, Bahadur shows that

$$-2n^{-1}\ln(1 - F(T_n)) \overset{n \to \infty}{\to} ab(\theta) =: c(\theta), \qquad \text{for all } \theta \neq \theta_0.$$

He calls $c(\theta)$ the approximate *slope* of the sequence $\{T_n\}$, and

$$\text{ARE}_{1,2}^{(B)}(\theta) = c^{(1)}(\theta)/c^{(2)}(\theta) \tag{3.6.1}$$

is called the approximate *relative efficiency* (in the Bahadur sense) of two standard sequences $\{T_n^{(1)};\ n \geq 1\}$ and $\{T_n^{(2)};\ n \geq 1\}$. The idea behind Bahadur's comparison of tests is to look at the rate of convergence (towards 0) of the level of the tests. (See also Bahadur, 1971.)

On the other hand, Pitman considers tests at level α and a sequence of alternatives which converges to the hypothesis at a certain rate (typically

$n^{-1/2}$). In this way he obtains a limiting power which is different from 1 (as $n \to \infty$), which he takes as the basis for his comparison between different tests.

Using the formalism of Noether (1955) we can compute the *asymptotic Pitman efficacy* E (3.2.6) which is a monotone increasing function of the asymptotic power by (3.2.7), and the *Pitman relative efficiency*

$$\mathrm{ARE}_{1,2}^{(P)} = E_1/E_2 \tag{3.6.2}$$

of two tests with efficacies E_1 and E_2.

Wieand (1976, p. 1005) gives a condition under which Bahadur and Pitman efficiencies coincide, namely

$$\lim_{\theta \to \theta_0} \mathrm{ARE}_{1,2}^{(B)}(\theta) = \lim_{\alpha \to 0} \mathrm{ARE}_{1,2}^{(P)}(\alpha, \beta), \tag{3.6.3}$$

where α and β are the level and the power. Wieand's condition is satisfied in several important cases.

P-Values

Let $G_n(\cdot\,; \theta)$ be the distribution function of the test statistic T_n under F_θ. Then, the *P-value* P_n is defined as

$$P_n := 1 - G_n(T_n; \theta_0). \tag{3.6.4}$$

Lambert and Hall (1982) investigate the asymptotic nonnull distribution of P-values and show that P-values are asymptotically lognormal (under the alternative), that is

$$\mathscr{L}_{F_\theta}\left(n^{1/2}\left[-n^{-1}\ln(P_n) - c_L(\theta)\right]\right) \to \mathscr{N}\left(0, \tau^2(\theta)\right),$$

for all $\theta \neq \theta_0$. This $c_L(\theta)$ is the *Bahadur half-slope*:

$$c_L(\theta) = c(\theta)/2 = \lim_{n \to \infty}\left(-n^{-1}\ln(P_n)\right) \quad \text{a.s. } (F_\theta). \tag{3.6.5}$$

Comparison with Lambert's Approach

Lambert (1981) defines an influence function for the testing problem in terms of P-values. Let $\{P_n; n \geq 1\}$ be any sequence of P-values that has a slope $\tilde{c}_L(H)$ under the distribution H. Lambert's influence function of $\{P_n\}$

at H is defined by

$$\mathrm{IF}_L(x; \{P_n\}, H) := \mathrm{IF}(x; \tilde{c}_L, H)$$

$$= \lim_{\substack{t \to 0 \\ >}} \left(\tilde{c}_L(H_{t,x}) - \tilde{c}_L(H)\right)/t, \qquad (3.6.6)$$

where we assume that $\{P_n\}$ has a slope $\tilde{c}_L(H_{t,x})$ at $H_{t,x} := (1 - t)H + t\Delta_x$, for every sufficiently small t.

Note that Lambert's influence function IF_L is not defined at the null hypothesis, where the $\mathrm{IF}_{\mathrm{test}}$ of Subsection 3.2a is calculated most often. The next proposition shows that $\mathrm{IF}_{\mathrm{test}}$ and IF_L are proportional at the alternative F_θ if the P-value depends on the data only through the test statistic, as is usually the case.

Theorem. If $\mathrm{IF}_{\mathrm{test}}(x; T, F_\theta)$ exists, and the Bahadur half-slope $\tilde{c}_L$ depends on the data only through the test statistic [i.e., $\tilde{c}_L(H) = d(T(H))$, where d is a differentiable function], then

$$\mathrm{IF}_L(x; \{P_n\}, F_\theta) = c_L'(\theta)\mathrm{IF}_{\mathrm{test}}(x; T, F_\theta).$$

Proof. Applying the definitions of $\mathrm{IF}_{\mathrm{test}}$ and ξ we have:

$$\mathrm{IF}_L(x; \{P_n\}, F_\theta) = \frac{\partial}{\partial t}\left[\tilde{c}_L((1 - t)F_\theta + t\Delta_x)\right]_{t=0}$$

$$= \frac{\partial}{\partial t}\left[d(T((1 - t)F_\theta + t\Delta_x))\right]_{t=0}$$

$$= d'(\xi(\theta))\mathrm{IF}(x; T, F_\theta)$$

$$= \frac{\partial}{\partial \theta}\left[d(\xi(\theta))\right]\mathrm{IF}_{\mathrm{test}}(x; T, F_\theta). \qquad \square$$

Remark 1. This theorem shows that IF_L and $\mathrm{IF}_{\mathrm{test}}$ are proportional. Therefore, they have the same qualitative behavior, as far as boundedness and continuity properties are concerned (if d' is continuous). Note that this theorem is a variant of Remark 3 of Subsection 3.2a, because the P-value is a "standardized" test statistic.

Remark 2. From the above theorem we obtain

$$\int \mathrm{IF}_L(x; \{P_n\}, F_\theta)^2 \, dF_\theta(x) = c_L'(\theta)^2 \int \mathrm{IF}_{\mathrm{test}}(x; T, F_\theta)^2 \, dF_\theta(x).$$

If also $\int \mathrm{IF}_L(x; \{P_n\}, F_\theta)^2 \, dF_\theta(x) = \tau^2(\theta)$ (Lambert, 1981, p. 651) and $\lim_{\theta \to \theta_0} \int \mathrm{IF}_{\text{test}}(x; T, F_\theta)^2 \, dF_\theta(x) = \int \mathrm{IF}_{\text{test}}(x; T, F_{\theta_0})^2 \, dF_{\theta_0}(x)$, the latter expression being equal to $1/E(T, F_{\theta_0})$ by Property 2 of Subsection 3.2b, then we can rewrite the asymptotic efficacy E in terms of c_L and τ:

$$E(T, F_{\theta_0}) = \lim_{\theta \to \theta_0} \left(c'_L(\theta)/\tau(\theta) \right)^2. \qquad (3.6.7)$$

3.6b. Eplett's Approach

In his 1980 paper, Eplett constructs a kind of influence curve for two-sample rank tests. Let us give the main idea of his definition (in our notation). In the case of two-sample location rank tests (3.3.9) he calculates the derivative of the asymptotic power

$$\beta(T, F) = 1 - \Phi\left(\lambda_{1-\alpha} - \Delta\sqrt{E(T, F)} \right) \qquad (3.6.8)$$

which is defined in Noether (1955) for the sequence of alternatives $\theta_N = \Delta/\sqrt{N}$ (see also Subsection 3.2b). Here, Φ is the standard normal cdf with density $\Phi' = \phi$, and $\lambda_{1-\alpha} = \Phi^{-1}(1 - \alpha)$. We shall call this derivative the *asymptotic power function* (*APF*), given by

$$\frac{\partial}{\partial t} \left[\beta(T, (1 - t)F + tG) \right]_{t=0} = \int \mathrm{APF}(x; T, F) \, dG(x). \qquad (3.6.9)$$

Applying the chain rule, we immediately find that

$$\mathrm{APF}(x; T, F) = -\tfrac{1}{2}\phi\left(\lambda_{1-\alpha} - \Delta\sqrt{E(T, F)} \right) \Delta E(T, F)^{3/2} \mathrm{CVF}_{\text{test}}(x; T, F)$$

$$(3.6.10)$$

so the APF is directly proportional to the change-of-efficacy function of Section 3.5 (the factors do not depend on x). In order to obtain a more simple expression, Eplett then calculates the derivative of

$$\tilde{\beta}(T, F) := \left[E(T, F)/(\lambda(1 - \lambda)) \right]^{1/2}. \qquad (3.6.11)$$

The resulting function (which is proportional to the APF) he calls the influence curve of the two-sample rank test, and throughout the paper it is treated as an analogue to the IF of an estimator. However, this function equals

$$-\tfrac{1}{2}\mathrm{CVF}_{\text{test}}(x; T, F)E(T, F)^{3/2}\big/(\lambda(1 - \lambda))^{1/2}, \qquad (3.6.12)$$

so it corresponds to the $\mathrm{CVF}_{\text{test}}$ and not to the $\mathrm{IF}_{\text{test}}$. The distinction is crucial, also in view of the different interpretations of large positive and large negative values (see the discussion preceding Definition 2 of Subsection 2.5a).

*3.7. M-TESTS FOR A SIMPLE ALTERNATIVE

Let us now discuss the case of testing against a simple alternative. That is to say, $H_0 = \{F_{\theta_0}\}$ is the null hypothesis and $H_1 = \{F_{\theta_1}\}$ is the alternative. The classical (nonrobust) test for this case is the likelihood ratio test (LRT), given by the Neyman–Pearson lemma. It is an M-test of type (3.3.2) with

$$\psi(x) = \ln\left(f_{\theta_1}(x)/f_{\theta_0}(x)\right) - \int \ln\left(f_{\theta_1}/f_{\theta_0}\right) dF_{\theta_0} \qquad (3.7.1)$$

so its $\mathrm{IF}_{\text{test}}$ can be computed from (3.3.3).

The censored likelihood ratio test (CLRT), invented by Huber (1965) as a robust alternative to the LRT, is also an M-test with

$$\psi(x) = \left[\ln\left(f_{\theta_1}(x)/f_{\theta_0}(x)\right)\right]_{k_2}^{k_1} - \int \left[\ln\left(f_{\theta_1}/f_{\theta_0}\right)\right]_{k_2}^{k_1} dF_{\theta_0}, \qquad (3.7.2)$$

where the notation means truncation at k_1 and k_2. Huber found this CLRT as the solution of a minimax problem. Let $\mathscr{P}_j$ be a neighborhood of F_{θ_j}, for $j = 0, 1$. Consider the following problem:

For a given $0 < \alpha < 1$ find a test which maximizes $\inf\{$ expected fraction of rejections at H; $H \in \mathscr{P}_1\}$, under the side condition $\sup$ $\{$ expected fraction of rejections at H; $H \in \mathscr{P}_0\} \leq \alpha$

$$(3.7.3)$$

The CLRT solves this problem if the $\mathscr{P}_j$'s are either ε-contamination (2.7.1), Prohorov neighborhoods (2.2.1), or total variation neighborhoods. From this exact finite-sample minimax result, Huber (1968) derived a corresponding approach for location estimators and showed that the Huber estimator is minimax in a finite-sample sense (for a normal model).

Note that the CLRT defined by (3.7.3) is the likelihood ratio test between a least favorable pair of distributions $Q_j \in \mathscr{P}_j$, $j = 0, 1$. The existence of such a least favorable pair can be studied in a general framework using 2-alternating capacities (Huber and Strassen, 1973; Bednarski, 1980, 1982).

It is difficult to determine the maximum level and the minimum power (within the neighborhoods) of the censored likelihood ratio test. Asymptotic values were given by Huber-Carol (1970) who obtained a nontrivial limiting level and power using the technique of *shrinking neighborhoods*. Rieder (1978) then extended this method to construct an asymptotic testing model by defining neighborhoods in terms of ε-contamination and total variation. In subsequent work he derived estimates from tests (Rieder, 1980) and studied the robustness properties of rank tests (Rieder, 1981). Rieder's asymptotic model was extended by Wang (1981) to the case where there are nuisance parameters.

Let us go into some detail. If the sample size grows in the setup introduced above, the power tends to 1 (as long as $\mathcal{P}_0$ and $\mathcal{P}_1$ are disjoint). As in Subsection 3.2b, the testing problem must get harder as n grows. Let us therefore consider the sequence of problems of testing $\theta = \theta_0$ against $\theta = \theta_0 + \Delta/\sqrt{n}$, and let the radius of $\mathcal{P}_{n0}$ and $\mathcal{P}_{n1}$ (defined as above) shrink at the same rate. The sequence of problems is then solved by a sequence of M-tests, given by

$$\psi^{(n)}(x) = \left[\frac{\sqrt{n}}{\Delta} \ln\left(f_{\theta_n}(x)/f_{\theta_0}(x) \right) \right]_{k_1(n)}^{k_2(n)}$$

$$- \int \left[\frac{\sqrt{n}}{\Delta} \ln\left(f_{\theta_n}/f_{\theta_0} \right) \right]_{k_1(n)}^{k_2(n)} dF_{\theta_0}$$

which tends to

$$\psi^*(x) = \left[\frac{\partial}{\partial \theta} [f_\theta(x)]_{\theta_0}/f_{\theta_0}(x) \right]_{k_1^*}^{k_2^*} - \int \left[\frac{\partial}{\partial \theta} [f_\theta]_{\theta_0}/f_{\theta_0} \right]_{k_1^*}^{k_2^*} dF_{\theta_0}(x)$$

$$= \left[\left(\frac{\partial}{\partial \theta} [f_\theta(x)]_{\theta_0}/f_{\theta_0}(x) \right) - a^* \right]_{d^*}^{c^*}. \qquad (3.7.4)$$

This expression is not a ψ-function of a CLRT any more, but a limiting case. Properties 1 and 2 of Subsection 3.2b apply with $\mathrm{IF}_{\mathrm{test}}(x; \psi^*, F_{\theta_0})$ given by (3.3.3). The same holds for the LRT, with

$$\psi^*(x) = \frac{\partial}{\partial \theta} [f_\theta(x)]_{\theta_0}/f_{\theta_0}(x). \qquad (3.7.5)$$

We see that here $\mathrm{IF}(x; \psi^*, F_{\theta_0})$ is proportional to $(\partial/\partial\theta)[\ln f_\theta(x)]_{\theta_0}$ so the LRT is asymptotically efficient by Property 3 of Subsection 3.2b, but often the gross-error sensitivity γ_{test}^* of this test is infinite, pointing to a lack of robustness.

Let us now apply Theorem 1 of Section 2.4, where ψ is replaced by ψ^*. The *optimal B-robust* M-test (i.e., the M-test with maximal asymptotic power subject to $\psi_{\text{test}}^* \leq c$) is given by

$$\tilde{\psi}^*(x) = \left[\left(\frac{\partial}{\partial \theta} \left[f_\theta(x) \right]_{\theta_0} / f_{\theta_0}(x) \right) - a \right]_{-b}^{b}, \tag{3.7.6}$$

which corresponds to a censored likelihood ratio test (3.7.4).

EXERCISES AND PROBLEMS

Section 3.1

1. Show that the Bartlett test is a generalization of the F-test by proving that they give the same result when $k = 2$. (*Hint*: Prove that $T > \frac{1}{2}(c^{1/2} + c^{-1/2})$ if and only if $T > c$ or $T < 1/c$.)

Subsection 3.2a

2. Compute the test influence function (Definition 1) of the z-test in the location model $\mathcal{N}(\theta, 1)$, at $\theta_0 = 0$. Also compute the IF of the χ^2-test in the scale model $\mathcal{N}(0, \sigma)$, at $\sigma_0 = 1$. Which test is more easily affected by outliers?

3. Compute the test influence function (Definition 2) of the z-test for location shift between two samples, at $F = \Phi$. Compute the IF of the F-test for comparing two variances, also at $F = \Phi$.

Subsection 3.2b

4. Verify Property 1 for the test influence functions of Exercise 2. Compute the asymptotic efficacy and the absolute asymptotic efficiency of these tests (at $F = \Phi$) by means of their IF_{test}.

5. Verify Properties 4 and 5 for the test influence functions of Exercise 3, and compute the asymptotic efficiency of these tests (at $F = \Phi$) by means of (3.2.10).

Subsection 3.2c

6. Because $\gamma_{test}^* = \infty$ for the two-sample F-test, what do you think would happen to the maximal asymptotic levels in Table 1 of Section 3.1 if asymmetric gross-error contamination were allowed?

7. Differentiate the equations of (3.2.17) with respect to ε, and make first-order Taylor expansions around $\varepsilon = 0$ [i.e., $g(\varepsilon) \cong g(0) + \varepsilon g'(0)$] for both. Do you recognize the resulting expressions?

8. Use the values in Tables 1 and 2 to approximate the maximal *finite-sample* level of the sign test and the Wilcoxon test (with nominal level 5%) when there are 1, 5, or 10 (possibly very asymmetric) outliers in a sample of 100 observations. Do the same for the minimal finite-sample power at $\theta = 0.3$, when testing $\theta = 0$ at the normal model $\mathcal{N}(\theta, 1)$.

9. Suppose you know that $IF_{test,1}(x; T, \Phi)$ is equal to (3.2.19) for the two-sample median test for shift and to (3.2.20) for the two-sample Wilcoxon test. What can you say about their asymptotic efficiency, maximal level, and minimal power?

Subsection 3.2d

10. Which location estimator is obtained from the one-sample z-test, using Definition 3? Which scale estimator is obtained from the one-sample χ^2-test for testing $\sigma = \sigma_0$? Are these functionals Fisher consistent at the normal model? Applying Property 7 to both.

11. Repeat the reasoning of Subsection 3.2d for two-sample estimators $D_{m,n}(X_1, \ldots, X_m; Y_1, \ldots, Y_n)$ of location shift, defined (similar to Definition 3) as the shift for which a two-sample test sees no difference. Also do the same for estimators of the ratio of scales σ_x/σ_y based on two-sample tests.

12. Which two-sample estimators are obtained (using the previous exercise) from the z-test for shift and the F-test for comparing two variances? Give their IF by means of the analog to Property 7, or directly. Note that these IF's are functions of two variables!

Subsection 3.3a

13. Show that the functionals (3.3.5) corresponding to rank statistics are not Fisher consistent (with a theoretical argument or example).

14. Derive the influence function (3.3.6) by means of Property 7.

Subsection 3.3b

15. Derive the influence function (3.3.11) by means of the two-sample analog to Property 7 (i.e., use the corresponding two-sample shift estimator).

16. (Extensive) Make plots of $IF_{test,1}(x; T, \Phi)$ for the quartile, Ansari–Bradley, Mood, and Klotz tests of scale. Use the influence functions to compute their asymptotic efficiencies at Φ. (For the second and third test, this leads to nontrivial integrals.)

Section 3.4

17. Construct the table with maximal asymptotic level and minimal asymptotic power, corresponding to Tables 1 and 2 of Subsection 3.2c, for the truncated normal scores test (3.4.4) with $b = 1.5$. (*Hint*: The necessary constants for computing E and γ^*_{test} can be found in Table 1 of Subsection 2.5d.) Does the truncated normal scores test behave rather like the sign test or like the Wilcoxon test, from this point of view?

18. Plot the robustified Klotz scores (3.4.7) and the corresponding $IF_{test,1}(x; T, \Phi)$ for a large and a small value of b. Describe what happens if $b \uparrow \infty$ and if $b \downarrow 0$.

Subsection 3.5

19. Make a plot of the change-of-efficacy function of the Wilcoxon test, and compare it to that of the Van der Waerden–Van Eeden test.

20. Derive (3.5.7) in analogy with (2.7.3), and use it to obtain (3.5.8).

21. Compare the exact minimal asymptotic power with the approximate minimal asymptotic power, both given in (3.5.8), for the sign test, Wilcoxon test, and the truncated normal scores test with $b = 1.4$, all in the case of *symmetric* gross-error contamination with $\varepsilon = 0.01$, 0.05, 0.10. (*Hint*: Use the entries of Table 1 of Section 2.7.) What happens for the Van der Waerden–Van Eeden test?

22. Consider the one-sample Wilcoxon test with nominal level $\alpha = 5\%$. What is the minimal asymptotic power of this test for $\Delta = 3.0$, when the underlying distribution F belongs to the *symmetric* gross-error neighborhood with $\varepsilon = 0.10$ around $F = \Phi$? Use this to approximate

the minimal *finite-sample* power at $\theta = 0.3$, when testing $\theta = 0$ in case there are at most ten (symmetrically located) outliers in a sample of 100 observations coming from a normal distribution $\mathcal{N}(\theta, 1)$. What is the (asymptotic or finite-sample) maximal level when F belongs to this symmetric gross-error neighborhood?

23. Sketch the graph of $-\ln\inf_{G \in \mathscr{P}_\varepsilon} E(T, G)$ as a function of ε and show how it contains κ^*, analogous to Figure 2 of Section 2.7. Draw this plot for the sign test, making use of the values in Exercise 76 of Chapter 2.

Subsection 3.6b

24. Compute the derivative with respect to ε (at $\varepsilon = 0$) of the approximation (3.5.8), and show that it is equal to the infimum of the asymptotic power function (3.6.10).

CHAPTER 4

Multidimensional Estimators

4.1. INTRODUCTION

The goal of this chapter is to show that the basic estimation principle used in the one-dimensional case—leading to the optimal estimator discussed in Section 2.4—applies to more general models with (finite-dimensional) vector-valued parameters. As was mentioned in Chapter 1, the resulting general estimation method may be viewed as the robustification of the maximum likelihood method, and therefore it bears a similar degree of generality.

After generalizing the concepts of influence function, sensitivities, and M-estimators in Section 4.2, we introduce the optimal B-robust estimators and give a step-by-step description of their application to specific models. Sections 4.4 and 4.5 bring up some general considerations about nuisance parameters and invariance structures. The complementary section includes a theoretical comment and discusses the problem how M-estimates can be computed practically.

As an example, the problem of estimating a location and a scale parameter simultaneously is discussed first. We will also consider a simple regression model with particular distributional assumptions and the gamma distribution with shape and scale parameter. The most important models—general multiple regression and covariance matrix estimation—are treated in the subsequent chapters. Stefanski et al. (1984) discuss optimal estimators for logistic regression, and Watson (1983) applies the method to circular data.

Testing problems have not been dealt with in this general multivariate framework. Other areas of research include the generalization of optimal redescending estimators and of the change-of-variance function. For the regression case, some results on these problems will be presented in Chapters 6 and 7.

4.2. CONCEPTS

4.2a. Influence Function

Let the setup of Section 2.1a be generalized to include observations in an arbitrary space $\mathscr{X}$ and vector-valued parameters, so that $\Theta \subset \mathbb{R}^p$, say. (Vectors $\theta \in \Theta$ will be denoted as column vectors.) Consider functionals T defined on a suitable subset of the set of probability measures on $\mathscr{X}$, taking values in Θ.

The influence function is defined exactly as in the one-dimensional case:

Definition 1. The (p-dimensional) *influence function* of a functional T at a distribution F is given by

$$\mathrm{IF}(x; T, F) := \lim_{h \downarrow 0} \left\{ \left(T\left[(1 - h)F + h\Delta_x \right] - T[F] \right) / h \right\}.$$

Under mild regularity conditions, we again obtain

$$\int \mathrm{IF}(x; T, F) \, dF(x) = 0, \tag{4.2.1}$$

where 0 is the p-dimensional null vector. More stringent conditions are needed to ensure that T is asymptotically normal with the (variance–) covariance matrix

$$V(T, F) := \int \mathrm{IF}(x; T, F) \mathrm{IF}(x; T, F)^T \, dF(x). \tag{4.2.2}$$

Now let $\{ F_\theta \}_\Theta$ be a parametric model with densities $f_\theta(x)$, and denote by

$$s(x, \theta) := \frac{\partial}{\partial \theta} \ln f_\theta(x) := \left(\frac{\partial}{\partial \theta^{(1)}} \ln f_\theta(x), \dots, \frac{\partial}{\partial \theta^{(p)}} \ln f_\theta(x) \right)^T$$

$$\tag{4.2.3}$$

the vector of likelihood scores, and by

$$J(\theta) := \int s(x,\theta)s(x,\theta)^T dF_\theta(x) \qquad (4.2.4)$$

the Fisher information matrix.

Theorem 1 (Asymptotic Cramér–Rao inequality). For Fisher-consistent estimators $(T(F_\theta) \equiv \theta)$,

$$d^T V(T, F_\theta)d \ge d^T J(\theta)^{-1}d \quad \text{for all } d \in \mathbb{R}^p$$

under suitable regularity conditions (see proof).

Proof (cf. Subsection 2.1b). Write, for fixed θ,

$$T(F_{\tilde\theta}) - T(F_\theta) = \int \text{IF}(x; T, F_\theta) f_{\tilde\theta}(x)\, dx + R(\tilde\theta - \theta),$$

where R is a remainder term. Regularity conditions are needed to ensure that $R(t) = o(t)$. Assuming Fisher consistency and differentiating with respect to $\tilde\theta$ at $\tilde\theta = \theta$, we get

$$\int \text{IF}(x; T, F_\theta)s(x,\theta)^T dF_\theta(x) = I, \qquad (4.2.5)$$

corresponding to (2.1.10) in the one-dimensional case, if differentiation and integration can be interchanged (which needs another weak regularity condition). Now, let $u(x) := [\text{IF}(x; T, F_\theta)^T, s(x,\theta)^T]^T$. The "covariance matrix" of u,

$$\int u(x)u(x)^T dF_\theta(x) = \begin{bmatrix} V(T, F_\theta) & I \\ I & J(\theta) \end{bmatrix},$$

is positive semidefinite. Let $d \in \mathbb{R}^p$ be arbitrary and $\tilde{d} := [d^T, (-J(\theta)^{-1}d)^T]^T$. Then

$$0 \le \tilde{d}^T \begin{bmatrix} V(T, F_\theta) & I \\ I & J(\theta) \end{bmatrix} \tilde{d} = d^T V(T, F_\theta)d - 2d^T J(\theta)^{-1}d + d^T J(\theta)^{-1}d$$

$$= d^T V(T, F_\theta)d - d^T J(\theta)^{-1}d. \qquad \square$$

4.2b.   Gross-Error Sensitivities

There are several ways to generalize of the gross-error sensitivity, the most direct one being

Definition 2.   The (unstandardized) *gross-error sensitivity* of an estimator (functional) T at a distribution F is

$$\gamma_u^*(T, F) := \sup_x \{ \| \mathrm{IF}(x; T, F) \| \},$$

where $\| \cdot \|$ denotes the Euclidean norm.

In many problems, the parametrization of the model distributions is to some extent arbitrary. In multiple regression, for example, the scales of the independent variables are most often a matter of arbitrary choice, and the scales of the parameters adjust to this choice. It is natural to postulate that a measure of robustness of parameter estimators should be invariant to scale transformations of individual parameter components at least—but the sensitivity defined above is not. Two possible standardizations overcome this defect.

The first one consists in measuring the IF, which is an asymptotic bias, in the metric given by the asymptotic covariance matrix of the estimator.

Definition 3.   If $V(T, F)$ exists, the *self-standardized sensitivity* is defined by

$$\gamma_s^*(T, F) := \sup_x \{ \mathrm{IF}(x; T, F)^T V(T, F)^{-1} \mathrm{IF}(x; T, F) \}^{1/2}$$

if $V(T, F)$ is nonsingular, else by ∞.

Lemma 1.   The squared γ_s^* is not less than the dimension of the parameter,

$$\gamma_s^*(T, F)^2 \geq p.$$

Proof.   Integrate the inequality $\gamma_s^{*2} \geq \mathrm{IF}^T V^{-1} \mathrm{IF} = \mathrm{trace}(V^{-1} \mathrm{IF} \cdot \mathrm{IF}^T)$ with respect to dF. $\square$

We shall see later on (Subsection 4.3c) that this bound is sharp.

Obviously, poor efficiency can be the reason for a low γ_s^* value. We will show, however, that it is possible to obtain quite efficient estimators with a low self-standardized sensitivity.

The second standardization is connected with a given model $\{F_\theta\}_\Theta$, which determines a natural "local metric" for the parameter space through the Fisher information:

Definition 4. If $J(\theta)$ exists (for all θ) the *information-standardized sensitivity* is given by

$$\gamma_i^*(T, F) := \sup\left\{ \mathrm{IF}(x; T, F)^T J[T(F)]\mathrm{IF}(x; T, F) \right\}^{1/2}.$$

γ_i^* compares the bias (as measured by the IF) with the scatter of the maximum likelihood estimator at the estimated model distribution instead of that of T itself at F.

We now verify the *invariance* of the standardized sensitivities with respect to differentiable one-to-one parameter transformations $\bar{\theta} = \beta(\theta)$ with nonsingular Jacobian

$$B(\theta) := \partial\beta(\theta)/\partial\theta := \begin{bmatrix} \partial\beta^{(1)}(\theta)/\partial\theta^{(1)} & \cdots & \partial\beta^{(1)}(\theta)/\partial\theta^{(p)} \\ \vdots & & \vdots \\ \partial\beta^{(p)}(\theta)/\partial\theta^{(1)} & \cdots & \partial\beta^{(p)}(\theta)/\partial\theta^{(p)} \end{bmatrix}.$$

$$(4.2.6)$$

The "transformed model" writes $\{\bar{F}_{\bar\theta}\}_{\bar\theta \in \bar\Theta}$ with $\bar{F}_{\bar\theta} := F_\theta$. The densities, scores, and information matrix transform to

$$\bar{f}_{\bar\theta}(x) = f_\theta(x),$$

$$\bar{s}(x, \bar\theta) = B(\theta)^{-T} \cdot s(x, \theta), \qquad (4.2.7)$$

$$\bar{J}(\bar\theta) = B(\theta)^{-T} \cdot J(\theta) \cdot B(\theta)^{-1}.$$

[A^{-T} stands for $(A^{-1})^T$.] If T is an estimator of θ, it is natural to examine $\bar{T} := \beta(T)$ as an estimator for the transformed parameter. We have

$$\mathrm{IF}(x; \bar{T}, F) = B(T(F)) \cdot \mathrm{IF}(x; T, F),$$

$$V(\bar{T}, F) = B(T(F)) \cdot V(T, F) \cdot B(T(F))^T. \qquad (4.2.8)$$

The invariance of the standardized sensitivities is easily verified by inserting these relations into the definitions.

Remark 1. For a fixed parameter value θ_*, the *information-standardized sensitivity* may be viewed as the unstandardized sensitivity in an "orthonormalized" parameter system, namely $\bar{\theta} = J^{1/2}(\theta_*) \cdot \theta$, where $J^{1/2}$ is any root of J, that is, any solution of $(J^{1/2})^T J^{1/2} = J$.

Examples for these concepts are given in the next two subsections.

4.2c. *M*-Estimators

M-estimators for the multiparameter case are defined just as those for a single parameter (Subsection 2.3a):

Definition 5. An *M-estimator* (functional) is defined through a function $\rho : \mathscr{X} \times \Theta \to \mathbb{R}$ as the value $T(F) \in \mathbb{R}^p$ minimizing $\int \rho(x, t) \, dF(x)$ over t, or through a function $\psi : \mathscr{X} \times \Theta \to \mathbb{R}^p$ as the solution for t of the vector equation

$$\int \psi(x, t) \, dF(x) = 0.$$

Note that a ψ-function may be left-multiplied by any nonsingular matrix not depending on x.

For the maximum likelihood estimator, $\rho(x, \theta) = -\ln\{f(x, \theta)\}$, and $\psi(x, \theta) = s(x, \theta)$.

Again, any M-estimator defined by a differentiable ρ-function corresponds to an M-estimator defined by a ψ-function, namely by $\psi(x, \theta) = \partial \rho(x, \theta)/\partial\theta$, as long as the latter is defined and unique. But the M-estimators defined by a ψ-function are now a much wider and thus more flexible class, since the $\psi^{(j)}(x, \theta)$ need not be the partial derivatives of any function of θ. We will therefore define M-estimators by ψ-functions in what follows.

The influence function is derived like Eq. (2.3.5),

$$\mathrm{IF}(x; T, F) = M(\psi, F)^{-1} \psi(x, T(F)) \tag{4.2.9}$$

with the $p \times p$ matrix M given by

$$M(\psi, F) := -\int \left[\frac{\partial}{\partial\theta} \psi(x, \theta) \right]_{T(F)} dF(x). \tag{4.2.10}$$

The asymptotic covariance matrix reads

$$V(T, F) = M(\psi, F)^{-1} \cdot Q(\psi, F) \cdot M(\psi, F)^{-T}, \qquad (4.2.11)$$

with

$$Q(\psi, F) := \int \psi(x, T(F)) \cdot \psi(x, T(F))^T \, dF(x).$$

(A basic reference for asymptotic normality of M-estimators is Huber, 1967; see also Huber 1981, Chapter 6.)

If we specialize to a *model* with scores, Fisher consistency obviously implies

$$\int \psi(x, \theta) \, dF_\theta(x) = 0 \quad \text{for all } \theta \in \Theta, \qquad (4.2.12)$$

and from (4.2.5) we get

$$M(\psi, F) = \int \psi(x, \theta) \cdot s(x, \theta)^T \, dF_\theta(x). \qquad (4.2.13)$$

This last result shows that the influence function for Fisher-consistent M-estimators at F_θ is determined by the $\psi(\cdot, \theta)$ alone, without any reference to other θ values, and without using derivatives of ψ. As in the one-dimensional case, this fact will be extremely useful for the construction of M-estimators with prespecified local robustness properties, since the ψ-function can be constructed separately for each value of θ.

In Sections 2.3 and 2.4 some *alternative* classes of *estimators* are also considered. These do not generalize as easily to multivariate models as M-estimators do. In any case, let us assume that such an alternative estimator T obeys some mild regularity conditions similar to those needed in the preceding subsection. Then, the M-estimator defined by $\psi(x, \theta) :=$ IF$(x; T, F_\theta)$ is asymptotically equivalent to it at the model distributions. For local considerations at the model (influence function, sensitivity, asymptotic covariance matrix), it therefore suffices to consider M-estimators. Only global properties, such as the breakdown point, as well as higher order local properties, such as a multiparameter analog to the change-of-variance function, will in general be different for a general estimator and the asymptotically equivalent M-estimator.

4.2d. Example: Location and Scale

Let the observation space be the real line, that is, $\mathcal{X} = \mathbb{R}$ and $m = 1$, and let Θ be the half plane,

$$\Theta = \left\{ \begin{pmatrix} \mu \\ \sigma \end{pmatrix} \middle| \sigma > 0 \right\}.$$

The location-scale model consists of the distributions

$$F_{\mu, \sigma} = \mathscr{L}(\mu + \sigma Z),$$

where Z is distributed according to a symmetric distribution F_0 with density f_0 and $\mathscr{L}$ stands for "distribution of". Note that μ and σ are not necessarily expectation and standard deviation.

We will restrict our attention to *equivariant estimators*:

$$T\{\mathscr{L}(a + bX)\} = \begin{pmatrix} a + b \cdot T^{(1)}\{\mathscr{L}(X)\} \\ |b| \cdot T^{(2)}\{\mathscr{L}(X)\} \end{pmatrix}, \quad \text{for all } a, b \in \mathbb{R}$$

Then, for any F_0 symmetric about 0, $T^{(1)}(F_0) = 0$, and, if $T^{(2)}(F_0) > 0$, the first component of the influence function is odd, whereas the second component is even. For symmetry reasons, V is diagonal,

$$V(T, F_0) = \text{diag}\left\{ V(T^{(1)}, F_0), V(T^{(2)}, F_0) \right\},$$

that is, the two components are asymptotically independent. For model distributions, we have

$$\text{IF}(x; T, F_{\mu, \sigma}) = \sigma \cdot \text{IF}\{(x - \mu)/\sigma; T, F_0\},$$

$$V(T, F_{\mu, \sigma}) = \sigma^2 \cdot V(T, F_0).$$

In summary, for symmetric F_0, asymptotic results can be obtained for the location and scale components independently. This makes things much easier than they are in the general case.

Any equivariant *M-estimator* may be given by its ψ-function at $\theta_0 :=$ $(0, 1)^T$ just by

$$\psi(x; \mu, \sigma) = \psi_0\{(x - \mu)/\sigma\} \quad \text{with} \quad \psi_0(-z) = \begin{pmatrix} -\psi_0^{(1)}(z) \\ \psi_0^{(2)}(z) \end{pmatrix}$$

(cf. Subsection 4.5b). The influence function may be written as

$$IF(z; T, F_0) = \begin{pmatrix} B_1(\psi_0^{(1)})^{-1} \cdot \psi_0^{(1)}(z) \\ B_2(\psi_0^{(2)})^{-1} \cdot \psi_0^{(2)}(z) \end{pmatrix}$$

with

$$B_j(\psi_0^{(j)}) := \int \psi_0^{(j)}(z) \cdot s_0^{(j)}(z) \, dF_0(z), \qquad j = 1, 2,$$

where $s_0(z)$ is the likelihood scores function at θ_0,

$$s_0(z) = \begin{pmatrix} -f_0'(z)/f_0(z) \\ -z \cdot f_0'(z)/f_0(z) - 1 \end{pmatrix}.$$

The asymptotic covariances vanish for symmetry reasons, and the asymptotic variances equal $V(T^{(j)}, F_0) = A_j(\psi_0^{(j)})/B_j(\psi_0^{(j)})^2$, where

$$A_j(\psi_0^{(j)}) := \int \psi_0^{(j)}(z)^2 \, dF_0(z), \qquad j = 1, 2.$$

Example 1. The famous couple *mean* $\overline{X}$ and sample *standard deviation* S are the maximum likelihood estimators for $F_0 = \Phi$ [up to a factor $(n - 1)/n$ for S^2]. They equal the M-estimator given by

$$\psi_0^{ML}(z) = \begin{pmatrix} z \\ z^2 - 1 \end{pmatrix}.$$

For $F_0 = \Phi$, we have

$$A_1(\overline{X}) = B_1(\overline{X}) = 1,$$

$$A_2(S) = B_2(S) = 2,$$

$$IF(z; T_{ML}, \Phi) = \begin{pmatrix} z \\ (z^2 - 1)/2 \end{pmatrix},$$

$$V(T_{ML}, \Phi) = J(\theta_0)^{-1} = \text{diag}(1, \tfrac{1}{2}).$$

As usual, the maximum likelihood estimator for the normal model has an unbounded influence function, and we would like to get a robust variant of it.

Example 2. In analogy to the Huber estimator for the location problem [see (2.3.15) and "Location" in Subsection 2.4b], a natural idea is to "bring in" the observations which are too far from the estimated location, where the critical distance is some multiple of the estimated scale parameter, and to determine these estimates implicitly by the requirement that they should be the classical estimators taken from the transformed observations. More precisely, this estimator is given by the ψ_0-function

$$\psi_{0,b}^{P2}(z) := \begin{pmatrix} \psi_b(z) \\ \psi_b(z)^2 - \beta \end{pmatrix}$$

with $\beta = 1$. It is easy to see that this results in an underestimation of σ^2 for standard normal observations. Therefore, Huber (1981, p. 137; 1964, p. 96) sets

$$\beta = \int \psi_b(z)^2 \, d\Phi(z) = 2\Phi(b) - 1 - 2b\varphi(b) + 2b^2(1 - \Phi(b))$$

$$= \chi_3^2(b^2) + b^2 \left[1 - \chi_1^2(b^2) \right]$$

in his "*Proposal 2*" in order to get Fisher consistency. (χ_k^2 is the Chi-squared distribution function with k degrees of freedom.) Figure 1 compares the transformations $z \mapsto \psi_0^{\mathrm{ML}}(z)$ and $z \mapsto \psi_{0,b}^{P2}(z)$.

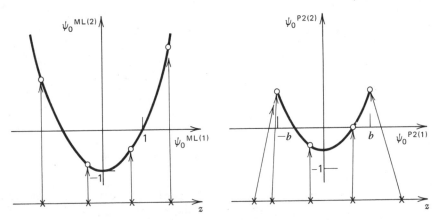

Figure 1. The functions ψ_0^{ML} and $\psi_{0,1.2}^{P2}$.

The integrals needed for IF and V are

$$A_1(\psi_b) = \int \psi_b(z)^2 \, d\Phi(z) = \beta,$$

$$B_1(\psi_b) = \int \psi_b(z) \, z \, d\Phi(z) = 2\Phi(b) - 1 = \chi_1^2(b^2),$$

$$A_2(\psi_b^2 - \beta) = \int \psi_b(z)^4 \, d\Phi(z) - \beta^2 = 3 \cdot \chi_5^2(b^2) + 1 - \chi_1^2(b^2) - \beta^2,$$

$$B_2(\psi_b^2 - \beta) = \int \psi_b(z)^2 z^2 \, d\Phi(z) - \beta = 3 \cdot \chi_5^2(b^2) + 1 - \chi_3^2(b^2) - \beta.$$

(Two useful formulas in this connection are

$$\int_{\|z\|^2 \le b^2} \|z\|^{2k} \, d\Phi_p(z) = p \cdot (p + 2) \cdot \cdots \cdot [p + 2(k - 1)] \cdot \chi_{p+2k}^2(b^2),$$

$$(4.2.14)$$

where Φ_p is the p-variate standard normal distribution, and

$$\chi_{p+2}^2(b^2) = \chi_p^2(b^2) - 2b^2 \cdot f_p(b^2)/p, \qquad (4.2.15)$$

where f_p is the density of χ_p^2.)

The sensitivities are clearly given by the influence vector for $z = b$ or $z = 0$:

$$\gamma^*(T_{P2}, \Phi)^2 = \max\left\{ c_\mu^2 \cdot b^2 + c_\sigma^2 \cdot (b^2 - \beta)^2, c_\sigma^2 \cdot \beta^2 \right\}$$

with

for γ_u^*: $c_\mu^2 = B_1(\psi_b)^{-2}$, $c_\sigma^2 = B_2(\psi_b^2 - \beta)^{-2}$;

for γ_s^*: $c_\mu^2 = A_1(\psi_b)^{-1}$, $c_\sigma^2 = A_2(\psi_b^2 - \beta)^{-1}$;

for γ_i^*: $c_\mu^2 = B_1(\psi_b)^{-2}$, $c_\sigma^2 = 2 \cdot B_2(\psi_b^2 - \beta)^{-2}$.

Example 3. *Median* and *median deviation* make another instructive example of a location and scale estimator pair. They can be defined by

$$\psi_0^{\text{Med}}(z) = \begin{pmatrix} \text{sign}(z) \\ \text{sign}(|z| - \beta) \end{pmatrix},$$

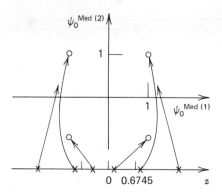

Figure 2. The function ψ_0^{Med}.

where $\beta = \Phi^{-1}(3/4) = 0.6745$ is introduced again to obtain Fisher consistency [cf. (2.3.20)]. Figure 2 sketches the transformation $z \mapsto \psi_0^{\mathrm{Med}}(z)$. The calculations

$$A_1(\mathrm{sign}) = A_2\left(\psi_0^{\mathrm{Med}(2)}\right) = 1,$$

$$B_1(\mathrm{sign}) = \int |z|\, d\Phi(z) = 2\varphi(0) = 0.798,$$

$$B_2\left(\psi_0^{\mathrm{Med}(2)}\right) = 4\beta \cdot \varphi(\beta) = 0.857$$

yield IF and the sensitivities

$$\gamma_u^* = 1.71, \qquad \gamma_s^* = 2, \qquad \gamma_i^* = 2.07.$$

The asymptotic efficiencies become $B_1(\mathrm{sign})^2 = 0.637$ for location, and $0.5 \cdot B_2(\psi_0^{\mathrm{Med}(2)})^2 = 0.368$ for scale.

Remark 2. In many applications, the *scale* is regarded as a *nuisance parameter*. Many location estimators need a scale value in order to be sensibly defined, and this value usually has to be estimated. In Section 4.4, we shall discuss more generally how to treat nuisance parameters.

Example 4. When scale is a nuisance parameter, it makes sense to estimate it as robustly as possible, even if quite inefficiently (cf. "Location" in Subsection 2.3b). The simplest scale estimator with breakdown point $\varepsilon^* = \frac{1}{2}$ is the rescaled MAD used in the last example, with the rather poor efficiency of 37%. It can be combined with a highly efficient estimator of location, like the Huber estimator with high bending point b.

Table 1.   Four Different Estimates of Location and Scale for Three Sets of Data

| | Data: 0.0, 0.8, 1.0, 1.2, 1.3, 1.3, 1.4, 1.8, and | | | | | |
| | 2.4, 4.6 | | 4.6, 4.6 | | 24.0, 46.0 | |
Estimator	Location	Scale	Location	Scale	Location	Scale
Mean and standard deviation (Ex. 1)	1.58	1.23	1.80	1.55	7.88	15.21
Huber's Proposal 2, $b = 2$ (Ex. 3)	1.46	0.95	1.80	1.61	6.10	11.04
Median and rescaled MAD[a] (Ex. 2)	1.30	0.74	1.30	0.74	1.30	0.74
Huber estimator, $b = 2$, with rescaled MAD[a] (Ex. 4)	1.41	0.74	1.47	0.74	1.47	0.74

[a] Defined by upper median of absolute residuals.

The values obtained by applying the estimators discussed in the four examples to Cushny and Peebles' data (Fig. 1 of Section 2.0) are given in the first pair of columns of Table 1. The location value for Huber's Proposal 2, with $b = 2$, is placed halfway between the bulk of the robust estimates shown in Figure 1b of Section 2.0 and the nonrobust arithmetic mean. When two observations are made into outliers (last two column pairs of Table 1), Proposal 2 follows the nonrobust classical estimate. In other words, it breaks down. This is clear from Huber (1981, Section 6.6), since the theoretical breakdown point for $b = 2$ is $\varepsilon^* = 0.19$. As a side remark, note that the scale estimate of Proposal 2 is even higher than the classical standard deviation for the second dataset because of the consistency correction. In summary, the Huber estimator with MAD scaling is more reliable than Proposal 2 if scale is regarded merely as a nuisance parameter.

Remark 3.   We found that the use of artificial, small data set like those shown in Table 1 is generally more instructive than Monte Carlo studies for exploring robustness properties.

Remark 4.   In this subsection, we have made heavy use of the *invariance* structure of the model and the considered estimators. The implications of such structures, which are present in most of the commonly used statistical models, will be discussed in a general setup in Section 4.5.

4.3. OPTIMAL ESTIMATORS

4.3a. The Unstandardized Case

Consider the problem of finding the most efficient estimator under the condition that the gross-error sensitivity must not exceed a given bound c. For the one-dimensional case, the problem was solved in Section 2.4. In the multiparameter case, the term *"efficient"* needs clarification. The usual partial ordering of variance–covariance matrices,

$$U \leq V \Leftrightarrow V - U \text{ positive semidefinite,} \qquad (4.3.1)$$

suggests that one should look for an estimator minimizing V in that sense. If the sensitivity is not restricted, the maximum likelihood estimator solves the problem. However, in the case of a restriction such an estimator does not exist, as will become clear later (Subsection 4.6a). On the other hand, if instead we measure efficiency by the trace of V, a solution can be found for the problem.

More precisely, consider a parametric model $(F_\theta)_\Theta$. θ_* shall be a fixed parameter value, and F_* the corresponding distribution F_{θ_*}. We restrict our attention to the class of "regular" functionals

$$\mathcal{T}_*^r := \{ T | \mathrm{IF}(.; T, F_*) \text{ exists, } (4.2.1 + 5) \text{ are satisfied,}$$

$$\text{and } V(T, F_*) \text{ exists} \}.$$

(Here, existence of V means that the integral $\int \mathrm{IF} \cdot \mathrm{IF}^T dF_*$ exists. Alternatively, one could require that T is asymptotically normal with covariance matrix $V = \int \mathrm{IF} \cdot \mathrm{IF}^T dF_*$.) Among the estimators with influence function bounded by c, that is, within

$$\mathcal{T}_{c*}^u := \{ T \in \mathcal{T}_*^r | \gamma_u^*(T, F_*) \leq c \},$$

we want to find one that minimizes trace$\{V(\cdot, F_*)\}$. Since the solution is a cornerstone to large parts of the book, we shall discuss this problem in a *heuristic* fashion before giving it a formal treatment.

If there was no restriction on the sensitivity, it is well known that the maximum likelihood estimator T_{ML} would solve the problem in general. Its influence function is, according to (4.2.9) and (4.2.13),

$$\mathrm{IF}(x; T_{\mathrm{ML}}, F_*) = J(\theta_*)^{-1} \cdot s(x, \theta_*),$$

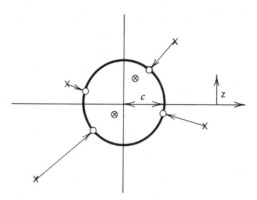

Figure 1. Sketch of the Huber function $z \mapsto h_c(z)$.

and is unbounded in many cases. Now, the solution for the one-dimensional problem suggests that we construct an M-estimator with a bounded ψ-function that is as similar as possible to the ψ-function $\psi_0(x, \theta_*) := s(x, \theta_*)$ of T_{ML} or any matrix multiple of it. In fact, the most meaningful version of the ψ-function defining an M-estimator is its influence function. Let us therefore examine $\bar{\psi}_0(x, \theta_*) := \mathrm{IF}(x; T_{\mathrm{ML}}, F_*) = J(\theta_*)^{-1}\psi_0(x, \theta_*)$. A modification of $\bar{\psi}_0(\cdot, \theta_*)$ is needed where its norm exceeds the bound c. In a first step, we can try to "cut down the influence" outside the hypersphere with radius c, still modifying it as little as possible. The function that transforms each point outside a hypersphere to its nearest point on it and leaves those inside alone is called the multidimensional *Huber function*,

$$h_c(z) := z \min(1, c/\|z\|), \qquad z \in \mathbb{R}^p \qquad (4.3.2)$$

(see Fig. 1).

Thus we consider an M-estimator given by a ψ-function that equals $h_c(\bar{\psi}_0(x, \theta_*))$ at $\theta = \theta_*$ as a candidate solution. But unfortunately, the modification has two undesirable side effects. First, Eq. (4.2.12) for Fisher consistency may no longer hold. This defect can be amended by shifting this new ψ-function, and we now consider

$$\psi_1(x, \theta_*) := h_c\left\{ J(\theta_*)^{-1}\psi_0(x, \theta_*)\right\} - a_1$$

with $a_1 := \int h_c\{\ldots\}\, dF_*(x)$. Second, however, ψ_1 does not coincide with the influence function of the respective estimator, since $M(\psi_1, F_*) \neq I$. This means that $\gamma^*(T_{\psi_1}, F_*)$ may be (and in fact is) greater than c. Remembering that $M(\bar{\psi}_0, F_*) = I$, and that we have not modified $\bar{\psi}_0$ a great deal, we can still hope that $M(\psi_1, F_*) \approx I$, and that we have therefore taken a step in the right direction. In a second step we might apply the same procedure to ψ_1, obtaining

$$h_c\{M(\psi_1, F_*)^{-1}\psi_1(x, \ell_*)\} - a_2$$

$$= h_c\{M(\psi_1, F_*)^{-1}(h_c[J(\theta_*)^{-1}\psi_0(x, \theta_*)] - a_1)\} - a_2.$$

The reader may convince himself or herself that the inner h_c function might modify ψ_0 in places where it does no longer seem necessary in order to get the influence function below its bound c. It should therefore be dropped. This results in

$$\psi_2(x, \theta_*) = h_c\{A_2[\psi_0(x, \theta_*) - \tilde{a}_1]\} - a_2$$

with suitable A_2 and $\tilde{a}_1$. Continuing this way, we can hope to eventually obtain a fixed point ψ_∞, which indeed defines the estimator for which we are looking. In fact, this procedure describes an algorithm for obtaining the solution (see Subsection 4.3d, step 4). Examining the way in which the limit $\psi_\infty =: \psi_c^u$ was constructed, the *solution* can be described more directly by the fixed point conditions as follows.

Consider a fixed θ_* and c. Let A be a $p \times p$ nonsingular matrix, and let a be a vector of dimension p. Then define

$$\psi_c^{A,a}(x) := h_c\{A[s(x, \theta_*) - a]\}$$

[where h_c is the Huber function given by (4.3.2)]. Find, if possible, A^* and a^* such that

$$\int \psi_c^{A^*, a^*}(x) s(x, \theta_*)^T dF_*(x) = I, \tag{4.3.3}$$

$$\int \psi_c^{A^*, a^*}(x) dF_*(x) = 0, \tag{4.3.4}$$

and let

$$\psi_c^u(x, \theta_*) := \psi_c^{A^*, a^*}(x).$$

If the construction is possible for all θ_*, ψ_c^u is said to be the ψ-function of the optimal unstandardized B-robust estimator T_c^u. (The superscript u of $\mathcal{T}_c^u$ and T_c^u is a mnemonic for "unstandardized".) Equation (4.3.3) may be changed into

$$A^* \cdot \int \left[s(x, \theta_*) - a^* \right] \left[s(x, \theta_*) - a^* \right]^T$$

$$\cdot \min \left\{ 1, c/\| A^* \left[s(x, \theta_*) - a^* \right] \| \right\} dF_*(x) = I$$

if (4.3.4) is valid. This shows that A^* must be symmetric.

The second equation, (4.3.4), ensures Fisher consistency, while the first one makes ψ_c^u coincide with the influence function of T_c^u by (4.2.9) and (4.2.13). Therefore, $\|\text{IF}(x; T_c^u, F_*)\| = \|h_c\{\ldots\}\| \leq c$ by the definition of h_c, and T_c^u will belong to $\mathcal{T}_{c*}^u$ under some regularity conditions.

These *conditions* seem difficult to treat in the general context; they should be ascertained for each specific model under consideration. Stahel (1981a, Sections A43 and B32) nevertheless gets some results for the general case. [Under regularity conditions for the model, a solution of (4.3.3) and (4.3.4) for fixed θ extends to a differentiable solution in a neighborhood, and $T_c^u \in \mathcal{T}_{c*}^u$.] Existence and uniqueness of A^* and a^* will be discussed in Subsection 4.3c. Note, however, that asymptotic normality should also be ascertained for the optimality property to be meaningful. Yet another problem of the kind is—even if the existence of a well-behaved ψ_c^u may be established—whether this leads to well-defined estimated values for given samples (or distributions). Even if all these problems have to be treated separately for specific models, the optimality question may be answered in the general context:

Theorem 1 (*Optimality of T_c^u*). If

    (i) $J(\theta_*)$ exists;
    (ii) A^* and a^*, defined above, exist;
    (iii) $T_c^u \in \mathcal{T}_*^r$;
    (iv) $\text{IF}(x; T_c^u, F_*) = U \cdot \psi_c^u(x, \theta_*)$ for some matrix U,

then $T_c^u \in \mathcal{T}_{c*}^u$, and it minimizes trace$\{V(\cdot, F_*)\}$ over $\mathcal{T}_c^u$. For all $T \in \mathcal{T}_{c*}^u$ which attain this minimum, we have

$$\text{IF}(x; T, F_*) = \psi_c^u(x, \theta_*) \quad \text{a.s.}(F_*).$$

Remark 1. The minimized criterion may also be interpreted as the asymptotic mean-squared error,

$$n \cdot E_*\{\|T - \theta_*\|^2\} \to \int \|\mathrm{IF}(x; T, F_*)\|^2 \, dF_*(x) = \mathrm{trace}\{V(T, F_*)\}.$$

Remark 2. Let us reiterate that the *optimality* of the estimator T_c^u in $\mathcal{T}_{c*}^u$ is *weaker* than the one possessed by the maximum likelihood estimator in $\mathcal{T}_{\infty*}^u$; the latter minimizes the asymptotic variance of the projection $d^T T$ on all $d \in \mathbb{R}^p$ simultaneously. (It attains the minimum in the partial ordering of V-matrices.) We will come back to this point in the next subsection.

Remark 3. Condition (iv) is a very weak regularity condition. We do not know of any (formal) M-estimator with continuous ψ-function where it fails.

Proof (of Theorem 1) (cf. Krasker, 1977, Theorem 4, and, for $p = 1$, Section 2.4.). Let $G(x)$ be the influence function of any estimator in $\mathcal{T}_{c*}^u$ at F_*. Then G must satisfy [see (4.2.5) and (4.2.1)]

$$\int G(x)s(x, \theta_*)^T \, dF_*(x) = I,$$

$$\int G(x) \, dF_*(x) = 0,$$

$$\|G(x)\| \le c \quad \text{for all } x,$$

and $V(T, F_*)$ equals $\int G(x)G(x)^T \, dF_*(x)$. In fact, the problem is to find the function G that minimizes the trace of the latter integral subject to the three given constraints.

Now, let A and a be a fixed but arbitrary nonsingular matrix and vector, respectively. Instead of minimizing trace$\{\int GG^T \, dF_*\}$, we can minimize the trace of

$$\int \{G(x) - A[s(x, \theta_*) - a]\}\{G(x) - A[s(x, \theta_*) - a]\}^T \, dF_*(x)$$

$$= \int GG^T \, dF_* - A \int [s - a]G^T \, dF_* - \int G[s - a]^T \, dF_* A^T$$

$$+ A \int [s - a][s - a]^T \, dF_* A^T$$

$$= \int GG^T \, dF_* - A - A^T + A[J(\theta_*) + aa^T]A^T,$$

since the trace is an additive function. [The existence of the integrals follows from assumption (i). Note that the extra terms on the right-hand side are related to Lagrangian multipliers.] The trace of the left-hand side may be written as

$$\int \| G(x) - A[s(x,\theta_*) - a] \|^2 \, dF_*(x).$$

It is obvious that $h_c\{A[s(x,\theta_*) - a]\}$ minimizes the integrand pointwise, hence also the integral, among all G fulfilling the third side condition. But, in general, the two first constraints are violated by this G, and we need the arbitrariness of A and a to satisfy them. More explicitly, if $A = A^*$ and $a = a^*$, this function coincides with $\psi_c^u(x,\theta_*)$ and thus satisfies the first two constraints by construction of A^* and a^*. If the conditions (ii) to (iv) hold, the matrix U in (iv) is the identity, and the estimator defined by ψ_c^u is in $\mathcal{T}_{c_*}^u$. Then the above argument shows that $\psi_c^u(\cdot,\theta_*)$ is the unique G minimizing the criterion under the three side conditions, and the latter are contained in the definition of $\mathcal{T}_{c_*}^u$. □

4.3b. The Optimal B-Robust Estimators

The problem just solved could be reformulated restricting one of the *standardized sensitivities*. The solution is given in Section 4.6a. However, the standardization of the sensitivities does not fit together with the use of trace(V) as the criterion for optimization; if the influence function is measured in the metric given, for example, by the inverse information matrix for defining the sensitivity, the same metric should be used for measuring the asymptotic mean-squared error, which in the light of Remark 1 is equivalent to the criterion. Remark 1 of Section 4.2b may help to see that *for the information-standardized sensitivity* and the respective criterion, *the problem essentially coincides with the unstandardized case*. For the self-standardized sensitivity, the situation is somewhat different, as will be explained below. In any case, if the standardization of the sensitivity matches the criterion, the optimal estimators have the same form as the solution for the unstandardized case just discussed.

Definition 1 (*Optimal B-robust estimators*). Let c be given. For nonsingular $p \times p$ matrices A and p-vectors a, let

$$\psi_c^{A,a}(x,\theta) := h_c\{A \cdot [s(x,\theta) - a]\}.$$

Assume that any one of the equations

(a) $$\int \psi_c^{A,a}(x,\theta) \cdot s(x,\theta)^T \, dF_\theta(x) = I,$$

(b) $$\int \psi_c^{A,a}(x,\theta) \cdot \psi_c^{A,a}(x,\theta)^T \, dF_\theta(x) = I, \qquad (4.3.5)$$

or

(c) $$\int \psi_c^{A,a}(x,\theta) \cdot s(x,\theta)^T \, dF_\theta(x) = J^{1/2}(\theta),$$

where $J^{1/2}$ is any differentiable root of J, $(J^{1/2})^T \cdot J^{1/2} = J$, together with

$$\int \psi_c^{A,a}(x,\theta) \, dF_\theta(x) = 0, \qquad (4.3.6)$$

has a differentiable solution $(A^*(\theta), a^*(\theta))$. Then the corresponding ψ-function $\psi_c^{A^*(\theta),a^*(\theta)}$ is, in each case, respectively, called the ψ-function (a) $\psi_c^u(x,\theta)$ of the optimal B_u-robust estimator T_c^u, (b) $\psi_c^s(x,\theta)$ of the optimal B_s-robust estimator T_c^s, or (c) $\psi_c^i(x,\theta)$ of the optimal B_i-robust estimator T_c^i.

The bunch of questions asked before Theorem 1 again arise and will be left open, and we just treat optimality questions here.

Theorem 1 states an optimality for the unstandardized case, and the arguments at the beginning of this subsection imply that the optimal B_i-*robust* estimator T_c^i minimizes

$$\text{trace}\{ J(\theta_*) \cdot V(\cdot, F_*) \}$$

over the class $\mathcal{T}_{c*}^i \subset \mathcal{T}_*^r$ with information-standardized sensitivity not exceeding c.

For the *self-standardized* case, only a weaker property can be ascertained, which in fact holds in all three cases.

Theorem 2 (*Admissibility*). Let $\psi^* = \psi_c^u$, or $\psi^* = \psi_c^s$, or $\psi^* = \psi_c^i$, and let T^* and $\mathcal{T}^*$ (for a fixed θ_*) be defined in the corresponding manner. If

(i) $J(\theta_*)$ exists.
(ii) ψ^* exists.

(iii) $T^* \in \mathcal{T}_*^r$.

(iv) $\mathrm{IF}(x; T^*, F_*) = U \cdot \psi^*(x, \theta_*)$ for some matrix U,

then $T^* \in \mathcal{T}^*$, and there is no estimator T in $\mathcal{T}^*$ with smaller asymptotic covariance matrix $V(T, F_*) \underset{\neq}{\leq} V(T^*, F_*)$, $(T^*$ is "admissible" in $\mathcal{T}^*$, see Subsection 4.6a. The notation $A \underset{\neq}{\leq} B$ stands for $B - A$ positive semidefinite and $A \neq B$.)

Proof. We first transform to $\bar{\theta} = B \cdot \theta$ with $B = I$ for the unstandardized, $B = J^{1/2}(\theta_*)$ for the information-standardized, and

$$B = \int \psi_c^s(x, \theta_*) \cdot s(x, \theta_*)^T \, dF_*(x) = M(\psi_c^s, F_*)$$

for the self-standardized case. Then

$$\mathrm{IF}(x; \bar{T}^*, F_*) = B \cdot B^{-1} \cdot \psi^*(x, \theta_*),$$

which shows that ψ^* coincides (in θ_*) with the ψ-function of the unstandardized optimal estimator for the transformed parameter and therefore minimizes $\mathrm{trace}\{ B \cdot V(T, F_*) \cdot B^T \}$ over the class of all estimators T with

$$\| \mathrm{IF}(x; \bar{T}, F_*) \| = \| B \cdot \mathrm{IF}(x; T, F_*) \| \leq c.$$

In this class, then, there is no estimator with smaller V matrix. [If $V_1 - V_2$ is positive semidefinite and nonzero, $\mathrm{trace}(B \cdot V_1 \cdot B^T) > \mathrm{trace}(B \cdot V_2 \cdot B^T)$.] This proves the unstandardized and information-standardized cases. For the self-standardized case, it is intuitively clear that, if it is impossible to get a smaller asymptotic covariance matrix without enlargening the raw bias, it is also impossible to achieve this without increasing the bias "divided" by the covariance. Formally, let T be $\in \mathcal{T}_{c*}^s$ with $V(T, F_*) \leq V(T_c^s, F_*)$. Then

$$V(\bar{T}, F_*) \leq V(\bar{T}_c^s, F_*) = I,$$

$$c^2 \geq \gamma_s^*(T, F_*)^2 = \gamma_s^*(\bar{T}, F_*)^2$$

$$= \sup_x \left\{ \mathrm{IF}(x; \bar{T}, F_*)^T \cdot V(\bar{T}, F_*)^{-1} \cdot \mathrm{IF}(x; \bar{T}, F_*) \right\}$$

$$\geq \sup_x \left\{ \| \mathrm{IF}(x; \bar{T}, F_*) \|^2 \right\} = \gamma_u^*(\bar{T}, F_*)^2.$$

Therefore, $\overline{T} \in \mathcal{T}_{c*}^{u}$, and by the first part of the proof, $V(\overline{T}, F_*) = V(\overline{T}_c^s, F_*)$; whence $V(T, F_*)$ cannot be smaller than $V(T_c^s, F_*)$.   □

The ψ-functions of the optimal B-robust estimators have, up to equivalence (see below), a special form which is worth being named.

Definition 2. A *W-estimator* is an M-estimator characterized by a weight function $W: \mathcal{X} \times \Theta \to \mathbb{R}$ and a centering function $C: \Theta \to \mathbb{R}^p$ through

$$\psi(x, \theta) = W(x, \theta) \cdot [s(x, \theta) - C(\theta)].$$

In words, the ψ-function of a W-estimator must have (for fixed θ) the same direction in $\mathbb{R}^p$ as the ψ-function s of the maximum likelihood estimator after subtraction of a centering vector, which is used to achieve Fisher consistency.

Although the ψ-functions given in Definition 1 are not of this form, they define W-estimators, since they may be replaced by

$$A^*(\theta)^{-1} \cdot \psi(x, \theta) = [s(x, \theta) - a^*(\theta)]$$

$$\cdot \min\{1, c/\|A^*(\theta) \cdot [s(x, \theta) - a^*(\theta)]\|\}.$$

4.3c.   Existence and Uniqueness of the Optimal ψ-Functions

Now, the questions of the existence and uniqueness of the ψ-functions defining the optimal estimators are considered. These problems should not be mixed up with the existence and uniqueness of a value of such an estimator for a given sample or distribution.

For the ψ-function of the optimal B_s-robust estimator, the existence problem has a satisfactory answer.

Theorem 3 (*Existence of ψ_c^s*).   If

$$F_*\{s(x, \theta_*) \in H\} = 0$$

holds for all ($p - 1$)-dimensional hyperplanes H, then the Eqs. 4.3.5(b) and (4.3.6) have a solution (at least one) if $c > \sqrt{p}$, and no solution if $c < \sqrt{p}$.

Proof.   The second assertion follows from Lemma 1 of Subsection 4.2b, whereas the first one is a consequence of a theorem by Maronna (1976), which we will mention in a slightly generalized form in Subsection 5.3c. This is seen by equating $z = s^{A,a}(x)$, $[w_\mu^\psi(r^2)]^2 = w_\eta^\psi(r^2) = \min(1, c^2/r^2)$, $w_\delta^\psi \equiv 1$. □

Clearly, if there is a solution of (4.3.5(b)) and (4.3.6) for $c = \sqrt{p}$, it defines the most B_s-robust estimator.

As to the *uniqueness of* ψ_c^s, it is easy to see that A^* is determined only up to premultiplication by an orthogonal matrix. A^* may therefore be required to be lower triangular with positive diagonal, for example.

The uniqueness of such a solution has not been proved yet in the general case. Maronna (1976) includes the respective conjecture, and Krasker and Welsch (1982) are able to derive it from Theorem 1 of Maronna (1976) in the regression context, since the vector a^* vanishes for symmetry reasons. For $p = 1$, Huber (1981, Section 11.1) gives the proof.

In the *unstandardized* and *information-standardized* cases, the uniqueness of A^* and a^* is clear from Theorem 1 and the arguments mentioned at the beginning of Subsection 4.3b (if $J^{1/2}$ is a fixed root of J). As to the existence, Bickel (1981, p. 22) shows for scalar θ ($p = 1$) that there is a solution as soon as $\mathcal{T}_{c*}^u$ is not empty, and Rieder (1983, personal communication) extended this result to multidimensional parameters. As a consequence, there must again be a critical c value above which there is a solution, and below which there is none. For the regression case, a lower bound for such a critical value is given in Subsection 6.3b.

4.3d. How to Obtain Optimal Estimators

In order to facilitate the derivation of optimal estimators in any parametric model the reader might come across, a step-by-step description for *constructing the* ψ-*functions* shall be given, which is then illustrated for two special models, the Γ distribution and a problem of simple regression with known carrier distribution and error variance. The most important cases, namely multiple regression and the estimation of covariance matrices, are treated in the following chapters.

Note that in this section we treat only how to determine the constants (A^* and a^*) in the formula for $\psi_c(\cdot, \theta_*)$ for an arbitrary, fixed value θ_*. This is to be distinguished clearly from calculating an estimated value for a given sample as discussed in Subsection 4.6b.

Step 1

Fix the value(s) of θ for which the ψ-function of the desired optimal estimator should be calculated.

If the model under consideration possesses some invariance structure, Step 1 reduces the set of parameter values at which the characterizing constants for the two standardized estimators must be calculated to a lower-dimensional space than p — for the most popular models even to a single value θ_0. We will discuss such structures in Section 4.5.

Example 1 (Γ *distribution*). Let X denote a random variable with a (two-parameter) Γ distribution, that is, with density proportional to $(x/\sigma)^{\alpha-1} \cdot \exp(-x/\sigma)$, $x \geq 0$. σ is a scale parameter. From our point of view, it is convenient to logarithmize scale parameters. For the sake of later conventions (Section 4.4), we denote $\tau := \ln(\sigma)$ as the first parameter component. Thus, the density is

$$f_{\tau,\alpha}(x) = [\sigma \cdot \Gamma(\alpha)]^{-1} \cdot (x/\sigma)^{\alpha-1} \cdot e^{-x/\sigma}, \qquad \sigma = e^{\tau}, \quad \alpha > 0,$$

where Γ is the Gamma function satisfying

$$\Gamma(\alpha + 1) = \alpha \cdot \Gamma(\alpha).$$

Note that $\mathscr{X} = \mathbb{R}$, $\Theta = \mathbb{R} \times \{x > 0\}$. We shall consider an arbitrary parameter value $(\tau_*, \alpha_*)^T$, although we could, for reasons just mentioned, specify $\tau_* = 0$. This example illustrates the general multiparameter situation, even though there are just two components.

Example 2 (*Simple linear regression*). Let $\mathscr{X} = \mathbb{R}^2$, $\Theta = \mathbb{R}^2$, $X^{(1)} \sim \mathscr{N}(0,4)$, and $X^{(2)} - (\theta^{(1)} + \theta^{(2)} \cdot X^{(1)})|X^{(1)} \sim \Phi$. Choose $\theta_* = 0$ as the fixed parameter value. This proves to be sufficient because of the high degree of invariance of the model (compare the location-scale case and Section 4.5). In order to illustrate this fact, we shall discuss the steps for general θ as well.

Step 2

Compute the likelihood scores in θ_* and the Fisher information $J(\theta_*)$.

Example 1. The scores are

$$s(x; \tau, \alpha) = \begin{pmatrix} z - \alpha \\ \ln(z) - \tilde{\Gamma}(\alpha) \end{pmatrix}, \qquad z := x/\sigma = x \cdot e^{-\tau},$$

where $\tilde{\Gamma}$ is the so-called ψ or digamma function, $\tilde{\Gamma}(\alpha) := d \ln(\Gamma(\alpha)]/d\alpha$, and satisfies

$$\tilde{\Gamma}(\alpha + 1) = \tilde{\Gamma}(\alpha) + \alpha^{-1}.$$

From $\int s(z; 0, \alpha) \, dF_{0,\alpha}(z) = 0$ we infer that

$$\int \ln(z) \, dF_{0,\alpha}(z) = \tilde{\Gamma}(\alpha).$$

This leads to

$$J(\tau, \alpha) = \begin{bmatrix} \alpha & 1 \\ 1 & K_\alpha - \tilde{\Gamma}(\alpha)^2 \end{bmatrix},$$

(independent of τ), where

$$K_\alpha := \int [\ln(z)]^2 \, dF_{0,\alpha}(z)$$

can be shown by partial integration (integrate $z^{\alpha-1}!$) to satisfy

$$K_{\alpha+1} = K_\alpha + \frac{2}{\alpha} \tilde{\Gamma}(\alpha).$$

Example 2. The scores are

$$s(z, 0) = \begin{pmatrix} z^{(2)} \\ z^{(1)} \cdot z^{(2)} \end{pmatrix},$$

and the information equals

$$J(0) = \begin{bmatrix} 1 & 0 \\ 0 & 4 \end{bmatrix}.$$

For general θ and given x, let $z := (x^{(1)}, x^{(2)} - \theta^{(1)} - \theta^{(2)} x^{(1)})^T$. Then, $s(x, \theta) = s(z, 0)$. It is easy to see that $J(\theta)$ is constant.

Step 3

The matrix A^* and the vector a^* to be calculated often allow simplifications justified by symmetry considerations. Guess at a special form for A^*

and $a*$, if applicable. If you find a solution of this form in Step 4, the guess is justified. Rewrite Eqs. (4.3.5) and (4.3.6) for the special problem, using

$$W^{A,a}(x) := \min\{1, c/\|A[s(x,\theta_*) - a]\|\},$$

$$M_k := \int [s(x,\theta_*) - a] \cdot [s(x,\theta_*) - a]^T \cdot W^{A,a}(x)^k dF_*(x), \qquad (4.3.7)$$

$$k = 1, 2,$$

in the form

(a) $A^{-1} = M_1$ for T_c^u (A is symmetric),

(b) $A^{-1} \cdot A^{-T} = M_2$ for T_c^s (A is lower triangular),   (4.3.8)

(c) $A^{-1} = M_1 \cdot J^{-1/2}(\theta_*)$ for T_c^i,

and

$$a = \int s(x,\theta_*) \cdot W^{A,a}(x) dF_*(x) \Big/ \int W^{A,a}(x) dF_*(x). \qquad (4.3.9)$$

Example 1.   The term $s(z,\theta_*) - a$ may be written as $(z - \tilde{a}^{(\tau)}, \ln(z) - \tilde{a}^{(\alpha)})^T$, say, which does not involve τ or α. In (4.3.9), a is replaced by $\tilde{a}$, while $s(z,\theta_*)$ becomes $[z, \ln(z)]^T$. No worthwhile further simplification of the equation seems possible. Because we chose τ rather than σ as a parameter, the unstandardized case (a) also leads to an estimator which is equivariant with respect to scale transformation.

Example 2.   We can guess that $a* = 0$ and $A* = \text{diag}(d_1, d_2)$. Let

$$B(b^2) := \int h_b(x) x \, d\Phi(x) = \chi_1^2(b^2) = 2\Phi(b) - 1,$$

$$\tilde{A}(b^2) := \int h_b(x)^2 \, d\Phi(x) = b^2 \cdot [1 - \chi_1^2(b^2)] + \chi_3^2(b^2),$$

[Cf. Subsection 4.2d; $h_b(x)$ is identical to $\psi_b(x)$ used there and earlier, since its dimension is 1]. Note that $V = (X^{(1)}/2)^2$ has a χ_1^2 distribution. The off-diagonal elements of M_k vanish, and the diagonal elements of the matrix equation lead to

(a)
$$d_1 = \left[\int B\{ c^2 / (d_1^2 + 4v\,d_2^2) \} \, d\chi_1^2(v) \right]^{-1},$$

$$d_2 = \left[\int 4v \cdot B\{ c^2 / (d_1^2 + 4v\,d_2^2) \} \, d\chi_1^2(v) \right]^{-1};$$

(b)
$$d_1 = \left[\int \tilde{A}\{ c^2 / (d_1^2 + 4v\,d_2^2) \} \, d\chi_1^2(v) \right]^{-2},$$

$$d_2 = \left[\int 4v \cdot \tilde{A}\{ c^2 / (d_1^2 + 4v\,d_2^2) \} \, d\chi_1^2(v) \right]^{-2};$$

(c)
$$d_1 = \left[\int B\{ c^2 / (d_1^2 + 4v\,d_2^2) \} \, d\chi_1^2(v) \right]^{-1},$$

$$d_2 = \left[\int 4v \cdot B\{ c^2 / (d_1^2 + 4v\,d_2^2) \} \, d\chi_1^2(v) \right]^{-1} \cdot 2.$$

For general θ, using $s(x, \theta) = s(z, 0)$ as in step 2, it turns out that the M_k are identical to those obtained for $\theta = 0$ (and given A and a). Therefore, the equations to be solved are identical for all θ, and thus A^* and a^* do not depend on θ.

Step 4

Calculations may done using the above equations for an iterative improvement, starting from $a = 0$ and $A = J^{1/2}(\theta_*)^{-T}$ for the optimal B_s- and B_i-robust estimators, and $A = J(\theta_*)^{-1}$ for the B_u-robust version (since these values solve the equations for $c = \infty$). Convergence has not been proved, but for the self-standardized case, Maronna's way of proving existence (cf. Subsection 4.3c) gives a strong hint. In the unstandardized case, the algorithm coincides with the construction given at the beginning of Subsection 4.3a.

Example 1. The calculation of J involves a numerical integration. Since exactness is not needed, one could alternatively use the recurrence relations for K_α and $\tilde{\Gamma}(\alpha)$ given above together with some values of K_α and $\tilde{\Gamma}(\alpha)$ for $0 < \alpha \le 1$. We did not calculate numerical values for A^* and a^* in this example. Each iteration step needs four numerical integrations of a similar nature as those appearing in the next example.

Example 2.   The starting values for T_c^s and T_c^i are $d_1 = 1$, $d_2 = \frac{1}{2}$, and for $c = 2.5$ one gets $d_1^2 = 1.689$, $d_2^2 = 0.4651$ after seven iterations for T_c^s, and $d_1^2 = 1.961$, $d_2^2 = 0.2036$, after six iterations for T_c^i.

Solution

The ψ-function in θ_* for the optimal estimator is

$$\tilde{\psi}_c(x, \theta_*) = \left[s(x, \theta_*) - a^* \right] \cdot \min\left\{ 1, c / \| A^* \cdot \left[s(x, \theta_*) - a^* \right] \| \right\},$$

where A^* and a^* are the results of Step 4, or any matrix multiple of it, such as ψ_c^* given in Definition 1.

Example 2 (*continued*).   For the optimal B_s-robust estimator with $c = 2.5$, the ψ-function is

$$\tilde{\psi}_c^s(z, 0) = \begin{pmatrix} z^{(2)} \\ z^{(1)} \cdot z^{(2)} \end{pmatrix}$$

$$\cdot \min\left\{ 1, 2.5 / \left[1.689 \cdot z^{(2)2} + 0.4651 \cdot z^{(1)2} \cdot z^{(2)2} \right]^{1/2} \right\}.$$

For general θ, the ψ-function is $\tilde{\psi}^s(x, \theta) = \tilde{\psi}^s(z, 0)$, where z is obtained from x as before. An identity of this kind holds for general parametric models with invariance structures, as Section 4.5 shows.

Remark 4.   The *choice of the sensitivity bound c* is a matter of the specific problem at hand, since it regulates the "degree of robustness" of the estimator. The bound may depend differentiably on θ_*. A natural choice is obtained by requiring a certain asymptotic relative efficiency compared with the maximum likelihood estimator at the model distributions. However, results for regression suggest that such an approach may lead to using estimators with very low robustness. In many applications, it may be wise to choose c near its lower bound, which, however, is known only for the self-standardized case. Useful guidance in the choice of c depends on the specific model, and will therefore be discussed in other places.

4.4.   PARTITIONED PARAMETERS

4.4a.   Introduction: Location and Scale

In many models, the parameter vector splits up in a natural way into a "*main* part" and a "*nuisance* part." The interest then focuses on good robustness and efficiency properties for the main part. We will again look for the estimator which is most efficient for the main part under a bound on the "sensitivity of the main part." The precise formulation is given in the next subsection.

The best known example is clearly the problem of estimating location when scale is unknown. We saw in Section 4.2d that the location component of the influence function and asymptotic variance does not depend on the scale estimator used, as long as equivariance holds and the model distributions are symmetric. The optimality problem for the location part is therefore unaffected by the uncertainty about scale; one may just use any equivariant scale estimator.

The last statement suggests that even a nonrobust scale estimator could be used. But if this was done in combination with, for example, a Huber estimator for the location part, the location component of the combined estimator would usually not be qualitatively robust. This can be seen by considering contamination with a point mass of fixed weight $\varepsilon > 0$ that moves to infinity; the scale estimate, and thus the (Huber) location estimate will move along with it. (Note that this represents another case of an estimator with bounded influence function which is not qualitatively robust.)

Taking the argument one step further, it is intuitively clear that the maximal bias of the location part in a contamination model (with fixed $\varepsilon > 0$) depends on the maximal (positive) bias of the scale part. Andrews et al. (1972, Section 6G) found in their simulation study that the *scale should be estimated most robustly in order to improve the robustness properties for the location estimator* (incurring minimal efficiency losses at the model.) Another reason for this guideline is furnished by considerations of breakdown aspects, as illustrated at the end of Subsection 4.2d.

The possibility of treating location and scale separately for asymptotic properties at the model is a consequence of the symmetry and invariance characteristics of the model. While some important models allow for such a simplification, the general case looks more complicated, and there is some efficiency loss for the main parameter if the nuisance parameter is unknown.

*4.4b.   Optimal Estimators

Notation.   For any p-dimensional vector a, let

$$a_{(1)} := \left(a^{(1)}, a^{(2)}, \ldots, a^{(q)} \right)^T, \qquad a_{(2)} := \left(a^{(q+1)}, \ldots, a^{(p)} \right)^T,$$

where $q < p$ is a fixed number. For the sake of compatibility of notation with testing problems (Chapter 7), $\theta_{(1)}$ will be the nuisance part of the parameter, while $\theta_{(2)}$ is the main part. The respective partition of matrices will be denoted by

$$A =: \begin{bmatrix} A_{(11)} & A_{(12)} \\ A_{(21)} & A_{(22)} \end{bmatrix}.$$

The *optimality problem* now consists of minimizing the trace of $V_{(22)}$ subject to a bound on $\|IF_{(2)}\|$. Remarkably enough, this leaves again some freedom for the nuisance estimation even in the general case, and that part can be optimized simultaneously.

Definition 1 (*Optimal B-robust estimators for partitioned parameters*). Let c_1 and c_2 be given. For nonsingular $p \times p$ matrices A and p-vectors a let

$$\psi^{A,a}_{c_1,c_2}(x,\theta)_{(i)} := h_{c_i}\{(A \cdot [s(x,\theta) - a])_{(i)}\}, \qquad i = 1,2. \quad (4.4.1)$$

Assume that any one of the equations

(a) $\qquad \int \psi^{A,a}_{c_1,c_2}(x,\theta) \cdot [s(x,\theta) - a]^T dF_\theta(x) = I,$

(b) $\qquad \int \psi^{A,a}_{c_1,c_2}(x,\theta) \cdot \psi^{A,a}_{c_1,c_2}(x,\theta)^T dF_\theta(x) = I \qquad (4.4.2)$

with lower triangular A, or

(c) $\qquad \int \psi^{A,a}_{c_1,c_2}(x,\theta) \cdot [s(x,\theta) - a]^T dF_\theta(x) = J^{1/2}(\theta),$

where $J^{1/2}(\theta)$ is the upper triangular root of $J(\theta)$, which is required to be differentiable, together with

$$\int \psi^{A,a}_{c_1,c_2}(x,\theta) \, dF_\theta(x) = 0 \qquad (4.4.3)$$

has a differentiable solution $(A^*(\theta), a^*(\theta))$. Then, the corresponding ψ-function $\psi^{A^*(\theta),a^*(\theta)}_{c_1,c_2}$ is called, in each case, respectively, the ψ-function (a) $\psi^{pu}_{c_1,c_2}$ of the optimal B^p_u-robust estimator, (b) $\psi^{ps}_{c_1,c_2}$ of the optimal B^p_s-robust estimator, or (c) $\psi^{pi}_{c_1,c_2}$ of the optimal B^p_i-robust estimator.

A sketch of the transformation $z := A(s - a) \mapsto \psi^{A,a}_{c_1,c_2}$ might help to see how the new construction just adapts the ideas of Section 4.3 to the new type of restriction on the influence function (Fig. 1).

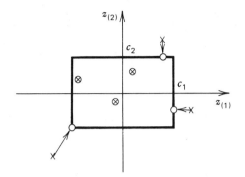

Figure 1. The transformation $z = A[s(x, \theta) - a] \mapsto \psi^{A, a}_{c_1, c_2}(x, \theta)$.

Theorem 1 (*Optimality* of the optimal B_u^p-robust estimator). If the conditions of Theorem 1 of Section 4.3 hold for $\psi^* = \psi^{pu}_{c_1, c_2}$, then the optimal B_u^p-robust estimator minimizes trace$\{V_{(22)}(T, F_*)\}$ over all $T \in \mathcal{T}_*^r$ with $\|\text{IF}(x; T, F_*)_{(2)}\| \le c_2$ and simultaneously trace$\{V_{(11)}(T, F_*)\}$ over all $T \in \mathcal{T}_*^r$ with $\|\text{IF}(x; T, F_*)_{(1)}\| \le c_1$.

Proof. Let $c_1 = \infty$. Then the argument used to prove Theorem 1 of Section 4.3 shows that ψ^{pu}_{∞, c_2} minimizes trace$(V_{(22)})$ under $\|\text{IF}_{(2)}\| \le c_2$. Note that in (4.4.2(a)), the term $-a$ in the brackets may be dropped because of (4.4.3) [cf. (4.3.3)]. Instead of solving (4.4.2) and (4.4.3) for A and a, we can solve for A and $\tilde{a} = Aa$, and write

$$\left\{ A[s(x, \theta) - a] \right\}_{(i)} = A_{(i)} s(x, \theta) - \tilde{a}_{(i)},$$

where $A_{(i)} := [A_{(i1)} \ A_{(i2)}]$. In this form (4.4.2(a)) (without the $-a$ term) and (4.4.3) decay into two independent systems of equations that determine $A_{(i)}$ and $\tilde{a}_{(i)}$, $i = 1$ and 2, respectively. The second system is

$$\int h_{c_2} \left[A_{(2)} s(x, \theta) - \tilde{a}_{(2)} \right] s(x, \theta)^T \, dF_\theta(x) = [0 \ I].$$

$$\int h_{c_2} \left[A_{(2)} s(x, \theta) - \tilde{a}_{(2)} \right] dF_\theta(x) = 0,$$

and does not contain c_1. Therefore $\text{IF}(x; T^{pu}_{c_1, c_2}, F_\theta)_{(2)} = \psi^{pu}_{c_1, c_2}(x, \theta)_{(2)}$ and $V(T^{pu}_{c_1, c_2}, F_\theta)_{(22)}$ do not depend on c_1 either. The same argument holds for the other part.   □

Remark 1. The optimal B_u^p-robust estimator treats the main and nuisance parts of the parameter in a symmetric manner.

Remark 2. Some efficiency for the main parameter is indeed lost if the nuisance part has to be estimated, unless $A_{(21)}^* = 0$. This can be seen by applying the uniqueness part of Theorem 1 of Subsection 4.3a to the estimation of the main parameter.

Remark 3. The efficiency of the main parameter is unaffected by a *transformation* of the nuisance parameter, that is, the optimal estimator for the parameter $\bar{\theta} = B \cdot \theta$ with

$$
B = \begin{bmatrix} B_{(11)} & B_{(12)} \\ 0 & I \end{bmatrix}
$$

has the same $V_{(22)}$ as the one for the original parameter.

The optimal B_l^p-robust estimator is justified by the fact that it coincides for each θ_* with the optimal B_u^p-robust one for a locally "orthonormalized" parameter system which retains the subspace containing the main parameter, namely $\bar{\theta} = J^{1/2}(\theta_*)\theta$. (Since $J^{1/2}$ is upper triangular, $\bar{\theta}_{(2)}$ is a function of $\theta_{(2)}$ only.) It is therefore optimal in a similar sense as the optimal B_u^p-robust estimator. The self-standardized variant may not be justified by such an argument, in spite of its name.

Example. Consider the Γ *distribution* as in Example 1 of Subsection 4.3d. The optimal B_s^p-robust estimator for the parameter partitioned into τ (nuisance) and α (main) is given by

$$
\psi(x; \tau, \alpha) = \begin{pmatrix} h_{c_1}\{A_{11}(z - a^{(1)})\} \\ h_{c_2}\{A_{12}(z - a^{(1)}) + A_{22}[\ln(z) - a^{(2)}]\} \end{pmatrix},
$$

where $z = x/e^\tau$ and A and a are a function of α and satisfy

$$
\int \psi(z; 0, \alpha) \cdot \psi(z; 0, \alpha)^T dF_{0,\alpha}(z) = I,
$$

$$
\int \psi(z; 0, \alpha) dF_{0,\alpha}(z) = 0.
$$

Remark 4. The experience with the location-scale problem mentioned above suggests that c_1 should be chosen as small as possible, while for c_2 Remark 4 of Subsection 4.3d applies.

*4.5. INVARIANCE

4.5a. Models Generated by Transformations

Three of the most important models in statistical data analysis, the location-scale, regression, and multivariate normal models, and some less important ones share a well-known structure:

Definition 1 (*Transformation model*). Let $\{\alpha_\theta\}_{\theta \in \Theta}$ be a parametrized group of transformations $\alpha_\theta \colon \mathcal{X} \to \mathcal{X}$, F_0 any fixed distribution on $\mathcal{X}$, and

$$F_\theta = \mathscr{L}\{\alpha_\theta(Z)\}, \quad \text{where } Z \sim F_0.$$

If $\theta \mapsto F_\theta$ is one–one, we call $\{F_\theta\}$ the model generated by F_0 and the transformations α_θ (cf. Fraser, 1968, and Barndorff-Nielsen et al., 1982).

As an example, the normal location-scale model is generated by the standard normal distribution and the transformations $\{\alpha_{\mu,\sigma} \colon z \mapsto \mu + \sigma z \mid \mu \in \mathbb{R}, \sigma > 0\}$.

The composition of the transformations α_θ induces a *group operation* $\Box$ in Θ by the definition

$$\theta = \theta' \Box \theta'' \Leftrightarrow \alpha_\theta(x) \equiv \alpha_{\theta'}\{\alpha_{\theta''}(x)\}. \tag{4.5.1}$$

The neutral element of this operation shall be denoted by θ_0, and the inverse by θ^-.

4.5b. Models and Invariance

Definition 2. A *model* $\{F_\theta\}_{\theta \in \Theta}$ is *invariant* with respect to a transformation $\alpha \colon \mathcal{X} \to \mathcal{X}$, if, for every $\theta \in \Theta$, there exists a $\bar{\theta} \in \Theta$ with

$$X \sim F_\theta \Rightarrow \alpha(X) \sim F_{\bar{\theta}}.$$

It is invariant with respect to a set $\mathscr{A}$ of transformations, if it is invariant for every $\alpha \in \mathscr{A}$. The transformation $\theta \mapsto \bar{\theta}$ corresponding to α shall be denoted by $\tilde{\alpha}$.

Clearly, a model generated by a group of transformations is invariant with respect to these transformations, since

$$\alpha = \alpha_{\theta_*} \Rightarrow \bar{\theta} = \theta_* \square \theta.$$

Some models are invariant with respect to a larger group. For example, the location-scale model with symmetric F_0 is also invariant with respect to multiplication by a negative factor.

At the end of Subsection 4.2b, we described the action of *parameter transformations* on density, scores, and Fisher information (4.2.7). Now, these transformations $\theta \mapsto \bar{\theta}$ bear strong connections to the corresponding x-transformation α. If $\bar{x} := \alpha(x)$ and $\bar{\theta} = \tilde{\alpha}(\theta)$,

$$\bar{f}_{\bar{\theta}}(x) = f_\theta(x) = f_{\bar{\theta}}(\bar{x}) \cdot \det\{ \partial\alpha(x)/\partial x \},$$

$$\bar{s}(x,\bar{\theta}) = s(\bar{x},\bar{\theta}).$$

If the model is generated by a group of transformations, and if $\alpha = \alpha_{\theta_*^-}$ and hence $\bar{\theta} = \tilde{\alpha}(\theta) = \theta_*^- \square \theta$, we get by fixing $\theta = \theta_*$

$$\bar{f}_{\theta_*}(x)\, dx = \bar{f}_{\theta_0}(x)\, dx = f_{\theta_0}(\bar{x})\, d\bar{x}$$

$$= f_{\theta_*}(x)\, dx$$

with $\bar{x} = \alpha_{\theta_*}^{-1}(x)$, and

$$\bar{s}(x,\bar{\theta}_*) = \bar{s}(x,\theta_0) = s(\bar{x},\theta_0)$$

$$= B^T s(x,\theta_*)$$

with

$$B := \left\{ [\partial\tilde{\alpha}(\theta)/\partial\theta]_{\theta_*} \right\}^{-1} = [\partial(\theta_*\square\theta)/\partial\theta]_{\theta_0}.$$

The distribution of the scores with respect to the transformed parameter is thus equal to the distribution of the scores under F_0 in the original parameter system. This will help to construct the optimal standardized estimators, since they are equivariant with respect to parameter transformations. In addition, we get

$$J(\theta_*) = B^{-T} J(\theta_0) B^{-1}.$$

4.5c. Equivariant Estimators

Definition 3. Let the model $\{F_\theta\}$ be invariant with respect to the transformation $\alpha\colon \mathscr{X} \to \mathscr{X}$. An *estimator* (functional) T is called *equivariant* with respect to α if

$$T\{\tilde{\alpha}(F)\} = \tilde{\alpha}\{T(F)\},$$

where $\tilde{\alpha}(F)$ is the distribution of $\alpha(X)$, $X \sim F$.

The influence function of such an estimator transforms properly:

$$\mathrm{IF}\{\alpha(x); T, \tilde{\alpha}(F)\} = [\partial\tilde{\alpha}(\theta)/\partial\theta]_{T(F)} \cdot \mathrm{IF}(x; T, F). \quad (4.5.2)$$

Consider now a model generated by a group of transformations, with respect to which T shall be equivariant. Then

$$\mathrm{IF}\{\alpha_{\theta_*}(z); T, F_{\theta_*}\} = [\partial(\theta_*\Box\theta)/\partial\theta]_{T(F_0)}\mathrm{IF}(z; T, F_0), \quad (4.5.3)$$

and the standardized gross-error sensitivities are constant over the model distributions.

M-estimators may be given by ψ-functions with a corresponding equivariance relation. Since

$$\int\psi(x, \theta)\, dF(x) = 0 \Leftrightarrow \int\psi(z, \theta_0)\, dF(x) = 0$$

with $z := \alpha_\theta^{-1}(x)$, the ψ-function

$$\tilde{\psi}(x, \theta) := \psi\{\alpha_\theta^{-1}(x); \theta_0\} =: \psi_0(z) \quad (4.5.4)$$

determines the same *M*-estimator as ψ.

For the optimal *B*-robust estimators, we get the following nice result:

Theorem 1 (*Equivariance of the optimal B_s- and B_i-robust estimators*). For a model generated by a group of transformations indexed by an open $\Theta \subset \mathbb{R}^p$, the optimal B_s- and B_i-robust estimators are equivariant with respect to the generating transformations.

Proof. See remark at the end of the last subsection. A formal proof for the information-standardized estimator is as follows: Let $A_0 := A^*(\theta_0)$,

$a_0 := a^*(\theta_0)$ be the solutions of (4.3.5(c)) and (4.3.6) for θ_0, such that

$$\int h_c\{A_0[s(z,\theta_0) - a_0]\} \cdot s(z,\theta_0)^T dF_0(z) = J^{1/2}(\theta_0).$$

With the notation used at the end of the last subsection, and $z = \bar{x}$, we get $s(z,\theta_0) = B^T \cdot s(x,\theta_*)$; $J^{1/2}(\theta_0) \cdot B^{-1}$ is a root $J^{1/2}(\theta_*)$ of $J(\theta_*) = B^{-T} \cdot J(\theta_0) \cdot B^{-1}$; $dF_0(z) = dF_{\theta_*}(x)$, and

$$\int h_c\{A_0[B^T s(x,\theta_*) - a_0]\}s(x,\theta_*)^T dF_{\theta_*}(x)B = J^{1/2}(\theta_*)B.$$

Therefore $A^*(\theta_*) = A_0 B^T$ and $a^*(\theta_*) = B^{-T}a_0$ are the solution of (4.3.5(c)) and (4.3.6) for θ_*, and

$$\psi_c^i(x,\theta_*) = h_c\{A^*(\theta_*)[s(x,\theta_*) - a^*(\theta_*)]\} = h_c\{A_0[s(z,\theta_0) - a_0]\}$$

$$= \psi_c^i(z,\theta_0),$$

which demonstrates the equivariance. Similar considerations apply in the self-standardized case.   □

Remark 1. In order to obtain the ψ-functions for the optimal B_s- of B_i-robust estimators, it suffices to determine $\psi_0 = \psi(\cdot,\theta_0)$ in view of (4.5.4) and the theorem (see Subsection 4.3d, Step 1). This is often simpler than to determine $\psi(\cdot,\theta)$ for general θ. What is more, it allows to calculate the coefficients $A(\theta_0)$ and $a(\theta_0)$ and tabulate or store them; the general $\psi(x,\theta)$ can then be obtained from an explicit formula. In the general case, since $A(\theta)$ and $a(\theta)$ must be known for all θ, they cannot be stored in any reasonable way, but have to be calculated every time they are needed. We come back to this point in Subsection 4.6b (Algorithm 4 and Remark 5).

Examples of these concepts are all models treated in this volume except for the Γ distribution (Example 1 of Section 4.3d) with respect to the parameter α. We therefore omit further instances.

*4.6. COMPLEMENTS

4.6a. Admissible *B*-Robust Estimators

In Section 4.3, we stated that there was no estimator which minimizes the asymptotic covariance matrix in the strong sense under a bound on any of the sensitivities. We therefore minimized just the trace of V in the unstandardized case. Here, we want to explore what other functionals of V we could consider for optimization while keeping the bound on one of the gross-error sensitivities fixed.

The partial ordering (4.3.1) of V matrices entails a natural notion of admissibility:

Definition 1. Given a class $\mathcal{T}$ of functionals, $T \in \mathcal{T}$ is called (asymptotically covariance) *admissible* within $\mathcal{T}$ at F if $V(T, F)$ exists and there is no functional $\tilde{T} \in \mathcal{T}$ which dominates it, that is, for which

$$V(\tilde{T}, F) \underset{\neq}{\leq} V(T, F).$$

The notion is clearly invariant with respect to regular differentiable parameter transformations: If T is admissible within $\mathcal{T}$ at F ($= F_\theta$), then $\overline{T} = \beta(T)$ is admissible within $\overline{\mathcal{T}} := \{\beta(T') | T' \in \mathcal{T}\}$ at F ($= \overline{F}_{\beta(\theta)}$). The standardized sensitivities lead to invariant classes of estimators, that is,

$$\overline{(\mathcal{T}_{c*}^s)} = \{\beta(T) | T \in \mathcal{T}_{c*}^s\} = \mathcal{T}_{c*}^s$$

(and analogously for $\mathcal{T}_{c*}^i$). If, therefore, T is admissible (at F_θ) under such a restriction, then $\overline{T}$ is admissible (at $\overline{F}_{\beta(\theta)}$) under the same restriction. For the unstandardized sensitivity, this property holds for transformations with orthogonal derivative $\partial\beta(\theta)/\partial\theta$ only (compare Subsection 4.2b).

The following sufficient condition helps to find admissible estimators:

Lemma 1. Let C be any nonsingular matrix. An estimator minimizing

$$\text{trace}\{C \cdot V(\cdot, F) \cdot C^T\}$$

within $\mathcal{T}$ is admissible within $\mathcal{T}$ at F.

The proof is straightforward.

Extending the idea of Subsection 4.3a, we obtain estimators which indeed minimize this criterion under a bound on any one of the sensitivities. The construction uses a generalization of the Huber function to an elliptical restriction:

$$\tilde{h}_{c,C}(z) = \min_{\|C^{-1}v\| \leq c}^{-1} (\|z - v\|).$$

($\min^{-1}$ denotes the argument for which the minimum is attained.) Clearly,

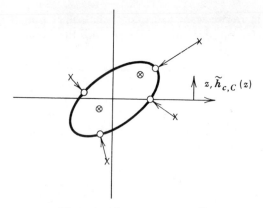

Figure 1. The transformation $\tilde{h}_{c,C}$.

$\tilde{h}_{c,C}(z) = z$ for $\|C^{-1}z\| \leq c$ and $\tilde{h}_{c,I}(z) = h_c(z)$. A sketch of $z \to \tilde{h}(z)$ in two dimensions will be helpful (Fig. 1). Now let

$$\psi_{c,C}^{A;a}(x,\theta) := C^{-1}\tilde{h}_{c,C}\{CA[s(x,\theta) - a]\} \qquad (4.6.1)$$

and determine A and a as in Definition 1 of Subsection 4.3b to get $\psi_{c,C}^u$, $\psi_{c,C}^s$, or $\psi_{c,C}^i$. Note that the ψ-functions of the estimators introduced in that definition are a special case, $\psi_c^\cdot = \psi_{c,I}^\cdot$.

Theorem 1. The M-estimators defined by $\psi_{c,C}^u$, $\psi_{c,C}^s$, and $\psi_{c,C}^i$ are admissible at F_* within $\mathcal{T}_{c*}^u$, $\mathcal{T}_{c*}^s$, and $\mathcal{T}_{c*}^i$, respectively, if they obey the conditions of Theorem 2 in Subsection 4.3b.

Proof. For the unstandardized case, a slight extension of the proof of Theorem 1 of Section 4.3 demonstrates that $\psi_{c,C}^u$ minimizes trace$\{CV(\cdot, F_*)C^T\}$ within $\mathcal{T}_{c*}^u$, and Lemma 1 may then be applied. From this result the admissibility of the estimator based on $\psi_{c,C}^i$ follows in the manner of the proofs of Sections 4.3 and 4.4 when considering the transformed parameter $\bar{\theta} = J^{1/2}(\theta_*) \cdot \theta$. The self-standardized case is proved as in Theorem 2 of Subsection 4.3b. $\qquad \square$

Note that $\psi_{c,C}^i$ does not minimize trace$\{\tilde{C}V\tilde{C}^T\}$ with $\tilde{C} = C$ but with $\tilde{C} = CJ^{1/2}(\theta_*)$. Specifically, $\psi_{c,C}^i$ with $C^{-1} = J^{1/2}(\theta_*)$ minimizes trace$\{V(\cdot, F_*)\}$ within $\mathcal{T}_{c*}^i$.

Stahel (1981a) gives further admissible estimators which may be interpreted as limits of those presented above. He conjectures that this wider class is "*essentially complete*" for $\mathcal{T}_{c*}^u$, $\mathcal{T}_{c*}^s$, or $\mathcal{T}_{c*}^i$, respectively; that is, to any admissible estimator there corresponds a member of the class which has the same asymptotic covariance matrix at F_*.

4.6b.   Calculation of M-Estimates

Given a sample $(x_1, x_2, \ldots, x_n)$ we have to solve

$$\text{ave}_i\{\psi(x_i, t)\} = 0$$

for t. [ave_i is defined as $(1/n)\Sigma_{i=1}^n$.] A Newton–Raphson step $t \to t + \Delta t$ uses the derivative matrix $\text{ave}_i\{\partial\psi(x_i, t)/\partial t\}$. If we approximate the average over the sample by the integral $\int \ldots dF_t$ over the tentatively estimated distribution we get

$$\Delta t = -\left\{\int \frac{\partial\psi(x, t)}{\partial t} dF_t(x)\right\}^{-1} \cdot \text{ave}_i\{\psi(x_i, t)\}$$

$$= \text{ave}_i\{\text{IF}(x_i; T, F_t)\}$$

for Fisher-consistent estimators. The last line suggests the name "influence algorithm." Another heuristic argument results from applying (2.1.7) in its multidimensional form to $G = F_n$ and $F = F_t$. For maximum likelihood estimators, the method is well known (Cox and Hinkley, 1974, p. 308) and often called the *scoring method*.
A more detailed formulation is the following algorithm.

Algorithm 1 (*Influence algorithm* for Fisher-consistent estimators).

1. Initial value t.
2. $M := \{\int\psi(x, t) \cdot s(x, t) dF_t(x)\}^{-1}$.
3. $\Delta t := M \cdot \text{ave}_i\{\psi(x_i, t)\}$, $\quad t := t + \Delta t$.
4. If $\Delta t \approx 0$, stop; else go to (2).

Remark 1. The *initial value* must be found in a way specific to the problem. If T should be robust, it is important that the initial value is already so. In some cases, a moments method using robust moment estimators like median and median deviation can be appropriate (for higher moments, see Hampel, 1975, p. 376 and the general method of Siegel, 1982).

Remark 2. In practice, one should bound the correction Δt ($\Delta t :=$ $h_k(\Delta t)$). Preliminary experience shows that even relaxation with a factor smaller than 1 may be needed in order to avoid oscillation in later steps.

Remark 3. It may be advisable to repeat step 3 several times before going back to 2 in order to reduce the number of integrations and inversions.

The method seems to be *widely applicable* in principle. Even for a T which is not an M-estimator, one might consider finding the value of the asymptotically equivalent M-estimator by the algorithm and to perform a specific step in the end. On the other hand, there are two major drawbacks. First, step 2 is feasible only for low dimension m of the observation space and low to moderate dimension p of the parameter. In specific problems, one should consider approximating M. Second, there is no proof of convergence or optimality known to us. Still, it performs similar to two specific algorithms in the estimation of covariance matrices (Stahel, 1981a).

On the whole, this and the following ideas may be considered in any problem until better specific methods are derived.

Let us turn to the special problem of obtaining the value of an *optimal B-robust estimator*. Here, the function $\psi(\cdot, t)$ is not given analytically. A matrix A^* and a vector a^* must be determined for each t that occurs during the iteration.

Algorithm 2 (*Influence algorithm for optimal B-robust estimators*).

1. Initial value for t.
    $A := J^{1/2}(t)^{-T}$ for the B_s- and B_i-robust estimators, $:= J(t)^{-1}$ for the B_u-robust estimator, and $a := 0$.
2. A and a as given by (4.3.8) and (4.3.9).
3. Iterate (2) until convergence.
4. $M := I$ for the B_u-robust estimator, $:= (AM_1)^{-1}$ [M_1 from (4.3.7)] for the B_s-robust estimator, and $:= J^{1/2}(t)$ for the B_i-robust estimator.
5. $\Delta t := M \operatorname{ave}_i\{h_c\{A[s(x_i, t) - a]\}\}$, $t := t + \Delta t$.
6. If $\Delta t \approx 0$, stop; else go to 2.

Remark 4. For the B_s-robust estimator, step 4 may be simplified to $M := A^T$ [replace M_1 by M_2 and use (4.3.8(b))]; although this leads to a somewhat different procedure.

Since the procedure includes two simultaneous iterations, some modifications might improve the numerical efficiency:

Algorithm 3.

1, 2. As before.
3. Optional—iterate step 2 a few times.
4, 5. As before.
6. If $\Delta t \approx 0$ and A and a have changed little in step 2, stop.
7. Repeat step 5 a few times, then go to step 2.

It is clear that Algorithms 2 and 3 have the same generality but suffer from the same drawbacks as the general influence algorithm.

In many models, there are *invariance* structures which typically allow to simplify the calculations substantially. Let the model be generated by a group of transformations as in Definition 1 of Section 4.5, and let T be equivariant with respect to all α_θ. Then, the influence function $\mathrm{IF}(\cdot\,; T, F_t)$ in step 2 of Algorithm 1 may be calculated from (4.5.3),

$$\mathrm{IF}(x_i; T, F_t) = \left[\partial(t \square \theta)/\partial\theta \right]_{\theta_0} \cdot \mathrm{IF}\left\{ \alpha_t^{-1}(x_i); T, F_0 \right\}.$$

But the influence function at F_0 may be calculated before the main iteration procedure and possibly also characterized by tabulated constants. Similar modifications apply to the calculation of A^* and a^* for the optimal B_s- and B_i-robust estimators, as Theorem 1 of Section 4.5 shows.

The formulas still require the calculation of $\left[\partial(t \square \theta)/\partial\theta \right]_{\theta_0}$ and α_t^{-1} for the values of t obtained during the iteration. But note that finding T for the sample $x_1, \ldots, x_n$ is equivalent to finding $\beta = \alpha_t^{-1}$ such that $\beta(x_1), \ldots, \beta(x_n)$ yield a T-value of θ_0, or

$$\mathrm{ave}_i\left\{ \psi\left[\beta(x_i), \theta_0 \right] \right\} = 0.$$

We assume that the constants needed to define β allow to recover t. If the Newton–Raphson method is modified as above, one finds the following procedures:

Algorithm 4 (*Influence algorithm for invariant problems*).

1. Determine constants needed for the calculation of $\mathrm{IF}(\cdot\,; T, F_0)$.
2. Find an initial value for t, and constants needed for $\beta(\cdot) = \alpha_t^{-1}(\cdot)$.

3. $z_i := \beta(x_i)$.

4. $\Delta t := \text{ave}_i\{\text{IF}(z_i; T, F_0)\}$,
$\Delta\beta := \alpha^{-1}_{\theta_0 + \Delta t}$ or an approximation to it,
$\beta := \Delta\beta \circ \beta$ (composite transformation).

5. If $\Delta t \approx 0$, go to step 6; else go to step 3.

6. Calculate the estimate t from $\alpha^{-1}_t = \beta$.

Remark 5. The calculation of the optimal B_s- and B_i-robust estimators is a special case here; the determination of A and a is needed just for step 1.

Remark 6. If the composition of transformations in step 4 is tedious, one might put $z_i := \Delta\beta(z_i)$ directly. Care is then needed to avoid numerical problems (accumulation of rounding errors) and to keep track of the estimate.

The algorithm was applied to covariance matrix estimation by Stahel (1981a). In fact, that specific method was the starting point for some of the general considerations of this subsection.

Remark 7. In Algorithms 2, 3, and 4, the calculation of Δt may be based on a different idea, not using influence functions. The general structure of the algorithms may then still be adequate.

EXERCISES AND PROBLEMS

Subsection 4.2a

1. (a) Compute the influence function of the empirical covariance matrix (with factor $1/n$) at a bivariate distribution with expectation 0 and covariance matrix Σ. (b) If $\Sigma = I$, in what regions of the plane are the components of the influence function positive, negative, and zero? (c) Compute the influence functions for trace and determinant of the empirical covariance matrix, and for the empirical correlation.

2. Consider the normal location model $\{\mathcal{N}_2(\mu, I)\}$ in the plane. Let T be a trimmed mean defined in the following way: Drop the observations that are among the 5% most extreme ones in any of the four axis directions. Sketch the transformation $x \mapsto \text{IF}(x; T, \mathcal{N}_2(0, I))$.

3. In Theorem 1 of Section 4.2, regularity conditions have been left unspecified. (a) Is it sufficient to assume that T is a von Mises

functional in order to have $R(t) = o(t)$ in the proof? (b) Give suffi-cient conditions that allow to do the required interchange of integra-tion and differentiation.

Subsection 4.2b

4. Let the vector-valued gross-error sensitivity γ_v^* be defined component-wise by

$$\gamma_v^{*(j)}(T, F) := \sup_x \left\{ \left| IF^{(j)}(x; T, F) \right| \right\}.$$

   Find order relations among γ_u^*, $\gamma_v^{*(j)}$, $\|\gamma_v^*\|$, and $\sup_j(\gamma_v^{*(j)})$. Discuss advantages and disadvantages of these quantities as measures of robustness.

5. Discuss the use of other metrics for alternative definitions of the gross-error sensitivity.

6. In the one-dimensional normal scale model $\{ \mathcal{N}(0, \sigma^2) \}$ consider the (rescaled) median deviation S_n (see "Location" and Example 4 of Subsection 2.3a).
   (a) Verify $\gamma_u^*(S_n, \Phi) = 1/[4\Phi^{-1}(0.75) \cdot \varphi\{\Phi^{-1}(0.75)\}] = 1.167$ and that $V(S_n, \Phi) = [\gamma_u^*(S_n, \Phi)]^2 = 1.360$. (b) Using these results, calculate γ_s^* and γ_i^*. (c) Consider $\tau = \ln(\sigma)$ instead of σ as the parameter and $T = \ln(S_n)$ as the estimator. How does this change affect the sensitivi-ties? (d) Derive the same quantities for $\mathcal{N}(0, 4)$ instead of Φ. Discuss the result (cf. Section 4.5).

Subsection 4.2c

7. (Short) Verify the formulas (4.2.14) and (4.2.15).

Subsection 4.3b

8. (Short) Explain why only Fisher-consistent estimators are considered in the optimality problem. What is the respective restriction in the Cramér–Rao inequality (for finite n)?

Subsection 4.3c

9. Consider a model with scalar parameter ($p = 1$) and a fixed distribu-tion F_θ. In the plane spanned by the axes $V(\cdot, F_\theta)$ and $\gamma_u^*(\cdot, F_\theta)$,

sketch the range of possible values. Draw the lines $\gamma_s^*(\cdot\,, F_\theta) = 2$ and $\gamma_i^*(\cdot\,, F_\theta) = 2$. Discuss the advantages of γ_s^* vs. γ_i^* as measures of robustness.

10. Consider the normal location-scale model $\{\mathcal{N}(\mu, \sigma^2)\}$. If $\gamma = \sigma^2$ is used instead of σ as the second parameter component, how does this change affect the optimal estimators in the three cases? Is there any benefit in using $\tau = \ln(\sigma)$ rather than σ? [*Hint*: Use (4.2.7).]

Subsection 4.3d

11. Consider the normal location-scale model $\{\mathcal{N}(\mu, \sigma^2)\}$. (a) Do steps 1 to 3 of Subsection 4.3d for this model. (b) Without performing actual calculations, write down the formulas for optimal ψ-functions. Sketch and discuss the qualitative behavior of the two components of the influence function. Sketch the figure corresponding to Figure 1 of Subsection 4.2d for an optimal estimator. (c) Do actual calculations for one of the three types of optimality and a fixed bound c. For the self-standardized case with $c = 2.5$, we obtained $A^* = \text{diag}(\sqrt{2.07}\,,$ $\sqrt{2.42}\,)$, $a^* = (0, -0.304)^T$ for $\mu = 0$, $\sigma = 1$ after 17 iterations of the algorithm given in Subsection 4.3d.

12. Derive optimal estimators for the Weibull distribution model, given by the densities

$$f_{\sigma, \gamma}(x) = \frac{\gamma}{\sigma}\left(\frac{x}{\sigma}\right)^{\gamma - 1} \exp\left\{\left(\frac{x}{\sigma}\right)^\gamma\right\}.$$

Section 4.4

13. Consider a location-scale model. Sketch a picture analogous to Figure 1 in Subsection 4.2d for an optimal B^p-robust estimator for arbitrary c_1 and c_2.

14. (Short) Convince yourself that in a location-scale model both Huber's Proposal 2 (cf. Subsection 4.2d) (with large enough constant b) and its variant A15 (cf. Section 1.4), where the scale is estimated by the median deviation, are optimal in the sense of Subsection 4.4b. Which modification of Proposal 2 is needed for small constants b? Note that these estimators are not optimal in the sense of Section 4.3.

15. In the simple regression example described as Example 2 in Subsection 4.3d consider the intercept $\theta^{(1)}$ as a nuisance parameter. (a) Give the equations that define the B^p-optimal ψ-functions in $\theta_0 = 0$. (b) Intro-

duce a scale model for the errors $X^{(2)} - \theta^{(2)}X^{(1)}$. Now the parameter falls into three "natural parts" (the three components), two of which are usually regarded as nuisance parameters. Sketch how the idea of Section 4.4 can be extended to such cases. (In practice, the distribution of $X^{(1)}$ can also be assumed to follow a parametric model, which gives rise to more nuisance parameters.) (c) In the model without scale, assume that $X^{(1)} \sim \mathcal{N}(\mu, 4)$ with $\mu \neq 0$ (but fixed). How does this affect the optimal ψ-functions?

Section 4.5

16. (Short) Give the basic distribution F_0 and the generating transformations $\{\alpha_\theta\}$ for (a) the normal location-scale model $\{\mathcal{N}(\mu, \sigma^2)\}$; (b) the standard linear regression model with random carriers (distribution K) and normal errors; and (c) the m-dimensional normal model $\{\mathcal{N}_m(\mu, \Sigma)\}$.

17. Verify the assertion about constant gross-error sensitivities stated near the beginning of Subsection 4.5c.

Subsection 4.6a

18. Let $p = 2$ and $C = \text{diag}(c_1, c_2)$. (a) Sketch the limit of the function $z \mapsto C^{-1}\tilde{h}_{c,C}(Cz)$ for $c_2 \to 0$. (b) Use this function to define an M-estimator in the same way as near (4.6.1). (c) Prove that this estimator is admissible. (d) Generalize the idea in as many ways as you can think of (for $p \geq 2$). (cf. Stahel, 1981a).

Subsection 4.6b

19. Adapt Algorithm 4 to the location-scale model. (Note that there are better algorithms in this case.)

CHAPTER 5

Estimation of Covariance Matrices and Multivariate Location

5.1. INTRODUCTION

The estimation of covariance matrices may be called the key to multivariate statistics. Robust estimators of these matrices open the door to the robustification of the classical normal-theory multivariate procedures.

A public start on the subject was presented by Gnanadesikan and Kettenring in 1972. Hampel (1973a) apparently was the first to mention the idea of affinely equivariant M-estimators for covariance matrices. The basic paper on these estimators is due to Maronna (1976). He treated the problems of existence and uniqueness, asymptotic distribution, and breakdown point for an important subclass of these estimators. An attempt by Schönholzer (1979) to extend his findings did not reach much further. Huber (1977b; 1981, Chapter 8) calculated the influence function and commented on computation and breakdown. He and Collins (1982) presented optimal estimators in a minimax sense. Stahel (1981a) derived the optimal B-robust estimators and added further remarks on breakdown properties. He also introduced an estimator with high breakdown point (see also Stahel, 1981b), which was found independently by Donoho (1982) (see also Donoho and Huber, 1983).

Most of these results will be presented in this chapter. After the formal introduction of the model, we will build up some machinery, which simplifies the derivation of many useful formulas and may give some deeper insight into the structure of the model. Then, optimal estimators will be given. Finally, we discuss breakdown aspects.

270

As pointed out above, the robustification of multivariate procedures can be achieved by introducing robust estimators of covariance matrices. This was done for discriminant analysis by Randles et al. (1978), Broffitt et al. (1980), and Campbell (1982). For specific problems, adapted methods may be more suitable. Devlin et al. (1975), Hampel (1975, p. 377) and others (see Huber, 1981, Sections 8.2–3) give such specific methods for simple correlations. Principal components were studied by Campbell (1980), Devlin et al. (1981), Ruymgaart (1981), Chen and Li (1981), and Li and Chen (1981).

Testing problems have received less attention. Muirhead and Waternaux (1980) examine the distribution of the classical sample covariance matrix of non-normal observations with known fourth moments, and apply the results to testing canonical correlation coefficients. The test statistics used in classical multivariate analysis are functions of the sample covariance matrix. If this estimate is replaced by a robust one, the resulting tests supposedly are robust, too. The null distribution of such robustified test statistics can be derived from the distribution of the robust covariance estimate used, and this was done for several statistics by Tyler (1981, 1982, 1983a).

Some authors suggest to estimate the elements of a covariance matrix separately from the one- and two-dimensional marginal distributions, notably by the correlation estimators mentioned above and some scale estimators. In general, the resulting matrices are not positive semidefinite, and their use should be restricted to situations where the individual elements are of primary interest.

Our discussion will be confined to the affinely equivariant M-estimators and another class of equivariant procedures.

5.2. THE MODEL

5.2a. Definition

Let F_0 be a spherically symmetric distribution in $\mathscr{X} = \mathbb{R}^m$ (i.e., F_0 shall be invariant under orthogonal transformations in $\mathbb{R}^m$). The model to be considered in this chapter comprises the distributions obtained by applying all affine transformations

$$\alpha_{A,a}(z) = Az + a, \qquad a \in \mathbb{R}^m, \quad A \text{ nonsingular } m \times m \text{ matrix}$$

to F_0. The resulting model distributions are often called elliptical distributions.

The pair A and a is not a suitable parameter for the model since A and $A\Gamma$ with orthogonal Γ lead to the same distribution. But $\Sigma = AA^T$ is the same for all A leading to the same distribution (and only for these), and the pair Σ and a could be used as a parametrization. Since the symmetric matrix Σ contains redundance, and the theory of Chapter 4 applies only to vectors, it will prove useful to introduce the following notation.

Notation 1.   If S is a symmetric matrix, let $\text{vecs}(S)$ be the vector

$$\text{vecs}(S) = \left(s_{11}/\sqrt{2}, \ldots, s_{mm}/\sqrt{2}, s_{21}, s_{31}, s_{32}, \ldots, s_{m,m-1}\right)^{T}. \quad (5.2.1)$$

The factor $1/\sqrt{2}$, applied to the diagonal elements, will simplify some formulas later on. A reason for this may be seen in the fact that

$$\| \text{vecs}(S) \|^{2} = \frac{1}{2}\Sigma_{ij}s_{ij}^{2} = \frac{1}{2}\text{trace}(SS^{T}). \quad (5.2.2)$$

Remark 1.   Searle (1978) and Henderson and Searle (1979) introduce a similar operator, omitting the factor (cf. McCulloch, 1982). Alternatively, the full matrices can be stacked columnwise to form a vector, and, later on, Kronecker products may be used (cf. Tyler, 1982; Magnus and Neudecker, 1979, and references therein). Our formalism is simpler and sufficient for our purposes.

The parameter vector of the model will consist of the two parts $\text{vecs}(\Sigma)$ and $\mu = a$.

Notation 2.   Instead of writing $\theta = [\text{vecs}(\Sigma)^{T}, \mu^{T}]^{T}$ we shall often use $\theta = (\Sigma, \mu)$ for short; F_{θ} will also be denoted as $F_{\Sigma,\mu}$.

Definition 1.   The *covariance–location model* generated by (a spherically symmetric) F_0 is

$$\left\{ F_{\Sigma,\mu} | \mu \in \mathbb{R}^{m}, \Sigma \text{ positive definite} \right\},$$

where $F_{\Sigma,\mu}$ is the distribution of

$$\alpha_{L,\mu}(Z) = LZ + \mu \quad \text{with } LL^{T} := \Sigma.$$

Technically, $\theta := [\text{vecs}(\Sigma)^{T}, \mu^{T}]^{T}$ is considered being the parameter.

Note that the dimension of θ is

$$p = m(m + 1)/2 + m. \quad (5.2.3)$$

All roots L of Σ yield the same distribution. We will choose L to be lower triangular with positive diagonal in order to have a unique transformation $\bar{\alpha}_{\Sigma,\mu} = \alpha_{L,\mu}$ for each parameter value (thus making $LL^T = \Sigma$ the Cholesky factorization). The covariance–location model is then the model generated by F_0 and the transformations $\bar{\alpha}_{\Sigma,\mu}$ in the sense of Subsection 4.5a.

The simplest and by far most relevant example is the normal distribution, where $F_0 = \mathcal{N}_m(0, I)$ and $F_{\Sigma,\mu} = \mathcal{N}_m(\mu, \Sigma)$. Σ is therefore called a (pseudo-) covariance matrix also in other cases.

Remark 2. The (rescaled) standard *normal distribution* is in fact the only spherically symmetric distribution for which the components of the random variable are independent. Since independence is a more natural concept than spherical symmetry, the generalization gained by allowing for different F_0 is spurious. A set of nearby elliptical distributions is certainly not a sufficiently rich neighborhood to be used for relevant robustness considerations. Remember that the approach based on influence functions intends to safeguard against all kinds of contamination. We feel that it is more relevant to examine the behavior in full but infinitesimal neighborhoods than in finite but "thin" ones. Nevertheless, there is a growing literature on elliptical distributions, most of which contain robustness considerations (e.g., Tyler, 1983b; Muirhead and Waternaux, 1980).

We shall allow for different elliptical models to be used as *central models* in parts of our discussion, even though this does not seem to have great relevance.

For later purposes, we characterize Σ by two numbers: Let λ_j be the eigenvalues of Σ. We call

$$\tau := \frac{1}{m} \ln \det (\Sigma) = \mathrm{ave}_j \ln(\lambda_j) \qquad (5.2.4)$$

the *log-size* parameter, and define the *shape* parameter Σ by the "standard deviation" of the logarithmized eigenvales,

$$\eta^2 := \mathrm{ave}_j \left[\ln (\lambda_j) - \tau \right]^2 . \qquad (5.2.5)$$

The introduction of the matrix elements σ_{ij} as the components of the parameter vector might seem somewhat unrelated to the structure of the situation. It is reassuring to see that $\|\mathrm{vecs}(\Sigma)\|$, τ, and η are invariant under orthogonal x-transformations, inducing $\Sigma \to \Gamma \Sigma \Gamma^T$ [consider (5.2.2)].

The model is *invariant* with respect not only to the generating transformations $\alpha_{L,\mu}$ (Subsection 4.5b), but to all affine transformations $\alpha_{A,a}$ with nonsingular A. For later use, we note that

$$\tilde{\alpha}_{A,a}\left(\begin{array}{c} \text{vecs}(\Sigma) \\ \mu \end{array}\right) = \left(\begin{array}{c} \text{vecs}(A\Sigma A^T) \\ A\mu + a \end{array}\right), \qquad (5.2.6)$$

which is linear in $\theta = (\Sigma, \mu)$, and therefore

$$\frac{\partial \tilde{\alpha}_{A,a}(\theta)}{\partial \theta}\left(\begin{array}{c} \text{vecs}(\Sigma_*) \\ \mu_* \end{array}\right) = \left(\begin{array}{c} \text{vecs}(A\Sigma_* A^T) \\ A\mu_* \end{array}\right) \qquad (5.2.7)$$

5.2b. Scores

The density of F_0 (with respect to Lebesgue measure), if it exists, must be a function of the (squared) radius only,

$$f_0(z) =: f^z(\|z\|^2),$$

and

$$f_{\Sigma,\mu}(x) = \det(\Sigma)^{-1/2} \cdot f^z(v)$$

with

$$v := (x - \mu)^T \Sigma^{-1}(x - \mu).$$

From

$$\frac{\partial v}{\partial \sigma_{ij}} = -\left[\Sigma^{-1}(x - \mu)(x - \mu)^T \Sigma^{-T}\right]_{ij},$$

$$\frac{\partial v}{\partial \mu} = -2\Sigma^{-1}(x - \mu), \qquad (5.2.8)$$

(cf. Dwyer, 1967; Tracy and Dwyer, 1969; McCulloch, 1982), and the general formula

$$\frac{\partial G(\Sigma)}{\partial[\text{vecs}(\Sigma)]} = \left(\sqrt{2}\,\frac{\partial G}{\partial \sigma_{11}}, \ldots, \sqrt{2}\,\frac{\partial G}{\partial \sigma_{mm}}, \frac{\partial G}{\partial \sigma_{21}} + \frac{\partial G}{\partial \sigma_{12}}, \ldots\right)$$

we get the scores

$$s\left[x, \left(\begin{array}{c} \text{vecs}(\Sigma) \\ \mu \end{array}\right)\right] = \left(\begin{array}{c} \text{vecs}\left[\Sigma^{-1}(x - \mu)(x - \mu)^T \Sigma^{-1}\omega^v(v) - \Sigma^{-1}\right] \\ \Sigma^{-1}(x - \mu)\omega^v(v) \end{array}\right)$$

$$(5.2.9)$$

with

$$\omega^v(v) := -2\, d\ln\{f^z(v)\}/dv.$$

This may be written as

$$s\left[x, \begin{pmatrix} \mathrm{vecs}(\Sigma) \\ \mu \end{pmatrix}\right] = \begin{pmatrix} \mathrm{vecs}[A^{-T}\Sigma^s(z)A^{-1}] \\ A^{-T}\mu^s(z) \end{pmatrix},$$

where A is a root of Σ $(AA^T = \Sigma)$, $z = A^{-1}(x - \mu)$, and

$$\begin{pmatrix} \mathrm{vecs}[\Sigma^s(z)] \\ \mu^s(z) \end{pmatrix} := s\left[z, \begin{pmatrix} \mathrm{vecs}(I) \\ 0 \end{pmatrix}\right] = \begin{pmatrix} \mathrm{vecs}[zz^T\omega^v(\|z\|^2) - I] \\ z\omega^v(\|z\|^2) \end{pmatrix}$$

$$(5.2.10)$$

are the scores for the neutral parameter $\theta_0 = (I, 0)$.

Note that f^z is not equal to the density f^v of the distribution F^v of the squared radius $V = \|Z\|^2$; instead $f^z(v)$ has to be multiplied by the surface of the hypersphere with squared radius v in order to get $f^v(v)$, and with the factor accounting for $\|Z\| \to V$. Thus,

$$f^v(v) = \left[\pi^{m/2}/\Gamma(m/2)\right]v^{m/2-1}f^z(v),$$

$$\omega^v(v) = -2\,d\ln\{f^v(v)\}/dv + (m-2)/v.$$

$$(5.2.11)$$

For the normal distribution, f^v is the χ^2_m density, and $\omega^v(v) \equiv 1$.

5.3. EQUIVARIANT ESTIMATORS

5.3a. Orthogonally Equivariant Vector Functions and d-Type Matrices

Before we derive some nice results from the invariance properties of the covariance–location model, it is useful to build up some algebraic machinery.

Let $a: \mathbb{R}^m \to \mathbb{R}^p$ be an arbitrary function. Each element of $\mathbb{R}^p$ can be written in the form $(\mathrm{vecs}(\Sigma)^T, \mu^T)^T$, corresponding to the meaning of the two parts of the parameter θ. Let us denote the respective parts of the function $a(z)$ by $\mathrm{vecs}[\Sigma^a(z)]$ and $\mu^a(z)$, such that

$$a(z) =: \begin{pmatrix} \mathrm{vecs}[\Sigma^a(z)] \\ \mu^a(z) \end{pmatrix}.$$

Definition 1. A function $a(z)$ is *orthogonally equivariant* if it obeys the functional equations

$$\Sigma^a(\Gamma z) = \Gamma \Sigma^a(z) \Gamma^T,$$

$$\mu^a(\Gamma z) = \Gamma \mu^a(z),$$

for all orthogonal matrices Γ.

Lemma 1. Orthogonally equivariant vector functions are characterized by three functions w_η^a, w_δ^a, w_μ^a: $\mathbb{R}^+ \to \mathbb{R}$ through

$$\Sigma^a(z) = zz^T w_\eta^a(\|z\|^2) - I w_\delta^a(\|z\|^2),$$

$$\mu^a(z) = z w_\mu^a(\|z\|^2).$$

$$(5.3.1)$$

On the other hand, any function $a(z)$ of this form is orthogonally equivariant. Furthermore,

$$\|a(z)\|^2 = \frac{1}{2}\left(1 - \frac{1}{m}\right)\left[v w_\eta^a(v)\right]^2 + \frac{1}{2m} u_\tau^a(v)^2 + v w_\mu^a(v)^2, \quad (5.3.2)$$

where $v := \|z\|^2$ and

$$u_\tau^a(v) := v w_\eta^a(v) - m w_\delta^a(v).$$

Proof. Let $z_* = (r, 0, 0, \ldots, 0)^T$, $r > 0$. Choose for Γ the change of sign of the kth component, $\gamma_{ij} = \delta_{ij}(1 - 2\delta_{ki})$, $k \geq 2$ to see first that $\mu^a(z_*)^{(k)} = 0$ for $k \geq 2$, and therefore, for some $w_\mu^a(r^2)$, we have $\mu^a(z_*) = (r \cdot w_\mu^a(r^2), 0, 0, \ldots, 0)^T$. Second, it shows that $\sigma_{ij}^a(z_*) = 0$ for $i \neq j$. If we now apply the interchange of two components k and l, $\gamma_{ij} = \delta_{ij} - \delta_{ij}\delta_{ik} - \delta_{ij}\delta_{il} + \delta_{ik}\delta_{jl} + \delta_{il}\delta_{jk}$, with $k, l \geq 2$, we see that the $\sigma_{kk}^a(z_*)$, $k \geq 2$, are equal. We may thus write $\Sigma^a(z_*)$ as

$$\Sigma^a(z_*) = \text{diag}\left\{r^2 \cdot w_\eta^a(r^2), 0, 0, \ldots, 0\right\} - I w_\delta^a(r^2)$$

with suitable w_η^a, w_δ^a: $\mathbb{R}^+ \to \mathbb{R}$. For $r = 0$, repeat the above considerations with $k = 1$ in order to see that w_μ^a, w_η^a, and w_δ^a may be extended to 0 $[w_\mu^a(0)$

and $w_\eta^a(0)$ being arbitrary]. In order to get $a(z)$ for general $z \neq 0$, choose a Γ with first column $z/\|z\|$, and $z_* = (\|z\|, 0, 0, \ldots, 0)^T$ and apply the equivariance relations of Definition 1.   □

Note that the scores function $s(z, \theta_0)$ has the mentioned form. We shall need to calculate integrals of the type

$$D = \int a(z) \cdot b(z)^T dF_0(z)$$

where a and b are of the form (5.3.1). If one writes down the elements of D, symmetry considerations show that D is determined by four numbers, d_ν^D, d_ρ^D, d_η^D, and d_μ^D, say, as

$$D = \left[\begin{array}{ccc} \left[\begin{array}{cccc} d_\nu^D & d_\rho^D & \cdots & d_\rho^D \\ d_\rho^D & d_\nu^D & \cdots & d_\rho^D \\ \vdots & \vdots & \vdots & \vdots \\ d_\rho^D & d_\rho^D & \cdots & d_\nu^D \end{array} \right] & 0 & 0 \\ 0 & \left[\begin{array}{cc} d_\eta^D & 0 \\ & \ddots \\ 0 & d_\eta^D \end{array} \right] & 0 \\ 0 & 0 & \left[\begin{array}{cc} d_\mu^D & 0 \\ & \ddots \\ 0 & d_\mu^D \end{array} \right] \end{array} \right] \begin{array}{l} \left.\vphantom{\begin{array}{c}a\\a\\a\\a\end{array}}\right\} m \\ \left.\vphantom{\begin{array}{c}a\\a\end{array}}\right\} \dfrac{m(m-1)}{2} \\ \left.\vphantom{\begin{array}{c}a\\a\end{array}}\right\} m \end{array}$$

$$(5.3.3)$$

We will show shortly that $d_\nu^D = d_\eta^D + d_\rho^D$. Since such matrices are nearly diagonal, we introduce the term "d-type matrix":

Definition 2. The *d-type matrix* D of order $m \geq 2$, given by the three numbers d_η^D, d_τ^D, and d_μ^D, is the matrix (5.3.3) with $d_\nu^D = d_\eta^D + d_\rho^D$, $d_\rho^D = (d_\tau^D - d_\eta^D)/m$.

Lemma 2. If $D = \int a(z) \cdot b(z)^T \, dF_0(z)$ with spherically symmetric F_0, then D is the d-type matrix given by

$$d_\eta^D = \left(1 + \frac{2}{m}\right)^{-1} \int \left(\frac{v}{m}\right)^2 w_\eta^a(v) w_\eta^b(v) \, dF^v(v),$$

$$d_\tau^D = \frac{1}{2m} \int u_\tau^a(v) u_\tau^b(v) \, dF^v(v),$$

$$d_\mu^D = \int \frac{v}{m} w_\mu^a(v) w_\mu^b(v) \, dF^v(v),$$

where F^v is the distribution of $V = \|Z\|^2$, $Z \sim F_0$.

Proof. Symmetry considerations show that D is of the form (5.3.3). Clearly, d_ρ^D equals

$$\frac{1}{2} \int \left[z_i^2 w_\eta^a(\|z\|^2) - w_\delta^a(\|z\|^2) \right] \left[z_j^2 w_\eta^b(\|z\|^2) - w_\delta^b(\|z\|^2) \right] dF_0(z)$$

$$(5.3.4)$$

with $i \neq j$, while d_ν^D is the same with $i = j$, i and j being arbitrary otherwise. Also,

$$d_\eta^D = \int z_i^2 z_j^2 w_\eta^a(\|z\|^2) w_\eta^b(\|z\|^2) \, dF_0(z), \qquad i \neq j, \qquad (5.3.5)$$

$$d_\mu^D = \int z_i^2 w_\mu^a(\|z\|^2) w_\mu^b(\|z\|^2) \, dF_0(z).$$

In the last equation, let i run from 1 to m and sum up to get the form of d_μ^D given in the lemma. Before working on d_η^D, note that, for any function G,

$$\int z_1^4 G(\|z\|^2) \, dF_0(z) = 3 \int z_1^2 z_2^2 G(\|z\|^2) \, dF_0(z), \qquad (5.3.6)$$

which is shown by calculating the following conditional expectations:

$$E_0\left(z_1^4 | z_1^2 + z_2^2\right) = \left(z_1^2 + z_2^2\right) \frac{1}{2\pi} \int_0^{2\pi} \cos(\alpha)^4 \, d\alpha = \frac{1}{8}\left(z_1^2 + z_2^2\right),$$

$$E_0\left(z_1^2 z_2^2 | z_1^2 + z_2^2\right) = \left(z_1^2 + z_2^2\right) \frac{1}{2\pi} \int_0^{2\pi} \cos(\alpha)^2 \sin(\alpha)^2 \, d\alpha = \frac{3}{8}\left(z_1^2 + z_2^2\right).$$

If we now sum up the right-hand side of (5.3.5) over all i and j (including $i = j$) and make use of (5.3.6) with $G(v) = w_\eta^a(v)w_\eta^b(v)$, we get d_η^D as stated in the lemma. By multiplying out (5.3.4) for $i = j$ and for $i \neq j$, and using (5.3.6) again, one verifies $d_\nu^D = d_\eta^D + d_\rho^D$. Finally, adding the integrals (5.3.4) over all i and j we get

$$m\left(d_\eta^D + md_\rho^D\right) = \frac{1}{2} \int \left[vw_\eta^a(v) - mw_\delta^a(v)\right]\left[vw_\eta^b(v) - mw_\delta^b(v)\right] dF^v(v),$$

and the left-hand side is md_τ^D by Definition 2. □

Lemma 3. d-type matrices (denoted A, B, D) have the following properties:

1. Multiplication is an Abelian group operation with

   $$D = A \cdot B \Leftrightarrow d_h^D = d_h^A \cdot d_h^B, \quad h = \eta, \tau, \mu,$$

   $$D = I \quad \Leftrightarrow d_h^D = 1, \quad h = \eta, \tau, \mu,$$

   $$D = A^{-1} \Leftrightarrow d_h^D = 1/d_h^A, \quad h = \eta, \tau, \mu.$$

2. $d_\eta^D, d_\tau^D, d_\mu^D$ are the eigenvalues of D, d_η^D and d_τ^D being the eigenvalues of the upper-left m rows and columns.

3. If Γ is an orthogonal $m \times m$ matrix with a column with all entries equal $(= 1/\sqrt{m})$, and

   $$\tilde{\Gamma} = \begin{pmatrix} \Gamma & 0 & 0 \\ 0 & I & 0 \\ 0 & 0 & I \end{pmatrix},$$

   then the transformation $D \to \tilde{\Gamma}^T D \tilde{\Gamma}$ makes all d-type matrices diagonal simultaneously.

Proof. (1) and (2) follow immediately from (3), which is easily proven. □

Lemma 4. d-type matrices transform orthogonally equivariant vector functions to orthogonally equivariant vector functions,

$$b(z) = D \cdot a(z) \Leftrightarrow$$

$$w_\eta^b(v) = w_\eta^a(v) \cdot d_\eta^D, \quad u_\tau^b(v) = u_\tau^a(v) \cdot d_\tau^D, \quad w_\mu^b(v) = w_\mu^a(v) \cdot d_\mu^D.$$

Proof. The jth component of $D \cdot a(z)$, $l \le j \le m$, is

$$\tfrac{1}{2}\left\{ d_\eta^D \left[z_j^2 w_\eta^a(\|z\|^2) - w_\delta^a(\|z\|^2) \right] + d_\rho^D \Sigma_i \left[z_i^2 w_\eta^a(\|z\|^2) - w_\delta^a(\|z\|^2) \right] \right\}$$

$$= \tfrac{1}{2}\left\{ d_\eta^D z_j^2 w_\eta^a(\|z\|^2) - \left[\left(d_\eta^D + m d_\rho^D \right) w_\delta^a(\|z\|^2) - d_\rho^D \|z\|^2 w_\eta^a(\|z\|^2) \right] \right\}.$$

Therefore, $b(z)$ is clearly of the form (5.3.1), with

$$w_\delta^b(v) = d_\tau^D w_\delta^a(v) - d_\rho^D v w_\eta^a(v),$$

such that

$$u_\tau^b(v) = v w_\eta^b(v) - m w_\delta^b(v)$$

$$= v w_\eta^a(v) d_\eta^D - m w_\delta^a(v) d_\tau^D + v w_\eta^a(v) \left(d_\tau^D - d_\eta^D \right)$$

$$= u_\tau^a(v) d_\tau^D. \qquad \qquad \square$$

5.3b. General Results

Equivariant estimators in the covariance location model must satisfy

$$T\{ \tilde{\tilde{\alpha}}_{A,a}(F) \} = \tilde{\alpha}_{A,a}\{ T(F) \} = \begin{pmatrix} \text{vecs}\left[A\hat{\Sigma}(F)A^T \right] \\ A\hat{\mu}(F) + a \end{pmatrix}$$

where $T(F) =: [\hat{\Sigma}(F), \hat{\mu}(F)]$. Inserting $F = F_0$, $a = 0$, and any orthogonal Γ for A, we obtain

$$\hat{\Sigma}(F_0) = \Gamma\hat{\Sigma}(F_0)\Gamma^T, \qquad \hat{\mu}(F_0) = \Gamma\hat{\mu}(F_0),$$

and therefore $T(F_0)$ must be proportional to θ_0,

$$\hat{\Sigma}(F_0) = \sigma_0 I, \qquad \hat{\mu}(F_0) = 0$$

for some $\sigma_0 \in \mathbb{R}$, (compare the proof of Lemma 1). Combining this equation with the first one, we get

$$\hat{\Sigma}(F_{\Sigma,\mu}) = \sigma_0 \Sigma, \qquad \hat{\mu}(F_{\Sigma,\mu}) = \mu,$$

and Fisher consistency holds if and only if $\sigma_0 = 1$. If $\hat{\tau}$ is the estimator of size obtained from $\hat{\Sigma}$ by (5.2.4), then clearly $\hat{\tau}(F_0) = \ln(\sigma_0)$.

The *influence function* of equivariant estimators is, according to (4.5.2) and (5.2.7),

$$\mathrm{IF}\big(Lz + \mu; T, F_{(LL^T),\mu}\big) = \begin{pmatrix} \mathrm{vecs}\big[L\Sigma^I(z)L^T\big] \\ L\mu^I(z) \end{pmatrix}, \qquad (5.3.7)$$

where

$$\begin{pmatrix} \mathrm{vecs}\big[\Sigma^I(z)\big] \\ \mu^I(z) \end{pmatrix} := \mathrm{IF}(z; T, F_0).$$

This equation still holds if L is not lower triangular, and inserting $L = \Gamma$, orthogonal, and $\mu = 0$ shows that $\mathrm{IF}(\cdot; T, F_0)$ is an orthogonally equivariant vector function. Now, the machinery of the preceding subsection produces the following results:

1. $\mathrm{IF}(\cdot; T, F_0)$ is of the form (5.3.1), with characterizing functions w_η^I, u_τ^I, and w_μ^I, say, which by (4.2.1) and (4.2.5) fulfill

$$\int u_\tau^I(v)\, dF^v(v) = 0,$$

$$\left(1 + \frac{2}{m}\right)^{-1} \int \left(\frac{v}{m}\right)^2 w_\eta^I(v)\omega^v(v)\, dF^v(v) = 1,$$

$$\frac{1}{2m} \int u_\tau^I(v)\big[v\omega^v(v) - m\big]\, dF^v(v) = 1,$$ \hfill (5.3.8)

$$\int \frac{v}{m} w_\mu^I(v)\omega^v(v)\, dF^v(v) = 1.$$

2. The norm of the influence function at F_0 is

$$\big\|\mathrm{IF}(z; T, F_0)\big\|^2 = \frac{1}{2}\left(1 - \frac{1}{m}\right)\big[vw_\eta^I(v)\big]^2 + \frac{1}{2m} u_\tau^I(v)^2 + vw_\mu^I(v)^2,$$

where $v := \|z\|^2$.

3. $zw_\mu^I(\|z\|^2)$ is the influence function of the location component $\hat{\mu}$ of T at F_0, and the last term $vw_\mu^I(v)^2$ of $\|\mathrm{IF}(z; T, F_0)\|^2$ is its squared norm;

$u_\tau^I(\|z\|^2)/(m\sigma_0)$ is the influence function of $\hat{\tau} = \ln\det(\hat{\Sigma})/m$, its square being essentially the second term of $\|\text{IF}(z; T, F_0)\|^2$. The first expression is the squared norm of the influence function $\text{vecs}[zz^T - (\|z\|^2/m)I]w_\eta^I(\|z\|^2)$ of $\sigma_0 \cdot \text{vecs}[\exp(-\hat{\tau})\hat{\Sigma} - I]$, and indeed equals the square of the influence function of $\hat{\eta}\sigma_0\sqrt{m/2}$. [For a proof, let $\Sigma(\varepsilon) := \sigma_0 I + \varepsilon\Sigma^I(r, 0, \ldots, 0)$, which is diagonal, and calculate the derivative of the corresponding shape parameter $\eta(\varepsilon)$ for $\varepsilon > 0$. Apply l'Hôpital's rule to obtain $\eta'(0)$.] In summary the influence function splits up into three orthogonal components which correspond (at F_0) to the three parts shape, size, and location of the parameter.

4.　The *asymptotic covariance matrix* of equivariant covariance–location estimators is

$$V(T, F_{\Sigma,\mu}) = \left[\partial\tilde{\alpha}_{L,\mu}(\theta)/\partial\theta\right]V(T, F_0)\left[\partial\tilde{\alpha}_{L,\mu}(\theta)/\partial\theta\right]^T, \qquad LL^T = \Sigma,$$

with the derivatives given in (5.2.7), and $V(T, F_0)$ is a d-type matrix with

$$d_\eta^V = \left(1 + \frac{2}{m}\right)^{-1}\int\left(\frac{v}{m}\right)^2 w_\eta^I(v)^2\, dF^v(v),$$

$$d_\tau^V = \frac{1}{2m}\int u_\tau^I(v)^2\, dF^v(v),$$

$$d_\mu^V = \int\frac{v}{m}w_\mu^I(v)^2\, dF^v(v).$$

Clearly, this is simply a formal description, asymptotic normality not being rigorously established.

5.　The *Fisher information* matrix at θ_0 is of d-type, given by

$$d_\eta^J = \left(1 + \frac{2}{m}\right)^{-1}\int\left(\frac{v}{m}\right)^2 \omega^v(v)^2\, dF^v(v),$$

$$d_\tau^J = \frac{1}{2m}\int[v\omega^v(v) - m]^2\, dF^v(v),$$

$$d_\mu^J = \int\frac{v}{m}\omega^v(v)^2\, dF^v(v).$$

[For the normal distribution, $J(\theta_0) = I$, since $\omega^v(v) \equiv 1$ and $F^v = \chi_m^2$.] This bounds the $V(., F_0)$ matrices for Fisher-consistent estimators to

$$d_h^V \geq 1/d_h^J, \qquad h = \eta, \tau, \mu.$$

6. The standardized *sensitivities* are

$$\gamma_s^*(T, F_{\Sigma,\mu})^2 = \sup_v \left\{ \tfrac{1}{2} \left(1 - \frac{1}{m} \right) \left[vw_\eta^I(v) / d_\eta^V \right]^2 \right.$$

$$\left. + \frac{1}{2m} \left[u_\tau^I(v) / d_\tau^V \right]^2 + v \left[w_\mu^I(v) / d_\mu^V \right]^2 \right\},$$

$$\gamma_i^*(T, F_{\Sigma,\mu})^2 = \sup_v \left\{ \tfrac{1}{2} \left(1 - \frac{1}{m} \right) \left[vw_\eta^I(v) \, d_\eta^J \right]^2 \right.$$

$$\left. + \frac{1}{2m} \left[u_\tau^I(v) \, d_\tau^J \right]^2 + v \left[w_\mu^I(v) \, d_\mu^J \right]^2 \right\}.$$

In view of 3, it seems natural to look at the suprema of the three components, or equivalently, at the sensitivites for $\hat{\eta}$, $\hat{\tau}$, and $\hat{\mu}$,

$$\gamma_s^*(\hat{\eta}, F_0) =: \gamma_\mu^s(T) := \sup_v \left[vw_\eta^I(v) \right] \Big/ \sqrt{m(m+2) d_\eta^V} \,,$$

$$\gamma_s^*(\hat{\tau}, F_0) =: \gamma_\tau^s(T) := \sup_v |u_\tau^I(v)| \Big/ \sqrt{2m d_\tau^V} \qquad (5.3.9)$$

$$\gamma_s^*(\hat{\mu}, F_0) =: \gamma_\mu^s(T) := \sup_v \left[\sqrt{v} \, w_\mu^I(v) \Big/ \sqrt{d_\mu^V} \right]$$

For the information-standardized sensitivities, replace d_h^V by $(d_h^J)^{-1}$, $h = \eta, \tau, \mu$.

5.3c. *M*-Estimators

Since any *M*-estimator may be given using its influence function as defining ψ-function, Eq. (5.3.7) and the considerations following it say that every equivariant *M*-estimator admits a ψ-function of the following type: $\psi_0(z) := \psi(z, \theta_0)$ is orthogonally equivariant, and $\psi(x, \theta)$ may be chosen as

$$\psi \left\{ x, \begin{pmatrix} \mathrm{vecs}(\Sigma) \\ \mu \end{pmatrix} \right\} = \psi_0 \{ L^{-1}(x - \mu) \} \qquad (5.3.10)$$

(where $LL^T = \Sigma$), since this ψ coincides with the influence function up to the matrix $\partial\tilde{\alpha}/\partial\theta$ [see (5.2.6) and (5.2.7)]. Conversely:

Lemma 5. An *M*-estimator of this type is affinely equivariant, that is, let ψ be of the described form. If $T = (\hat{\Sigma}, \hat{\mu})$ is a solution of $\int \psi(x, T) \, dF(x) = 0$, then $\overline{T} = ([A\hat{\Sigma}A^T], A\hat{\mu} + a)$ is a solution of $\int \psi(\overline{x}, \overline{T}) \, d\overline{F}(\overline{x}) = 0$ with $\overline{F} = \tilde{\alpha}_{A,a}(F)$.

Proof. With $\hat{L}\hat{L}^T = \hat{\Sigma}$ we have $\int \psi_0 \{\hat{L}^{-1}(x - \hat{\mu})\} \, dF(x) = 0$, and $\int \psi_0 \{\hat{L}^{-1}[A^{-1}(\bar{x} - a) - \hat{\mu}]\} \, d\bar{F}(\bar{x}) = 0$ by the definition of $\bar{F}$. This may be rewritten as $\int \psi_0 \{\Gamma^T \tilde{L}^{-1}[\bar{x} - (a + A\hat{\mu})]\} \, d\bar{F}(\bar{x}) = 0$ with orthogonal Γ and lower triangular $\tilde{L}$. Orthogonal equivariance of ψ_0 yields $\int \psi_0 \{\tilde{L}^{-1}[\bar{x} - (a + A\hat{\mu})]\} \, d\bar{F}(\bar{x}) = 0 = \int \psi(\bar{x}, \bar{T}) \, d\bar{F}(\bar{x})$ with $\bar{T} = (\tilde{L}\tilde{L}^T, A\hat{\mu} + a)$. Finally, $\tilde{L}\tilde{L}^T = A\hat{L}\Gamma\Gamma^T\hat{L}^T A^T = A\hat{\Sigma}A^T$. $\qquad\square$

A more explicit description of the affinely equivariant M-estimators is as follows: Let $w_\eta^\psi, w_\delta^\psi, w_\mu^\psi : \mathbb{R}^+ \to \mathbb{R}$ be three scalar functions. Solve

$$\int zz^T w_\eta^\psi (\|z\|^2) \, dF(x) = I \int w_\delta^\psi (\|z\|^2) \, dF(x),$$

$$\int z w_\mu^\psi (\|z\|^2) \, dF(x) = 0, \qquad (5.3.11)$$

where $z := L^{-1}(x - \mu)$, for L and μ, thus obtaining $\hat{L}$ and $\hat{\mu}$. Then estimate Σ by $\hat{L}\hat{L}^T$ and μ by $\hat{\mu}$. This estimator coincides with the M-estimator determined by the ψ-function (5.3.10), where ψ_0 is given by w_η^ψ, w_δ^ψ, and w_μ^ψ through (5.3.1).

Equation (5.3.11) is equivalent to

$$\int \left(zz^T - I \cdot \|z\|^2/m \right) w_\eta^\psi (\|z\|^2) \, dF(x) = 0,$$

$$\int u_\tau^\psi (\|z\|^2) \, dF(x) = 0, \qquad (5.3.12)$$

$$\int z w_\mu^\psi (\|z\|^2) \, dF(x) = 0$$

(where u_τ^ψ is defined as in Lemma 1). One component equation of the first line is redundant. In spite of the spurious complication, this alternative system gives more insight into the structure of covariance matrices. The three lines involve only one of the functions w_η^ψ, u_τ^ψ, and w_μ^ψ each. They reflect the split mentioned in result 3 of Subsection 5.3b, as is shown in result 3 which follows.

Remark 1. Different w_δ^ψ functions can be used to define the same M-estimator. In fact, w_δ^ψ can be replaced by

$$\tilde{w}_\delta^\psi (v) \doteq w_\delta^\psi (v) + \text{const} \cdot u_\tau^\psi (v).$$

On the other hand, w_η^ψ, u_τ^ψ, and w_μ^ψ are uniquely determined up to scalar factors.

Example 1. Huber (1977b; 1981, Section 8.10) develops the following *minimax approach*. He considers a contamination "neighborhood" of the normal model in the set of all elliptical models (cf. Remark 2 of Subsection 5.2a!). Then he poses a minimax problem, as described in Section 2.7, for the estimation of location and (pseudo-) covariance separately. The solution for the location part has no simple analytic form. For the covariance part, he obtains $w_\delta^\psi(v) = 1$ and

$$w_\eta^\psi(v) = \begin{cases} a/v, & v < a \\ 1, & a \le v \le b \\ b/v, & v > b, \end{cases}$$

with suitable constants a, b depending on the size ε of the neighborhood and on the dimension m. The lower cutpoint a vanishes for small ε and m. When it is positive, it gives rise to an awkward discontinuity at zero (at $x_i = \hat{\mu}$, that is).

This estimator is not consistent. Huber recommends to multiply the estimate obtained from w_η^ψ and $w_\delta^\psi \equiv 1$ by the appropriate constant τ in order to obtain the correct value for the normal distribution.

Let us complete the definition of the estimator by setting the weight function for location equal to the familiar $w_\mu^\psi(v) = \min(1, \sqrt{b_\mu/v})$ for the location part.

A numerical example is shown in Figure 1. The influence of the two outliers in the sample is reduced successfully, and the robust estimate suggests independence of the two variables while the ordinary estimate indicates a negative correlation. In this case, the outliers can be detected by other means as well. In higher dimensions, a robust estimator can be the only means of getting at them.

The *asymptotic expressions* for equivariant M-estimators may be written down using the formulas of the preceding subsections. Sufficient conditions for consistency and asymptotic normality, which render these expressions meaningful, are given by Maronna (1976) and Schönholzer (1979), see also Huber (1981).

1. $\hat{\Sigma}(F_0) = \sigma_0 I$ [if $\hat{\Sigma}(F_0)$ is unique], where σ_0 is obtained from

$$\int u_\tau^\psi(v/\sigma_0) \, dF^v(v) = 0. \tag{5.3.13}$$

T_ψ is Fisher consistent (or there is such a root) for all θ if and only if $\int u_\tau^\psi(v) dF^v(v) = 0$.

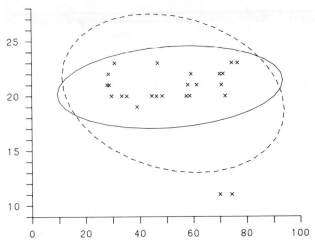

Figure 1. Scatterplot of two variables, together with ellipses of constant density containing 95% of the probability of a normal distribution as estimated by (a) the ordinary mean and empirical covariance matrix, and (b) the robust estimator defined in Example 1 with $a = 0$, $b = 2 = b_\mu$. The data are given in the first two columns of Table 4 of Subsection 7.5d.

2. The matrices $M(T_\psi, F_0)$, $Q(T_\psi, F_0)$, and $V(T_\psi, F_0)$ given by (4.2.11) and (4.2.13) are d-type matrices with characterizing numbers

$$d_\eta^M = \left(1 + \frac{2}{m}\right)^{-1} \int \left(\frac{v}{m}\right)^2 w_\eta^\psi(v)\omega^v(v)\, dF^v(v),$$

$$d_\tau^M = \frac{1}{2m} \int u_\tau^\psi(v)\left[v\omega^v(v) - m\right] dF^v(v),$$

$$d_\mu^M = \int \frac{v}{m} w_\mu^\psi(v)\omega^v(v)\, dF^v(v);$$

$$d_\eta^Q = \left(1 + \frac{2}{m}\right)^{-1} \int \left(\frac{v}{m}\right)^2 w_\eta^\psi(v)^2\, dF^v(v),$$

$$d_\tau^Q = \frac{1}{2m} \int u_\tau^\psi(v)^2\, dF^v(v),$$

$$d_\mu^Q = \int \frac{v}{m} w_\mu^\psi(v)^2\, dF^v(v);$$

$$d_h^V = d_h^Q / \left(d_h^M\right)^2, \qquad h = \eta, \tau, \mu.$$

3. The influence function is of the form (5.3.1) with

$$w_\eta^I(v) = w_\eta^\psi(v)/d_\eta^M, \qquad u_\tau^I(v) = u_\tau^\psi(v)/d_\tau^M, \qquad w_\mu^I(v) = w_\mu^\psi(v)/d_\mu^M.$$

The question of *existence and uniqueness* (or domain) of such estimators, that is, of a solution of Equation (5.3.11) for a given F, is of clear significance in this context, and it is even more important since it is connected to the existence and uniqueness of ψ-functions defining the optimal B_s-robust estimator in general models (see Subsection 4.3c).

Theorem 1 (*Existence* of affinely equivariant M-estimators). The following conditions guarantee the existence of a solution of (5.3.11):
Condition for w_δ^ψ:

 (i) $w_\delta^\psi(v)$ is constant, $w_\delta^\psi(v) \equiv w_\delta^* > 0$.

Conditions for w_η^ψ:

 (ii) $w_\eta^\psi(v)$ is continuous for $v > 0$.
 (iii) $w_\eta^\psi(v) \geq 0$.
 (iv) $w_\eta^\psi(v)$ is nonincreasing for $v > 0$.
 (v) $\psi_\eta(v) := v w_\eta^\psi(v)$ is bounded.
 (vi) ψ_η is nondecreasing.
 (vii) $\lim_{v \downarrow 0} \{\psi_\eta(v)\} = 0$.

Conditions for w_μ^ψ:

(viii) $w_\mu^\psi(v)$ is continuous for $v > 0$.
 (ix) $w_\mu^\psi(v) \geq 0$.
 (x) w_μ^ψ is bounded.
 (xi) $\sqrt{v}\, w_\mu^\psi(v)$ is bounded.

Condition connecting w_δ^ψ, w_η^ψ, and w_μ^ψ:

(xii) There is a $v^* > 0$ with $[0 \leq v \leq v^* \Rightarrow w_\mu^\psi(v) > 0$ and $\psi_\eta(v)$ strictly increasing in $v]$ and $\psi_\eta(v^*) > m w_\delta^*$.

Conditions on the distribution:

(xiii) $F\{X \in H\} \leq 1/m$ and $< \frac{1}{2}$ for all $(m-1)$-dimensional hyperplanes H of $\mathbb{R}^m$.

(xiv) $F\{ X \in H \} \le 1 - m \cdot w_\delta^* / \sup_v \{ \psi_\eta(v) \}$ for all H.

Condition (vii) may be replaced by

(xv) $\lim_{v \downarrow 0} \{ \psi_\eta(v) \} < m w_\delta^*$.

(xvi) $F\{ X = x \} = 0$ for all $x \in \mathbb{R}^m$ (which excludes sample distributions).

Maronna (1976) needed somewhat stronger conditions than (vii), (xiv), and (xii). The proofs of our conditions are given by Stahel (1981a), who just collects the results of Schönholzer (1979) (see Huber, 1981) and adds a detail.

The uniqueness of the solution can be proved only under very restrictive assumptions which exclude sample distributions. Indeed, for model distributions, the following result applies.

Theorem 2 (*Existence and uniqueness for elliptical distributions*). If (1) F has a density of the form $f(z) = f^z(\|z\|^2)$ with strictly decreasing f^z and (2) $w_\eta^\psi(v) \ge 0$ and $w_\mu^\psi(v) \ge 0$ for all $v \ge 0$, then (5.3.11) has a unique solution if and only if $\int u_\tau^\psi(v/\sigma) \, dF^v(v) = 0$ has a unique solution σ_0 for σ. The solution of (5.3.11) is then $T(F) = (\sigma_0 I, 0)$. The same is true if (1) and (2) are changed to (1') nonincreasing f^z and (2') $w_\eta^\psi(v) > 0$ and $w_\mu^\psi(v) > 0$ for all $v \ge 0$.

Proof. The existence part is obvious, and the only thing to show is that every solution must have the form $(\hat{\sigma}I, 0)$. This is proved in Stahel (1981a) by extending the idea of proof of Maronna's theorem 3(i) to the covariance part. □

Remark 2. By equivariance, the existence and uniqueness extends to all model distributions.

Remark 3. Schönholzer (1979) shows that, under quite severe restrictions on the w functions, existence and uniqueness carry over to a Prohorov neighborhood of the model distributions and thus to large samples from a model distribution.

A simple *algorithm* for computing estimates is the reweighing method of Maronna (1976), that has been described in another context in Subsection

4.3d, step 4. For that matter, note that (5.3.11) may be written as

$$\hat{\Sigma} = \frac{\int (x - \hat{\mu})(x - \hat{\mu})^T w_\eta^\psi(v)\, dF(x)}{\int w_\delta^\psi(v)\, dF(x)},$$

$$\hat{\mu} = \frac{\int x w_\mu^\psi(v)\, dF(x)}{\int w_\mu^\psi(v)\, dF(x)},$$

$$(5.3.14)$$

where $v := (x - \hat{\mu})^T \hat{\Sigma}^{-1}(x - \hat{\mu})$. The algorithm implied by these formulas may be slow and numerically unstable. Huber (1977b), Huber (1981, Section 8.11), and Stahel (1981a) discuss some more sophisticated ideas. (The "accelerated" algorithm described in Maronna, 1976, does not seem to be reliable.) More research is needed, however, in this area. The program library ROBETH described briefly in Subsection 6.4c implements the reweighing algorithm and the algorithm of Huber (1977b) (see Marazzi, 1980d).

Our next goal is to derive the optimal estimators. Since the two standardized versions must be *equivariant* (with respect to the generating transformations at least; see Subsection 4.5c), and they are *W-estimators*, it is useful to have a brief look at this class: The translation vector $C(\theta)$ allowed for in Definition 2 of Subsection 4.3b should help to achieve Fisher consistency. In the present model, at $\theta = \theta_0$, it suffices to allow for $C(\theta_0) = (c_0 I, 0)$. The general ψ_0-function of an equivariant, Fisher-consistent *W*-estimator is thus characterized by the weight function $w: \mathbb{R}^+ \to \mathbb{R}$ and $\beta \ (= c_0 + 1)$ through

$$w_\eta^\psi(v) = w(v)\omega^v(v) = w_\mu^\psi(v), \qquad u_\tau^\psi(v) = [v \cdot \omega^v(v) - m\beta]w(v),$$

$$\beta = \frac{\int (v/m)w(v)\omega^v(v)\, dF^v(v)}{\int w(v)\, dF^v(v)}.$$

5.4. OPTIMAL AND MOST *B*-ROBUST ESTIMATORS

5.4a. Full Parameter

In order to derive optimal *B*-robust estimators of the parameter (Σ, μ) in the sense of Section 4.3, the steps described in Subsection 4.3d will be carried out. The machinery of Subsection 5.3a will help to achieve great simplifications. The unstandardized estimator will not be derived, since we consider equivariance to be essential in this context. For the sake of simplicity of the formulas, we choose the normal model instead of a general F_0, since it is by far the most useful one, and since generalization to other F_0 is easy.

Step 1. Theorem 1 of Subsection 4.5c shows that it suffices to consider θ_0.

Step 2. The scores and information matrix are given by (5.2.10) and result (5) of Subsection 5.3b.

Step 3. It is natural to guess that A in $W^{A,a}$ [defined in (4.3.7)] will be a d-type matrix and a has the form $([\beta - 1]\text{vecs}(I), 0)$ [as $C(\theta_0)$ had at the end of the last subsection], since this makes $A(s - a)$ an orthogonally equivariant vector function. (The letters A and a are used here with a different meaning than in the preceding sections, but in agreement with Section 4.3.) Using $v := \|z\|^2$, we obtain

$$\left\| A\left[s(z, \theta_0) - a \right] \right\|^2$$

$$= \frac{1}{2}\left(1 - \frac{1}{m}\right)\left[v d_\eta^A \right]^2 + \frac{1}{2m}\left[(v - m\beta)\, d_\tau^A \right]^2 + v\left(d_\mu^A \right)^2 =: r(v)^2,$$

$$(5.4.1)$$

say, a quadratic function in v. Next,

$$W^{A,a}(z) = \min\{1, c/r(v)\} =: w(v) \qquad (5.4.2)$$

and the M_k, $k = 1, 2$, are d-type matrices determined by

$$d_\eta^{M_k} = \left(1 + \frac{2}{m}\right)^{-1} \int \left(\frac{v}{m}\right)^2 w(v)^k\, dF^v(v),$$

$$d_\tau^{M_k} = \frac{1}{2m} \int (v - m\beta)^2 w(v)^k\, dF^v(v), \qquad (5.4.3)$$

$$d_\mu^{M_k} = \int \frac{v}{m} w(v)^k\, dF^v(v).$$

The final equations reduce to

(b) $$\left(d_h^A\right)^2 = 1/d_h^{M_2}, \qquad h = \eta, \tau, \mu \quad \text{for } T_c^s,$$

or   (5.4.4)

(c) $$d_h^A = 1/d_h^{M_1}, \qquad h = \eta, \tau, \mu \quad \text{for } T_c^i,$$

and

$$\beta = \frac{\int (v/m) w(v)\, dF^v(v)}{\int w(v)\, dF^v(v)} \tag{5.4.5}$$

[case (a) being disregarded].

Note that the general equations, which consist of a p-vector and a $p \times p$ matrix equation reduce here to four scalar equations, independently of m or p!

Step 4. For some selected m and c, calculations were done using the last equations for iterative improvement of the starting values $\beta = d_\eta^A = d_\tau^A = d_\mu^A = 1$.

Results are given in Table 1. (Stahel, 1981a, presents some more numbers.)

Solution. In summary, the optimal B_s- and B_i-robust estimators equal $T(F) = (\text{vecs}(\hat{L}\hat{L}^T)^T, \hat{\mu}^T)^T$, where $\hat{L}$ and $\hat{\mu}$ are the solutions of

$$\int \left(zz^T - m\beta \cdot I\right) w\left(\|z\|^2\right) dF(x) = 0,$$

                                                                      (5.4.6)

$$\int zw\left(\|z\|^2\right) dF(x) = 0,$$

with

$$z = L^{-1}(x - \mu)$$

for L and μ, and w and β are given by (5.4.2) and (5.4.5).

The ψ-function for the optimal self-standardized estimator T_c^s exists in view of the general Theorem 3 of Subsection 4.3c if $c > c_* := \sqrt{p}$, and is nonexistent for $c < c_*$. [Do not mix up this existence with the existence of an estimated value for a given (sample) distribution F.] It is intuitively clear that there is an analogous constant c_* for T_c^i. What about $c = c_*$?

Table 1. Optimal B-Robust Estimators

Dimension	m	2	2	2	3	3	5	5	10	10	20
T_c^s Characteristic	$c = \gamma_s^*$	5.29	3.04	2.24	5.32	3.00	6.55	4.47	9.73	8.06	15.17
constants	β	0.970	0.803	0.517	0.952	0.598	0.946	0.704	0.943	0.825	0.906
	$(d_\eta^A)^2$	1.18	2.70	29.50	1.30	27.95	1.36	27.12	1.45	28.90	36.39
	$(d_\tau^A)^2$	1.26	3.06	18.74	1.45	16.75	1.55	16.15	1.70	19.09	28.03
	$(d_\mu^A)^2$	1.06	1.67	10.64	1.12	12.30	1.17	14.98	1.27	20.41	30.15
Efficiencies	$1/d_\eta^V$	0.950	0.800	0.663	0.950	0.719	0.950	0.784	0.950	0.860	0.917
	$1/d_\tau^V$	0.929	0.784	0.560	0.933	0.590	0.934	0.639	0.938	0.728	0.829
	$1/d_\mu^V$	0.984	0.877	0.644	0.978	0.694	0.973	0.760	0.966	0.845	0.911
Corner	v^*	6.76	2.54	0	6.44	0	7.74	0	11.09	0	0
	Pr	0.034	0.281	1	0.092	1	0.171	1	.351	1	1
Sensitivity	γ_i^*	5.46	3.42	2.99	5.48	3.90	6.74	5.60	9.99	9.45	16.65
T_c^i Characteristic	$c = \gamma_i^*$	5.43	3.38	—	5.46	—	6.72	—	9.98	—	—
constants	β	0.970	0.806	—	0.952	—	0.946	—	0.943	—	—
	$(d_\eta^A)^2$	1.24	3.32	—	1.37	—	1.43	—	1.53	—	—
	$(d_\tau^A)^2$	1.35	3.85	—	1.55	—	1.66	—	1.82	—	—
	$(d_\mu^A)^2$	1.08	1.88	—	1.15	—	1.20	—	1.32	—	—

Clearly, for $c \downarrow c_*$, the set where $w(v) = c/r(v)$ [instead of 1; see (5.4.2)] tends towards the whole space. Also, $w(v) \downarrow 0$ pointwise. But $d_\eta^A w(v)$ has a nonzero limit, which we call $w^*(v)$, and which may be found as follows: Let, for given d_τ, d_μ,

$$w^*(v) = c_* \Big/ \left\{ \frac{1}{2}\left(1 - \frac{1}{m}\right) v^2 + \frac{1}{2m}\left[(v - m\beta)\, d_\tau\right]^2 + v\, d_\mu^2 \right\}^{1/2},$$

and find the solution $(d_\tau^*, d_\mu^*, \beta)$ of

$$d_\tau^k = 1 \Big/ \left\{ \frac{1}{2m} \int (v - m\beta)^2 w^*(v)^k \, dF^v(v) \right\},$$

$$d_\mu^k = 1 \Big/ \int \frac{v}{m} w^*(v)^k \, dF^v(v),$$

$$\beta = \frac{\int (v/m) w^*(v) \, dF^v(v)}{\int w^*(v) \, dF^v(v)} \tag{5.4.7}$$

with $k = 2$ for $T_{c_*}^s$ and $k = 1$ for $T_{c_*}^i$. Then β and d_η^A, $d_\tau^A = d_\tau^* d_\eta^A$, and $d_\mu^A = d_\mu^* d_\eta^A$ for any $d_\eta^A \geq \sup_v\{w^*(v)\}$ are solutions of (5.4.4) and (5.4.5) for $c = c_*$. Solutions $(d_\tau^*, d_\mu^*, \beta)$ were found empirically for $T_{c_*}^s$; hence $T_{c_*}^s$ and probably also $T_{c_*}^i$ seem to exist. We did not try hard to prove their existence. The constants β, $d_\eta^A = \sup_v\{w^*(v)\}$, $d_\tau^A = d_\tau^* d_\eta^A$, and $d_\mu^A = d_\mu^* d_\eta^A$ are the limits of the respective constants for $c \downarrow c_*$. They are included in Table 1.

Table 1 also contains the "efficiencies," that is, the reciprocals of the numbers d_η^V, d_τ^V, and d_μ^V characterizing the (d-type) asymptotical variance–covariance matrix V of the estimators, the "corner point" v^*, where the weight function is 1 for the last time, the probability "Pr" that $w(v) < 1$, and the sensitivities. The sensitivity bounds c are selected either to give selected values of the efficiency for the shape parameter, or equal to the extreme value c_*. The differences between the two standardizations are minimal. The efficiencies and the corner point v^* for T_c^i, which are not given in the table, most often coincide within the precision of the calculations with the respective numbers for the T_c^s which has the same efficiency for shape and is given above it in the table.

Remark 1. In Huber's minimax solution (Example 1 of Subsection 5.3c), observations that are too close to the (estimated) location point are "brought out" in some cases. Do the optimal *B*-robust estimators do something of the kind? In the light of (5.4.1) and (5.4.2), it might happen that $r(0) > c$, and that points near the estimated center would therefore be downweighed. Empirically, such a case was not found. If the location part was not estimated, which would make the third term in (5.4.1) disappear, then downweighing near the center would occur for low bounds c.

5.4b. Partitioned Parameter

Since the parameter, the influence function and the asymptotic covariance matrix all split up into three natural parts, we may look for estimators which minimize d_h^V under the restriction $\gamma_h^s \le c_h$ or $\gamma_h^i \le c_h$, $h = \eta$, τ, and μ with given bounds c_h [cf. (5.3.9)]. This adapts the general problem discussed in Section 4.4 to the covariance–location model. The solution is quite satisfactory: There is an estimator which is optimal for all three criteria simultaneously (this is true in the general case, see Theorem 1 of Section 4.4), and, as in the location-scale case, if we focus on one part, we do not lose any asymptotic efficiency by the estimation of the other parts.

The functions w_η, u_τ, and w_μ, that define the estimator, are determined by separate implicit equations. w_η is given by

$$w_\eta(v) = \min\left\{1, c_\eta\sqrt{m(m+2)}\Big/[vd_\eta]\right\} =: \min(1, v_\eta^*/v),$$

$$d_\eta^k = \left(1 + \frac{2}{m}\right)\Big/\int\left(\frac{v}{m}\right)^2 w_\eta(v)^k \, dF^v(v),$$

(5.4.8)

with $k = 2$ for the self- and $k = 1$ for the information-standardized variant. u_τ is of the form $u_\tau(v) = w_\tau(v)(v - m\beta)$, w_τ being given by

$$w_\tau(v) = \min\left\{1, c_\tau / \left[\sqrt{1/2m}\,\|v - m\beta\|\,d_\tau\right]\right\},$$

$$d_\tau^k = 2m \Big/ \int (v - m\beta)^2 w_\tau(v)^k \, dF^v(v), \qquad (5.4.9)$$

$$\beta = \frac{\int (v/m) w_\tau(v) \, dF^v(v)}{\int w_\tau(v) \, dF^v(v)}.$$

Finally, w_μ is calculated from

$$w_\mu(v) = \min\left\{1, c_\mu / \left(\sqrt{v}\,d_\mu\right)\right\} =: \min\left\{1, \sqrt{v_\mu^*/v}\right\},$$

$$\qquad (5.4.10)$$

$$d_\mu^k = 1 \Big/ \int \frac{v}{m} w_\mu(v)^k \, dF^v(v).$$

Every optimal B_s^p-robust estimator is also an optimal B_i^p-robust one—for different c_h—as in problems with a scalar parameter. Tables 2 and 3 show the characterizing figures for some of these estimators. They include the "corner points" v_η^* where the weight falls below 1 and the probabilities "Pr" for $w_h(v) < 1$. For size estimation with low sensitivity, we have $w_\sigma(v) < 1$ also for small v, $v < v_\tau^0$ say, and the table includes $\tilde{\text{Pr}} := F^v(V < v_\tau^0)$.

Remark 2. The estimators given by (5.4.7)–(5.4.10) are defined in much simpler terms than those of the preceding subsection. Their optimality is at

Table 2. Optimal Estimators of Shape and Location

Dimension	m	2	2	2	3	3	5	5	10	10	20
Corner	$v_\eta^* = v_\mu^*$	5.66	2.45	0	6.41	0	7.87	0	11.32	0	0
	Pr	0.059	0.293	1	0.093	1	0.164	1	0.333	1	1
Shape: Efficiency	$1/d_\eta^V$	0.950	0.800	0.500	0.950	0.600	0.950	0.714	0.950	0.833	0.90
Sensitivities	γ_η^s	2.28	1.47	1	1.90	1	1.54	1	1.24	1	1
	γ_η^i	2.33	1.65	1.41	1.95	1.29	1.58	1.18	1.27	1.10	1.05
Location: Efficiency	$1/d_\mu^V$	0.993	0.956	0.785	0.993	0.849	0.992	0.905	0.990	0.951	0.97
Sensitivities	γ_μ^s	2.45	1.86	1.41	2.62	1.73	2.94	2.24	3.60	3.16	4.47
	γ_μ^i	2.46	1.91	1.60	2.63	1.88	2.95	2.35	3.61	3.24	4.53

Table 3. Optimal Estimators for Size

Dimension	m	2	2	2	3	3	5	5	10	10	20
Corner	v_τ^*	7.09	3.98	1.39	8.47	2.37	11.11	4.35	17.36	9.34	19.34
Points	v_τ^0	0	0	1.39	0	2.37	0	4.35	2.17	9.34	19.34
	β	0.971	0.863	0.693	0.972	0.789	0.974	0.870	0.976	0.934	0.967
	Pr	0.029	0.136	1	0.037	1	0.049	1	0.072	1	1
	P̌r	0	0	0.5	0	0.5	0	0.5	0.05	0.5	0.5
Efficiency	$1/d_\tau^V$	0.950	0.800	0.480	0.950	0.528	0.950	0.569	0.950	0.602	0.619
Sensitivities	γ_τ^s	2.89	1.71	1	2.55	1	2.22	1	1.92	1	1
	γ_τ^i	2.96	1.91	1.44	2.61	1.38	2.28	1.33	1.97	1.29	1.27

least as intuitive, and they are more flexible. In *practical application*, they are therefore preferable.

As in the previous subsection, let us now discuss the *most robust estimator* of the type just treated. The limit of the optimal B_s^p-robust shape estimator for $c_\eta \downarrow 1$ is given by

$$w_\eta^*(v) = 1/v$$

(which is the limit of $w_\eta(v)/v^*$); for $c_\tau \downarrow 1$, we eventually obtain

$$w_\tau^*(v) = |v - m\beta^*|^{-1}, \qquad m\beta^* = \text{med}(F^v),$$

$$u_\tau^*(v) = \text{sign}(v - m\beta^*);$$

and for $c_\mu \downarrow \sqrt{m}$,

$$w_\mu^*(v) = 1/\sqrt{v}.$$

Let us call the estimator given by w_η^*, w_τ^*, and w_μ^* the *median estimator*. It may be described in words: Find the affine transformation $z = \hat{L}^{-1}(x - \hat{\mu})$ for which the transformed observations show the following picture—if the z_i are projected onto the hypersphere with radius $\sqrt{m\beta^*}$, the projected points must have ordinary covariance matrix $\beta^* I$ and mean 0, and half of the points must lie within the hypersphere, the other half outside. Then estimate Σ by $\hat{L}\hat{L}^T$ and μ by $\hat{\mu}$ as before. In fact, this coincides with a suggestion of Hampel (1975, p. 377) and with the limiting case of Huber's minimax estimators (Example 1 of Subsection 5.3c). It also shares the defect of the discontinuity at the (estimated) center. The characterizing numbers for this estimator are included in Tables 2 and 3.

The asymptotic covariance matrix is given by

$$d_\eta^V = 1 + 2/m,$$

$$d_\tau^V = \left[2m\left(d_\tau^M\right)^2\right]^{-1}, \qquad d_\tau^M = 2\beta^* f^v(m\beta^*),$$

$$d_\mu^V = \frac{m}{2}\left[\Gamma\left(\frac{m+1}{2}\right)\middle/\Gamma\left(\frac{m}{2}\right)\right]^{-2}.$$

For large m, some formulas of Abramowitz and Stegun (1972, 6.1.41, 26.4.17, 6.1.37) show that efficiency losses tend to $1 - 2/\pi = 0.36$ for the size, and only $2/m$ for shape and $1/2m$ for location.

Remark 3. In moderate to high dimensions, it therefore seems appropriate to use a not quite extreme estimator of the type (5.4.8) and (5.4.10) introduced above in *applications*. It achieves almost as good robustness properties as possible while avoiding the discontinuity problem, and efficiency losses are tiny except for the size parameter. If the latter is of special interest, a less extreme u_τ may be adequate. The next section, however, shows that even the most robust M-estimator has poor breakdown properties in high dimensions.

5.5. BREAKDOWN PROPERTIES OF COVARIANCE MATRIX ESTIMATORS

5.5a. Breakdown Point of M-Estimators

The breakdown point was defined in Subsection 2.2a roughly as the smallest fraction of contamination to a distribution for which—by varying the contamination—the estimated value may be carried over all bounds. More precisely, the following statement is true:

Lemma 1. Let $T_n \to T(F)$ in probability (for all F), and let F_* and ε be fixed. If there is a sequence of distributions $\{H_k\}$ for which $\{T(G_k)\}$ with $G_k := (1 - \varepsilon)F_* + \varepsilon H_k$ is not contained in any compact proper subset of Θ, then $\varepsilon^*(\{T_n\}, F_*) \leq \varepsilon$.

Proof. If $\varepsilon^* > \varepsilon$, then there would exist K_ε with $G_k\{T_n \in K_\varepsilon\} \to 1$ as $n \to \infty$ for all k since $\pi(F_*, G_k) \le \varepsilon$ (the same is true for any well-known distance other than π). But this contradicts the assertion that $T(G_k) \notin K_\varepsilon$ for some k. $\qquad\qquad\qquad\qquad\qquad\qquad\qquad\qquad\qquad\qquad\qquad\square$

For ease of communication, we introduce

Definition 1. If a sequence $\{\theta_k\}$ is not contained in any compact proper subset K of Θ, we say the $\{\theta_k\}$ *diverges to the edge.*

We will find a "wicked sequence" $\{H_k\}$ to derive an upper bound on the breakdown point of affinely equivariant M-estimators. The crucial distribution G_* to look at is constructed as follows: Let $\tilde{F}_0$ be the projection of F_0 to the $(m-1)$-dimensional space $\{x | x^{(1)} = 0\}$; for example, $\tilde{F}_0$ is the $(m-1)$-dimensional standard normal distribution in this space. Denote by $\tilde{F}^v$ the distribution of $\|X\|^2$ when $X \sim \tilde{F}_0$; for example, the χ^2_{m-1}; and by H_* the symmetrized distribution of $\|X\|$ considered as a distribution on the first x axis. Let

$$G_* := \left(1 - \frac{1}{m}\right)\tilde{F}_0 + \frac{1}{m}H_*.$$

This crucial distribution is sketched in Figure 1.

Lemma 2. Let an M-estimator be given by (5.3.11) or (5.3.12). If the equation

$$\int u_\tau^\psi(v/\sigma)\, d\tilde{F}^v(v) = 0$$

has a solution σ_* for σ, then $\hat{\Sigma} = \sigma_* I$, $\hat{\mu} = 0$ is a solution for $T(G_*)$.

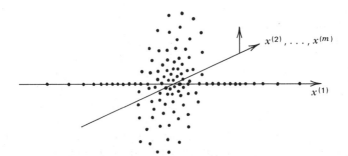

Figure 1. Sketch of the crucial distribution G_*.

Proof.   Using the form (5.3.12) of the defining equations and integrating first with constant $\|x\|$, the result is readily clear.   □

Theorem 1 (*Breakdown* bound for affine equivariant *M-estimators*). For a T which fulfills the assumptions of Lemma 2 and is weakly continuous at G_*, we have

$$\varepsilon^*\big(\{T_n\}, F_{\Sigma,\mu}\big) \leq 1/m \quad \text{for all } \Sigma, \mu.$$

Proof.   Let $\Sigma = I$, $\mu = 0$, and

$$G_k(\cdot) := \left(1 - \frac{1}{m}\right) \cdot F_0(\cdot) + \frac{1}{m} H_*(\cdot/s_k)$$

with $s_k \to \infty$. Transform the observations x like $x^{(1)} \mapsto x^{(1)}/s_k$ to get the distribution $\tilde{G}_k$. Note that $\tilde{G}_k \to G_*$, and the continuity assumption implies $\hat{\sigma}_{11}(\tilde{G}_k) \to \sigma^2_*$. By equivariance,

$$\hat{\sigma}_{11}(G_k) = s_k^2 \cdot \hat{\sigma}_{11}(\tilde{G}_k) \to \infty,$$

and Lemma 1 provides the result for θ_0. For other parameter values, it follows from the equivariance of T.   □

Remark 1.   In usual situations, the equation in Lemma 2 will really have a solution: If $\tilde{F}^v$ is replaced by F^v, then there must be a solution by Theorem 2 of Subsection 5.3c, and F^v will usually be quite similar to $\tilde{F}^v$ for moderate to large m.

Remark 2 (*historical*).   Maronna (1976) detected the low breakdown point of "monotone" M-estimators of covariance matrices (as described by the conditions of Theorem 1 of Subsection 5.3.c). Huber (1977b) generalized this result to nonmonotone estimators. The assumptions of his theorem include a questionable continuity property, however.

Remark 3.   An extreme *example* in which violation of this continuity leads to a *high breakdown point* is also unusual in the sense of Remark 1: Let F_0 be the uniform distribution on the set $\{x|1 - \Delta v \leq \|x\|^2 \leq 1\}$, and let T be defined by (5.3.11) with $w_\eta^\psi(v) = 2$ Indicator $(v; 1 - \Delta v \leq v \leq 1)$ and $w_\delta^\psi(v) \equiv 1/m$. The breakdown point for T approaches $\frac{1}{2}$ for $\Delta v \to 0$.

Whereas the theorem disqualifies all equivariant M-estimators, there are trivial nonequivariant procedures with breakdown $\frac{1}{2}$: Reject all observations showing up as univariate outliers in any of the m coordinates and calculate any covariance matrix (and location) estimator from the rest. Instead of being a favorable procedure, such a rule shows a *weakness* of the classical concept of *breakdown point* when applied to the covariance–location model and gives rise to more general considerations.

*5.5b. Breakdown at the Edge

Most frequently, the covariance–location model serves to describe dependence relations between variables or helps detecting so-called "multivariate outliers." Therefore, if the majority of the points in the sample happen to be near a hyperplane, it is highly desirable that this structure be detected. This idea leads to the following concept which applies to other models as well:

Definition 2.   The point of (gross error) *breakdown at the edge* for an estimator T and a given model is

$$\varepsilon_e^*(T) = \sup\{\varepsilon \leq 1| \text{ if } \{\theta_k\} \text{ diverges to the edge then}$$

$$\left(G_k = (1 - \varepsilon)F_{\theta_k} + \varepsilon H_k \Rightarrow \{T(G_k)\} \text{ diverges to the edge}\right)\}.$$

In the context of the *covariance–location model* we require the following: If $\{G_k\}$ is a sequence of distributions, each of which is contained in an ε-"neighborhood" of a model distribution F_{θ_k}, $\theta_k = (\Sigma_k, \mu_k)$, and if Σ_k tends to a singular matrix, then the estimated covariance matrix should also tend to a singular matrix.

For equivariant estimators in a *model generated by transformations*, this new kind of breakdown coincides with the first one. For a precise statement, let us recall that the group structure of the transformations $\{\alpha_\theta\}$ induces an operation $\square$ in Θ by

$$\theta = \theta' \square \theta'' \Leftrightarrow \alpha_\theta(\cdot) = \alpha_{\theta'}\{\alpha_{\theta''}(\cdot)\}$$

[see (4.5.1)]. θ^0 corresponds to the identity element, and θ^- to the inverse ($\theta^- \square \theta = \theta^0$).

Theorem 2 (*Equality of breakdown points* for equivariant estimators).   In a model generated by a group of transformations let the operation $\square$ have the following property: For all $K, K' \subsetneqq \Theta$ compact, there is a $K'' \subsetneqq \Theta$ compact, such that $\{\theta \square \theta' | \theta \in K, \theta' \in K'\} \subset K''$ and $\{\theta^- | \theta \in K\} \subset K''$. Then for any estimator equivariant under the generating transformations, the breakdown point at the edge coincides with the ordinary (gross error) breakdown point.

Proof (cf. Theorem 1). To G_k there corresponds $\overline{G}_k = \tilde{\alpha}_{\theta_k}^{-1}(G_k) = (1 - \varepsilon) \cdot F_0 + \varepsilon \cdot \tilde{\alpha}_{\theta_k}^{-1}(H_k)$, and $T(G_k) = \theta_k \square T(\overline{G}_k)$. If θ_k diverges while $T(G_k)$ stays within a compact proper subset K of Θ, then $T(\overline{G}_k)$ diverges, so that $\varepsilon^* \leq \varepsilon_e^*$. On the other hand, let $\{G_k = (1 - \varepsilon) \cdot F_0 + \varepsilon \cdot H_k\}$ be such that $T(G_k)$ diverges. Then $\theta_k = T(G_k)^-$ diverges while $T\{\tilde{\alpha}_{\theta_k}(G_k)\} = \theta^0$ does not. Therefore $\varepsilon_e^* \leq \varepsilon^*$. $\square$

In the *covariance–location model*, the operation is given by (5.2.6) and the condition of the theorem is easily verified since Σ-matrices belonging to compact subsets of the parameter space have determinants bounded away from 0 and ∞.

This result evokes an argument for looking at equivariant estimators with a high (ordinary) breakdown point. Clearly, the coordinate-dependent procedure mentioned at the end of the last subsection does not improve the breakdown point at the edge, but the question whether there are other nonequivariant estimators with a high breakdown point at the edge remains open.

5.5c. An Estimator with Breakdown Point $\frac{1}{2}$

While univariate outliers in one of the coordinates were eliminated in the procedure mentioned at the end of Subsection 5.5a we now downweigh all observations sticking out in any projection:

Definition 3. For each direction $d \in \mathbb{R}^m$, $\|d\| = 1$, let $L_d(F)$ and $S_d(F)$ be a (one-dimensional) location and scale estimator (functional) of the distribution of the projection $d^T X$, $X \sim F$, respectively (with the corresponding equivariance properties), and

$$r(x; F) := \sup_d \{|d^T x - L_d(F)| / S_d(F)\}.$$

Then the *projection estimator* corresponding to L, S, and a weight function $w \colon \mathbb{R}^+ \to \mathbb{R}$ is defined as the ordinary weighted covariance–location estimator with weights $w\{r(x; F)^2\}$,

$$\hat{\Sigma}(F) = \frac{c\int w\{r(x; F)^2\}(x - \hat{\mu})(x - \hat{\mu})^T dF(x)}{\int w\{r(x; F)^2\} dF(x)},$$

$$(5.5.1)$$

$$\hat{\mu}(F) = \frac{\int w\{r(x; F)^2\} x \, dF(x)}{\int w\{r(x; F)^2\} dF(x)},$$

where c is a fixed constant used to achieve Fisher consistency at the normal distribution.

The equivariance of such estimators follows from the definition of $r(x; F)$.

Theorem 3. Suppose that:

(i) The location-scale estimator $(L, S)^T$ has breakdown point $\frac{1}{2}$ at the projection of F_0 onto a straight line.
(ii) w is continuous, positive, and bounded, and $w(r^2)r^2$ is bounded.
(iii) The support of F_0 is the whole $\mathbb{R}^m$.

Then, the *gross-error breakdown point* of the projection estimator defined by L, S, and w is $\frac{1}{2}$ for all model distributions.

Proof. Because of affine equivariance, it suffices to consider F_0 as the model distribution. Let $\varepsilon < \frac{1}{2}$, and denote by G the distributions in the respective gross-error neighborhood around F_0. Because of (i), $L_d(G)$ is bounded and $S_d(G)$ is bounded away from 0 uniformly for all d and G. By (ii), $w[r(x; G)^2]$ is bounded away from 0 on any ball $\{x \mid \|x\| < c\}$, uniformly in G. Because of (iii), then, the denominator $\int w(r^2) \, dG$ in (5.5.1) is bounded away from 0. Also, since the estimator is a weighted classical estimator, the estimated covariance matrix may not be singular. It remains to show that the numerators in (5.5.1) are bounded. Clearly,

$$\|x\| \le \left| d_x^T x - L_{d_x}(G) \right| + \left| L_{d_x}(G) \right| \le S_{d_x}(G)r(x; G) + \left| L_{d_x}(G) \right|,$$

where $d_x = x/\|x\|$ $(x \ne 0)$. Since $S_d(G)$ and $L_d(G)$ are uniformly bounded over all d and G,

$$w\left[r(x; G)^2\right]\|x\|^2 \le c_1 w\left[r(x; G)^2\right]r(x; G)^2 + c_2 w\left[r(x; G)^2\right]r(x; G) + c_3$$

with suitable constants c_1, c_2, c_3, and similarly for $w(r^2)\|x\|$. Condition (ii) then makes sure that the numerators are bounded.   □

Remark 4. The theorem is the asymptotic version of Donoho's calculation of the *finite-sample breakdown point* (Donoho, 1982; Donoho and Huber, 1983).

The maximization involved in the definition of $r(x; F)$ poses a serious *computational problem*. Heuristic considerations show that there may be many local maxima even if the global maximum is much higher than these. (This is especially true if L and S are the median and the median deviation.) Since, therefore, ordinary nonlinear programming cannot solve the problem, one could try to evaluate the function to be maximized at all points of a fine "grid" on the hypersphere. In higher dimensions, such a grid would need very many points to be accurately fine, however.

A practicable solution for finite sample sizes n seems nevertheless possible (cf. the idea of Siegel, 1982): Choose m indices $(i_1, i_2, \ldots, i_m)$ randomly from $\{1, 2, \ldots, n\}$ and find d perpendicular to the hyperplane through the observations with these indices. Repeat this construction q times and treat the maximum over the q directions as the global maximum. A local maximization procedure could be used to improve on this preliminary solution. In the situation mentioned in Subsection 5.5a (Fig. 1), where the "good" observations lie near a hyperplane, the procedure will detect the structure if the indices (i_k) select "good" observations exclusively for at least one of the q choices. The probability for this is

$$p^* \approx 1 - \left[1 - (1 - \varepsilon)^m \right]^q,$$

where ε is the proportion of contamination in the sample, if n/m is large. The number q of choices necessary for $p^* = 0.95$ is given in Table 1.

Note that this approximation retains the affine equivariance of the exact estimator.

First results on the performance of the procedure for finite samples are given by Stahel (1981b).

It may be guessed that the *efficiency* of such estimators is as poor as that of related regression estimators (cf. Subsection 6.4a). They should be used as a starting value for the iteration cycles of a (nonunique) M-estimator with possibly redescending influence function.

Remark 5. The idea behind the projection estimators (and the regression method of Subsection 6.4a) can be used also in other multivariate problems. It is called "projection pursuit" (see Huber, 1985).

Table 1. Number of Choices Necessary for $p^* = 0.95$

m	ε			
	0.05	0.1	0.3	0.5
2	2	2	5	11
3	2	3	8	23
5	3	4	17	95
10	4	7	105	3067
20	7	24	3753	3141252

Remark 6. Another affine equivariant location-covariance method with 50% breakdown point is the *minimal volume ellipsoid* (MVE) estimator (Rousseeuw, 1983b). One determines the ellipsoid with smallest volume which covers (at least) 50% of the data, and uses its center as a location estimate (by inflating the ellipsoid, one finds appropriate confidence ellipsoids). A somewhat more efficient variant is the *minimal covariance determinant* (MCD) estimator, given as the mean of the 50% data points for which the determinant of the empirical covariance matrix is minimal. These are not projection pursuit estimators, but their computational cost is of the same order of magnitude.

EXERCISES AND PROBLEMS

Subsection 5.2a

1. (Short) Show that the rescaled standard normal distributions $\{\mathcal{N}_m(0, \sigma I)\}$ are the only spherically symmetric distributions for which the components $X^{(j)}$ are independent. For simplicity, consider distributions with densities (with respect to Lebesque's measure) only.

2. (Short) Let $U(A) := \partial \tilde{\alpha}_{A,a}(\theta)/\partial\theta$ for short. (The right-hand side is independent of θ and a.) Show that $\tilde{\alpha}_{A,a}^{-1} = \tilde{\alpha}_{A^{-1},-A^{-1}a}$ and therefore $U(A)^{-1} = U(A^{-1})$, and that $U(A)^T = U(A^T)$.

Subsection 5.2b

3. Verify (5.2.8) and (5.2.9), using the explicit formula that expresses Σ^{-1} and $\det(\Sigma)$ in terms of determinants of submatrices.

4. (Short) Verify (5.2.11).

5. (Short) Calculate f^v and ω^v for the case of (1) a normal distribution and (2) another spherically symmetric distribution of your choice.

6. In the bivariate normal model with zero location, replace the parameter $\theta = (\sigma_{11}/\sqrt{2}, \sigma_{22}/\sqrt{2}, \sigma_{12})^T$ by $\bar{\theta} = (\ln\sqrt{\sigma_{11}}, \ln\sqrt{\sigma_{22}}, \rho)^T$, where $\rho = \sigma_{12}/\sqrt{\sigma_{11}\sigma_{22}}$. Using (5.2.9) and (4.2.7), show that the scores for $\bar{\theta} = (0, 0, \rho)^T$ are

$$\bar{s}\left\{ x, \begin{pmatrix} 0 \\ 0 \\ \rho \end{pmatrix} \right\} = \begin{pmatrix} \left(1 - \rho^2\right)^{-1} x^{(1)}\left(x^{(1)} - \rho x^{(2)}\right) - 1 \\ \left(1 - \rho^2\right)^{-1} x^{(2)}\left(x^{(2)} - \rho x^{(1)}\right) - 1 \\ \left(1 - \rho^2\right)^{-1}\left[\left(1 - \rho^2\right)^{-1}\left(x^{(1)} - \rho x^{(2)}\right)\left(x^{(2)} - \rho x^{(1)}\right) - \rho\right] \end{pmatrix}$$

[or derive $\bar{s}(x, \bar{\theta})$ for general $\bar{\theta}$].

Subsection 5.3a

7. (Short) Let S and $\tilde{S}$ be symmetric matrices and let $\tilde{D}$ be the upper left corner of order $p(p + 1)/2$ of a d-type matrix. Evaluate $\text{vecs}(S)^T \tilde{D} \text{vecs}(\tilde{S})$.

Subsection 5.3b

8. Verify that the asymptotic covariance between the (i, j)th and the (k, l)th element of an equivariantly estimated variance–covariance matrix for a general model distribution is

$$V_{(ij),(kl)} = 2\sigma_{ij}\sigma_{kl} d_\rho^V + \left(\sigma_{jk}\sigma_{il} + \sigma_{ik}\sigma_{jl}\right) d_\eta^V$$

if $i \neq j$ and $k \neq l$, and that the same formula holds for $i = j$ or $k = l$ or both up to a factor of $\sqrt{2}$ or 2. [*Hint*: Let

$$S^{(kl)} := \begin{cases} \left(\delta_{ik}\delta_{lj} + \delta_{il}\delta_{kj}\right)_{i,j} & \text{if } k \neq l \\ \left(\delta_{ik}\delta_{kj} \cdot \sqrt{2}\right)_{i,j} & \text{if } k = l \end{cases}$$

and use result (4) of Subsection 5.3b and the results of Exercises 2 and 7.]

9. In the identity

$$\operatorname{cov}(X^{(1)}, X^{(2)}) = \tfrac{1}{4}\left[\operatorname{var}(X^{(1)} + X^{(2)}) - \operatorname{var}(X^{(1)} - X^{(2)})\right] \quad (*)$$

replace var by the square of a robust scale estimator S to get a robust covariance estimator C.
1. Express the influence function of C at the bivariate normal distribution $\mathcal{N}_2(0, \Sigma)$ in terms of the influence function $\operatorname{IF}(x; S, \Phi)$ of the scale estimator at the standard normal distribution.
2. In order to obtain a correlation estimator R, Gnanadesikan and Kettenring (1972) suggest to replace the $X^{(j)}$ in $(*)$ by $\overline{X}^{(j)} = X^{(j)}/S(X^{(j)})$. Solve part (1) for R.
3. Huber (1981, Section 8.2) notes that $|R|$ may exceed 1. He suggests using

$$R^* := \frac{S(\overline{X}^{(1)} + \overline{X}^{(2)})^2 - S(\overline{X}^{(1)} - \overline{X}^{(2)})^2}{S(\overline{X}^{(1)} + X^{(2)})^2 + S(\overline{X}^{(1)} - \overline{X}^{(2)})^2}$$

instead of R. Do part (1) again for R^*.

10. Consider the model of classical discriminant analysis: Let $P(K = 1) = p$ and $P(K = 0) = 1 - p$ and $X|K \sim \mathcal{N}_m(\mu_0 + K(\mu_1 - \mu_0), \Sigma)$. Derive the influence functions of: (1) an estimator $(\operatorname{vecs}(\hat{\Sigma})^T, \hat{\mu}_0^T, \hat{\mu}_1^T)^T$ obtained from an affinely equivariant estimator as discussed in Subsection 5.3b, at $\Sigma = I$, $\mu_0 = 0$, $\mu_1 = (1, 0, 0, \ldots)^T$; and (2) the respective estimator of the Mahalonobis distance $(\mu_1 - \mu_0)^T \Sigma^{-1}(\mu_1 - \mu_0)$. [*Hint*: Use (5.2.8).]

Subsection 5.3c

11. (Research) Can Eq. (5.3.12) be used instead of (5.3.11) in order (1) to develop an improved algorithm and (2) to prove the existence of estimated values under conditions different from those of Theorem 2 of Subsection 5.3c (e.g., W–estimators)?

12. Adapt Algorithm 4 of Subsection 4.6b to the covariance–location model.

13. (Short) Verify that the only W-estimator that is also a maximum likelihood estimator (for a suitable F_0) is the classical estimator with constant weight function.

Subsection 5.4a

14. (Research) Does the most robust estimator defined by (5.4.7) exist?

Subsection 5.4b

15. Consider Huber's minimax solution (Example 1 of Subsection 5.3c). Show that, if $a = 0$, the minimax solution is optimal in the sense of Subsection 5.4b [*Hint*: First reconcile Huber's way of making the estimator consistent with introducing β in (5.4.9).] Note that these estimators fulfill the conditions of Theorem 1 of Subsection 5.3c, which shows that there exists an estimate.

16. Continuing the previous problem, show that for $a = b$, the minimax estimator equals the median estimator. If $0 < a < b$, the minimax estimator is not optimal B^p-robust. The optimal B^p-robust estimators avoid the discontinuity problem of the minimax estimators with $a > 0$ [except for those with extreme w_η^ψ function $w_\eta^\psi(v) = 1/v$].

Section 5.4

17. (Extensive) Derive an optimal robust estimator of the Mahalonobis distance which treats all the other parameters in the model as nuisance parameters (Section 4.4).

18. (Research) Derive optimal redescending B-robust estimators in analogy with Section 2.6. (For a minimax approach, see Collins, 1982.)

Section 5.5

19. For any seven points x_i in the plane, determine their "outlyingness" $r(x_i; F_n)$ as defined by $L =$ median, $S =$ MAD, (1) approximately, using as the "grid" of directions the axes and the diagonals and (2) by guessing the direction d_i that determines $r(x_i; F_n)$ for each point i.

20. (Short) Show that in the plane ($m = 2$) one outlier, added to four fixed observations, causes the breakdown of any projection estimator based on $L =$ median and $S =$ MAD.

21. Find the finite-sample breakdown point of the projection estimators considered in Theorem 3 (cf. Donoho, 1982; Donoho and Huber, 1983).

CHAPTER 6

Linear Models: Robust Estimation

6.1. INTRODUCTION

6.1a. Overview

The linear regression model is one of the most widely used tools in statistical analysis and the least-squares method is a very popular estimation technique for this model. But in spite of its mathematical beauty and computational simplicity, this estimator suffers a dramatic lack of robustness. Indeed, one single outlier can have an arbitrarily large effect on the estimate. The goal of this chapter is to investigate the robustness properties of a wide class of regression estimators generalizing least-squares and to propose robust alternative estimators. The testing problem and a proposal for a robust model selection are discussed in Chapter 7.

Section 6.2 is devoted to the Huber-estimator. We derive its influence function and by means of the fundamental concepts of influence of the residual and influence of position in factor space, we show that more refined estimators are needed in order to cope with outlying points in the factor space. General M-estimators are introduced in Section 6.3. We compute their influence function and their change-of-variance function, discuss different sensitivities, and solve Hampel's optimality problem. Finally, in Section 6.4 we discuss the breakdown point of regression estimators and give some numerical results on the asymptotic efficiency of bounded influence estimators. A short account on the available computer programs is also included.

Since the linear model is a special case of the general parametric model, the results of this chapter can in principle be derived from those of Chapter

4. However, because of its theoretical and practical importance, it is worthwhile to adapt the formulas and to discuss explicitly the results in this case. Moreover, the equivariance properties of the linear model simplify the form of the optimal estimators and allow a nice and intuitive interpretation of the results that help to understand the general theory. Therefore, this chapter can be read before Chapters 4 and 5.

6.1b. The Model and the Classical Least-Squares Estimates

We consider the following linear model.

Let $\{(x_i, y_i) : i = 1, 2, \ldots, n\}$ be a sequence of independent identically distributed random variables such that

$$y_i = x_i^T \theta + e_i, \qquad i = 1, \ldots, n,$$

where $y_i \in \mathbb{R}$ is the ith observation,

   $x_i \in \mathbb{R}^p$ is the ith row (written as a column vector) of the design matrix $X_{n \times p}$,

   $\theta \in \Theta \subset \mathbb{R}^p$ is a p-vector of unknown parameters ($p \geq 1$),

   $e_i \in \mathbb{R}$ is the ith error.

We suppose that Θ is open and convex and that e_i is independent of x_i and has a symmetric distribution $G(e/\sigma)$, where $\sigma > 0$ is a scale parameter, with density g with respect to the Lebesgue measure.

Let $K(x)$ be the distribution function of x_i, with the density k with respect to the Lebesgue measure. We denote by $f_\theta(x, y)$ the joint density of (x_i, y_i), that is,

$$f_\theta(x, y) = \sigma^{-1} g\big((y - x^T \theta)/\sigma\big) k(x),$$

and by $F_\theta(x, y)$ the corresponding distribution function.

In the usual formalization, one considers a linear model with observations $y_1, \ldots, y_n$ and fixed carriers $x_1, \ldots, x_n$. However, it is convenient to consider a random-carriers model, especially for dealing with extreme design vectors (which might be wrong). The results derived for this case can be interpreted and applied in a fixed-carriers model.

Classical estimation and test procedures in linear models are based on the well-known method of least squares. This method was first published by Legendre in 1805, but the question of the priority between Gauss and Legendre over this discovery is still being discussed (Stigler, 1981). Consider for a moment σ as fixed.

A *least-squares* (*LS*) *estimate* T_n^{LS} of θ is any statistic that minimizes the Euclidean norm of the residuals:

$$\Gamma(\theta) = \sum_{i=1}^{n} \left((y_i - x_i^T \theta)/\sigma \right)^2,$$

that is, T_n^{LS} is defined by

$$\Gamma(T_n^{LS}) = \min\{ \Gamma(\theta) | \theta \in \Theta \}.$$

Its optimality is stated by the following result.

Theorem [Gauss–Markov (GM); see Scheffé, 1959, p. 14]. Under the assumptions

$$Ee_i = 0, \qquad i = 1, \ldots, n, \qquad \text{(6.1.GM1)}$$

$$\text{cov}(e_i, \ldots, e_n) = \sigma^2 I, \qquad \text{(6.1.GM2)}$$

every estimable function $b^T\theta$ has a unique linear estimate which has minimum variance in the class of all unbiased linear estimates. It is given by $b^T T_n^{LS}$, where T_n^{LS} is any least-squares estimate. If in addition the errors are normally distributed, then this estimator has minimum variance among all unbiased estimators.

Some remarks about this result are in order.

Linearity is a drastic restriction: many maximum likelihood estimators [e.g., under the logistic and all *t*-distributions (including the Cauchy) of the errors] are not linear. Rejection of outliers is also a nonlinear operation. In fact, the least-squares estimator is optimal in the class of all unbiased estimators, only if the errors are normally distributed. Therefore, the restriction to linear estimators can be justified only by normality (or simplicity). The normal model is never exactly true and in the presence of small departures from the normality assumption on the errors, the least-squares procedures (estimators and tests) lose efficiency drastically (Huber, 1973a, 1977c; Hampel, 1973a, 1978a, 1980; Schrader and Hettmansperger, 1980; Ronchetti, 1982a, 1982b). Thus, one would prefer procedures which are only nearly optimal at the normal model but which behave well in a certain neighborhood of it.

Figure 1 shows the effect of an outlying observation P_1 on the least-squares fit. The data are taken from Ezekiel and Fox (1959, pp. 57–58). The

Table 1. Water Flow at Two Points on Kootenay River in January (units of cfs)

Year	Newgate (y)	Libby (x)
1931	19.7	27.1
32	18.0	20.9
33	26.1	33.4
34	44.9	77.6
1935	26.1	37.0
36	19.9	21.6
37	15.7	17.6
38	27.6	35.1
39	24.9	32.6
1940	23.4	26.0
41	23.1	27.6
42	31.3	38.7
43	23.8	27.8

Permission Ezekiel and Fox (1959).

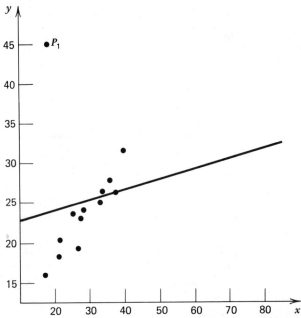

Figure 1. LS fit for the data of Table 1 (point corresponding to year 1934 changed to P_1).

water flow at two different points (Libby, Mont. and Newgate, B.C.) on Kootenay River in January was measured. Table 1 shows the data. For illustration purposes, we changed the point corresponding to the year 1934 to (20.0, 44.9) and denoted it by P_1.

6.2. HUBER-ESTIMATORS

In this section we assume the linear model of Section 6.1 with normal errors $(G = \Phi)$. The results generalize readily to other model's distribution functions G satisfying the conditions (F1) and (F2) of Subsection 2.5a.

In 1973 Huber (see Huber, 1973a, 1977c) extended his results on robust estimation of a location parameter to the case of linear regression. Basically, he proposed to compute weighted least-squares estimates with weights (redefined iteratively) of the form

$$w_i = \min\{1, c/|r_i|\}, \qquad (6.2.1)$$

where r_i is the ith residual and c is a positive constant. The weights thus are not fixed, but depend on the estimate. More generally, Huber proposed M-estimators T_n defined by

$$\Gamma(T_n) = \min\{\Gamma(\theta)|\theta \in \Theta\}, \qquad (6.2.2)$$

where

$$\Gamma(\theta) = \sum_{i=1}^{n} \rho\left((y_i - x_i^T\theta)/\sigma\right), \qquad (6.2.3)$$

for some function $\rho: \mathbb{R} \to \mathbb{R}^+$ and for a fixed σ. If ρ has a derivative $(\partial/\partial r)\rho(r) = \psi(r)$, T_n satisfies the system of equations (with the p-vectors x_i)

$$\sum_{i=1}^{n} \psi\left((y_i - x_i^T T_n)/\sigma\right)x_i = 0. \qquad (6.2.4)$$

Remark 1. The least-squares estimator is defined by the function $\rho(r) = r^2/2$ and the L_1-estimator by $\rho(r) = |r|$ [see Bloomfield and Steiger, 1983 for a nice account of L_1-estimation]. The Huber-estimator defined by the weights (6.2.1) may be obtained putting $\rho(r) = \rho_c(r)$ in (6.2.3) [$\psi(r) = \psi_c(r)$; cf. (2.3.15)], and is in fact the maximum likelihood estimator when the errors are distributed according to the distribution with density proportional to $\exp(-\rho_c(r))$ (cf. Section 2.7).

Remark 2. In practice, one usually has to estimate the scale parameter σ along with θ. A possible way to do this is to minimize, with given β,

$$\sum_{i=1}^{n} \left[\rho\left(\left(y_i - x_i^T\theta \right)/\sigma \right) + \beta \right] \sigma$$

with respect to θ and σ. Taking derivatives we obtain the following equations

$$\sum_{i=1}^{n} \psi\left(\left(y_i - x_i^T T_n \right)/\hat{\sigma} \right) x_i = 0, \tag{6.2.5}$$

$$\sum_{i=1}^{n} \chi\left(\left(y_i - x_i^T T_n \right)/\hat{\sigma} \right) = 0, \tag{6.2.6}$$

where $\psi(r) = \rho'(r)$ and $\chi(r) = r\psi(r) - \rho(r) - \beta$. The choice $\psi(r) = \psi_c(r)$, $\chi(r) = (\psi_c(r))^2 - \beta$ corresponds to Huber's Proposal 2 (see Huber, 1981, p. 137). Alternatively, one can consider ψ and χ unrelated and solve (6.2.5) and (6.2.6) simultaneously. The choice $\chi(r) = \text{sign}(|r| - \tilde{\beta})$ with an appropriate $\tilde{\beta}$, defines Hampel's median deviation (Hampel, 1974). For details see Subsection 2.5e.

Remark 3. Note that M-estimation based on a dispersion function of the form (6.2.3) is related to a metric. Let

$$\nu(u, v) := \sum_{i=1}^{n} \rho\left(\left(u^{(i)} - v^{(i)} \right)/\sigma \right),$$

$$u = \left(u^{(1)}, \ldots, u^{(n)} \right)^T, \qquad v = \left(v^{(1)}, \ldots, v^{(n)} \right)^T.$$

McKean and Schrader (1980) show that in most cases there exists a monotone function f such that $f(\nu(u, v))$ is a metric on $\mathbb{R}^n$. This f exists for $\rho = \rho_c$ (Huber-estimator) and for ρ defining Hampel's three-part redescending estimator (Hampel, 1974) (see Subsection 2.6a).

In order to study the robustness properties of the Huber-estimator T^{Hu} we compute its influence function at the model distribution $F_\theta(x, y)$ with density (ignoring the scale)

$$f_\theta(x, y) = \phi(y - x^T\theta)k(x).$$

Using (4.2.9) with

$$F = F_\theta, \qquad \psi(x, \theta) = \psi_c(y - x^T\theta)$$

we obtain

$$\text{IF}(x, y; T^{\text{Hu}}, F_\theta) = \psi_c(y - x^T\theta) \cdot M^{-1}x, \qquad (6.2.7)$$

where

$$M = (E\psi_c') \cdot (Exx^T) = \left(\int \psi_c'(r)\, d\Phi(r)\right) \cdot \left(\int xx^T\, dK(x)\right).$$

We see that this influence function depends on y only through $r := y - x^T\theta$. Following Hampel (1973a, 1978b) we can rewrite IF as a product of two factors, namely the (scalar) *influence of the residual* (IR) and the (vector-valued) *influence of position in factor space* (IP):

$$\text{IF}(x, x^T\theta + r; T^{\text{Hu}}, F_\theta) =: \text{IT}(x, r; T^{\text{Hu}}, F_\theta)$$

$$= \text{IR}(r; T^{\text{Hu}}, \Phi) \cdot \text{IP}(x; T^{\text{Hu}}, K), \qquad (6.2.8)$$

where

$$\text{IR}(r; T^{\text{Hu}}, \Phi) = \psi_c(r) / (E\psi_c'),$$

$$\text{IP}(x; T^{\text{Hu}}, K) = (Exx^T)^{-1}x.$$

The factorization (6.2.8) is unique if we define IR as the influence function of the corresponding M-estimator of location (defined by ψ_c) (see 2.3.12). IT is called the *total influence* on the estimator at F_θ.

The influence of the residual $\text{IR}(r; T^{\text{Hu}}, F_\theta)$ is bounded. This is an improvement of least-squares estimators from the robustness point of view. But still, the influence of position in factor space is unbounded. Thus a single x_i, which is an outlier in the factor space, will almost completely determine the fit (Hampel, 1973a). In this sense the Huber-estimator and all the estimators defined through (6.2.4), including the L_1-estimator, are only the first step in the robustification of a regression estimator. In order to cope with problems caused by outlying points in the factor space, we need more refined estimators.

Figure 1 shows the effect of a leverage point P_2 on the least-squares *and* the Huber-estimator. (The data are the same as those of Figure 1 in Subsection 6.1b with P_1 replaced by P_2.)

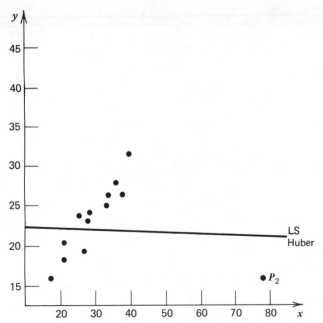

Figure 1. LS and Huber's fit for the data of Table 1 (Section 6.1) (point corresponding to year 1934 changed to P_2).

Remark 4. The influence function for fixed-carriers models can be derived as follows (see Huber, 1983). Let $x_1, \ldots, x_n$ be fixed and denote by $G_{\theta i}$ the distribution of $y_i = x_i^T \theta + e_i$. The defining equation for the estimator T is

$$\sum_{i=1}^{n} \int \psi\left(y_i - x_i^T T(H_1, \ldots, H_n)\right) x_i \, dH_i(y_i) = 0 \qquad (6.2.9)$$

and Fisher consistency requires $T(G_{\theta 1}, \ldots, G_{\theta n}) = \theta$. Then, by putting $H_{j;\, y} = (1 - \varepsilon)G_{\theta j} + \varepsilon \Delta_y$ in (6.2.9) and by computing the derivative at $\varepsilon = 0$, we get

$$\mathrm{IF}_j\left(y; T; G_{\theta 1}, \ldots, G_{\theta n}\right) = \psi\left(y - x_j^T \theta\right)$$

$$\cdot \left[\sum_{i=1}^{n} \int \psi'\left(y_i - x_i^T \theta\right) x_i x_i^T \, dG_{\theta i}(y_i) \right]^{-1} \cdot x_j;$$

$$(6.2.10)$$

see Exercise 7.

6.3. *M*-ESTIMATORS FOR LINEAR MODELS

6.3a. Definition, Influence Function, and Sensitivities

Definition 1. An *M*-estimator T_n for linear models is defined implicitly by the (vector) equation

$$\sum_{i=1}^{n} \eta\big(x_i, (y_i - x_i^T T_n)/\sigma\big)x_i = 0 \qquad (6.3.1)$$

(see Maronna and Yohai, 1981), where the function $\eta: \mathbb{R}^p \times \mathbb{R} \to \mathbb{R}$ satisfies the following conditions:

(6.3.ETA1′) (i) $\eta(x, \cdot)$ is continuos on $\mathbb{R} \setminus \mathscr{C}(x; \eta)$ for all $x \in \mathbb{R}^p$, where $\mathscr{C}(x; \eta)$ is a finite set. In each point of $\mathscr{C}(x; \eta)$, $\eta(x, \cdot)$ has finite left and right limits.

   (ii) $\eta(x, \cdot)$ is odd and $\eta(x, r) \geq 0$ for all $x \in \mathbb{R}^p$, $r \in \mathbb{R}^+$.

(6.3.ETA2′) For all x, the set $\mathscr{D}(x; \eta)$ of points in which $\eta(x, \cdot)$ is continuous but in which $\eta'(x, r) := (\partial/\partial r)\eta(x, r)$ is not defined or not continuous, is finite.

We shall also need the following regularity conditions:

(6.3.ETA3′) (i) $M := E\eta'(x, r)xx^T = \int \eta'(x, r)xx^T d\Phi(r)\, dK(x)$ exists and is nonsingular.

   (ii) $Q := E\eta^2(x, r)xx^T = \int \eta^2(x, r)xx^T d\Phi(r)\, dK(x)$ exists and is nonsingular.

These conditions correspond to the conditions on ψ given in Subsection 2.5a.

The form (6.3.1) restricts the *M*-estimators, as compared with the general formula $\sum_{i=1}^{n} \psi(x_i, y_i; T_n) = 0$ (cf. Definition 5, Subsection 4.2c) in two ways: $\psi(x_i, y_i; \theta)$ must have the same direction as x and the scalar $\eta(x, r)$ depends on θ only through $r = y - x^T\theta$. Both restrictions are justified by equivariance considerations.

There have been several proposals for choosing η. For a stimulating discussion we refer to the papers by Hampel (1978b), Welsch (1979), Krasker and Welsch (1982), Huber (1983), and to the Ph.D thesis of Hill (1977). All known proposals of η may be written in the form

$$\eta(x, r) = w(x) \cdot \psi(r \cdot v(x)),$$

for appropriate functions $\psi \colon \mathbb{R} \to \mathbb{R}$ and $w \colon \mathbb{R}^p \to \mathbb{R}^+$, $v \colon \mathbb{R}^p \to \mathbb{R}^+$ (weight functions). Huber (1973a) uses $w(x) = 1$, $v(x) = 1$, and Mallows's and Andrews's proposals (see Hill, 1977) set $v(x) = 1$ and $w(x) = 1$, respectively. Hill and Ryan (see Hill, 1977) proposed $v(x) = w(x)$ and, finally, Schweppe (see Merrill and Schweppe, 1971; Handschin et al., 1975) suggested choosing $v(x) = 1/w(x)$.

Let now $\sigma = 1$ for simplicity.

The functional $T(F)$ corresponding to the M-estimator defined by (6.3.1) is the solution of

$$\int \eta(x, y - x^T \cdot T(F)) x \, dF(x, y) = 0.$$

Define

$$M(\eta, F) := \int \eta'(x, y - x^T \cdot T(F)) x x^T dF(x, y). \qquad (6.3.2)$$

Then the influence function of T at a distribution F (on $\mathbb{R}^p \times \mathbb{R}$) is given by

$$\mathrm{IF}(x, y; T, F) = \eta(x, y - x^T \cdot T(F)) \cdot M^{-1}(\eta, F) x. \qquad (6.3.3)$$

Maronna and Yohai (1981) show, under certain conditions, that these estimators are consistent and asymptotically normal with asymptotic covariance matrix

$$V(T, F) = \int \mathrm{IF}(x, y; T, F) \cdot \mathrm{IF}^T(x, y; T, F) \, dF(x, y)$$

$$= M^{-1}(\eta, F) \cdot Q(\eta, F) \cdot M^{-1}(\eta, F), \qquad (6.3.4)$$

where

$$Q(\eta, F) := \int \eta^2(x, y - x^T \cdot T(F)) x x^T dF(x, y)$$

[cf. (4.2.2) and also Yohai and Maronna, 1979]. At the model distribution $F = F_\theta$ we obtain

$$\mathrm{IF}(x, y; T, F_\theta) = \eta(x, y - x^T\theta) \cdot M^{-1}x, \qquad (6.3.5)$$

$$V(T, F_\theta) = M^{-1}QM^{-1}, \qquad (6.3.6)$$

where

$$M := M(\eta, F_0) = E\{\eta'(x, r)xx^T\},$$

$$Q := Q(\eta, F_0) = E\{\eta^2(x, r)xx^T\}.$$

Note that M and Q, and thus $V(T, F_\theta)$, do not depend on θ. Moreover, the sensitivities (see Subsection 4.2b) equal

$$\gamma_u^*(T, F_\theta) = \sup_{x, y} |\eta(x, y - x^T\theta)| \||M^{-1}x\|,$$

$$\gamma_s^*(T, F_\theta) = \sup_{x, y} |\eta(x, y - x^T\theta)| \cdot (x^TQ^{-1}x)^{1/2},$$

$$\gamma_i^*(T, F_\theta) = J(G) \cdot \sup_{x, y} |\eta(x, y - x^T\theta)| \cdot (x^TM^{-1}(Exx^T)M^{-1}x)^{1/2}$$

where $J(G) = \int(g'/g)^2\, dG$ $\quad (J(\Phi) = 1)$.

For linear models one may also consider the *influence on the fitted value* (self-influence; see Hampel, 1978b) given by

$$x^T \cdot \text{IF}(x, y; T, F_\theta)$$

with the corresponding sensitivity

$$\gamma_f^*(T, F_\theta) := \sup_{x, y} |x^T \cdot \text{IF}(x, y; T, F_\theta)|$$

$$= \sup_{x, y} |\eta(x, y - x^T\theta)| \cdot x^TM^{-1}x.^\ddagger$$

Note that γ_s^*, γ_i^*, and γ_f^* are invariant with respect to nonsingular linear transformations of the parameter space which, for linear models, correspond to a change of the basis in the space of the design vectors x.

‡One can also define the *influence on the predicted value at* x_0,

$$x_0^T\text{IF}(x, y; T, F_\theta),$$

with the corresponding sensitivity

$$\gamma_p^*(T, F_\theta, x_0) := \sup_{x, y} |x_0^T\text{IF}(x, y; T, F_\theta)|$$

$$= \sup_{x, y} |\eta(x, y - x^T\theta)| |x_0^TM^{-1}x.$$

Definition 2. We say that the estimator T defined by η is B_u-$(B_{s^-}, B_{i^-}, B_{f^-})robust$ if and only if $\gamma_u^*(\gamma_s^*, \gamma_i^*, \gamma_f^*)$ is finite.

6.3b. Most *B*-Robust and Optimal *B*-Robust Estimators

Consider the linear model with normal errors (see Section 6.1) and let $\{\eta\}$ be the class of *M*-estimators for linear models defined in Subsection 6.3a.

In this subsection we want to find the most *B*-robust estimator in $\{\eta\}$, that is, the estimator with the smallest sensitivity γ^* at the model. Furthermore, we want to apply the optimality results of Section 4.3 to obtain optimal *B*-robust estimators for linear models.

The following proposition gives lower bounds for the unstandardized and the self-standardized gross-error sensitivity at the model, see Ronchetti and Rousseeuw (1985).

Proposition 1.

(i)   Always $\gamma_u^* \geq p(\pi/2)^{1/2}/E\|x\|$. If $E[xx^T/\|x\|]$ is a scalar matrix, then $\eta(x, r) = \text{sign}(r)/\|x\|$ reaches this lower bound, so we call this estimator most B_u-robust.

(ii) Always $\gamma_s^* \geq p^{1/2}$. If $E[xx^T/\|x\|^2]$ is a scalar matrix, then $\eta(x, r) = \text{sign}(r)/\|x\|$ reaches this lower-bound, so we call this estimator most B_s-robust.

(Here a scalar matrix is a real multiple of the identity matrix.)

Proof. First, we have $M = E[y\eta(x, y)xx']$, so $I_p = E[y\eta(x, y) M^{-1}xx']$. Taking the trace, $p = E[y\eta(x, y)x'M^{-1}x] \leq E[|y| \cdot |\eta(x, y)| \cdot \|x\| \cdot \|M^{-1}x\|] \leq \gamma_u^* \cdot E[|y| \cdot \|x\|] = \gamma_u^* \cdot 2\Phi'(0)E[\|x\|]$, from which the desired inequality follows. If $E[xx'/\|x\|]$ is a scalar matrix and $\eta(x, y) = \text{sign}(y)/\|x\|$, then $M = 2\Phi'(0)E[xx'/\|x\|] = 2\Phi'(0)E[\|x\|](1/p)I_p$ so $\gamma_u^* = p(\pi/2)^{1/2}/E[\|x\|]$.

Second, we have $Q = E[\eta^2(x, y)xx']$, so $I_p = E[\eta^2(x, y)xx'Q^{-1}]$. Taking traces, $p = E[\eta^2(x, y)x'Q^{-1}x] = E[(|\eta(x, y)|(x'Q^{-1}x)^{1/2})^2] \leq \gamma_s^{*2}$. If $E[xx'/\|x\|^2]$ is a scalar, then $\eta(x, y) = \text{sign}(y)/\|x\|$ satisfies $Q = E[xx'/\|x\|^2] = (1/p)I_p$ (because its trace equals 1); hence $\gamma_s^* = \sup|\eta(x, y)|(x'Q^{-1}x) = \sup(1/\|x\|)\|x\|\sqrt{p} = \sqrt{p}$.   □

In the case $p = 1$ or K is radially symmetric, that is,

(6.3.R) K depends only on $\|x\|$, then $E[xx^T/\|x\|]$ and $E[xx^T/\|x\|^2]$ are indeed scalar matrices. When $p = 1$ (regression line through the origin), Proposition 1 gives the most B_u- (and B_s-) robust estimator $T_n = \text{med}\{y_i/x_i\}$ (see Krasker, 1980, p. 1340). When also $x \equiv 1$ (location problem), the solution reduces to the sample median. Proposition 1 improves Krasker's lower bound (1980, Proposition 2) for γ_u^*, which he gave for a special family of estimators with which we shall deal later.

Note that the most B_u- (and B_s-) robust estimator of Proposition 1 has a Mallows-type η-function, but can also be considered as a member of the Schweppe class by writing sign(r) as sign($r \cdot \|x\|$).

The optimality results follow directly from Section 4.3. From Theorem 1 of that section we get the following result.

Proposition 2. The estimator within $\{\eta\}$ which minimizes the trace of the asymptotic covariance matrix V [the trace of $J(\Phi)Exx^TV$ respectively] under the condition of a sufficiently large bound c on the sensitivity γ_u^* (γ_i^* respectively) is defined by an η-function of Schweppe's form,

$$\eta(x, r) = w(x) \cdot \psi_c(r/w(x)).\qquad(6.3.7)$$

The optimal weight function $w(x)$ is given by

$$w(x) = \|Ax\|^{-1},\qquad(6.3.8)$$

where the matrices A are defined implicitly by the equations

$$E\left[\psi_c'(r\|Ax\|)xx^T\right] =$$

u: $\qquad E\left[(2\Phi(c/\|Ax\|) - 1)xx^T\right] = A^{-1},\qquad(6.3.9)$

i: $\qquad E\left[(2\Phi(c/\|Ax\|) - 1)xx^T\right] = A^{-1}(Exx^T)^{1/2},\qquad(6.3.10)$

where $(Exx^T)^{1/2}$ is any root of Exx^T.

More precisely, if c is large enough that the solution A of (6.3.9) exists, then (6.3.7) satisfies the stated optimality. There is a lower bound on c (given by Proposition 1(i)), and Krasker (1980) shows the existence for sufficiently large c, but one does not know how large c must be. Assuming either $p = 1$ or (6.3.R), one can show that this estimator exists if and only if $c > p(\pi/2)^{1/2}/E\|x\|$ (Ronchetti and Rousseeuw, 1985, Lemma 1).

The optimal estimator in the unstandardized case (optimal B_u-robust) is called the *Hampel–Krasker estimator* (see Hampel, 1978b; Krasker, 1977, 1980).

Applying Theorem 2 of Section 4.3 we can state now an admissibility result for the self-standardized case.

Proposition 3.   For each $c > p^{1/2}$, the estimator defined by (6.3.7) and (6.3.8), where the matrix A is determined implicitly by the equation

$$E\left[\left(\|Ax\|^{-2}\psi_c^2(r\|Ax\|)\right)xx^T\right] = (A^TA)^{-1}, \qquad (6.3.11)$$

is admissible (see Definition of Section 4.6) among all estimators of the class $\{\eta\}$ satisfying $\gamma_s^* \leq c$. [The left-hand side of (6.3.11) can be integrated with respect to the error distribution yielding

$$E\left[\left(2c^2/\|Ax\|^2 - 1 + 2(1 - c^2/\|Ax\|^2)\Phi(c/\|Ax\|)\right.\right.$$

$$\left.\left. - 2(c/\|Ax\|)\Phi'(c/\|Ax\|)\right)xx^T\right].]$$

This estimator will be called *Krasker–Welsch estimator*.

Krasker and Welsch (1982) discuss extensively the self-standardized case and show that the estimator given by Proposition 3 satisfies the first-order necessary condition for efficiency, that is, if there is an estimator that minimizes the asymptotic covariance matrix in the strong sense within $\{\eta\}$ under the condition of a bounded self-standardized sensitivity, this is, of the form given above. Note, however, that Bickel (1984b) suggests that the Krasker–Welsch estimator might not be strongly optimal (cf. also Ruppert, 1985).

Remark 1.   Welsch (1981) conjectures that we can obtain the same kind of result if we solve the optimality problem with respect to the "fitted-value sensitivity." The optimal B_f-robust estimator is defined by a η-function of the form (6.3.7) but with weight function

$$w(x) = \|Ax\|^{-2}, \qquad (6.3.12)$$

where A is determined implicitly by the equation

$$E\left[\left(2\Phi(c/\|Ax\|^2) - 1\right)xx^T\right] = (A^TA)^{-1}. \qquad (6.3.13)$$

Now we want to give the solution of Hampel's optimality problem in a subclass of general *M*-estimators for linear models, namely, the class of Mallows estimators. There is a particular interest in Mallows's class of estimators because they may provide good diagnostic information (Welsch, 1981). Moreover, they have bounded local-shift sensitivity, as opposed to Schweppe-type estimators. Mallows estimators are defined by a η-function of the form

$$\eta(x, r) = w(x) \cdot \psi(r),$$

where $w: \mathbb{R}^p \to \mathbb{R}^+, \psi: \mathbb{R} \to \mathbb{R}$ are appropriate functions. The optimal *B*-robust estimator within this class is defined by the η-function

$$\eta(x, r) = w(x) \cdot \psi_c(r), \tag{6.3.14}$$

where $w(x)$ is given by

u,s,i:
$$w(x) = w_b(\|Bx\|), \tag{6.3.15}$$

f:
$$w(x) = w_b(\|Bx\|^2), \tag{6.3.16}$$

$$w_b(t) := \psi_b(t)/t = \min(1, b/|t|),$$

and b, c are some positive constants. Note that here b and c are not uniquely defined through the bound on the sensitivity. The matrices B are implicitly defined according to the four cases by the following equations

u:
$$Ew_b(\|Bx\|)xx^T = B^{-1}, \tag{6.3.17}$$

s:
$$Ew_b^2(\|Bx\|)xx^T = (B^TB)^{-1}, \tag{6.3.18}$$

i:
$$Ew_b(\|Bx\|)xx^T = B^{-1}(Exx^T)^{1/2}, \tag{6.3.19}$$

f:
$$Ew_b(\|Bx\|^2)xx^T = (B^TB)^{-1}. \tag{6.3.20}$$

These results can be derived analogously to the general case.

Figure 1 shows how the Huber-estimator, the optimal Mallows estimator [defined by (6.3.14), (6.3.15), and (6.3.17)], and the Hampel–Krasker estimator downweigh the observations.

Figure 2 shows the Hampel–Krasker and the optimal Mallows estimator for the data set of Figure 1 (Section 6.2).

Huber:

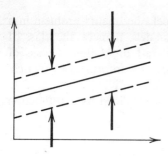

$$\sum_{i=1}^{n} \psi_c(r_i) x_i = 0$$

Huber *does not* bound the influence of position.

Mallows:

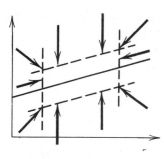

$$\sum_{i=1}^{n} \psi_c(r_i) w_i x_i = 0$$

Mallows *downweighs leverage points regardless of the magnitude of the corresponding residual.*

Hampel–Krasker:

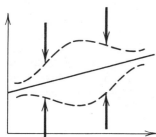

$$\sum_{i=1}^{n} \psi_c\!\left(\frac{r_i}{\tilde{w}_i}\right) \tilde{w}_i x_i = \sum_{i=1}^{n} \psi_{c\tilde{w}_i}(r_i) x_i = 0$$

Hampel–Krasker *downweighs leverage points only if the corresponding residual is large.*

Figure 1. Bounding the influence of the residual and the influence of position in factor space.

Remark 2. Mallows (1983) pointed out that one should impose constraints on both the gross-error sensitivity *and* the local-shift sensitivity. In the case of simple regression through the origin ($p = 1$), the estimator that minimizes the asymptotic variance within the class $\{\eta\}$, under the side conditions of a bounded gross-error sensitivity and a bounded local-shift sensitivity is given by

$$\eta(x, r) = w_b(\|x\|)\psi_{c(\|x\|)}(r),$$

where $c(\|x\|) = \tilde{c} \cdot \max(1,\ b/\|x\|)$, and the constants $b, \tilde{c}$ are determined uniquely through the bounds on the sensitivities.

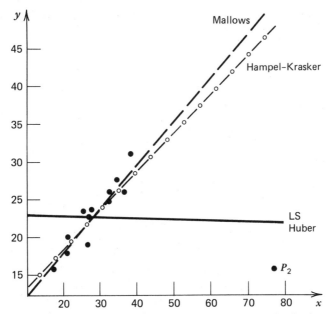

Figure 2. The LS, Huber-, Mallows, and Hampel–Krasker estimators for the data of Figure 1 (Section 6.2).

Remark 3. Equations (6.3.9)–(6.3.11), (6.3.13), and (6.3.17)–(6.3.20) are of the form (5.3.11). Thus, the solution of each equation determines a robust covariance matrix in the factor space which is used in the corresponding function [(6.3.8), (6.3.12), (6.3.15), and (6.3.16)] to assign a weight to each *x* according to its outlyingness from the center.

6.3c. The Change-of-Variance Function; Most *V*-Robust and Optimal *V*-Robust Estimators

In this subsection we extend the notion of change-of-variance function to the regression model and in analogy with the gross-error sensitivity we define an unstandardized and a self-standardized change-of-variance sensitivity. We find the most *V*-robust estimators for regression and we show that the optimal *B*-robust estimators solve the related optimality problem with respect to the change-of-variance sensitivity; therefore, they are also optimal *V*-robust estimators.

The proofs of the propositions in this subsection use similar techniques as in the location case (see Section 2.5). Therefore we drop most of them in our exposition. For further details see Ronchetti and Rousseeuw, 1985; Ronchetti, 1982b.

Consider the class of M-estimators given in Subsection 6.3a and let $V(T, F)$ be the asymptotic covariance matrix of an M-estimator T defined by (6.3.4).

Definition 3. We define the *change-of-variance function* of an M-estimator corresponding to η at the model distribution F_θ by the matrix

$$\text{CVF}(x, y; T, F_\theta) := \left[(\partial/\partial\varepsilon) V\left(T, (1 - \varepsilon) F_\theta + \varepsilon\Delta_{(x,y)}\right) \right]_{\varepsilon=0},$$

for all x, y where this expression exists.

Let $\mathscr{D}(x; \eta) \subset \mathbb{R}$ be the finite set of points where $\eta'(x, \cdot)$ is not defined (see 6.3.ETA2'). After some elementary calculations applying Definition 3, we obtain

$$\text{CVF}(x, y; T, F_\theta) = M^{-1}QM^{-1} + \eta^2(x, y - x^T\theta) \cdot M^{-1}xx^TM^{-1}$$

$$-\eta'(x, y - x^T\theta) \cdot (M^{-1}xx^TM^{-1}QM^{-1}$$

$$+ M^{-1}QM^{-1}xx^TM^{-1}) \qquad (6.3.21)$$

and taking traces

$$\text{trace CVF}(x, y; T, F_\theta) = \text{trace} V(T, F_\theta) + \eta^2(x, y - x^T\theta) \cdot \|M^{-1}x\|^2$$

$$-2\eta'(x, y - x^T\theta) \cdot (M^{-1}x)^T M^{-1}Q(M^{-1}x),$$

$$(6.3.22)$$

for all (x, y) such that $y - x^T\theta \notin \mathscr{D}(x; \eta)$.

In analogy with γ^* we propose the following measures of sensitivity.

Definition 4. We call

$$\kappa_u^*(T, F_\theta) := \sup\{\text{trace CVF}(x, y; T, F_\theta)/$$

$$\text{trace} V(T, F_\theta) | y - x^T\theta \notin \mathscr{D}(x; \eta)\} \qquad (6.3.23)$$

the *unstandardized change-of-variance sensitivity* and

$$\kappa_s^*(T, F_\theta) := \sup\left\{ \text{trace} \left[\text{CVF}(x, y; T, F_\theta) \cdot V^{-1}(T, F_\theta) \right] \right|$$

$$y - x^T\theta \notin \mathscr{D}(x; \eta) \right\} \qquad (6.3.24)$$

the *self-standardized change-of-variance sensitivity*. Applying (6.3.22) and (6.3.24) we obtain

$$\kappa_u^*(T, F_\theta) = \sup\left\{ 1 + \eta^2(x, y - x^T\theta) \cdot \| M^{-1}x \|^2 / \text{trace} V(T, F_\theta) \right.$$

$$- 2\eta'(x, y - x^T\theta) \cdot (M^{-1}x)^T M^{-1}Q \, (M^{-1}x) / \text{trace} V(T, F_\theta) \Big|$$

$$y - x^T\theta \notin \mathscr{D}(x; \eta) \right\} \qquad (6.3.25)$$

and

$$\kappa_s^*(T, F_\theta) = \sup\left\{ p + \eta^2(x, y - x^T\theta) \cdot x^T Q^{-1} x - 2\eta'(x, y - x^T\theta) \cdot x^T M^{-1}x \Big|$$

$$y - x^T\theta \notin \mathscr{D}(x; \eta) \right\}. \qquad (6.3.26)$$

Note that both definitions (6.3.23) and (6.3.24) extend the change-of-variance of sensitivity of the one-dimensional location case (see Section 2.5) to the regression case and that κ_s^* is invariant with respect to nonsingular linear transformations of the parameter space.

As for the location case (cf. Subsections 2.5a), the change-of-variance sensitivity is defined by means of the supremum of the change-of-variance function. Note, however, that the infimum of that function though less important than the supremum, also has some meaning and importance, especially if stability of confidence intervals is desired.

Again the following equality holds:

$$\int \text{CVF}(x, y; T, F_\theta) \, dF_\theta(x, y) = 0.$$

Remark 4. The change-of-variance function can be also computed at distributions F outside the model F_θ. Moreover, Definitions 3 and 4 can be applied to general parametric models and one gets formulas similar to (6.3.21), (6.3.25), and (6.3.26).

Definition 5. We say that the estimator defined by η is V_u- (V_s-)*robust* if and only if κ_u^* (κ_s^*) is finite.

As in the location case we have the following relation between B- and V-robustness.

Proposition 4. V_u-robustness implies B_u-robustness. V_s-robustness implies B_s-robustness. In fact, $\gamma_u^* \leq [(\kappa_u^* - 1)\text{MSE}]^{1/2}$ and $\gamma_s^* \leq [\kappa_s^* - p]^{1/2}$, where $\text{MSE} := \text{trace}\, V(T, F_\theta)$.

Proof. First, assume that $\kappa_u^* := \kappa_u^*(\eta) < \infty$. Let us suppose that there exists (x_0, y_0) such that $|\eta(x_0, y_0)| \cdot \|M^{-1}x_0\| > [(\kappa_u^* - 1)\text{MSE}]^{1/2}$. Without loss of generality, $y_0 \notin \mathscr{D}(x_0, \eta)$ and $y_0 > 0$. We also have $x_0 \neq 0$, and we put $C := \mathscr{C}(x_0, \eta)$, $D := \mathscr{D}(x_0, \eta)$, and $\psi(y) := \eta(x_0, y)\|M^{-1}x_0\|$ for all $y \notin C$. *Case A:* If $(M^{-1}x_0)'QM^{-1}(M^{-1}x_0) = 0$, then $1 + \psi^2(y_0)/\text{MSE} - 2\psi'(y_0)(M^{-1}x_0)'QM^{-1}(M^{-1}x_0)/(\|M^{-1}x_0\|\text{MSE}) > 1 + (\kappa_u^* - 1) = \kappa_u^*$, contradicting (6.3.25). *Case B:* Suppose that $(M^{-1}x_0)'QM^{-1}(M^{-1}x_0) > 0$. If $\psi'(y_0) \leq 0$, then $1 + \psi^2(y_0)/\text{MSE} - 2\psi'(y_0)(M^{-1}x_0)'QM^{-1}(M^{-1}x_0)/(\|M^{-1}x_0\|\text{MSE}) > 1 + (\kappa_u^* - 1) = \kappa_u^*$, a contradiction. Therefore $\psi'(y_0) > 0$; hence there exists $\varepsilon > 0$ such that $\psi'(t) > 0$ for all t in $[y_0, y_0 + \varepsilon)$, and thus $\psi(y) > \psi(y_0)$ for all y in $(y_0, y_0 + \varepsilon]$. We now show that $\psi(y) > \psi(y_0)$ for all $y > y_0$, $y \notin C$. Suppose the opposite were true; then $y_0 + \varepsilon \leq y' := \inf\{y > y_0; \ y \notin C$ and $\psi(y) \leq \psi(y_0)\} < \infty$. As in points of C only upward jumps are allowed (otherwise, $\kappa_u^* = \infty$), we have $y' \notin C$. [Take $c \in C$, $c > y_0$. If $\psi(c -) < \psi(y_0)$ then $y' < c$, and if $\psi(c -) \geq \psi(y_0)$ then $\psi(c +) > \psi(y_0)$, so $c \neq y'$.] Hence ψ is continuous at y', so $\psi(y') = \psi(y_0)$. Clearly, $\psi(y) > \psi(y_0)$ for all y in $(y_0, y')\setminus C$. Now there exists a point y'' in $(y_0, y')\setminus(C \cup D)$ such that $\psi'(y'') \leq 0$, because otherwise we could show that $\psi(y') > \psi(y_0)$ by starting in y_0 and going to the right, using the upward jumps in points of C. Since we must also have $\psi^2(y'') > \psi^2(y_0)$, it would hold that $1 + \psi^2(y'')/\text{MSE} - 2\psi'(y'')(M^{-1}x_0)'QM^{-1}(M^{-1}x_0)/(\|M^{-1}x_0\|\text{MSE}) > 1 + \psi^2(y'')/\text{MSE} > 1 + (\kappa_u^* - 1) = \kappa_u^*$, a contradiction. We conclude that $\psi(y) > \psi(y_0)$ for all $y > y_0$, $y \notin C$. We now proceed to the final contradiction. Because $C \cup D$ is finite, we can assume from now on that $[y_0, +\infty) \cap (C \cup D)$ is empty. Then $\psi^2(y) - 2\psi'(y)(M^{-1}x_0)'QM^{-1}(M^{-1}x_0)/\|M^{-1}x_0\| \leq b^2$ for all $y \geq y_0$, where $b := [(\kappa_u^* - 1)\text{MSE}]^{1/2}$. Because $\psi^2(y) > \psi^2(y_0) > b^2$, this gives $\psi'(y)(M^{-1}x_0)'QM^{-1}(M^{-1}x_0)/\|M^{-1}x_0\| \geq \frac{1}{2}[\psi^2(y) - b^2] > 0$. Therefore, $\psi'(y)/[\psi^2(y) - b^2] \geq d := \|M^{-1}x_0\|/[2(M^{-1}x_0)'QM^{-1}(M^{-1}x_0)] > 0$. For all $y \geq y_0$, we define $R(y) = (1/b)\coth^{-1}(\psi(y)/b)$. It follows that R is well defined and differentiable with derivative $\psi'(y)/[\psi^2(y) - b^2]$ on $[y_0, \infty)$. Therefore, $R(y) - R(y_0) \geq d(y - y_0)$ for all $y \geq y_0$, which implies that $\coth^{-1}(\psi(y)/b) \leq b[dy_0 - R(y_0) - dy]$. The left-hand side of the latter inequality is strictly positive because $\psi(y)/b > 1$, but the right-hand side tends to $-\infty$ when $y \to \infty$, which is clearly impossible. *Case C*: Now suppose that $(M^{-1}x_0)'QM^{-1}(M^{-1}x_0) < 0$. As in case *B*, we see that $\psi'(y_0) < 0$; hence there exists $\varepsilon > 0$ such that $\psi'(t) < 0$ for all t in $(y_0 - \varepsilon, y_0]$, and thus $\psi(y) > \psi(y_0)$ for all y in $[y_0 - \varepsilon, y_0)$. As in case *B*, we show $\psi(y) > \psi(y_0)$ for all $y < y_0$ (this time making use of $-\infty < y' := \sup\{y < y_0; y \notin C$ and $\psi(y) \leq \psi(y_0)\} \leq y_0 - \varepsilon$ and the fact that only downward jumps of ψ are allowed in points of C). However, this is a contradiction because ψ has to be negative for negative y. As we have reached a contradiction in cases *A*, *B*, and *C*, we can conclude that $\gamma_u^* \leq [(\kappa_u^* - 1)\mathrm{MSE}]^{1/2}$.

Second, assume that $\kappa_s^*(\eta) < \infty$ and that there exists (x_0, y_0) such that $|\eta(x_0, y_0)|(x_0'Q^{-1}x_0)^{1/2} > (\kappa_s^*(\eta) - p)^{1/2}$ where $y_0 \notin \mathscr{D}(x_0, \eta)$ and $y_0 > 0$. The proof then follows the same lines as in the first part, but now $\psi(y) := \eta(x_0, y)(x_0'Q^{-1}x_0)^{1/2}$; cases *A*, *B*, and *C* are based on the sign of $x_0'M^{-1}x_0$, $b := (\kappa_s^* - p)^{1/2}$; and $d := (x_0'Q^{-1}x_0)^{1/2}/(2x_0'M^{-1}x_0)$. ☐

Corollary. Suppose η is nondecreasing in r. Then,

(i) If $QM^{-1} \geq 0$, V_u- and B_u-robustness are equivalent.
(ii) If $M > 0$, V_s- and B_s-robustness are equivalent.

The following proposition gives the most V_u- and most V_s-robust estimators. It is analogous to Proposition 1.

Proposition 5

(i) Always $\kappa_u^* \geq 2$. If $E[xx^T/\|x\|]$ is a scalar matrix, then $\eta(x, r) = \mathrm{sign}(r)/\|x\|$ reaches this lower bound, so we call this estimator most V_u-robust.

(ii) Always $\kappa_s^* \geq 2p$. If $E[xx^T/\|x\|^2]$ is a scalar matrix, then $\eta(x, r) = \mathrm{sign}(r)/\|x\|$ reaches this lower bound, so we call this estimator most V_s-robust.

The optimality results with respect to the change-of-variance sensitivities are given in the next two propositions.

Proposition 6. Assume $p = 1$ or (6.3.R), in which case attention is restricted to functions $\eta(x, r) = \eta^*(\|x\|, r)$. Let T_c^{HK} be the

Hampel–Krasker estimator. Then, for each $k > 2$, there exists a $c > 0$ such that $\kappa_u^*(T_c^{HK}, F_\theta) = k$, and the Hampel–Krasker estimator minimizes the trace of the asymptotic covariance matrix among all estimators satisfying $\kappa_u^* \le k$. Any other solution of this extremal problem is equivalent to this estimator.

Proposition 7. Let T_c^{KW} be the Krasker–Welsch estimator. Then, for each $k > 2p$, there exists a c such that $\kappa_s^*(T_c^{KW}, F_\theta) = k$, and the Krasker–Welsch estimator is admissible among all estimators satisfying $\kappa_s^* \le k$.

Remark 5. In analogy to the location case, we conjecture that, if one looks for estimators with finite change-of-variance sensitivity *and* finite rejection point, the η-function defining the most efficient estimator satisfying those conditions, is of Schweppe's form (see Subsection 6.3a) with $\psi(y) = \chi_{r,k}(y)$ [see (2.6.9) and Fig. 7 in Subsection 2.6c]. We call this M-estimator for regression tanh-*regression estimator*.

6.4. COMPLEMENTS

6.4a. Breakdown Aspects

By means of infinitesimal methods we have derived in the last section optimal robust estimators for regression. In this subsection we want to discuss the breakdown-point aspects. Let us recall briefly the definition of the breakdown point ε^*. In words, it is the smallest percentage of contamination in the data that may cause the estimator to take on arbitrarily large values. The exact definition, which is asymptotic in nature, can be found in Subsection 2.2a along with a finite-sample version of this concept.

In the case of the least-squares estimator we find $\varepsilon^* = 0$. The first step in the robustification of the least-squares estimator is the Huber-estimator (see Section 6.2). Its breakdown point depends on the design (the distribution of the x's). It is never greater than 25% and is arbitrarily close to zero for longer-tailed designs. The most robust limiting case of the Huber-estimator, the L_1-estimator (that minimizes the sum of absolute deviations) has breakdown point 25% for uniform x's, less than 24% for normal x's, and arbitrarily close to zero for longer-tailed and asymmetric designs (cf. Hampel, 1975, p. 379). This is due to the fact that these estimators protect against outlying y_i but cannot cope with outliers in the factor space which have a large influence (leverage points) on the fit. In fact the L_1-straight line goes *right through* a leverage point if this is sufficiently distant (Exercise 12). For the M-estimators, treated in Section 6.3, the breakdown point cannot exceed $1/p$ where p is the dimension of the parameter space (see Theorem 1 of Section 5.5 and Maronna, 1976 and Maronna et al., 1979). Thus the breakdown point becomes very low when the number of parameters is moderately large.

Various other resistant estimators have been proposed by Theil (1950), Brown and Mood (1951), Sen (1968), Andrews (1974), Tukey (1970/71, Chapter 10), and Johnstone and Velleman (1985), but none of them reaches more than $\varepsilon^* = 30\%$ in the case of simple regression ($p = 2$). Jaeckel (1972) proposed some estimates which are defined by minimization of a robust dispersion measure of the residuals. For this purpose he uses L-estimators for scale (see Subsection 2.3b). As these estimators are translation invariant, they do not determine the constant term of the regression model. Moreover, the breakdown point of L-estimators for scale is at most 25%, which is reached at the interquartile range. However, this objection could be removed by replacing the scale estimator by another one having a larger breakdown point. Indeed, Hampel (1975, p. 380) suggested an estimator based on minimization of $\text{MAD}(r_i)$.

Let us now discuss two recent proposals for regression analysis with *maximal* breakdown point. Clearly, $\varepsilon^* = 50\%$ is the best we can hope to accomplish, because for larger amounts of contamination no equivariant estimator can distinguish between the "good" and the "bad" parts of the sample. Probably the first regression estimator with $\varepsilon^* = 50\%$ to be worked out was Siegel's (1982) repeated median algorithm.

For any p observations $(x_{i_1}, y_{i_1}), \ldots, (x_{i_p}, y_{i_p})$ in general position denote by $\theta(i_1, \ldots, i_p)$ the unique parameter vector determined by them. Siegel's estimator is then defined as

$$T_n^{(j)} = \operatorname*{med}_{i_1}\left(\ldots \left(\operatorname*{med}_{i_{p-1}}\left(\operatorname*{med}_{i_p} \theta^{(j)}(i_1, \ldots, i_p)\right)\right) \ldots \right), \qquad (6.4.1)$$

where $T_n^{(j)}$ denotes the jth component of T_n. As this estimator is defined coordinatewise, it is not equivariant with respect to linear transformations in the factor space. In general, the number of computations necessary for determining this estimator is of order $O(n^p)$. In the case of simple regression, Siegel's estimate of the slope reduces to

$$\hat{\beta} = \operatorname*{med}_{i} \operatorname*{med}_{j \neq i} \frac{y_i - y_j}{x_i - x_j}. \qquad (6.4.2)$$

Note that the computation of (6.4.1) poses prohibitive computational problems in higher dimensions.

Let us now present the other proposal (Hampel, 1975, p. 380; Rousseeuw, 1984). Many authors have robustified the least-(sum-of)-squares technique by replacing the squared residuals r_i^2 by something else, such as $|r_i|$ or $\rho(r_i)$, but have not touched the summation sign. However, the *least median of squares* estimator (LMS) is defined as the value T_n that minimizes

$$m_\theta = \text{med}\left\{ y_i - x_i^T\theta \right\}^2 \qquad (6.4.3)$$

over θ. The finite-sample breakdown point equals $([n/2] - p + 2)/n$. Therefore, asymptotically $\varepsilon^* = 50\%$. In the one-dimensional case, the LMS estimate is the midpoint of the shortest half of the sample, and in the case of simple regression the LMS line corresponds to the narrowest strip covering half of the points.

This proposal presents computational problems as well, but computer programs are available (see Subsection 6.4c). Algorithms for the computation of LMS estimates are discussed in Leroy and Rousseeuw (1984) and Steele and Steiger (1984).

A disadvantage of the LMS is its lack of efficiency when the errors are really normally distributed. Very recently, two variants of the LMS have been proposed which are more efficient: the *least trimmed squares (LTS) estimator* (Rousseeuw, 1984) and the *S-estimator* (Rousseeuw and Yohai, 1984).

The efficiency problem can also be overcome by calculating a one-step M-estimator with a redescending ψ-function like the one of the hyperbolic tangent estimator (2.6.9) or of the biweight (2.6.4) and using the LMS estimator as starting point. In the presence of leverage points, one might prefer to start with a LMS estimator and calculate a one-step M-estimator using a η-function corresponding to a bounded influence estimator [(6.3.7) or (6.3.14)].

As an example, let us consider the famous "stackloss data" of Brownlee (1965) which describe the operation of a plant for the oxidation of ammonia to nitric acid. We have selected this example because it is a set of real data and it has been examined by a great number of statisticians (Draper and Smith, 1966, p. 204; Daniel and Wood, 1971; Andrews, 1974; and many others) with the help of several methods. Summarizing their findings, it can be said that most people concluded that observations 1, 3, 4, 21 were outliers. However, this remains concealed to people only using least squares. Figure 1 shows a residual plot based on least squares (this is also called an

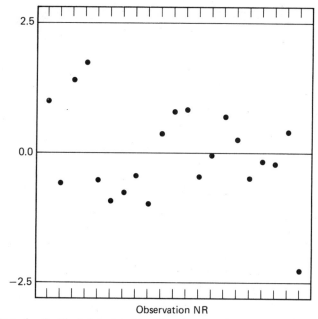

Figure 1. Residual plot of stackloss data, based on least-squares regression.

"index plot" because the horizontal axis lists the index of the observation). No residual falls outside the ± 2.5 $\hat{\sigma}$ band; only the residual of observation 21 looks perhaps slightly suspect. On the other hand, Figure 2 contains a residual plot of the same data but using least median of squares. In this plot, constructed from a *robust* regression, the reputed outliers 1, 3, 4, and 21 indeed lie far from the confidence strip.

This example shows that the main purpose of the LMS is to *detect* outliers. Therefore, we consider it as a useful diagnostic tool in data analysis. Also, the LMS could very well be used as a high-breakdown starting value for the iterative computation of M-estimators, especially with redescending ψ.

*6.4b. Asymptotic Behavior of Bounded Influence Estimators

In this subsection we summarize the results obtained by Maronna et al. (1979) comparing the asymptotic efficiency of the Hampel–Krasker estimator (see Proposition 2 of Subsection 6.3b) and the optimal B-robust Mallows estimator given in Subsection 6.3b. More numerical results can be

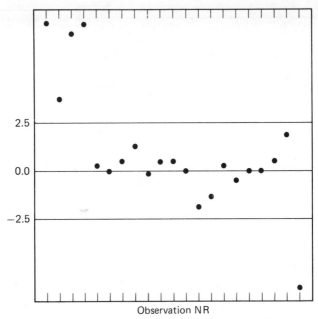

Figure 2.   Residual plot of stackloss data, based on least median of squares regression.

found in Hill (1977). We consider the simple model of a straight line through the origin:

$$y = x\theta + e.$$

In this case the estimators are of the form:

HK: $\eta(x, r) = |x|^{-1}\psi_{c_1}(r|x|)$ (Hampel–Krasker estimator),

MA: $\eta(x, r) = w_b(|x|)\psi_{c_2}(r)$ (optimal B-robust Mallows estimator).

Note that for this model the Hampel–Krasker estimator is the same as the Krasker–Welsch estimator.

Let G and K be the distributions of e and x, respectively. We choose contaminated normal distributions

$$(1 - \varepsilon)\Phi(t) + \varepsilon\Phi(t/s)$$

with the parameters

ε	0	0.1	0.1	0.05
s	1	3	5	10

The constants c_1, b, and c_2 are chosen so that both estimators have the same efficiency (with respect to least squares), when $x \sim N(0, 1)$ and $e \sim N(0, 1)$. For 95% efficiency we obtain the values

$$c_1 = 2.56, \qquad b = 1.73, \qquad c_2 = 1.6.$$

Note that b and c_2 are not uniquely determined.

Table 1. **Relative Deficiency of HK with Respect to MA**

K	G	$\varepsilon = 0$ $s = 1$	$\varepsilon = 0.1$ $s = 3$	$\varepsilon = 0.1$ $s = 5$	$\varepsilon = 0.05$ $s = 10$
$\varepsilon = 0$ $s = 1$		1.00	1.07	1.13	1.13
$\varepsilon = 0.1$ $s = 3$		0.98	1.02	1.08	1.07
$\varepsilon = 0.1$ $s = 5$		0.98	0.98	1.03	1.04
$\varepsilon = 0.05$ $s = 10$		0.94	0.94	0.98	1.00

Table 1 gives the relative deficiency of HK with respect to MA [as var(HK)/as var(MA)]. There are no spectacular differences. HK is a bit better than MA when the distribution of e has moderate tails and that of x has heavy tails and vice versa.

From Table 2 one can see the optimal robustness properties of HK. For the Mallows estimator the asymptotic variance can be written as $VR \cdot VX$, where VX depends only on w_b and on the distribution K of x and $VR = A/B^2$, with $A = E\psi_{c_2}^2$, $B = E\psi'_{c_2}$. Table 3 shows these factors.

Table 2.   Unstandardized Gross-Error
Sensitivities of HK and MA

K	G HK	$\varepsilon = 0$ $s = 1$ MA
$\varepsilon = 0$ $s = 1$	2.94	3.40
$\varepsilon = 0.1$ $s = 3 \cdot$	2.22	2.66
$\varepsilon = 0.1$ $s = 5$	1.73	2.15
$\varepsilon = 0.05$ $s = 10$	1.64	2.07

Table 3.   Factors of the
Asymptotic Variance for
MA

	VR	VX
$\varepsilon = 0$ $s = 1$	1.03	1.02
$\varepsilon = 0.1$ $s = 3$	1.30	0.65
$\varepsilon = 0.1$ $s = 5$	1.40	0.46
$\varepsilon = 0.05$ $s = 10$	1.23	0.37

As we can see from these results, the differences among bounded influence estimators are small. The message is that the important thing is to
treat leverage points, especially because in general they are difficult to detect
using standard methods. However, especially in view of their low breakdown in higher dimensions, bounded influence estimators are not the unique
and final answer of any analysis but rather should be used together with
other techniques.

6.4c. Computer Programs

Robust estimates are defined by means of implicit equations and their calculation requires numerical methods. At the moment four computer packages are available.

LINWDR

LINWDR is a computer program for robust M-estimators of regression and it is a modified version of the least-squares program LINWOOD (Daniel and Wood, 1971/80). It was developed by Dutter (1976, 1977, 1979, 1983).

The algorithm used for the computation of the Huber-estimator is the so-called H-algorithm (see Huber and Dutter, 1974; Huber 1977a; Dutter 1979) that is based on the following idea (cf. also Section 6.2). First, give a starting value for the parameter estimate and compute the fitted values $\hat{y}_i$ and the residuals r_i. Then, transform the residuals according to $z_i = \psi(r_i)$ and compute the pseudo-observations $y_i^* = \hat{y}_i + z_i$. Now update the estimate of θ by applying least squares to y_i^* and continue until convergence is reached. (Note that usually one has to update simultaneously the scale parameter.) Figure 3 shows this procedure for the data set given in Figure 1 of Subsection 6.1b. By using a robust starting value and performing only one step of this procedure, one obtains the so-called one-step Huber estimate whose properties have been investigated by Bickel (1975).

ROBETH

ROBETH, a subroutine package containing routines for robust linear regression as well as robust covariance matrix M-estimators, has been written at the ETH, Zürich and is continuously extended at the University of Lausanne (see Marazzi, 1980b, c, d). ROBETH computes many of the most recent procedures for robust regression, including bounded influence estimators and robust tests for linear models. Some nonparametric procedures (e.g., the Hodges and Lehmann estimator and associated confidence intervals) are also included.

The computation of the estimators is based on iteratively reweighting techniques (see Holland and Welsch, 1977) which have been adapted by including weights with respect to the x_i's. Basically, one writes (6.3.1) as

$$\sum_{i=1}^{n} w(x_i, r_i) r_i x_i = 0, \qquad (6.4.4)$$

where $r_i = (y_i - x_i^T T_n)/\sigma$, $w(x, r) = \eta(x, r)/r$, and then solves iteratively (6.4.4) using a weighted least-squares program. From a numerical point of

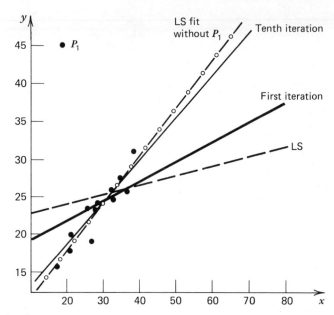

Figure 3.   Computation of the Huber-estimator for the data of Figure 1 in Section 6.1b.

view, the main property of these routines is the use of Householder transformations (cf. Lawson and Hanson, 1974) which allow to achieve numerical stability.

The ongoing development of ROBETH includes new robust techniques (e.g., the choice of variable in regression) and the development of the system control program ROBSYS, based on a problem-oriented language. It facilitates the use of the robust regression subroutines of ROBETH by avoiding the need of FORTRAN programming. It allows some data manipulation, assembles the appropriate ordered sets of subroutines, and supplies organized and graphical output.

TROLL

TROLL is a large interactive statistical system for modeling and data analysis. It was developed at MIT over the years 1966–1971 and extended at the National Bureau of Economic Research's Computer Research Centre during 1972–1977. The system contains a package called BIF for the computation of bounded influence regression estimators. The algorithm is basically the same as in ROBETH.

Differences between BIF and ROBETH include the computation of robust distances in factor space, the choice of cutoff parameters, and the convergence criterion (see Samarov and Welsch, 1982; Peters et al., 1982).

PROGRESS

PROGRESS is a program for robust regression, running on both mainframe computers and micros. It was developed at the universities of Brussels and Delft (Leroy and Rousseeuw, 1984). It performs high-breakdown multiple regression by means of the LMS method (6.4.3). The algorithm is based on a repeated drawing of random subsamples of p data points, followed by a calculation of the median-squared residual with respect to the regression equation corresponding to these p points. The regression plane with the smallest median-squared residual is then treated as an initial solution. The number of subsamples is determined by the requirement that the probability of finding at least one combination consisting of p "good" points should be close to one, an idea which was already used in multivariate location by Stahel (1981b). In order to improve statistical efficiency, a one-step re-weighted least-squares estimate is computed with weights based on the LMS.

The output of PROGRESS consists of results concerning LS, LMS, and reweighted LS based on LMS. The program gives the regression coefficients with their standard deviation and t-value, their variance–covariance matrix, an estimate for the scale parameter σ, the determination coefficient (R squared), and residual plots of two types (in order to identify the outliers in an easy way, and to detect possible inadequacies of the model). For instance, Figures 1 and 2 are output of PROGRESS. Also, the program contains two options for dealing with missing values.

*6.4d. Related Approaches

In this subsection we summarize briefly some approaches related to bounded influence regression.

Diagnostic techniques have a connection with bounded influence regression. For instance, Welsch (1982) proposes various finite-sample sensitivities (that are approximations to the sensitivities computed from the influence function) and uses them to measure the influence of single (and small subsets of) observation(s) on the least-squares estimates, predicted values, and so on. A complete account of these techniques can be found in Belsley et al. (1981), Cook and Weisberg (1980, 1982) and Atkinson (1982).

Multiple outliers pose serious problems to diagnostic techniques because single deletion does not remove the influence of leverage whereas multiple deletion of k points leads to computational complexity $O\left(\binom{n}{k}\right)$. Therefore, in such a situation the diagonal elements of the hat matrix corresponding to leverage points might not be large. Similar techniques are used by Pregibon (1981) for logistic regression.

Huber (1983) investigates bounded influence regression by means of minimax techniques. For the model of simple regression through the origin ($p = 1$), he conjectures that the Krasker–Welsch estimator might be the minimax strategy of the statistician in a certain contamination game, in which the minimax strategy of Nature would selectively put most of the contamination on the leverage points. Therefore, Huber feels that bounded influence regression might be overly pessimistic. This conjecture was found incorrect by Rieder (1983, personal communication) who showed that the solution of the contamination game proposed by Huber does not equal the Krasker–Welsch estimator.

More heuristically, there are also indications, as the often use of regression diagnostic, that bounded influence regression might be in fact too optimistic!

Samarov (1983) uses the influence function and the change-of-variance function to approximate the mean-square error over a gross-error model [cf. (2.7.3) and (2.7.7) and also Subsection 1.3e, and Hampel et al., 1982] and derives minimax estimators for regression.

Other approaches to robust regression include configural polysampling techniques (Tukey, 1981; Pregibon and Tukey, 1981; O'Brien, 1984) and rank methods (Jurečková, 1971, 1977). The problem of asymmetric error distributions is discussed in Carroll (1979) and Collins et al. (1985). Related work can be found in Carroll and Ruppert (1980, 1982), Ruppert and Carroll (1980), Beran (1982), Morgenthaler (1979), and Marazzi (1980a).

Work for related models has been done by Shoemaker (1982) and Rocke (1983) who consider robust estimators in random-effects models. Kelly (1984) uses the influence function for the detection of influential observations in the error-in-variables model and Milasevic (1983, personal communication) proposes rank estimates for this model which turn out to have a bounded and continuous influence function.

EXERCISES AND PROBLEMS

Section 6.1b

1. Discuss critically the assumptions and the results of the Gauss–Markov theorem.

Section 6.2

2. Show if ρ is not convex (or ψ is not monotone) there may be multiple solutions to the analogs of the normal equations, corresponding to local minima (and maxima) of the minimization problem.

3. Show in this case there may even be multiple global solutions to the minimization problem.

4. (Short) Show the Huber-estimator defined by (6.2.4) with $\psi = \psi_c$ can be computed using iteratively reweighted least squares with weights (6.2.1).

5. (Research) Give an M-estimator for which there exists no monotone function f such that $f(\nu(u, v))$ is a metric in $\mathbb{R}^n$, where ν is defined in Remark 3.

6. Derive (6.2.7) from (6.2.4) with $\psi = \psi_c$.

7. Compare (6.2.7) with (6.2.10) and interpret the results.

8. Why are leverage points so hard to detect with least squares?

9. What are the possible interpretations for a leverage point not close to the line through the other points?

10. Show if the point P_2 in Figure 1 is sufficiently far out, the largest least-squares residuals belong to "ordinary data."

11. Show the same is true for Huber-type regression.

12. Show the regression line which minimizes $\sum_{i=1}^{n}|y_i - \hat{y}_i|$ (the L_1-norm), which is a limiting case of Huber-type regression, goes right through P_2 in Figure 1 if all other points are on the other side of $\overline{X}$.

Section 6.3a

13. (Short) What are the reasons for considering γ_s^* and γ_i^*?

14. Compute the influence function (6.3.3) for the proposals discussed in Subsection 6.3a. What is the relation between the weight function, the η-function, and the influence function?

15. Show that Mallows-type regression puts separate bounds on influence of residual and influence of position in factor space; that Huber-type regression only bounds the former; and that Schweppe-type regression bounds the product.

16. Explain some differences between robust parameter fitting and robust prediction by comparing the corresponding sensitivities.

Section 6.3b

17. (Extensive) Give lower bounds for γ_i^*, γ_f^*, and γ_p^*.

18. Show that the Hampel–Krasker and the Krasker–Welsch estimators can be considered as Huber-estimators with variable "corner" c in ψ_c.

19. To Proposition 2, How can you express the trace of the asymptotic covariance matrix V? Why minimizing trace(V) is a better criterion than minimizing det(V)?

20. What happens to the efficiency if a "good" leverage point (very close to the line through the other points) is downweighted as in Mallows-type regression?

21. What happens to the Schweppe-type estimate and its efficiency if a "good" leverage point is moved a bit in y-direction?

22. What are the advantages and disadvantages of Mallows-type estimators as compared to Schweppe-type estimators?

23. (Extensive) Prove the conjecture in Remark 1.

24. (Extensive) Solve Hampel's optimality problem with respect to the predicted value sensitivity γ_p^*.

25. Discuss the (close) relation between "downweighting" and "Winsorizing" or "Huberizing" residuals in regression.

26. (Research) Prove Mallows's result (see Remark 2) for $p > 1$.

27. Explain where robust covariance matrix estimates come in in multiple regression.

28. What are your (personal) main criteria for choosing a robust regression method?

Section 6.3c

29. (Extensive) Compute the change-of-variance function of an M-estimator for a general parametric model and derive (6.3.25) as a special case. (Use 4.2.11)

30. (Extensive) Prove Proposition 5.

31. (Research) Prove the conjecture in Remark 5.

Section 6.4a

32. Compute the breakdown point for the L_1-estimator when the x's are (1) uniform and (2) normal.

33. Show that the breakdown point of Tukey's resistant line (see Tukey, 1970/71, Chapter 10; Johnstone and Velleman, 1985) is $\frac{1}{6}$.

34. Show that the repeated median estimator has breakdown point 50%.

35. Pick a one-dimensional and a two-dimensional data set and compute the LMS estimator.

Section 6.4d

36. For each of the following data sets fit a straight line by means of (1) least squares, (2) least squares after the rejection of outliers (use different rejection rules), and (3) bounded-influence regression. Discuss the results and the differences. In (2), Which points are considered outliers by different rejection rules?

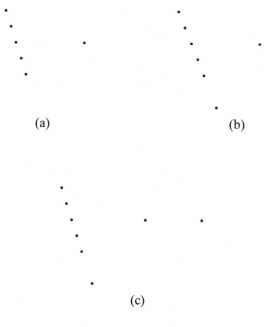

        (a)   (b)

        (c)

37. Compute the influence function and the breakdown point of a regression estimator of your choice.

CHAPTER 7

Linear Models: Robust Testing

7.1. INTRODUCTION

7.1a. Overview

This chapter is devoted to robust testing in linear models. We follow the approach used by Ronchetti (1982a, 1982b) in his Ph.D. thesis and present the ideas and the infinitesimal techniques developed in that work for constructing a robust version of the classical F-test for linear models.

We introduce a general class of test procedures for linear models and investigate their asymptotic properties (influence function and asymptotic distribution) (Section 7.2). Then, solving the optimality problem for infinitesimal robustness of tests, we derive the optimal robust test procedure in this class. The robustification of the F-test is achieved in two steps. First, the optimal robust tests with respect to the influence of the residual only (ρ_c-test, Subsection 7.3a) are derived. Second, taking into account the influence of position in factor space, one imposes the condition of a bounded total influence, obtaining optimal bounded influence tests which are the natural counterpart to optimal bounded influence estimators (Subsections 7.3b and 7.3c). In this framework we discuss in Subsection 7.3d a robust procedure for the selection of the variables.

An alternative class of procedures, $C(\alpha)$-type tests for linear models, is introduced in Section 7.4. It is shown that they are locally equivalent to the former ones at the parametric model, having the same influence function and the same asymptotic efficiency. Moreover, an interesting connection between the optimal robust $C(\alpha)$-type test and an asymptotically minimax test is shown (Subsection 7.4d).

Finally, in Section 7.5, we compare the asymptotic behavior of several test procedures under different distributions and we give an example based on real data that shows the excellent performance of the optimal robust test.

The present discussion is devoted to distributional robustness and we do not study other departures from the assumptions in the linear model. For instance, Andersen et al. (1981) investigate the behavior of the test procedures in a two-way analysis of variance model with correlated errors. A complete discussion of the effects of unsuspected serial correlations on tests and confidence intervals is given in Section 8.1.

7.1b. The Test Problem in Linear Models

Consider the general linear model of Section 6.1 and suppose we want to test the linear hypothesis

$$\ell_j(\theta) = 0, \qquad j = q + 1, \ldots, p,$$

where $\ell_{q+1}, \ldots, \ell_p$ are $(p - q)$ linearly independent estimable functions and $0 < q < p$.

Through a transformation of the parameter space we can reduce this hypothesis to

$$H_0: \theta^{(q+1)} = \cdots = \theta^{(p)} = 0. \tag{7.1.1}$$

Let Θ_ω be the subspace of Θ obtained by imposing the condition H_0. If the errors are normally distributed, the classical F-test for testing H_0 is optimal in several senses (see Lehmann, 1959 and for a summary Seber, 1980, pp. 34–37). It is defined by the critical region

$$\left\{ T > F_{p-q, n-p; 1-\alpha} \right\},$$

where α is the level of the test, $F_{p-q, n-p; 1-\alpha}$ is the $(1 - \alpha)$ quantile of the F-distribution with $(p - q)$ and $(n - p)$ degrees of freedom, and T is the test statistic. T is defined by

$$T = \left\| (X \cdot (T_\Omega)_n - X \cdot (T_\omega)_n)/\hat{\sigma} \right\|^2 / (p - q)$$

$$= \sum_{i=1}^{n} \left[\left((y_i - x_i^T(T_\omega)_n)/\hat{\sigma} \right)^2 - \left((y_i - x_i^T(T_\Omega)_n)/\hat{\sigma} \right)^2 \right] / (p - q),$$

where $(T_\Omega)_n$ and $(T_\omega)_n$ are the least-squares estimates of θ in the full (Θ) and reduced model (Θ_ω), respectively and

$$\hat{\sigma}^2 = \sum_{i=1}^{n} \left(y_i - x_i^T(T_\Omega)_n \right)^2 / (n - p)$$

is the least-squares (unbiased) estimate of σ^2.

The F-test is equivalent to the corresponding likelihood ratio test (see Subsection 7.3a for details). Since the likelihood ratio criterion is based on a very general and natural principle (for the theoretical motivations see, e.g., Akaike, 1973), we shall work with it in order to construct a robust version of the F-test.

From a robustness point of view, the classical test procedures based on the least-squares estimates in the full and reduced model have similar problems as the estimators themselves. Although the F-test (for testing H_0) is relatively robust with respect to the level (being an approximation of the permutation test; see Scheffé, 1959, p. 313), it does lose power rapidly in the presence of departures from the normality assumption on the errors (Hampel, 1973a, 1980; Schrader and Hettmansperger, 1980).

Whereas robust estimation theory in linear models has recently received more and more attention, the test problem has been somewhat neglected. On the other hand, the need for robust test procedures is obvious because one cannot estimate the parameters robustly and apply unmodified classical test procedures for testing hypotheses about them.

Only recently, Schrader and McKean (1977) and Schrader and Hettmansperger (1980) proposed a new class of tests for linear models that we shall discuss in Subsections 7.2c and 7.3a. These tests are based on Huber estimates in the full and reduced model. But, as was mentioned above, this is only the first step for robustifying the F-test. Like Huber estimates, these tests do not overcome problems caused by situations where the fit is mostly determined by outlying points in the factor space. In order to construct tests which fit bounded influence estimation, Ronchetti (1982a, 1982b) introduced a new class of tests and studied their asymptotic properties. In this way he could derive optimal bounded influence tests which are the counterpart of optimal bounded influence estimators.

Throughout the next sections we assume the general linear model defined in Section 6.1 with normal errors and the corresponding test problem.

7.2. A GENERAL CLASS OF TESTS FOR LINEAR MODELS

7.2a. Definition of a τ-Test

The aim of this section is to define a class of tests that can be viewed as an extension of the log–likelihood ratio test for linear models.

Let us first introduce the class of functions

$$\tau : \mathbb{R}^p \times \mathbb{R} \to \mathbb{R}^+, \qquad (x, r) \to \tau(x, r)$$

with the following properties:

(7.2.TAU) $\quad \tau(x, r) \not\equiv 0$, $\tau(x, r) \geq 0$ for all $x \in \mathbb{R}^p$, $r \in \mathbb{R}$, and $\tau(x, 0)$ $= 0$ for all $x \in \mathbb{R}^p$. $\tau(x, \cdot)$ is differentiable for all $x \in \mathbb{R}^p$. Let $\eta(x, r) := (\partial / \partial r) \tau(x, r)$.

(7.2.ETA1) $\quad$ (i) $\quad \eta(x, \cdot)$ is continuous and odd for all $x \in \mathbb{R}^p$.
$\qquad$ (ii) $\quad \eta(x, r) \geq 0$ for all $x \in \mathbb{R}^p$ and $r \in \mathbb{R}^+$.

(7.2.ETA2) $\quad \eta(x, \cdot)$ is differentiable on $\mathbb{R} \setminus \mathscr{D}(x; \eta)$ for all $x \in \mathbb{R}^p$ where $\mathscr{D}(x; \eta)$ is a finite set. Let

$$\eta'(x, r) := (\partial / \partial r) \eta(x, r) \qquad \text{if } x \in \mathbb{R}^p, r \in \mathbb{R} \setminus \mathscr{D}(x; \eta)$$

$$:= 0 \qquad \text{otherwise,}$$

and assume

$$\sup_r |\eta'(x, r)| < \infty \text{ for all } x \in \mathbb{R}^p.$$

We shall also assume the following regularity conditions:

(7.2.ETA3) $\quad$ (i) $\quad M := E\eta'(x, r)xx^T$ exists and is nonsingular.
$\qquad$ (ii) $\quad Q := E\eta^2(x, r)xx^T$ exists and is nonsingular.

These conditions are stronger than those given in Subsection 6.3a. Here $\mathscr{C}(x; \eta)$ is the empty set, that is, we consider only continuous functions η.

Definition 1 (τ-tests). Let τ satisfy the conditions given above, and define the corresponding M-estimators $(T_\Omega)_n$ and $(T_\omega)_n$ in the full and reduced model, respectively, by

$$\Gamma((T_\Omega)_n) = \min\{\Gamma(\theta)|\theta \in \Theta\}, \qquad (7.2.1)$$

$$\Gamma((T_\omega)_n) = \min\{\Gamma(\theta)|\theta \in \Theta_\omega\}, \qquad (7.2.2)$$

with

$$\Gamma(\theta) := \sum_{i=1}^{n} \tau(x_i, (y_i - x_i^T\theta)/\sigma). \qquad (7.2.3)$$

A test based on the test statistic

$$S_n^2(x_1, \ldots, x_n; y_1, \ldots, y_n) = \frac{2}{p-q}\frac{1}{n}[\Gamma((T_\omega)_n) - \Gamma((T_\Omega)_n)]$$

$$(7.2.4)$$

is called a τ-test. "Large" values of S_n^2 are significant.

(In order to give a critical region we shall give the asymptotic distribution of S_n^2 under H_0; see Subsection 7.2b.) The estimation of σ is discussed in the next remark.

Notation. For any vector $x \in \mathbb{R}^p$ denote by $\tilde{x}$ the vector $(x^{(1)}, \ldots, x^{(q)}, 0, \ldots, 0)^T$.

$(T_\omega)_n$ and $(T_\Omega)_n$ fulfill the equations

$$\sum_{i=1}^{n} \eta(x_i, r_{\omega i})\tilde{x}_i = 0 \qquad (7.2.5)$$

$$\sum_{i=1}^{n} \eta(x_i, r_{\Omega i})x_i = 0, \qquad (7.2.6)$$

where

$$r_{\omega i} := (y_i - x_i^T(T_\omega)_n)/\sigma, \qquad r_{\Omega i} := (y_i - x_i^T(T_\Omega)_n)/\sigma.$$

Note that $(T_\omega)_n = ((T_\omega)_n^{(1)}, \ldots, (T_\omega)_n^{(q)}, 0, \ldots, 0)^T$ (under H_0 the last $(p-q)$ components of θ equal 0!).

Table 1.   Relation Between M-Estimators (Defined in Subsection 6.3a) and τ-Tests

$\tau(x, r)$	$\eta(x, r)$	Estimator Corresponding to η
$r^2/2$	r	Least squares
$\rho(r)$	$\psi(r)$	Huber
$w(x) \cdot \rho(r)$	$w(x) \cdot \psi(r)$	Mallows
$w(x) \cdot \rho(r/w(x))$	$\psi(r/w(x))$	Andrews
$\rho(r \cdot w(x))$	$w(x)\psi(r \cdot w(x))$	Hill and Ryan
$w^2(x) \cdot \rho(r/w(x))$	$w(x)\psi(r/w(x))$	Schweppe

Examples.   Some choices of τ are of the form $\tau(x, r) = w(x)\rho(r \cdot v(x))$ for certain functions w, v, and ρ. They correspond to the estimators given in Subsection 6.3a, when ψ is set to $\psi(r) = (\partial/\partial r)\rho(r)$, as Table 1 demonstrates. The conditions (7.2.TAU), (7.2.ETA1), (7.2.ETA2), and (7.2.ETA3) can then be written in terms of w, v, and ρ.

Actually, one usually has to estimate the scale parameter σ. A suitable way is to estimate σ in the full model, taking the median deviation or the Proposal 2 estimate of Huber (see Remark 2, Section 6.2). More precisely, for a given real function χ, one has to solve (7.2.6) and

$$\sum_{i=1}^{n} \chi(r_{\Omega i}) = 0 \qquad (7.2.7)$$

for $(T_\Omega)_n$ and σ, and then (7.2.5) for $(T_\omega)_n$, using σ. Since the estimation of σ does not affect the asymptotic results presented later (provided it is robust and consistent), from now on we put $\sigma = 1$ for simplicity.

7.2b.   Influence Function and Asymptotic Distribution

Let S, T_ω, and T_Ω be the functionals corresponding to S_n, $(T_\omega)_n$, and $(T_\Omega)_n$ (see Definition 1), that is

$$S(F) = \left\{ 2(p - q)^{-1} \int \left[\tau\left(u, v - u^T T_\omega(F)\right) \right. \right.$$

$$\left. \left. - \tau\left(u, v - u^T T_\Omega(F)\right) \right] dF(u, v) \right\}^{1/2} \qquad (7.2.8)$$

where F is an arbitrary distribution function on $\mathbb{R}^p \times \mathbb{R}$ and T_ω, T_Ω fulfill the system of equations

$$\int \eta\big(u, v - \tilde{u}^T T_\omega(F)\big)\tilde{u}\, dF(u, v) = 0,$$

$$\int \eta\big(u, v - u^T T_\Omega(F)\big)u\, dF(u, v) = 0.$$

[Note that $T_\omega^{(j)}(F) = 0$, for $j = q + 1, \ldots, p$ and for all F, and $S_n = S(F_n)$, $(T_\omega)_n = T_\omega(F_n)$, and $(T_\Omega)_n = T_\Omega(F_n)$, where F_n is the empirical distribution function of (x_i, y_i), $i = 1, \ldots, n$.] The square root of the usual statistic is taken in order to obtain a nonzero influence function (cf. Ronchetti and Rousseeuw, 1980). The next proposition gives the influence functions of T_ω, T_Ω, and S at the null hypothesis. [Note that, under the null hypothesis, $\theta = \tilde{\theta} = (\theta^{(1)}, \ldots, \theta^{(q)}, 0, \ldots, 0)^T$, so $F_{\tilde{\theta}}$ is the model distribution under the null hypothesis.]

Proposition 1. Assume (7.2.TAU), (7.2.ETA1), (7.2.ETA2), and the following conditions:

(7.2.IF1) $h(\alpha) := \int \eta(x, y - x^T\alpha)x\, dF_{\tilde{\theta}}(x, y)$ exists for all $\alpha \in \Theta \subset \mathbb{R}^p$. $(\partial/\partial\alpha)h(\alpha)$ exists and is continuous.

(7.2.IF2) $h(\tilde{\theta}) = 0$.

Then, the influence functions of T_ω, T_Ω, and S exist and equal:

(i) $\mathrm{IF}(x, y; T_\omega, F_{\tilde{\theta}}) = \eta(x, y - x^T\tilde{\theta}) \cdot (\tilde{M})^+ x$,

(ii) $\mathrm{IF}(x, y; T_\Omega, F_{\tilde{\theta}}) = \eta(x, y - x^T\tilde{\theta})M^{-1}x$,

(iii) $\mathrm{IF}(x, y; S, F_{\tilde{\theta}}) = \big|\eta(x, y - x^T\tilde{\theta})\big| \cdot \{[x^T(M^{-1} - (\tilde{M})^+)x]/(p - q)\}^{1/2}$,

where

$$\tilde{M} := \begin{bmatrix} M_{(11)} & 0 \\ 0 & 0 \end{bmatrix} \quad \text{and} \quad (\tilde{M})^+ \text{ denotes the pseudoinverse of } \tilde{M},$$

$$(\tilde{M})^+ = \begin{bmatrix} M_{(11)}^{-1} & 0 \\ 0 & 0 \end{bmatrix},$$

$M_{(11)}$ being the upper-left corner of M of order $q \times q$ (cf. Subsection 4.4b).

Proof. Assertions (i) and (ii) of this proposition follow from Theorem G11.1 of Stahel (1981a, p. 116), with $P = F_{\hat{\theta}}$. His conditions "a", "b", "c" follow from (7.2.IF1), (7.2.IF2), and condition "d" from (7.2.ETA3) (i). Finally, condition "e" follows from (7.2.ETA2), since $\{(x, y)|y - x^T\theta \in \mathcal{D}(x; \eta)\}$ is a regular hyperplane in his sense (Stahel, 1981a, p. 12). To show (iii), denote by $\Delta_{(x, y)}$ the distribution on $\mathbb{R}^p \times \mathbb{R}$ that puts mass 1 at (x, y) and define the following ε-contaminated distribution

$$F_{\hat{\theta}; \varepsilon, (x, y)} := (1 - \varepsilon) F_{\hat{\theta}} + \varepsilon\Delta_{(x, y)}.$$

After a straightforward computation we get

$$S^2(F_{\hat{\theta}}) = 0 \quad [\text{by } (7.2.\text{IF2})], \quad \left[(\partial/\partial\varepsilon) S^2(F_{\hat{\theta}; \varepsilon, (x, y)})\right]_{\varepsilon=0} = 0,$$

and

$$\left[(\partial^2/\partial\varepsilon^2) S^2(F_{\hat{\theta}; \varepsilon, (x, y)})\right]_{\varepsilon=0}$$

$$= 2(p - q)^{-1}\left[-\eta(x, y - x^T\tilde{\theta})\tilde{x}^T \cdot \text{IF}(x, y; T_\omega, F_{\hat{\theta}})\right.$$

$$\left. + \eta(x, y - x^T\tilde{\theta})x^T \cdot \text{IF}(x, y; T_\Omega, F_{\hat{\theta}})\right]$$

$$= 2(p - q)^{-1} \cdot \eta^2(x, y - x^T\tilde{\theta}) \cdot (x^T M^{-1}x - x^T(\tilde{M})^+ x).$$

Using l'Hôpital's rule twice, we obtain

$$\text{IF}(x, y; S, F_{\hat{\theta}}) = \lim_{\varepsilon \to 0}\left(S(F_{\hat{\theta}; \varepsilon, (x, y)}) - S(F_{\hat{\theta}})\right)/\varepsilon$$

$$= \left(\lim_{\varepsilon \to 0} S^2(F_{\hat{\theta}; \varepsilon, (x, y)})/\varepsilon^2\right)^{1/2}$$

$$= \left(\tfrac{1}{2} \cdot \left[(\partial^2/\partial\varepsilon^2) S^2(F_{\hat{\theta}; \varepsilon, (x, y)})\right]_{\varepsilon=0}\right)^{1/2}. \qquad \square$$

The following proposition will yield an alternative formula for the influence function of S.

Proposition 2. Let M be given by (7.2.ETA3) (i). Then:

(i) There exists a nonsingular lower triangular ($p \times p$) matrix U with positive diagonal elements such that

$$UU^T = M \quad \text{and} \quad U^T(M^{-1} - (\tilde{M})^+)U = I - \tilde{I}.$$

Moreover,

$$M_{(22.1)} = M_{(22)} - M_{(12)}^T M_{(11)}^{-1} M_{(12)} = U_{(22)} U_{(22)}^T.$$

(ii)   The function

$$m : \mathbb{R}^p \to \mathbb{R}^+, \qquad x \to m(x) := \left(x^T \big(M^{-1} - (\tilde{M})^+ \big) x \right)^{1/2}$$

is a pseudonorm in $\mathbb{R}^p$, that is, m satisfies the conditions $\forall \alpha \in \mathbb{R}$, $x, y \in \mathbb{R}^p$; $m(0) = 0$, $m(\alpha x) = |\alpha| m(x)$, and $m(x + y) \le m(x) + m(y)$.

Proof.

(i)   First, observe that the matrix M is positive definite, since using (7.2.ETA1) (i) and (ii) and performing an integration by parts we have

$$M = \int \eta'(x, r) \, d\Phi(r) x x^T \, dK(x) = \int \eta(x, r) r \, d\Phi(r) x x^T \, dK(x),$$

where the inner integral $\int \eta(x, r) r \, d\Phi(r)$ is positive for all $x \in \mathbb{R}^p$. Let $UU^T = M$ be the Choleski decomposition of M. Then $U^T M^{-1} U = I$ and

$$U^T (\tilde{M})^+ U = \begin{bmatrix} U_{(11)}^T M_{(11)}^{-1} U_{(11)} & 0 \\ 0 & 0 \end{bmatrix} = \tilde{I},$$

since $U_{(11)} U_{(11)}^T$ is the Choleski decomposition of $M_{(11)}$. The formula for $M_{(22.1)}$ follows from a straightforward calculation.

(ii)   From (i), we get

$$x^T \big(M^{-1} - (\tilde{M})^+ \big) x = \left\| (U^{-1} x)_{(2)} \right\|^2. \qquad \square$$

Combining this result with (iii) of Proposition 1, we find the following formulas.

Corollary 1.   If the conditions of Proposition 1 hold, then:

(i)

$$\text{IF}(x, y; S, F_{\tilde{\theta}}) = (p - q)^{-1/2} \cdot \left\| \eta \big(x, y - x^T \tilde{\theta} \big) \cdot (U^{-1} x)_{(2)} \right\|.$$

(ii)

$$\int (\text{IF}(x, y; S, F_{\hat{\theta}}))^2 \, dF_{\hat{\theta}}(x, y)$$

$$= (p - q)^{-1} \cdot \text{trace}\left[Q(M^{-1} - (\tilde{M})^+) \right]$$

$$= (p - q)^{-1} \cdot E\left\{ \eta^2(x, r) \cdot \left\| (U^{-1}x)_{(2)} \right\|^2 \right\}.$$

Remark 1. $m(\cdot)$ defines a pseudodistance in the space of the x's. It measures how well observations at x can distinguish between the full and the reduced model.

In order to compute the asymptotic distribution of the test statistic defining the τ-test, we first perform a von Mises expansion of the functional S^2 defined by (7.2.8). Following von Mises (1947), we get

$$S^2(H_1) = S^2(F_{\hat{\theta}}) + \left[(\partial/\partial\varepsilon) S^2(H_\varepsilon) \right]_{\varepsilon=0}$$

$$+ \tfrac{1}{2} \left[(\partial^2/\partial\varepsilon^2) S^2(H_\varepsilon) \right]_{\varepsilon=0} + \cdots,$$

where

$$H_\varepsilon := (1 - \varepsilon) F_{\hat{\theta}} + \varepsilon H_1.$$

We perform the same kind of calculations as in the proof of Proposition 1 and obtain

$$S^2(F_{\hat{\theta}}) = 0,$$

$$\left[(\partial/\partial\varepsilon) S^2(H_\varepsilon) \right]_{\varepsilon=0} = 0,$$

$$\left[(\partial^2/\partial\varepsilon^2) S^2(H_\varepsilon) \right]_{\varepsilon=0} = 2(p - q)^{-1} \int \left(\zeta^T(u, v) \cdot (M^{-1} - (\tilde{M})^+) \right)$$

$$\cdot \int \zeta(x, y) \, dH_1(x, y) \right) d(H_1 - F_{\hat{\theta}})(u, v)$$

$$= 2(p - q)^{-1} \left(\int \zeta^T(u, v) \, dH_1(u, v) \right)$$

$$\cdot (M^{-1} - (\tilde{M})^+) \cdot \left(\int \zeta(u, v) \, dH_1(u, v) \right),$$

where

$$\zeta(u,v) := \eta(u, v - u^T\tilde{\theta})u.$$

Thus, we have

$$S^2(H_1) = (p-q)^{-1}\left(\int \zeta^T(u,v)\, dH_1(u,v)\right) \cdot \left(M^{-1} - (\tilde{M})^+\right)$$

$$\cdot \left(\int \zeta(u,v)\, dH_1(u,v)\right) + \dots,$$

and evaluating S^2 at the empirical distribution function $H_1 = F_n$, we obtain the following expansion

$$(p-q)nS_n^2 = (p-q)nS^2(F_n)$$

$$= V_n^T(\tilde{\theta}) \cdot \left(M^{-1} - (\tilde{M})^+\right) \cdot V_n(\tilde{\theta}) + \dots,$$

where

$$V_n(\theta) := n^{-1/2} \sum_{i=1}^{n} \eta\left(x_i, y_i - x_i^T\theta\right) \cdot x_i.$$

Ronchetti (1982b, Section 4.3) shows under the given conditions that the statistics

$$nS_n^2,$$

$$(p-q)^{-1}V_n^T(\tilde{\theta}) \cdot \left(M^{-1} - (\tilde{M})^+\right) \cdot V_n(\tilde{\theta}),$$

and

$$W_n^2 := (p-q)^{-1}n^{1/2}\left((T_\Omega)_n^T\right)_{(2)} M_{(22.1)} n^{1/2}\left((T_\Omega)_n\right)_{(2)}$$

have the same asymptotic distribution at the model. More precisely, under the sequence of alternatives

$$H_n: \theta^{(j)} = n^{-1/2}\left(\Delta_{(2)}\right)^{(j)}, \qquad j = q+1, \dots, p$$

where $\Delta := (\Delta^{(1)}, \dots, \Delta^{(p)})^T$, the asymptotic distribution of all three statistics is the distribution of

$$\sum_{j=q+1}^{p} \left(\lambda_j^{1/2}N_j + \left(C^T\Delta_{(2)}\right)^{(j)}\right)^2, \qquad (7.2.9)$$

where $N_{q+1}, \ldots, N_p$ are independent univariate standard normal variables, $\lambda_{q+1} \geq \cdots \geq \lambda_p > 0$ are the $(p - q)$ positive eigenvalues of $Q \cdot (M^{-1} - (\check{M})^+)$ (Q and M are defined in 7.2.ETA3) and C is a Choleski root of $M_{(22.1)}$ that satisfies

$$CC^T = M_{(22.1)} \qquad (7.2.10)$$

$$C^T(M^{-1}QM^{-1})_{(22)}C = \Lambda_{(22)} = \operatorname{diag}(\lambda_{q+1}, \ldots, \lambda_p). \qquad (7.2.11)$$

Remark 2. In Subsection 7.5b we shall discuss the computational aspects of the distribution given by (7.2.9).

Remark 3. A related result on the distribution of the likelihood ratio test statistic when the data do not come from the parametric model under consideration was obtained by Kent (1982) (cf. also Foutz and Srivastava, 1977).

Remark 4. The test defined by the test statistic W_n^2 is invariant with respect to linear transformations of the parameter space which leave the hypothesis H_0 invariant, that is, under $\theta \to D\theta$, where D is a $(p \times p)$ nonsingular matrix with $D_{(21)} = 0$.

Corollary 2. Let be $c > 0$. Then, under the given conditions, the following equivalence holds:

$$\sup_{x,y} |IF(x, y; S, F_\theta)|$$

$$= \sup_{x,y} |IF(x, y; W, F_\theta)| \leq c$$

$$\Leftrightarrow \sup_{x,y} \left\| IF(x, y; (T_\Omega)_{(2)}, F_\theta) \right\|_{M_{(22.1)}} \leq c \cdot (p - q)^{1/2},$$

where $\| \cdot \|_{M_{(22.1)}}$ denotes the norm in $\mathbb{R}^{p-q}$ generated by $M_{(22.1)}$.

The proof follows easily from the definitions of S_n, W_n, and $((T_\Omega)_n)_{(2)}$. The expression

$$\left\| IF(x, y; (T_\Omega)_{(2)}, F_\theta) \right\|_{M_{(22.1)}}^2$$

$$= IF^T(x, y; (T_\Omega)_{(2)}, F_\theta) \cdot M_{(22.1)} \cdot IF(x, y; (T_\Omega)_{(2)}, F_\theta)$$

$$= IF^T(x, y; (T_\Omega)_{(2)}, F_\theta) \cdot \left((M^{-1})_{(22)}\right)^{-1} \cdot IF(x, y; (T_\Omega)_{(2)}, F_\theta)$$

defines a standardized sensitivity for the estimator $((T_\Omega)_n)_{(2)}$. This will be compared with other standardizations for estimators (see Subsection 7.3c, Remark 4).

7.2c. Special Cases

In this subsection we study the cases where the distribution (7.2.9) reduces to

$$\lambda \cdot \chi^2_{p-q}(\delta^2),$$

where $\chi^2_{p-q}(\delta^2)$ denotes the χ^2-distribution with $(p-q)$ degrees of freedom and noncentrality parameter δ^2. In each case we shall give the standardization factor λ and the noncentrality parameter δ^2. Note that the asymptotic power of the corresponding test is a monotone increasing function of δ^2.

Case 1. $\tau(x, r) = \rho(r)$

In this case we have

$$\lambda_j = A/B = (p-q)^{-1}\text{trace}\left[Q\left(M^{-1} - (\tilde{M})^+\right)\right]$$

$$= \int\left(\text{IF}(x, y; S, F_{\tilde\theta})\right)^2 dF_{\tilde\theta}(x, y) =: \lambda \qquad j = q+1,\ldots,p,$$

$$\delta^2 = \left(B^2/A\right)\Delta^T_{(2)}\left(Exx^T\right)_{(22.1)}\Delta_{(2)},$$

where

$$A := E\psi^2, \quad B := E\psi', \quad \rho' = \psi$$

and Q and M turn out to be

$$Q = A \cdot Exx^T, \qquad M = B \cdot Exx^T.$$

This class of tests was proposed by Schrader and McKean (1977) and Schrader and Hettmansperger (1980) and carries out M-estimation in a natural way.

Let us look at the influence function of these tests. Using (iii) of Proposition 1, we have

$$\text{IF}(x, y; S, F_{\hat{\theta}}) = \left(\left|\psi(y - x^T\tilde{\theta})\right|/B^{1/2}\right)$$

$$\cdot\left(\left(x^T\left((Exx^T)^{-1} - (E\tilde{x}\tilde{x}^T)^+\right)x\right)/(p - q)\right)^{1/2}.$$

$$(7.2.12)$$

Using again the notions of influence of residual and influence of position in factor space (Section 6.2), we can rewrite (7.2.12) as

$$\text{IF}(x, y; S, F_{\hat{\theta}}) = \text{IT}(x, r; S, F_{\hat{\theta}}) = \text{IR}(r; S, \Phi) \cdot \text{IP}(x; S, K),$$

$$(7.2.13)$$

with

$$\text{IR}(r; S, \Phi) = |\psi(r)|/B^{1/2},$$

$$\text{IP}(x; S, K) = \left(\left(x^T\left((Exx^T)^{-1} - (E\tilde{x}\tilde{x}^T)^+\right)x\right)/(p - q)\right)^{1/2}.$$

Remark 5. One can standardize the test statistic S by dividing it by $B^{1/2}$. While this operation does not change the test, IR becomes the absolute value of the influence function of a location M-estimator defined by ψ.

Remark 6. If we choose a bounded function ψ, the influence of the residual is bounded too, but the total influence IT is not because $\text{IP}(x; S, K) \to \infty$, when $\|x\| \to \infty$. This justifies the consideration of the more general class of τ-tests.

Case 2. $q = p - 1$

In this case we obtain the asymptotic distribution $\lambda \cdot \chi_1^2(\delta^2)$ with

$$\lambda = \lambda_p = E\left\{\eta^2(x, r) \cdot \left|(U^{-1}x)^{(p)}\right|^2\right\},$$

$$\delta^2 = \left((u_{pp})^2/\lambda_p\right)(\Delta^{(p)})^2 = \left(M_{(22.1)}/\lambda_p\right)(\Delta^{(p)})^2,$$

where the matrix U is given by Proposition 2.

We will find a standardization of IF$(x, y; S, F_\theta)$ which extends in a natural way Definition 1 of Subsection 3.2a to the τ-tests. Although it is not necessary to standardize the influence function of the test statistic in order to solve the optimality problem for infinitesimal robustness of tests (see Section 3.4), we shall show that in this case such a standardization exists. Define

$$\xi_\tau(\theta) := S^2(F_\theta)$$

$$= 2(p - q)^{-1} \int \big(\tau\big(x, y - x^T T_\omega(F_\theta)\big)$$

$$- \tau\big(x, y - x^T\theta\big)\big) f_\theta(x, y) \, dx \, dy,$$

where

$$f_\theta(x, y) = \phi\big(y - x^T\theta\big) \cdot k(x).$$

Assume that both estimators T_ω, T_Ω are Fisher consistent at the reduced and full model

$$T_\omega(F_{\tilde\theta}) = \tilde\theta, \qquad T_\Omega(F_\theta) = \theta$$

and denote by a dot the differentiation with respect to θ. Then, after some straightforward calculations we obtain the vector

$$\dot\xi_\tau(\tilde\theta) = 0$$

and the matrix

$$\ddot\xi_\tau(\tilde\theta) = 2(p - q)^{-1} \cdot \int \eta\big(x, y - x^T\tilde\theta\big) \cdot \big(-\dot T_\omega^T(F_\theta)\tilde x + x \big)$$

$$\cdot \dot f_\theta^T(x, y) \, dx \, dy, \tag{7.2.14}$$

and, finally, in the partitioned notation

$$\dot\xi_\tau(\tilde\theta) = 0,$$

$$\big(\ddot\xi_\tau(\tilde\theta)\big)_{(11)} = \big(\ddot\xi_\tau(\tilde\theta)\big)_{(12)} = \big(\ddot\xi_\tau(\tilde\theta)\big)_{(21)} = 0,$$

$$\big(\ddot\xi_\tau(\tilde\theta)\big)_{(22)} = M_{(22.1)} = 2(p - q)^{-1} U_{(22)} U_{(22)}^T. \tag{7.2.15}$$

If $q = p - 1$, $U_{(22)} = u_{pp}$; hence

$$(\ddot{\xi}_\tau(\tilde{\theta}))_{(22)} = 2u_{pp}^2.$$

Therefore, we propose in this case to standardize the influence function of the test statistic as follows:

$$\mathrm{IF}_{\mathrm{test}}(x, y; S, F_{\tilde{\theta}}) := \mathrm{IF}(x, y; S, F_{\tilde{\theta}})/\left(\tfrac{1}{2}\ddot{\xi}_\tau(\tilde{\theta})\right)_{pp}^{1/2}$$

$$= \mathrm{IF}(x, y; S, F_{\tilde{\theta}})/u_{pp}. \tag{7.2.16}$$

The significance of this definition will become clear in Section 7.4. Note that (7.2.16) is consistent with Definition 1 of Subsection 3.2a in the sense that it has the following properties:

(i)

$$\mathrm{IF}_{\mathrm{test}}(x, y; S, F_\theta) = \left|\lim_{\theta^{(p)} \to 0}\mathrm{IF}\left(x, y; S^2, F_\theta\right)/(\xi_\tau(\theta))^{(p)}\right|,$$

so (7.2.16) is the natural extension of (3.2.3).

(ii) $\mathrm{IF}_{\mathrm{test}}(x, y; S, F_{\tilde{\theta}})$ is invariant to monotone differentiable transformations of the test statistic.

(iii)

$$\delta^2 = \left((u_{pp})^2/\lambda_p\right) \cdot (\Delta^{(p)})^2$$

$$= \left\{\int (\mathrm{IF}_{\mathrm{test}}(x, y; S, F_{\tilde{\theta}}))^2 \, dF_{\tilde{\theta}}(x, y)\right\}^{-1} \cdot (\Delta^{(p)})^2.$$

Case 3. The Density of x is Spherically Symmetric with Respect to $x_{(2)}$

If

$$(7.2.\mathrm{SPH}) \qquad \begin{cases} k(x) = k_{\mathrm{SPH}}\left(x_{(1)}, \|x_{(2)}\|\right) \\ \tau(x, r) = \tau_{\mathrm{SPH}}\left(x_{(1)}, \|x_{(2)}\|; r\right), \end{cases}$$

then

$$\lambda_j = \lambda^{(Q)}/\lambda^{(M)} =: \lambda, \qquad j = q + 1, \ldots, p,$$

$$\delta^2 = \left((\lambda^{(M)})^2/\lambda^{(Q)}\right) \cdot \|\Delta_{(2)}\|^2,$$

where

$$\lambda^{(Q)} := (p - q)^{-1} E\left\{ \eta^2(x, r) \cdot \|x_{(2)}\|^2 \right\},$$

$$\lambda^{(M)} := (p - q)^{-1} E\left\{ \eta'(x, r) \cdot \|x_{(2)}\|^2 \right\}.$$

In this case we have

$$U_{(22)} = \left(\lambda^{(M)}\right)^{1/2} \cdot I_{(22)}$$

and a natural standardization of the influence function of the test statistic is

$$\mathrm{IF}_{\text{test}}(x, y; \tau, F_\theta) = \mathrm{IF}(x, y; S, F_\theta)/\left(\lambda^{(M)}\right)^{1/2}$$

$$= \mathrm{IF}(x, y; S, F_\theta)/u_{pp}.$$

7.3. OPTIMAL BOUNDED INFLUENCE TESTS

In this section we solve the optimality problem for infinitesimal robustness of tests (see Section 3.4), that is, we find a τ-test which maximizes the asymptotic power, subject to a bound on the influence function of the test statistic at the null hypothesis. Note that the test which maximizes the asymptotic power without any conditions on the influence function, is given by $\tau(x, r) = r^2/2$.

Using the results of Section 7.2, we shall give the solution to this problem for Huber's regression (Subsection 7.3a), Mallows's regression (Subsection 7.3b), and in the general case (Subsection 7.3c).

Throughout this section we assume the linear model of Section 6.1 and the corresponding test problem of Section 7.1.

7.3a. The ρ_c-Test

Proposition 1.   Assume $\tau(x, r) = \rho(r)$. Then the test which maximizes the asymptotic power, under the side condition of a bounded influence of the residual

$$\sup_r \mathrm{IR}(r; S, \Phi) \leq b,$$

is given by Huber's function $\rho_c(r)$, where $c = c(b)$ (cf. Subsection 2.4b).

Proof. (Ronchetti and Rousseeuw, 1980). Since the asymptotic power is a monotone increasing function of the noncentrality parameter

$$\delta^2 = \left[(E\psi')^2 / E\psi^2 \right] \cdot \Delta_{(2)}^T (Exx^T)_{(22.1)} \Delta_{(2)}$$

(see Subsection 7.2c), the assertion follows easily using Hampel's Lemma 5 (see Subsection 2.4a).   □

Remark 1. The ρ_c-test can be viewed as a likelihood ratio test when the error distribution has a density g_0 proportional to $\exp(-\rho_c(r))$. This distribution minimizes the Fisher information within the gross-error model $\mathscr{P}_\varepsilon(\Phi)$ ("least favorable distribution:" see Section 2.7). Note that a test of the same type (a likelihood ratio test under a least favorable distribution) is used by Carroll (1980, p. 73) who proposes a robust method for testing transformations to achieve approximate normality.

Remark 2. The ρ_c-test is asymptotically equivalent to a test proposed by Bickel (1976, p. 167) who applies the classical F-test to transformed observations (see Schrader and Hettmansperger, 1980; cf. also Huber, 1981, p. 197).

7.3b. The Optimal Mallows-Type Test

As with estimation, there is a need for tests with bounded total influence. The optimal test under this restriction will be given in the next subsection. Here we consider the class of tests corresponding to Mallows-type estimators only, for the same reasons as in Subsection 6.3b.

Proposition 2. Consider the class of tests defined by $\tau(x, r) = w(x) \cdot \rho(r)$. Assume either ($q = p - 1$) or (7.2.SPH) of Subsection 7.2c. Then, the test which maximizes $M_{(22.1)}/\lambda_p$ or $(\lambda^{(M)})^2/\lambda^{(Q)}$ respectively, and therefore the asymptotic power, under the side condition of a bounded influence function,

$$\sup_{x,y} \left(\left| \eta(x, y - x^T\tilde{\theta}) \right| / u_{pp} \right) \cdot \left((x^T (M^{-1} - (\tilde{M})^+)x) / (p - q) \right)^{1/2} \le b_{\tilde{\theta}},$$

is defined by the function

$$\tau_M(x, r) = w_c \left(\|z_{(2)}\| \cdot (p - q)^{-1/2} \right) \cdot \rho_{c'}(r), \tag{7.3.1}$$

where

$$w_c(t) := \psi_c(t)/t, \qquad z = U_M^{-1}x;$$

U_M is a lower triangular matrix defined implicitly by

$$Ew_c\left(\|z_{(2)}\| \cdot (p - q)^{-1/2}\right)zz^T = I;$$

and c, c' are positive constants such that $cc' < b_\theta$. Under (7.2.SPH) we have: $z_{(2)} = x_{(2)} \cdot (p - q)^{1/2}$.

Proof.   See Ronchetti (1982a, b).

7.3c.   The Optimal Test for the General M-Regression

In order to state the optimality result, we first have to show the existence of a certain matrix.

Proposition 3.   Let $c > 0$. If Exx^T is finite and nonsingular, then:

(i)   For sufficiently large $c > 0$ there exists a symmetric and positive definite ($p \times p$) matrix $M_S(c)$ which satisfies the equation

$$E\left(\left(2\Phi\left(c \cdot (p - q)^{1/2}/\left|x^T(M^{-1} - (\tilde{M})^+)x\right|^{1/2}\right) - 1\right)xx^T\right) = M.$$

$$(7.3.2)$$

(ii) M_S converges to Exx^T when $c \to \infty$.

(iii)   Denote by U_S the lower triangular matrix with positive diagonal elements such that $U_S \cdot U_S^T = M_S$ and define

$$\eta_S(x, r) := \left(\|z_{(2)}\|/(p - q)^{1/2}\right)^{-1} \cdot \psi_c\left(r\|z_{(2)}\|/(p - q)^{1/2}\right),$$

with $z = U_S^{-1}x$. Then, $M_S = E\eta_S'(x, r)xx^T$.

Proof.   Assertions (i) and (ii) can be shown using the same techniques as in Krasker (1980, Proposition 1, p. 1338), noting that

$$\|\tilde{M}\| \leq \|M\|, \text{ where } \|M\| := \sum_{i, j}m_{ij}^2.$$

(iii) follows from Proposition 2 of Section 7.2 using (7.3.2).   □

Remark 3. The subscript S for M_S indicates that M_S is the matrix M corresponding to $\eta_S(x, r)$; this function is of the Schweppe form (see Subsection 6.3a).

Proposition 4. Assume either $(q = p - 1)$ or (7.2.SPH). Then, the test that solves the optimality problem for infinitesimal robustness within the class of τ-tests, is defined by a function of the form

$$\tau_S(x, r) = \left(\|z_{(2)}\| / (p - q)^{1/2} \right)^{-2} \cdot \rho_c\left(r\|z_{(2)}\| / (p - q)^{1/2} \right)$$

$$= \rho_{\bar{c}(x)}(r),$$

where $\bar{c}(x) := c \cdot (p - q)^{1/2} / \|z_{(2)}\|$, $z = U_S^{-1}x$, and U_S is the lower triangular matrix (which exists because of Proposition 3) for which

$$E\left(\left(2\Phi\left(c \cdot (p - q)^{1/2} / \|z_{(2)}\| \right) - 1 \right) zz^T \right) = I. \tag{7.3.3}$$

Proof (Ronchetti, 1982b). We show the assertion under the condition $(q = p - 1)$. The proof is similar under (7.2.SPH). Put

$$M(\eta) = E\eta'(x, r)xx^T, \qquad Q(\eta) = E\eta^2(x, r)xx^T,$$

and let $\lambda_p(\eta)$ be the positive eigenvalue of

$$Q(\eta) \cdot \left(M^{-1}(\eta) - (\tilde{M})^+(\eta) \right).$$

Moreover, denote by $U(\eta)$ the Choleski root of $M(\eta)$ as given by Proposition 2 of Section 7.2: $U(\eta)$ is a lower triangular matrix with positive diagonal elements. We have to solve the following problem.

For a given $b > 0$, find a test which maximizes $M_{(22.1)}/\lambda_p$, under the side condition

$$\sup_{x, r}\left(|\eta(x, r)| / u_{pp} \right) \cdot \left(x^T(M^{-1} - (\tilde{M})^+)x \right)^{1/2} \le b. \tag{7.3.4}$$

Since $U^T(M^{-1} - (\tilde{M})^+)U = I - \tilde{I}$ by Proposition 2 of Section 7.2, we obtain

$$\lambda_p(\eta) = (U^{-1}QU^{-T})_{pp} = E\left\{ \eta^2(x, r)\left|(U^{-1}x)^{(p)}\right|^2 \right\}, \tag{7.3.5}$$

$$M_{(22.1)}(\eta) = \left(u_{pp}(\eta) \right)^2. \tag{7.3.6}$$

Moreover, (7.3.4) becomes

$$\sup_{x,r}\left(\left|\eta(x,r)\right|/u_{pp}\right)\cdot\left|\left(U^{-1}x\right)^{(p)}\right| \le b. \qquad (7.3.7)$$

Choose $c > 0$ such that $b = c/(U_S)_{pp}$, where U_S is defined by Proposition 3(iii) [this c exists because $c/(U_S)_{pp}$ is continuous in c, and $c^2/(U_S)^2_{pp} \ge c^2/\text{trace}\, M_S \ge c^2/E\|x\|^2 \to \infty$, when $c \to \infty$], and restrict attention, without loss of generality, to those η for which

$$u_{pp}(\eta) = (U_S)_{pp}. \qquad (7.3.8)$$

(Multiplication of the test statistic by a positive constant does not change the test!)

Combining (7.3.5)–(7.3.8), the original problem reduces to minimize (7.3.5), under the conditions (7.3.8) and

$$\sup_{x,r}\left|\eta(x,r)\right|\cdot\left|\left(U^{-1}x\right)^{(p)}\right| \le c. \qquad (7.3.9)$$

As in other optimality proofs in this book, the main trick is to replace the integrand in (7.3.5) by an expression which makes pointwise minimization obvious. We have

$$E\eta^2(x,r)\cdot\left|\left(U^{-1}x\right)^{(p)}\right|^2 = E\left(\eta(x,r)\cdot\left(U^{-1}x\right)^{(p)} - r\cdot\left(U_S^{-1}x\right)^{(p)}\right)^2$$

$$-Er^2\cdot\left|\left(U_S^{-1}x\right)^{(p)}\right|^2 + 2,$$

since

$$E\left(\eta(x,r)\cdot r\cdot\left(U^{-1}x\right)^{(p)}\cdot\left(U_S^{-1}x\right)^{(p)}\right)$$

$$= \left(U^{-1}\cdot E\left(\eta(x,r)\cdot r\cdot xx^T\right)\cdot U_S^{-1}\right)_{pp},$$

and integrating by parts, this equals

$$\left(U^{-1}\cdot E\left(\eta'(x,r)xx^T\right)\cdot U_S^{-1}\right)_{pp} = \left(U^{-1}MU_S^{-1}\right)_{pp} = \left(U^{-1}UU^TU_S^{-1}\right)_{pp}$$

$$= \left(U^TU_S^{-1}\right)_{pp} = u_{pp}/(U_S)_{pp} = 1,$$

where in the last equations we have used (7.3.8) and $U_{(12)} = 0$. Thus, minimizing (7.3.5), under the conditions (7.3.8) and (7.3.9) is equivalent to minimizing

$$E\left\{\left(\eta(x,r)\cdot(U^{-1}x)^{(p)} - r\cdot(U_S^{-1}x)^{(p)}\right)^2\right\},$$

subject to (7.3.9).Let

$$h(x,r) = \eta(x,r)\cdot(U^{-1}x)^{(p)}, \qquad (7.3.10)$$

and let $\mathcal{H}$ be the set of all functions $h(x,r)$ that can be obtained in this way by varying η. We have to minimize

$$E\left\{\left(h(x,r) - r\left(U_S^{-1}x\right)^{(p)}\right)^2\right\}$$

under the side condition

$$|h(x,r)| \le c \text{ for all } x, r$$

within $\mathcal{H}$. If we drop the "within $\mathcal{H}$" for a moment, pointwise minimization yields the solution

$$h(x,r) = \psi_c\left(r\cdot\left(U_S^{-1}x\right)^{(p)}\right).$$

Now, since this $h(x,r)$ is of the form (7.3.10) with

$$\eta(x,r) = |z_{(2)}|^{-1} \cdot \psi_c\left(r|z_{(2)}|\right)$$

where $z = U_S^{-1}x$, it also solves the minimization problem within $\mathcal{H}$.

Any other solution defines a test which has the same influence function and the same asymptotic power and in this sense is locally equivalent to η_S.

□

Remark 4 (*Test based on quadratic forms of robust estimators*). Corollary 2 of Section 7.2 shows that the unstandardized sensitivity of the test statistic S_n equals a standardized sensitivity of the estimator $((T_\Omega)_n)_{(2)}$ [the last $p - q$ components of the M-estimator $(T_\Omega)_n$ in the full model]. Alternatively, one can define other standardizations of $((T_\Omega)_n)_{(2)}$. For instance, the choice $(V_{(22)})^{-1}$, where $V_{(22)}$ is the $(p - q) \times (p - q)$ lower-right block of the asymptotic covariance matrix V of $(T_\Omega)_n$, defines the self-standardized sensitivity γ_s^* of $((T_\Omega)_n)_{(2)}$ (cf. Subsections 4.2b and 6.3a). This is the unstandardized sensitivity of the quadratic form

$$R_n := \left[((T_\Omega)_n)_{(2)}^T \cdot (V_{(22)})^{-1} \cdot ((T_\Omega)_n)_{(2)} \right]^{1/2} \qquad (7.3.11)$$

which can be used as an alternative test statistic.

Note that R_n^2 is the Wald test statistic when $(T_\Omega)_n$ is the least-squares estimator. Its influence function is unbounded. Therefore, one can look for a function $\psi(x, y; \theta)$ defining an M-estimator $(T_\Omega)_n$ (see Definition 5, Subsection 4.2c) which maximizes the asymptotic power of the test defined by R_n, under the condition of a bounded influence function for R_n. This is equivalent to find $\psi(x, y; \theta)$ which minimizes $V_{(22)}$ under a bound on the self-standardized sensitivity of $((T_\Omega)_n)_{(2)}$. From the theory of robust estimation (see Subsections 4.4b and Section 6.3), we obtain only the following admissibility result.

Proposition 5. For a given $c \geq \sqrt{p - q}$, let $C_c(\psi)$ be the class of tests defined by (7.3.11) and such that $\sup_{x, y} |\mathrm{IF}(x, y; R, F_\theta)| \leq c$. Then, the test defined by the following function ψ^* is admissible within $C_c(\psi)$:

$$\psi_{(1)}^*(x, y; \theta) = r \cdot z_{(1)},$$

$$\psi_{(2)}^*(x, y; \theta) = \left(z_{(2)} / \|z_{(2)}\| \right) \cdot \psi_c \left(r \|z_{(2)}\| \right),$$

where $\psi_{(1)}^*$ and $\psi_{(2)}^*$ denote the first q and last $(p - q)$ components of ψ^*, respectively; $r = y - x^T \theta$, $z = Dx$, and D is defined implicitly by the matrix equation

$$E\psi^*(x, y; \theta) r x^T = I.$$

Under the conditions of the propositions given in this section, it is easy to see that the test defined by ψ^* is asymptotically equivalent (i.e., it has the same asymptotic power and the same influence function) to the corresponding optimally robust tests given in this section. Thus, under those conditions, ψ^* is in fact strongly optimal.

Remark 5 (*Finite-sample approximations to the optimal weights*). The η-functions defining optimally bounded influence tests and optimally bounded influence estimators are of the same form [see Propositions 2 and 4 of this section and (6.3.14) and (6.3.7)]; there is a difference only in the weights. The optimal weights for the test take into account that (after standardization) only the last $(p - q)$ components are of interest for the testing problem.

Welsch (1981) provides finite-sample approximations to the optimal weights for estimators. He takes $n - 1$ observations $[(x_i, y_i)$ omitted], computes an approximate influence function for (x_i, y_i), and in this way obtains the matrix

$$(n - 1)^{-1} X^T(i) \cdot X(i)$$

that can be viewed as an approximation to the matrix M. [$X(i)$ denotes the design matrix X without the ith row x_i^T.] Using this approximation to the matrix M we can obtain finite-sample approximations to the optimal weights for the tests:

$$w_c\left(v_i^{-1}\right) \cdot \rho_{c'}(r_i), \qquad \text{(Mallows's regression)}$$

$$\rho_{c \cdot v_i}(r_i) \qquad \text{(general case)},$$

where

$$v_i := \left((n - 1) \cdot (p - q)^{-1} \cdot \left(h_{ii}^{(\Omega)}/\left(1 - h_{ii}^{(\Omega)}\right) - h_{ii}^{(\omega)}/\left(1 - h_{ii}^{(\omega)}\right)\right)\right)^{-1/2}$$

and $h_{ii}^{(\cdot)}$ is the element (i, i) of the *hat matrix* $H = X(X^TX)^{-1}X^T$ of the corresponding model (see Hoaglin and Welsch, 1978). More empirical investigations about the quality and the properties of these approximations are necessary.

Remark 6.   Note that from these optimal robust tests one can easily derive robust confidence regions for the parameters and a robust version of stepwise regression. These procedures can be easily implemented in a package for robust regression. Especially, they are integrated in ROBETH (see subsection 6.4c). However, we would like to stress again the point that bounded influence regression should not be used blindly and should not be considered the only and final answer of any data analysis (cf. also the concluding remark in Subsection 6.4b).

Remark 7.   The theory developed up to now allows us to construct asymptotic critical regions for optimal bounded influence tests. However, the optimal test statistics that define these tests can be used as test statistics for permutation tests. This would guarantee, on one side, an *exact* level α-test (a property of permutation tests) and, on the other side, a high robustness of efficiency (a property of S_n^2). This idea has been applied for constructing a confirmatory test in connection with a Swiss hail experiment; cf. Hampel et al. (1983); Lambert (1985).

7.3d. A Robust Procedure For Model Selection

The Information Criterion is a powerful tool for choosing among different models which can be used to fit a given data set. If we denote by L_p the log–likelihood of the model with p parameters, this amounts to choose the model which minimizes (for a given fixed α)

$$\text{AIC}(p; \alpha) = -2L_p + \alpha p. \qquad (7.3.12)$$

The choice $\alpha = 2$ is based on information theoretic reasons and maximizes approximately the entropy of the model (see Akaike, 1973). Other choices for α are possible, (see, for instance, Bhansali and Downham, 1977).

If we apply (7.3.12) to the regression model of Section 6.1 with normally distributed errors, we obtain

$$\text{AIC}(p; \alpha) = K(n, \hat{\sigma}) + R_p/\hat{\sigma}^2 + \alpha \cdot p, \qquad (7.3.13)$$

where $K(n, \hat{\sigma})$ is a constant depending on the marginal of the x_i's, $\hat{\sigma}^2$ is some estimate of σ^2 and $R_p = \sum_{i=1}^n (y_i - x_i^T \hat{\theta}_p)^2$ is the residual sum of squares with respect to the least-squares estimate $\hat{\theta}_p \cdot \text{AIC}(p; 2)$ is equivalent to Mallows's C_p statistic (see Mallows, 1973). Since the AIC statistic for regression models is a direct consequence of the normality assumption on the errors' distribution [see (7.3.13)], we cannot use it in this form with robust estimators and robust tests. In this subsection we propose a robust selection procedure for regression that, first, allows us to choose the model which fits the *majority* of the data taking into account that the errors might not be exactly normally distributed, and second, that can be used consistently with new robust estimators and tests.

For a given constant α and a given function ρ, choose the model that minimizes

$$\text{AICR}(p; \alpha, \rho) = 2 \sum_{i=1}^n \rho(r_{i;p}) + \alpha p, \qquad (7.3.14)$$

where $r_{i;\,p} = (y_i - x_i^T T_{n;\,p})/\hat{\sigma}$, $\hat{\sigma}$ is some robust estimate of σ, and $T_{n;\,p}$ is the M-estimator defined by

$$\sum_{i=1}^{n} \psi(r_{i;\,p}) x_i = 0,$$

with $\psi(r) = d\rho/dr$.

The extension of AIC to AICR is the exact counterpart of that of maximum likelihood estimation to M-estimation (cf. Subsection 2.3a). In particular, if we choose ρ as Huber's function

$$\begin{aligned}\rho_c(r) &= r^2/2 && \text{if } |r| \le c \\ &= c|r| - c^2/2 && \text{otherwise,}\end{aligned}$$

then $T_{n;\,p}$ is Huber's estimator and AICR(p; α, ρ_c) is the generalized Akaike statistic computed under the least favorable errors' distribution (see Section 2.7).

In this case a robust estimate for σ can be obtained using Huber's Proposal 2 or Hampel's median absolute deviation in the model with all parameters (see Remark 2, Section 6.2).

Based on a result due to Stone (1977), Ronchetti (1982b, 1985) proposes to choose $\alpha = 2E\psi^2/E\psi' < 2$, whereas Hampel (1983b) obtains $\alpha = E\psi^2/E\psi' + E\psi^2/(E\psi')^2$ (which differs little from 2 for $\psi = \psi_c$ and c between 1.3 and 1.6) "by adding the average decrease of $\sum_{i=1}^{n}\rho(r_i)$ and the average increase of the total mean-square error of fit due to a superfluous parameter under normality."

For details we refer to Ronchetti (1982b, 1985) who proposed this procedure and discussed its properties (in particular, a relationship between AICR and robust tests); Hampel (1983b) who gives a general discussion on robust model selection (and extends AICR for bounded influence regression), and Martin (1980) who used a similar idea for autoregressive models.

Finally, note that an alternative robust procedure for the selection of variables can be obtained by replacing the F-test in the classical stepwise regression by the optimal robust test derived in this section. This procedure can be viewed as a robust version of stepwise regression.

*7.4. $C(\alpha)$ – TYPE TESTS FOR LINEAR MODELS

Consider the linear model of Section 6.1 with normal errors $(G = \Phi), \sigma = 1$, and the related test problem (Section 7.1).

Though $C(\alpha)$ tests can be defined for testing more than one linear hypothesis on θ (see Bühler and Puri, 1966), we focus here on the generalization of Neyman's original definition (see Neyman, 1958, 1979), that is, we shall be concerned only with testing a single parameter ($q = p - 1$). We define the class of $C(\alpha)$ – *type* tests and we compute their influence function and their asymptotic power. As far as these properties are concerned, we show that these tests are equivalent to the tests which have been introduced in Section 7.2. Moreover, we shall show the connection between our optimal test and an asymptotically minimax test. Details are provided by Ronchetti (1982b).

7.4a.   Definition of a $C(\alpha)$ – Type Test

The class of tests we want to define will be based on a function which satisfies the conditions (7.2.ETA1), (7.2.ETA2), (7.2.ETA3) (i) in Section 7.2.

Definition 1.   The class of $C(\alpha)$-type tests for linear models is defined by means of the test statistics

$$Z_n(\tilde{\theta}; \eta) := n^{-1/2} \sum_{i=1}^{n} \eta\left(x_i, y_i - x_i^T\tilde{\theta}\right) \cdot \left(U^{-1}x_i\right)^{(p)}$$

$$= n^{-1/2}u_{pp}^{-1} \cdot \sum_{i=1}^{n} \eta\left(x_i, y_i - x_i^T\tilde{\theta}\right) \cdot \left[x_i^{(p)} - U_{(21)} \cdot U_{(11)}^{-1}(x_i)_{(1)}\right],$$

$$(7.4.1)$$

where η satisfies the conditions (7.2.ETA1) and (7.2.ETA2) and U is the lower triangular matrix with positive diagonal elements such that $UU^T = M = E\eta'(x, r)xx^T$. "Large" absolute values of Z_n are significant.

Remark 1. $C(\alpha)$ tests were introduced by Neyman (1958) in a general framework. His basic motivation was to "seek for tests that have a compromise but a clear property of optimality and that are relatively easy to deduce" (Neyman, 1979, p. 1). Wang (1981, p. 1100) derived a (asymptotically minimax) robust version of the optimal $C(\alpha)$ test and Ronchetti (1982b) proposed the class of $C(\alpha)$-type tests and found the optimal bounded influence test in that class.

Remark 2. If we put $\eta(x, r) = -g'(r)/g(r)$ in (7.4.1), where g is the density of the error distribution, Z_n becomes the test statistic of the optimal $C(\alpha)$ test obtained by Neyman (1958, p. 228). In this sense the tests defined by (7.4.1) can be called $C(\alpha)$-type tests.

Remark 3. The test statistic Z_n depends on the unknown nuisance parameters $\theta^{(1)}, \ldots, \theta^{(p-1)}$. In Subsection 7.4b we shall discuss the properties of a Studentized version $Z_n((T_\omega)_n; \eta)$, $(T_\omega)_n$ being a suitable estimate of $\tilde{\theta}$.

Remark 4. In order to determine asymptotic critical regions of the test, we shall compute the asymptotic distribution of Z_n in the next Subsection.

7.4b. Influence Function and Asymptotic Power of $C(\alpha)$-Type Tests

Let Z be the functional corresponding to the test statistic Z_n, that is,

$$Z(F) = \int \eta(x, y - x^T\tilde{\theta}) \cdot (U^{-1}x)^{(p)} dF(x, y).$$

Then $Z(F_n) = n^{-1/2}Z_n$, where F_n is the empirical distribution function.

Proposition 1. Let $F_{\tilde{\theta}}$ be the model distribution under the null hypothesis. If (7.2.ETA1) and (7.2.ETA2) hold, then

(i) The influence function of Z is given by

$$\text{IF}(x, y; Z, F_{\tilde{\theta}}) = \eta(x, y - x^T\tilde{\theta}) \cdot (U^{-1}x)^{(p)}.$$

(ii)

$$\left|\text{IF}(x, y; Z, F_{\tilde{\theta}})\right| = \text{IF}(x, y; S, F_{\tilde{\theta}}) = \text{IF}(x, y; W, F_{\tilde{\theta}}),$$

where S and W are the functionals defining the test statistics of the tests introduced in Section 7.2.

Proof. Left as an exercise.

From this proposition we see that the influence function of the optimal $C(\alpha)$ test when the error distribution is normal, is $(y - x^T\theta)(U^{-1}x)^{(p)}$ and therefore unbounded in y and x. The goal of this section is to find a

$C(\alpha)$-type test (defined by η) that maximizes the asymptotic power and has a bounded influence function. Let us first compute the asymptotic power.

Proposition 2. Besides (7.2.ETA1), (7.2.ETA2), and (7.2.ETA3) (i) assume

(7.4.AS) $$E\{\eta^2(x,r) \cdot \|x\|^2\} < \infty.$$

Put

$$\lambda_p := E\eta^2(x,r) \cdot \left|(U^{-1}x)^{(p)}\right|^2.$$

Then, under the sequence of alternatives

$$H_{(n)} := \theta = \tilde{\theta} + n^{-1/2}\Delta, \qquad (7.4.2)$$

where

$$\Delta = (0, \ldots, 0, \Delta^{(p)})^T,$$

the test statistic Z_n has asymptotically a normal distribution with mean $u_{pp} \cdot \Delta^{(p)}$ and variance λ_p.

Moreover, the asymptotic power of the $C(\alpha)$-type test defined by Z_n is given by

$$1 - \Phi\left(\Phi^{-1}(1-\alpha) - \Delta^{(p)} \cdot u_{pp}/\lambda_p^{1/2}\right).$$

Proof. It suffices to compute $E_\theta Z_n$ under the sequence (7.4.2). Define

$$\xi(\theta) := \int \eta(x, y - x^T\theta) \cdot (U^{-1}x)^{(p)} \cdot f_\theta(x, y)\, dx\, dy,$$

where

$$f_\theta(x, y) = \phi(y - x^T\theta) \cdot k(x).$$

Then, for $j = 1, \ldots, p - 1$, we have

$$\left[(\partial/\partial\theta^{(j)})\xi(\theta)\right]_{\tilde{\theta}} = -\int \eta'(x, y - x^T\tilde{\theta})x^{(j)} \cdot (U^{-1}x)^{(p)}\, dF_{\tilde{\theta}}(x, y)$$

$$+ \int \eta(x, y - x^T\tilde{\theta}) \cdot (U^{-1}x)^{(p)} \cdot \left[(\partial/\partial\theta^{(j)})f_\theta(x, y)\right]_{\tilde{\theta}}\, dx\, dy.$$

$$(7.4.3)$$

Using

$$(U^{-1}x)^{(p)} = u_{pp}^{-1} \cdot \left(- U_{(21)}U_{(11)}^{-1}x_{(1)} + x^{(p)} \right), \qquad (7.4.4)$$

we get

$$E\left\{ \eta'(x,r)x_{(1)}^T \cdot (U^{-1}x)^{(p)} \right\}$$

$$= u_{pp}^{-1} \cdot \left(- U_{(21)} \cdot U_{(11)}^{-1} \cdot E\left\{ \eta'(x,r)x_{(1)}x_{(1)}^T \right\} + E\left\{ \eta'(x,r)x_{(1)}^T x^{(p)} \right\} \right)$$

$$= u_{pp}^{-1} \cdot \left(- U_{(21)} \cdot U_{(11)}^{-1} \cdot U_{(11)} \cdot U_{(11)}^T + M_{(12)}^T \right) = 0, \qquad (7.4.5)$$

and the first term of the right-hand side of (7.4.3) vanishes. In a similar way one can prove that also the second term of the right-hand side of (7.4.3) equals 0. Moreover,

$$\left[(\partial/\partial\theta^{(p)})\xi(\theta) \right]_{\tilde{\theta}}$$

$$= \int \eta\left(x, y - x^T\tilde{\theta} \right) \cdot (U^{-1}x)^{(p)} \cdot \left[(\partial/\partial\theta^{(p)})f_\theta(x,y) \right]_{\tilde{\theta}} dx\, dy$$

$$= E\left\{ \eta(x,r) \cdot (U^{-1}x)^{(p)} \cdot r \cdot x^{(p)} \right\}.$$

Integration by parts converts this to

$$E\left\{ \eta'(x,r) \cdot (U^{-1}x)^{(p)} \cdot x^{(p)} \right\} = u_{pp}^{-1} \cdot \left(- U_{(21)} \cdot U_{(11)}^{-1} \cdot M_{(12)} + m_{pp} \right)$$

$$= u_{pp}^{-1}\left(u_{pp}^2 \right) = u_{pp} \qquad (7.4.6)$$

using the same argument again.

Finally, denoting by a dot the differentiation with respect to θ, we obtain

$$\left(\dot{\xi}(\tilde{\theta}) \right)^{(j)} = 0, \qquad \text{for } j = 1, \ldots, p - 1, \qquad (7.4.7)$$

$$\left(\dot{\xi}(\tilde{\theta}) \right)^{(p)} = u_{pp}, \qquad (7.4.8)$$

and by (7.2.ETA1) (i),

$$\xi(\tilde{\theta}) = 0.$$

Therefore, using a Taylor expansion we get

$$E_\theta Z_n = n^{1/2}\xi(\theta) = n^{1/2}\xi^T(\tilde{\theta}) \cdot (\theta - \tilde{\theta}) + o(\|\theta - \tilde{\theta}\|)$$

$$= u_{pp}\Delta^{(p)} + o(n^{-1/2}).$$

This completes the proof.   □

Remark 5. As in Subsection 7.2c, Case 2, the suitable standardization of the influence function is

$$\text{IF}_{\text{test}}(x, y; Z, F_{\tilde{\theta}}) := \text{IF}(x, y; Z, F_{\tilde{\theta}})/(\dot{\xi}(\tilde{\theta}))^{(p)}$$

$$= \text{IF}(x, y; Z, F_{\tilde{\theta}})/u_{pp} \qquad (7.4.9)$$

(cf. 7.2.16). Note that $\xi(\theta) = E_\theta n^{-1/2} Z_n = Z(F_\theta)$. Thus, (7.4.9) extends the one-dimensional definition in a natural way. Again from Proposition 2 it follows that the asymptotic power of the $C(\alpha)$-type test defined by η is a monotone increasing function of $(\int \text{IF}_{\text{test}}(x, y; Z, F_{\tilde{\theta}}))^2 \, dF_{\tilde{\theta}}(x, y))^{-1}$.

Remark 6. The test defined by the test statistic Z_n depends on the unknown nuisance parameter $\tilde{\theta}$. By techniques similar to those used by Wang (1981) one can show that the result of Proposition 2 holds if we substitute $\tilde{\theta}$ by a suitable ($n^{1/2}$-consistent) estimate $(T_\omega)_n$ (see Wang, 1981, p. 1099). Moreover, the result of Proposition 1 still holds. To see this, suppose that the influence function of T_ω exists and define

$$Z(F) = \int \eta(x, y - x^T \cdot T_\omega(H)) \cdot (U^{-1}x)^{(p)} \, dF(x, y).$$

Then,

$$\text{IF}(x, y; Z, F_{\tilde{\theta}}) = \left[(\partial/\partial\varepsilon) Z((1 - \varepsilon)F_\theta + \varepsilon\Delta_{(x, y)}) \right]_{\varepsilon = 0}$$

$$= \eta(x, y - x^T\tilde{\theta}) \cdot (U^{-1}x)^{(p)} + \int \eta(s, v - s^T\tilde{\theta}) \cdot (U^{-1}s)^{(p)} \, dF_{\tilde{\theta}}(s, v)$$

$$- \left(\int \eta'(s, v - s^T\tilde{\theta}) \cdot (U^{-1}s)^{(p)} \cdot \tilde{s}^T \, dF_{\tilde{\theta}}(s, v) \right) \cdot \text{IF}(x, y; T_\omega, F_{\tilde{\theta}}).$$

$$(7.4.10)$$

The second term on the right-hand side vanishes, in view of (7.4.4) and (7.2.ETA1) (i), and so does the third term by (7.4.5).

7.4c. Optimal Robust $C(\alpha)$-Type Tests

From the point of view of the asymptotic properties (influence function and asymptotic efficiency) $C(\alpha)$-type tests defined by η, τ-tests, and tests based on the quadratic form W_n^2 with the same η are equivalent (cf. Subsections 7.2b and 7.2c with Subsection 7.4b). Applying Proposition 4 of Section 7.3 we have the following result.

Corollary. The test which solves the optimality problem for infinitesimal robustness (see Section 3.4) within the class of $C(\alpha)$-type tests is defined by a function of the form:

$$\eta_s(x, r) = \left|z^{(p)}\right|^{-1}\psi_c\left(r \cdot \left|z^{(p)}\right|\right) = \psi_{c/\left|z^{(p)}\right|}(r),$$

where $z = U_s^{-1}x$, U_s is the lower triangular matrix defined implicitly by

$$E\left(2\Phi\left(c/\left|z^{(p)}\right|\right) - 1\right)zz^T = I,$$

and c is a positive constant depending on the bound on the influence function.

7.4d. Connection with an Asymptotically Minimax Test

In this subsection we describe the connections between this optimal robust $C(\alpha)$-type test and a test introduced by Wang (1981). Wang studies the testing problem using minimax techniques. He considers the following situation.

Suppose we are given a parametric model $(\mathscr{X}, \mathscr{A}, \{F_\theta : \theta \in \Theta \subset \mathbb{R}^p\})$ (see Subsection 2.1a) and suppose we want to test a hypothesis on one component of the parameter θ, say $\theta^{(p)} = 0$, the other $(p - 1)$ components being nuisance parameters. Then, using the technique of shrinking neighborhoods, Wang is able to derive an asymptotically minimax test and to extend the work of Rieder (1978) to the case where the model distribution is indexed by nuisance parameters. Let us write this result in the situation of the linear model. Define

$$r := y - x^T\tilde{\theta}, \qquad a := \left(a^{(1)}, \ldots, a^{(p)}\right)^T,$$

$$d\left(x, a_{(1)}\right) := x^{(p)} + a_{(1)}^T x_{(1)} = x^{(p)} + \sum_{j=1}^{p-1} a^{(j)}x^{(j)},$$

$$\Lambda\left(a_{(1)}, \tilde{\theta}\right) = \Lambda^*\left(x, y; a_{(1)}, \tilde{\theta}\right) := r \cdot d\left(x, a_{(1)}\right). \qquad (7.4.11)$$

For a given $\varepsilon > 0$, $\delta_1 > 0$, define $V_0(a_{(1)}, \tilde{\theta})$ and $V_1(a_{(1)}, \tilde{\theta})$ implicitly by means of the equations

$$E\left[\max\left\{\left(\Lambda\left(a_{(1)}, \tilde{\theta}\right) - V_1\left(a_{(1)}, \tilde{\theta}\right)\right), 0\right\}\right] = \varepsilon/\delta_1,$$

(7.4.12)

$$E\left[\max\left\{\left(V_0\left(a_{(1)}, \tilde{\theta}\right) - \Lambda\left(a_{(1)}, \tilde{\theta}\right)\right), 0\right\} + \Lambda\left(a_{(1)}, \tilde{\theta}\right)\right] = \varepsilon/\delta_1.$$

(7.4.13)

Moreover, let $a_{(1)} = \alpha_{(1)}(\tilde{\theta})$ be the solution to the equation

$$E\left(\left[\Lambda\left(a_{(1)}, \tilde{\theta}\right)\right]_{V_0(a_{(1)}, \tilde{\theta})}^{V_1(a_{(1)}, \tilde{\theta})} \cdot r \cdot x_{(1)}\right) = 0,$$

(7.4.14)

and define

$$v^2(\tilde{\theta}) := E\left(\left[\Lambda\left(\alpha_{(1)}, \tilde{\theta}\right)\right]_{V_0(\alpha_{(1)}, \tilde{\theta})}^{V_1(\alpha_{(1)}, \tilde{\theta})}\right)^2.$$

(7.4.15)

Proposition 3. The test defined by the critical region

$$\left\{Y_n(\tilde{\theta}) \geq \Phi^{-1}(1 - \alpha) + \varepsilon V_1\left(\alpha_{(1)}(\tilde{\theta}), \tilde{\theta}\right)/v(\tilde{\theta})\right\},$$

(7.4.16)

where

$$Y_n(\tilde{\theta}) := n^{-1/2} \cdot \left(v(\tilde{\theta})\right)^{-1} \cdot \sum_{i=1}^{n} \left[\Lambda^*\left(x_i, y_i; \alpha_{(1)}(\tilde{\theta}), \tilde{\theta}\right)\right]_{V_0(\alpha_{(1)}, \tilde{\theta})}^{V_1(\alpha_{(1)}, \tilde{\theta})},$$

is an asymptotically minimax test at level α.

Proof. See Wang (1981), pp. 1099 and 1104.

The test defined by (7.4.16) will be called the *W-test*. In order to have a better performance *at the model*, Wang (1981, p. 1105) proposes a modification of the test (7.4.16) and defines a test by means of the following critical region

$$\left\{Y_n(\tilde{\theta}) \geq \Phi^{-1}(1 - \alpha)\right\}.$$

(7.4.17)

The test defined by (7.4.17) will be called the *modified W-test*. Then we have the following result.

Proposition 4. The modified W-test is a $C(\alpha)$-type test. It is equivalent to the optimal robust $C(\alpha)$-type test defined by η_s (given by the corollary), that is, both tests have the same influence function and the same asymptotic power.

Proof (Ronchetti, 1982b). We prove that $Y_n(\tilde{\theta}) = \lambda_p^{-1/2}(\eta_s) \cdot Z_n(\tilde{\theta}; \eta_s)$. Then, the assertion follows, using Proposition 2 and Remark 5. First, consider Eq. (7.4.14). Using (7.4.11) we have

$$0 = E\Big([\Lambda]_{V_0}^{V_1} \cdot r \cdot x_{(1)}\Big) = E\Big([r \cdot d]_{V_0}^{V_1} \cdot r \cdot x_{(1)}\Big)$$

$$= \int d(x, \alpha_{(1)}) \cdot x_{(1)} \cdot \left(\int [r]_{V_0/|d|}^{V_1/|d|} \cdot r \, d\Phi(r)\right) dK(x),$$

and (7.4.14) becomes

$$\int \Big(\Phi\big(V_1/|d(x, \alpha_{(1)})|\big) - \Phi\big(V_0/|d(x, \alpha_{(1)})|\big)\Big) \cdot d(x, \alpha_{(1)}) x_{(1)} \, dK(x) = 0.$$

$$(7.4.18)$$

Now, combining (7.4.12) and (7.4.13) and noting that $E\Lambda = 0$, we obtain

$$\int (\Lambda - V_1) \cdot 1_{\{\Lambda \geq V_1\}} + \int (\Lambda - V_0) \cdot 1_{\{\Lambda \leq V_0\}} = 0,$$

and performing these integrations we get, finally, by (7.4.18)

$$V_0 = -V_1 (:= -V). \tag{7.4.19}$$

For a given positive constant c, define η_s and U_S as in the corollary. Moreover, choose

$$V = c \cdot (U_S)_{pp}. \tag{7.4.20}$$

Then

$$\alpha_{(1)}^T = -(U_S)_{(21)} \cdot \big((U_S)_{(11)}\big)^{-1} \tag{7.4.21}$$

and

$$d(x, \alpha_{(1)}) = x^{(p)} - (U_S)_{(21)} \cdot \big((U_S)_{(11)}\big)^{-1} x_{(1)}$$

$$= \big(U_S^{-1} x\big)^{(p)} \cdot (U_S)_{pp}. \tag{7.4.22}$$

[The left member of (7.4.18) becomes

$$
E\left[\left(2\Phi\left(c/\left|(U_S^{-1}x)^{(p)}\right|\right) - 1\right) \cdot (U_S^{-1}x)^{(p)} \cdot (U_S)_{pp} \cdot x_{(1)}\right]
$$

$$
= \left[E\eta_S'(x,r) \cdot (U_S^{-1}x)^{(p)} \cdot x_{(1)}\right] \cdot (U_S)_{pp} = 0
$$

by (7.4.5).]

Moreover, by (7.4.15), (7.4.19), (7.4.20), and (7.4.22) we have

$$
v^2(\tilde\theta) = E\left([r \cdot d]_{-V}^{+V}\right)^2 = E\left(d^2(x;\alpha_{(1)}) \cdot \psi_{V/|d|}^2(r)\right)
$$

$$
= (U_S)_{pp}^2 \cdot E\left(\psi_{c/|z^{(p)}|}^2(r) \cdot \left|z^{(p)}\right|^2\right)
$$

$$
= (U_S)_{pp}^2 \cdot \lambda_p(\eta_S), \tag{7.4.23}
$$

where

$$
z^{(p)} = \left(U_S^{-1}x\right)^{(p)} \quad \text{and} \quad \lambda_p(\eta_S) = E\eta_S^2(x,r) \cdot \left|z^{(p)}\right|^2.
$$

Finally, we get

$$
Y_n(\tilde\theta) = n^{-1/2} \cdot \left((U_S)_{pp} \cdot \lambda_p^{1/2}(\eta_S)\right)^{-1} \cdot \sum_{i=1}^{n} [r \cdot d]_{-V}^{+V}
$$

$$
= n^{-1/2} \cdot \lambda_p^{-1/2}(\eta_S) \cdot \sum_{i=1}^{n} \left((U_S)_{pp}\right)^{-1} \cdot \psi_{c(U_S)_{pp}}\left(r \cdot z^{(p)} \cdot (U_S)_{pp}\right)
$$

$$
= n^{-1/2} \cdot \lambda_p^{-1/2}(\eta_S) \cdot \sum_{i=1}^{n} \psi_{c/|z^{(p)}|}(r) \cdot z^{(p)}
$$

$$
= \lambda_p^{-1/2}(\eta_S) \cdot Z_n(\tilde\theta; \eta_S). \qquad \qquad \square
$$

7.5. COMPLEMENTS

*7.5a. Computation of Optimal η Functions

In this subsection we discuss a computational aspect involved in the calculation of the η_S functions of optimal bounded influence tests—the computation of the matrix U_S.

Our first problem is to determine the lower triangular matrix with positive diagonal elements $U = U_S$ which solves the equation (see Proposition 4 of Section 7.3)

$$E\left[\left(2\Phi\left(c \cdot (p - q)^{1/2}/|z_{(2)}|\right) - 1\right)zz^T\right] = I, \qquad (7.5.1)$$

where $z = U^{-1}x$. Note, that if the x_i's are given and fixed, one has to estimate the expected value over K in (7.5.1) with an average over $\{x_1, \ldots, x_n\}$. Define

$$\overline{M}(U) := E\left[\left(2\Phi\left(c \cdot (p - q)^{1/2}/|z_{(2)}|\right) - 1\right)xx^T\right].$$

Then (7.5.1) is equivalent to

$$\overline{M}(U) = UU^T. \qquad (7.5.2)$$

We can find the solution U_S of (7.5.2) using the following algorithm (cf. Algorithm 2, Subsection 4.6b).

Algorithm

1. Define: $M_0 := Exx^T$. Let U_0 be the Choleski decomposition of M_0 (i.e., $U_0 U_0^T = M_0$ and U_0 lower triangular with positive diagonal elements). $U := U_0$.
2. Define $\overline{M}_1 := \overline{M}(U)$ and let U_1 be the Choleski decomposition of $\overline{M}_1$.
3. IF $\quad \|U_1 - U\| < \varepsilon$, GOTO 4; otherwise $U := U_1$ and GOTO 2.
4. END.

Empirical experience shows a rather quick convergence. We use this algorithm for computing the matrix U in Subsections 7.5c and 7.5d.

*7.5b. Computation of the Asymptotic Distribution of the τ-Test Statistic

In order to determine P-values we have to compute the distribution function of $\sum_{j=1}^m \lambda_j N_j^2$, where the N_j's are m independent normally distributed random variables and $\lambda_1 \geq \lambda_2 \geq \cdots \geq \lambda_m > 0$ are m positive real numbers (see Subsection 7.2b with $m := p - q$).

Let us denote this distribution function by $F_m(z; \lambda_1, \ldots, \lambda_m)$. Johnson and Kotz (1970, p. 149 ff.) discuss different approximations to F_m, especially a power-series expansion, a representation of F_m as a mixture of χ^2-distributions, and an expansion in terms of Laguerre polynomials. All these expansions seem to converge slowly (see Grenander et al., 1959, p. 377). The problem is discussed by Grenander et al. (1959) who propose the following idea. Consider the integral equation for F_m:

$$z \cdot F_m(z; \lambda_1, \ldots, \lambda_m) = \int_0^z \alpha(z - y) \cdot F_m(y; \lambda_1, \ldots, \lambda_m)\, dy, \quad (7.5.3)$$

where

$$\alpha(t) := 1 + \tfrac{1}{2} \sum_{j=1}^{m} \exp\left(-t/(2\lambda_j)\right).$$

Equation (7.5.3) can be discretized and one obtains

$$F_m(\ell \cdot \Delta; \lambda_1, \ldots, \lambda_m) =$$

$$\left(\ell - \tfrac{1}{2}\alpha(0)\right)^{-1} \cdot \sum_{\nu=1}^{\ell-1} \alpha((\ell - \nu) \cdot \Delta) \cdot F_m(\nu \cdot \Delta; \lambda_1, \ldots, \lambda_m), \quad (7.5.4)$$

for $\ell = 2, 3, \ldots$. The approximation based on (7.5.4) has not been tried out yet for τ-tests. For further approximations, see Solomon and Stephens (1977).

Remark. It would be interesting to try and use small-sample asymptotic techniques in order to find a better approximation to the distribution function F_m for small samples. These methods are very promising and have been used successfully in other situations, for example, for getting an approximation to the distribution of M-estimators (see Hampel, 1973b; Field, 1982; Field and Hampel, 1982; Daniels, 1954); cf. Section 8.5.

*7.5c.   Asymptotic Behavior of Different Tests for Simple Regression

In this subsection we present the results obtained by Ronchetti (1982b) comparing the asymptotic behavior of different tests for different distributions. In the exposition we apply the methods which have been used by Maronna et al. (1979) for comparing different estimators (see Subsection 6.4b). We consider simple regression:

$$y = x^T\theta + e,$$

where $x = (1, x^{(2)})^T$ and $\theta = (\theta^{(1)}, \theta^{(2)})^T$.

We want to test the hypothesis

$$H_0 : \theta^{(2)} = 0.$$

Let G and K be the distributions of e and $x^{(2)}$, respectively. For $x^{(2)}$ and e we choose the following contaminated normal distributions

$$(1 - \varepsilon)\Phi(t) + \varepsilon\Phi(t/s)$$

with the following parameters

ε	0	0.1	0.1	0.05
s	1	3	5	10

We want to study the asymptotic behavior of the optimally bounded influence tests:

$$\tau_M(x, r) = w_{c_1}(|z^{(2)}|) \cdot \rho_{c_2}(r) \qquad \text{(see Proposition 2 of Section 7.3),}$$

$$\tau_S(x, r) = \rho_{c_3/|z^{(2)}|}(r) \qquad \text{(see Proposition 4 of Section 7.3 and Corollary of Section 7.4).}$$

Moreover, in order to investigate the behavior of a test defined by means of a redescending η-function $(\eta(x, r) = (\partial/\partial r)\tau(x, r))$, we consider also the following procedure:

$$\tau_{MT}(x, r) = w_{c_4}(|z^{(2)}|) \cdot \tilde{\rho}(r; c_5, \kappa^*, c_6, A_5, B_5),$$

where $\tilde{\rho}$ is the ρ-function defining the hyperbolic tangent estimator (see Fig. 1 and Subsection 2.6c). $\tilde{\rho}$ is defined by

$$\begin{aligned}
\tilde{\rho}(r; c_5, \kappa^*, c_6, A_5, B_5) &= r^2/2 & \text{if } 0 \le |r| \le c_6 \\
&= (c_6)^2/2 + b(c_6) - b(r) & c_6 \le |r| \le c_5 \\
&= (c_6)^2/2 + b(c_6) - b(c_5) & c_5 \le |r|,
\end{aligned}$$

where

$$b(r) := 2(A_5/B_5) \cdot \ln\left[2 \cdot \cosh\left\{\left[((\kappa^* - 1)B_5^2/A_5)^{1/2}/2\right] \cdot (c_5 - |r|)\right\}\right].$$

Note that c_6, A_5, B_5 are computed implicitly in terms of c_5 and κ^*.

For each test we compute the standardized sensitivities (at $G = \Phi$)

$$u_{22}^{-1} \cdot \sup_{x, r}|\eta(x, r)| \cdot |z^{(2)}|$$

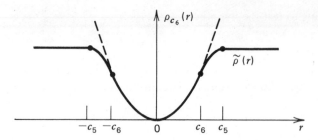

Figure 1. The ρ-function defining hyperbolic tangent estimators ($\tilde{\rho}$) and Huber's ρ-function (ρ_{c_6}).

and the efficacies

$$M_{22.1}/\lambda_2 = u_{22}^2/\lambda_2$$

under different distributions.

Note that the efficacy of the tests defined by τ_M and τ_{MT} factorizes; in these cases we have

$$M_{22.1}/\lambda_2 = DX \cdot DR,$$

where DX depends only on w_c and K and $DR = B^2/A$, with

$$A = E\psi_c^2, \quad B = E\psi_c', \quad \psi_c(r) = (\partial/\partial r)\rho_c(r).$$

The constants are chosen so that all the tests have the same asymptotic efficiency, when $x^{(2)} \sim N(0,1)$ and $e \sim N(0,1)$. Note that the constants for τ_M and τ_{MT} are not uniquely determined.

Table 1 describes the calibration as well as the standardized sensitivities of the tests.

Table 1. Calibration and Standardized Sensitivities at $x^{(2)} \sim N(0,1)$, $e \sim N(0,1)$

Test	Constants	DX	DR	Efficiency	u_{22}	A	B	Standardized Sensitivities
τ_M	$c_1 = 1.80$, $c_2 = 1.58$	0.98	0.97	0.95	0.900	0.809	0.886	3.16
τ_{MT}	$c_4 = 1.80$, $c_5 = 4.68$, $\kappa^* = 4.50$	0.98	0.97	0.95	0.910	0.842	0.905	3.37
	$c_6 = 1.70$, $A_5 = 0.84$, $B_5 = 0.90$							
τ_S	$c_3 = 2.67$	—	—	0.95	0.930	—	—	2.87

Table 2. Asymptotic Efficacies

K	Test	G $\varepsilon = 0$ $s = 1$	$\varepsilon = 0.1$ $s = 3$	$\varepsilon = 0.1$ $s = 5$	$\varepsilon = 0.05$ $s = 10$	Standardized Sensitivities
$\varepsilon = 0$ $s = 1$						
	τ_M	0.95	0.75	0.71	0.80	3.16
	τ_{MT}	0.95	0.77	0.77	0.87	3.37
	τ_S	0.95	0.78	0.72	0.78	2.87
$\varepsilon = 0.1$ $s = 3$						
	τ_M	1.52	1.21	1.13	1.28	2.70
	τ_{MT}	1.53	1.23	1.23	1.40	2.91
	τ_S	1.54	1.27	1.16	1.25	2.34
$\varepsilon = 0.1$ $s = 5$						
	τ_M	2.44	1.94	1.81	2.04	2.33
	τ_{MT}	2.44	1.97	1.97	2.24	2.51
	τ_S	2.51	2.06	1.86	2.00	1.94
$\varepsilon = 0.05$ $s = 10$						
	τ_M	2.99	2.38	2.23	2.51	2.23
	τ_{MT}	3.00	2.42	2.42	2.76	2.38
	τ_S	3.24	2.61	2.32	2.50	1.85

Table 3. Factorization for the τ_M- and τ_{MT}-Test

	$\varepsilon = 0$ $s = 1$	$\varepsilon = 0.1$ $s = 3$	$\varepsilon = 0.1$ $s = 5$	$\varepsilon = 0.05$ $s = 10$
DR				
τ_M	0.970	0.771	0.722	0.814
τ_{MT}	0.973	0.785	0.785	0.894
DX				
	0.979	1.569	2.512	3.087

From Table 2 one can obtain similar conclusions as for estimators (see Subsection 6.4b):

1. τ_{MT} is better than τ_M for all distributions under consideration.
2. τ_S is better than τ_{MT} when the distribution of e has moderate tails.
3. τ_S has the better standardized sensitivities (computed at $G = \Phi$).

Table 3 gives the factors for the τ_{M^-} and τ_{MT^-}test.

7.5d. A Numerical Example

The data to this example are taken from Draper and Smith (1966, 1981). We have the following variables:

Y = response or number of pounds of steam used per month,
X_8 = average atmospheric temperature in the month (in °F),
X_6 = number of operating days in the month.

Table 4 shows the data.
We consider the linear model

$$Y = \alpha + \beta_8 X_8 + \beta_6 X_6 + e,$$

and we want to test the hypothesis

$$H_0 : \beta_6 = 0.$$

The factor space is given by Figure 2 and the observations are plotted in Figure 3a.

From Figure 2 we see that there exist two outliers in the factor space. We want to study the behavior of the $\log_{10} P$-values of the F-, ρ-, and optimal τ-test when the observation (y_7) corresponding to the point ($X_8 = 74.4$, $X_6 = 11$) varies between 0 and 20. (Its actual value is 6.36.)
The tests under study are defined by the following functions:

Test	$\tau(x, r)$
F	$r^2/2$
ρ_{c_1}	$\rho_{c1}(r)$
Optimal τ	$\rho_{c_2/|z^{(2)}|}(r)$ $[z^{(2)} = (U_S^{-1}x)^{(2)}]$

Table 4

X_8	X_6	Y
35.3	20	10.98
29.7	20	11.13
30.8	23	12.51
58.8	20	8.40
61.4	21	9.27
71.3	22	8.73
74.4	11	6.36
76.7	23	8.50
70.7	21	7.82
57.5	20	9.14
46.4	20	8.24
28.9	21	12.19
28.1	21	11.88
39.1	19	9.57
46.8	23	10.94
48.5	20	9.58
59.3	22	10.09
70.0	22	8.11
70.0	11	6.83
74.5	23	8.88
72.1	20	7.68
58.1	21	8.47
44.6	20	8.86
33.4	20	10.36
28.6	22	11.08

Permission Draper and Smith (1966).

The scale parameter σ was estimated in the full model using Huber's Proposal 2.

The constants c_1 and c_2 are chosen such that the corresponding tests have a given efficiency, say 0.95, at the normal model (i.e., when x_i and e_i are normally distributed). We obtain the following values:

$$c_1 = 1.345, \qquad c_2 = 2.67.$$

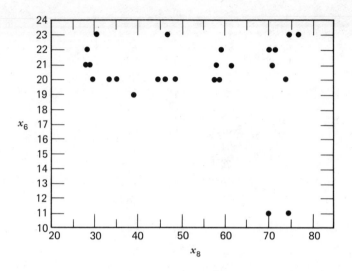

Figure 2. The variables X_6 and X_8. Permission Draper and Smith (1966).

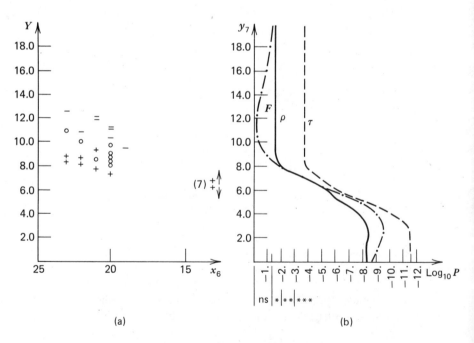

Figure 3. (a) $(-)$ $X_8 < 40$, $(+)$ $X_8 > 60$, (0) otherwise; and (b) $\log_{10} P$-values against y_7.

Figure 3b shows the $\log_{10} P$-value of the F-, ρ-, and τ-test when the observation (7) moves from 0 to 20. The plot shows that the F-test becomes even not significant for $8.7 \le y \le 18.7$, and again significant (but in the wrong direction!) for $y > 18.7$. This is due to the fact that the least-squares line is attracted by one single observation y_7. The ρ_c-test shows a better behavior. It is significant everywhere (at least at the 5% level for $y \ge 8.7$). For $y \ge 8.7$ the robust line (computed using the Huber-estimator) is not affected by y_7 and the values of the ρ_c-test statistic remains constant. Finally, the τ-test shows the best behavior, being strongly significant everywhere.

EXERCISES AND PROBLEMS

Section 7.1

1. Summarize the optimality properties (and the conditions under which these hold) of the F-test.

Section 7.2

2. Describe the relations and differences between the F-test, ρ-test, and τ-test.

Section 7.3

3. (Short) Why is the ρ_c-test only the first step for robustifying the F-test?

4. To Remark 4; show that the unstandardized sensitivity of R_n equals the self-standardized sensitivity of $((T_\Omega)_n)_{(2)}$.

5. (Research) Investigate the properties of the approximations discussed in Remark 5.

6. (Research) Show that the use of a robust test statistic in a permutation test leads to robustness of efficiency for the test.

7. What is the relation between AICR and the test statistic (7.2.4) with $\tau(x, r) = \rho(r)$?

8. (Extensive) Apply AIC and AICR to the data set given in Subsection 7.5d.

Section 7.4

9. Prove Proposition 1.

Section 7.5

10. (Research) Use (7.5.4) to approximate the asymptotic distribution of τ-tests.

11. What are the reasons for using a redescending η-function (like that defining the hyperbolic tangent estimator) to define an estimator or a test for regression?

12. Discuss the results of the numerical example. What happens with the LS estimate and the P-value of the F-test as y_7 increases? What about the ρ- and τ-test?

13. (Extensive) Compute the $\log_{10} P$-value of the F-, ρ-, and τ-test for the data in Figure 1 (Section 6.2) by moving the point P_2 along the vertical axis. Discuss the result.

CHAPTER 8

Complements and Outlook

8.1. THE PROBLEM OF UNSUSPECTED SERIAL CORRELATIONS, OR VIOLATION OF THE INDEPENDENCE ASSUMPTION

8.1a. Empirical Evidence for Semi-systematic Errors

The occurrence of violations of the famous "i.i.d." (independent identically distributed) assumption, even in physical and technical data, is acknowledged by eminent applied statisticians. There are old experiences in this direction in various subject matter sciences. Physicists have too often found that standard errors assigned to absolute physical constants later turned out to be too small; some physicists are used to dismissing the classical standard error of a mean as "merely statistical" [Jeffreys (1939) 1961, p. 301; p. 270]. This mistrust unfortunately seems to be justified to some extent. Thus, Youden (1972) cites a sequence of 15 determinations of the astronomical unit (mean distance earth–sun) from 1895 to 1961, where each new value was outside the experimenter's estimate of spread for the previous one!

It would be too simpleminded to attribute this result altogether to (constant) systematic errors, although such systematic errors also occur. A nice example is the older determinations of the velocity of light during this century, most of which gave about 299, 776 km/sec (± a few km/sec), while a minority gave about 299, 792 km/sec. It turned out (cf., e.g., Ahnert, 1961, p. 136) that the majority (!) of the data was wrong due to a systematic delay of the light caused by one common aspect of the methods, which was absent in the smaller group based on otherwise quite heterogeneous methods.

But very often one does not find a clear-cut separation of measurement series into a few different constants, corresponding to different systematic errors; rather the "systematic errors" seem to vary themselves in an often smooth but rather irregular way. The astronomer Newcomb (1895, p. 103) was aware of the existence of such errors and called them "semi-systematic." Karl Pearson (1902) carried out a large experiment himself [already alluded to in Subsection 1.2b, (iii)] in order to check the common assumption that astronomical observations can be decomposed into a random part and a constant systematic error caused by the observer and called the "personal equation." He found that the "personal equation" was not constant, but changed in rather peculiar ways, even with correlations among different observers. Gosset ("Student", 1927) discusses several series of chemical measurements of the same quantity, obtained with greatest care, which should be ideal examples for i.i.d. observations. He also found slow fluctuations and long-range dependencies and called these errors "semi-constant" errors. Elsewhere [cf. Jeffreys (1939) 1961, p. 298] he remarks that he has not encountered any determination which is not influenced by the date on which it is made. A similar experience with engineering data (C. Daniel, orally) is in the background of page 59 in Daniel and Wood (1971). It may even be suspected that some doubtful "discoveries" in astronomy, such as that of a planet of a nearby sun on the basis of slight wiggles of the latter's position, could have been caused by semi-systematic errors. Another striking example for such nonperiodic fluctuations is the sequence of 2885.5 (sic!) determinations of the velocity of light by Michelson et al. (1935). Two more recent examples (H. Graf, 1983 and orally; Graf et al., 1984) are series of 500 and 600 weight determinations of the same weight by the U.S. National Bureau of Standards, a section of which is given in Freedman et al. (1978, p. 91); one series looks clearly "out of control" and not like i.i.d. data, while the other looks like a marginal case; but even the most homogeneous subset of about 300 measurements shows weak yet highly significant ($P < 1\%$) serial correlations.

Such violations of the assumption of independence even in science and engineering data have consequences for standard errors and related quantities. The growing awareness of the problem among top data analysts in recent years is reflected by such chapter headings as "Hunting Out the Real Uncertainty. How $\sigma/\sqrt{n}$ Can Mislead" in Mosteller and Tukey (1977) or "Random Sampling and the Declaration of Independence" in Box et al. (1978). Already Jeffreys [(1939) 1961, p. 297] noted that in his examples means of 25–30 observations fluctuated like means of only 2–15 independent observations. The strong effects of even mild serial correlations are discussed, for example, in Scheffé (1959, Chapter 10). There remained, however, the question of the true correlation structures in real data or, more realistically, of simple correlation models which capture the most important correlation features of real data.

Important pioneering work on this question was done for agricultural data by H. Fairfield Smith (1938) almost 50 years ago. He noted that in a large uniformity trial the log variances of the average yield over neighboring plots against the log numbers of plots were very close to a falling straight line, but its slope of -0.7 was clearly above -1, the correct value under independence of the yields of neighboring plots. He then studied the corresponding plots for about 40 uniformity trials; almost all of them gave very neat straight lines, whose slopes ranged fairly uniformly between -0.05 and -0.80, with some clustering around -0.4. These findings, which are further discussed in Whittle (1956), strongly suggest power laws of the form $cn^{-\alpha}$, $0 < \alpha < 1$, for the variance of the mean of n identically distributed observations. A somewhat different example, which in retrospect also produces such a law (with beautiful straight lines clearly flatter than the Poisson line in log–log coordinates), is the standard error of estimates of the density of wireworms against the density given in Yates and Finney (1942) and reproduced as "error graph" in Section 7.20 of Yates [(1949) 1981]. (In passing it may be noted that Fisher and Mackenzie, 1922 boldly fit a $\tanh(d/d_0)^{-3/5}$ curve to a few rainfall correlations between points in distance d, which asymptotically would lead to the same kind of law for the variability of the mean rainfall.) The problem is that none of the customary stationary stochastic processes, like ARMA models, yields such a law; it can only be approximated for some transient stretch of time with a large number of parameters, but asymptotically the variance of the mean of all these processes with a "finite memory" decreases like n^{-1}.

8.1b. The Model of Self-Similar Processes for Unsuspected Serial Correlations

Now there are in fact stationary stochastic processes for which the variance of the mean decreases like $n^{-\alpha}$ for any α between 0 and 1 (in fact, any α between 0 and 2), namely the increment processes of so-called self-similar processes. They were invented more than 40 years ago in a rather abstract probabilistic setting by Kolmogorov (1940, 1941) and found early use in turbulence theory and later also in some other areas of physics. D. R. Cox (orally; cf. also Whittle, 1956) derived them as a model for yarn diameter variation in textile industry. It was B. B. Mandelbrot (1965; Mandelbrot and van Ness 1968; Mandelbrot and Wallis 1968, 1969a, 1969b; cf. also his later papers and his 1977 book) who introduced and established self-similar processes in subfields of statistical applications, first in hydrology, and later, more generally, for geophysical records. He was the first to give a satisfactory model for the "Hurst phenomenon" of the floods of the river Nile which had puzzled many probabilists (cf. also Feller, 1951). Briefly, the heights of the yearly floods were correlated ("seven fat years and seven lean years" of the Bible, hence the term "Joseph-effect") in a way which can be modeled by the increments of a self-similar process whose mean converges at rate $n^{-0.2}$ approximately. A related result had been found empirically by Hurst (1951, 1956, Hurst et al., 1965) for the cumulative ranges of the Nile data.

Mandelbrot called the self-similarity parameter or parameter of long-range persistence of correlations H in honor of Hurst, with $H = 1 - \alpha/2$. H can vary between 0 and 1 for stationary increments; $H = \frac{1}{2}$ corresponds to independence, $H > \frac{1}{2}$ to positive correlations, $H < \frac{1}{2}$ to negative correlations. In Mandelbrot and Wallis (1969b), estimates of H are given for about 70 geophysical data series such as river flows, rainfall, earthquake frequencies, tree-ring indices, thickness of geological layers, and also sunspot numbers. The estimates range fairly evenly over the whole range between 0.5 and 1 and are in most cases clearly above 0.5, giving strong evidence that nontrivial self-similar processes (or else more complicated processes) are needed for the description of most geophysical records.

Self-similar processes $Z(t)$ on the real line are characterized by the property that if the time scale is changed by a factor $a > 0$, all finite-dimensional distributions are changed by a factor a^H. From this simple scaling property follow many important results; thus for the variance of the increments $\text{Var}(Z(t) - Z(0)) = t^{2H} \text{Var}(Z(1) - Z(0))$ whence the variance of the mean, and the autocovariance function at lag $s \geq 1$ for the discretized increment process, becomes proportional to $(s - 1)^{2H} - 2s^{2H} + (s + 1)^{2H}$, which behaves asymptotically like s^{2H-2} for $H > \frac{1}{2}$. In some ways Gaussian self-similar processes are as simple and may be expected to become as important as the classical Gauss–Markov processes; both have only one correlation parameter, and while in one case the serial correlations decrease exponentially fast, they decrease hyperbolically in the other case.

Most mathematical properties of self-similar processes are very difficult to work with, however, which seems to be one reason for the late interest in them. A relatively simple way of defining self-similar processes is as "fractional integrals" over white noise; thus, the Gaussian self-similar processes, called "fractional Brownian motion," can be "defined" heuristically with Gaussian white noise $dB(t)$ by

$$Z(t)`` = "\int_{-\infty}^{t} (t - s)^{H-1/2}\, dB(s)/\Gamma(H + 1/2).$$

More precisely,

$$Z(t,\omega) - Z(0,\omega) = \left(\int_{-\infty}^{0} \left((t-s)^{H-1/2} - (-s)^{H-1/2} \right) dB(s,\omega) \right.$$

$$\left. + \int_{0}^{t} (t-s)^{H-1/2} dB(s,\omega) \right) / \Gamma(H + 1/2).$$

$H = \frac{1}{2}$ yields ordinary Brownian motion. The "derivative" or "increment process," called "fractional Gaussian noise," is a Schwartz distribution; however, in practice one will usually work with the increments of the discretized process (restricted to integer time points). For some further properties of self-similar processes, cf. for example, Mandelbrot and van Ness (1968); for pictures of computer simulations of the increment processes, see Mandelbrot and Wallis (1969a), Graf (1983), and Graf et al. (1984); for processes and pictures in higher dimensions, see Mandelbrot (1977). In recent years, there has been an increasing interest by probabilists in self-similar processes, with many mathematical papers, for example, by Dobrushin, by Major, and by Taqqu (cf. the references under these names).

8.1c. Some Consequences of the Model of Self-Similar Processes

One of the dangers of the autocorrelation function of self-similar processes is that although individual correlations may all be very small and hence nonsignificant when tested, their sum tends to infinity, implying infinite memory and nonnegligible correlation even between the distant past and the distant future. On the other hand, such unusual features are just what is needed to describe the empirical findings for semi-systematic errors.

One of Student's (1927) series of 135 supposedly i.i.d. data (nitrogen content of aspartic acid) was analyzed in depth by Hampel and J.-D. Tacier (cf. Hampel, 1979; Hampel et al., 1982; Graf et al., 1984). After the first six data points, the 99%-confidence interval based on the first n observations (assuming independence) does not contain the known true value anymore. Moreover, two 99%-confidence intervals based on about the first 50 and on almost all data do not even overlap! The variance of the means of nonoverlapping subsamples of size k again decreases like a power law different from k^{-1}. A recent estimate for H obtained by H. Graf via the power spectrum yields $H = 0.83 \pm 0.06$, with $\alpha = 0.34$ (cf. Hampel et al., 1982). Again adopting a self-similar process as the simplest possible model consistent with the data, we have to multiply the standard error and the confidence interval for the mean of all 135 observations by a factor of about 4.5 (this

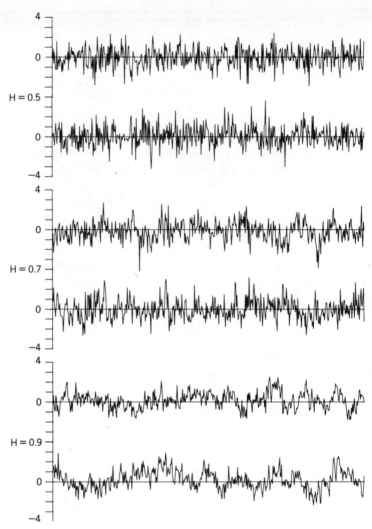

Figure 1. Simulated realizations of standardized fractional Gaussian noise for three different values of the self-similarity parameter H. Each realization is plotted in two lines and consists of 1024 points, which have been connected with straight lines. The case $H = .5$ is Gaussian white noise, while $H > .5$ means long-range dependence.

392

means, the variance of the mean of these 135 data points is about 20 times as large as naively estimated by assuming the n^{-1} rule)! The result is shocking, but it yields one reward: the corrected 99%-confidence interval based on all data contains the true value again. At least in this example, there is therefore no necessity to invoke a systematic error for the observed discrepancy between true and observed mean; this can be explained solely by our model for semi-systematic errors.

The strange behavior of Pearson's series can also be best modeled by self-similar processes. Assuming such a model to be true, the ratios γ^2 of the variance of the means of groups of 25 consecutive observations and of the variance of individual observations given in Jeffreys [(1939) 1961, p. 315, cf. above] allow simple and crude estimates for H for the six series by equating γ^2 with $25^{-\alpha}$; these estimates range fairly evenly with increasing γ^2 from $H = 0.58$ via 0.63, 0.64, 0.69, 0.84 to 0.91. For the two NBS weight series, Graf obtained by more sophisticated methods $H = 0.54$ (barely not significantly different from 0.5, with a P-value of 6%) and $H = 0.79$. Thus even for such high-quality series from physics, astronomy, and chemistry, long-range correlations seem to be the rule, with H anywhere between 0.5 and 1.

While for geophysical (including meteorological) and agricultural data, correlations were to be expected and only the form of the correlation function came as a surprise, the widespread occurrence of the same type of correlations among supposedly independent data under the most ideal conditions of exact sciences must sound revolutionary to many. It is contradicted by all elementary and almost all advanced textbooks of statistics which claim that the standard error of the mean in such situations goes to zero like $n^{-1/2}$. However, it is only the quantification of the widespread qualitative experience of top statisticians and scientists cited above—including the skeptical physicists who do not believe in statistics. In fact, if statisticians insist on such purely mathematical deductions from unwarranted assumptions as the $n^{-1/2}$ rule for the standard error, without regard to real data, they only disqualify themselves in the eyes of any sufficiently reasonable and experienced scientist. It is surprising how much blindness still exists against facts of which Jeffreys [(1939) 1961, pp. 297/298] wrote already many decades ago, in using the term "internal correlation" for the correlation structure of semi-systematic errors: "Internal correlation habitually produces such large departures from the usual rule that the standard error of the mean is $n^{-1/2}$ times that of one observation that the rule should never be definitely adopted until it has been checked. In a series of observations made by the same observer, and arranged in order of time, internal correlation is the normal thing, and at the present state of knowledge hardly needs a significance test any longer. It practically reduces to a problem of estimation. The question of significance arises only when special measures have been taken to eliminate the correlation and we want to know whether they have been successful."

Of course these facts have to be seen in the right perspective; they do not imply that *all* results derived under the independence assumption are nonsensical or wrong. For one thing, in special situations there appear to exist even very long sequences indistinguishable from sequences of independent data. An example are some 20 million dice throws done with special precautions by Longcor (see Iversen et al., 1971); even runs of odd or even of length 20 or more occurred with frequencies roughly expected under independence. Another aspect is that the relative (as opposed to the absolute) qualities of point estimates (at least those based on the empirical cumulative distribution function, thus disregarding the time order of the data) seem to be not too much affected by long-term serial correlations, at least for series which are not too long. (For the limiting case of very long series, J. Beran recently obtained quite different results; cf. Beran and Künsch, 1985). This makes it possible to discuss robust point estimation in a relevant way while disregarding the correlations. All confidence intervals and standard errors have to be adjusted, of course—probably roughly by the same factor. But a further aspect is that this effect (as far as long-term correlations are involved) is usually moderately small for small sample sizes; to compute some numbers: the adjustment factor is, in the simplest (though here perhaps inadequate) model, $n^{H-1/2}$. For $n = 5$, this is below 1.2, 1.5, and 2 for H up to 0.61, 0.75, and 0.93, respectively. For $n = 20$, the corresponding values of H are 0.56, 0.64, and 0.72; an H of 0.9 leads already to a factor of 3.3 for the confidence interval or 11 for the variance of the mean. There is of course the difficult problem of estimating H from small samples, so it is only somewhat fortunate that the effects are not as big yet. For $n = 100$, even $H = 0.6$ leads to a factor 1.6 for the standard error, and $H = 0.9$ to a factor 6.3; hence for supposedly independent data series of similar or greater length, a study of the correlation structure and an estimate of H should always be made. (For short series, the short-term correlations are relatively more important, and at least the first few correlations should be individually estimated.) It may be surmised that the observation that the chi-square test virtually always rejects the null hypothesis for sufficiently large samples (Berkson, 1938) is not only due to the null hypothesis being almost never exactly true, but also due to the use of a wrong critical region.

The main reason that statistics has been successfully applied in many fields despite these semi-systematic errors, seems, however, that one is very often not interested in absolute constants, but rather in contrasts. In contrasts, especially if determined in a split-plot fashion, the semi-systematic errors will largely cancel out. Another way of treating them in the determination of contrasts is randomization, whereby they become a part of the proper (independent) random error. The fundamental distinction between contrasts and absolute constants is a basic part of knowledge of experienced applied statisticians (cf., e.g., remarks by Box, by C. Daniel, and by others), though it is not reflected at all in the usual formalism of mathematical statistics.

8.1d. Estimation of the Long-Range Intensity of Serial Correlations

There appear to be at present four or five main approaches towards the estimation of the parameter H of a self-similar process: via the correlogram, the power spectrum, the variance–time function, the R/S (rescaled cumulative range) method by Mandelbrot, and variants of maximum likelihood. The correlogram is not recommended because of its many pitfalls (cf., e.g., Mandelbrot, 1972), such as artifacts caused by the correlations of the correlations; in the present context it is particularly unsuitable because it is not the small individual correlations which matter but their large sum. The variance–time function considers the variance of the means of nonoverlapping subsamples as function of the subsample size. It needs an iterative adjustment of the variance of the grand mean, but for approximately normal data it gives a rather direct picture of what is happening (cf. the graphs in Hampel et al., 1982 or Graf et al., 1984). The robustification of this method with regard to outliers apparently has not yet been studied. A closely related approach is to analyze the components of variance of a hierarchical splitup of the data sequence by means of ANOVA methods.

The method favored and used extensively by Mandelbrot is R/S-analysis, where R/S is the range of the cumulative process divided by the standard deviation of the increment process for some time span $(s, s + t)$. This quantity is inspired from standard techniques in hydrology, where R is the ideal capacity of a reservoir. The plot of log R/S against log t for many different time spans clusters around a straight line with slope H (in first approximation). This method could be shown to be extremely robust against changes of the marginal distribution; even for very long-tailed and skew distributions it gives good estimates of H (cf., e.g., Mandelbrot and Wallis, 1969c, or Mandelbrot and Taqqu, 1979). On the other hand, there are indications that for near-normal data this method is less efficient than others and that the relative efficiency even decreases with the sample size, suggesting a slower convergence rate (cf., e.g., Graf, 1983).

The power spectrum of a self-similar process has the property that in first approximation the regression of log power spectrum on log frequency gives a straight line with slope $1 - 2H$ (with approximately i.i.d. errors following roughly the double-exponential extreme-value distribution). This approach should lead to fairly robust methods which are in general more efficient than R/S. A crude version of this estimation approach has been implemented by D. L. Mohr (1981); a more refined version, taking care of various distributional approximations, robustifications, and bias corrections, is given by H. Graf (1983) [cf. also Graf et al. (1984)]. There are indications (cf. J. Beran, 1984) that his nonrobustified estimator is asymptotically as efficient as maximum likelihood under a normal self-similar process. Graf's robustified estimators, while of course losing some efficiency under this strict parametric model, allow, for example, downweighting of spikes and of high-frequency noise in the spectrum. Also, maximum likelihood estimation has been tried (cf. McLeod and Hipel, 1978) which soon becomes rather cumbersome with increasing length of the sequence and may possibly be quite nonrobust, although for Gaussian processes this and the spectrum-based methods are probably most efficient. But there is an approximate maximum likelihood estimator (Fox and Taqqu, 1984; J. Beran, 1984; cf. also Künsch, 1981) which is fairly simple also for large n, can be robustified, is flexible in incorporating additional short-term correlations, and for which there exists already an asymptotic theory. It certainly will be worthwhile to work out general computer programs for it and its robustified versions. Otherwise, for general purposes the refined power-spectrum methods seem to be best at present, except for extremely long-tailed marginal distributions (e.g., without first moments) where R/S might be the safest choice.

8.1e. Some Further Problems of Robustness against Serial Correlations

So far, we have modeled only the most salient features of observed semi-systematic errors by means of self-similar processes. The true correlation functions are unknown and will remain so. It is, however, to be expected that the simple model of self-similar processes has to be refined somewhat for a more accurate analysis. Most likely seems to be a combination of a self-similar process with a short-memory ARMA process with just a few parameters, to account for special behavior of the first few lags. Another question is whether self-similar processes are always sufficient to describe the long-range correlations, or whether and when other refinements become necessary there, too. We might also consider non-Gaussian distributions as basic parametric models and study various estimators under the corresponding non-Gaussian self-similar processes.

The problem of unsuspected serial correlations in supposedly i.i.d. data is just a special case of the problem of deviations from an assumed correlation structure. For example, it could be that one entertains an AR model with three correlation parameters, but that the true correlation structure is "slightly different." This robustness problem for time series is to be distinguished sharply from the hitherto treated robustness problems in time series, namely those caused by outliers and other changes in the marginal distributions (cf. Section 8.3). On a still higher level would be the combination of both types of robustness problems.

Another, still more general statistical problem is the case of higher-dimensional sets of observations, such as yields of plots in agriculture, or data with some other neighborship structure, as along a tree or network, when it can be surmised that neighboring data are more similar than distant ones. The question is again how the correlations decay in such a structure, and by what statistical methods they can be treated. There is already quite some probability theoretic work on higher-dimensional self-similar processes, but very little seems to be known about the statistical problems involved.

8.2. SOME FREQUENT MISUNDERSTANDINGS ABOUT ROBUST STATISTICS

8.2a. Some Common Objections against Huber's Minimax Approach

It will not be attempted to repeat all the discussions of misunderstandings about robust statistics which the interested reader can find, for example, in Hampel (1973a, 1978a, 1980). There are also already many such discussions earlier in this book, notably in Sections 1.1, 1.2, and in some parts of 1.3, especially in Subsections 1.2e and 1.3f. Specific discussions of the rejection of outliers can be found in Section 1.4, as well as in Hampel (1985). The present section will concentrate on four major topics of discussion in some depth.

We shall first discuss specifically a number of objections against Huber's 1964 paper, some of which are just partly justified and some of which consist of rather basic though widespread misunderstandings. This subsection can be read after Subsection 1.3b.

There are several reasons for starting with a discussion of this paper by Huber. The theory contained in it was the first general theory of robust statistics, became historically prominent, and to some extent has set the stage for the later theories. Many objections against it—whether justified or not—can be and are equally well held against the theory in this book. And at least some statisticians seem to have had merely a superficial glance at Huber's 1964 paper (or nowadays perhaps at his 1981 book), get distorted impressions and misconceptions from it, and believe they now can judge robust statistics.

1. "The asymptotic variance is not the actual variance." True; but the asymptotic variance (together with asymptotic normality) yields the only *simple* description of the distribution of the estimate; *and* in the cases considered it is an excellent description down to $n = 20$ and good even down to $n = 10$, as is known from Monte Carlo studies, for example, the large one in Andrews et al. (1972). [For $n < 10$, one needs more precise "asymptotic" methods of describing a distribution, such as those discussed in the survey article by Field and Hampel, 1982, which yield highly accurate approximations even in the tails down to about $n = 3$. Besides, there are arguments going back at least to Fisher (1922, p. 350), a strong proponent of the variance, that in general it is not the variance which should be considered as main criterion for finite samples, but rather the intrinsic accuracy or information, which reduces to the ("in-")variance in the cases of normality and asymptotic normality.]

2. "The scale parameter has to be known." True; but it can (and should) be estimated simultaneously. However, it should be noted that scale estimation changes the neighborhoods, since distributions with the same estimated scale, not with the same background scale of the normal part, are grouped together. On the other hand, it is no disadvantage that scale has to be estimated simultaneously, unless one is satisfied with a pure location estimator in the spirit of nonparametric statistics, which describes only one aspect of the data and not (at least approximately) the full data set. Contrariwise, scale estimation also allows rejection of outliers in dependence of their distance from the majority of the data. This cannot be achieved by L-estimators (linear functions of the order statistics) or R-estimators (estimators derived from rank tests) (cf. also Section 1.4 and Subsection 8.2c).

3. "The estimate is only defined by an implicit equation and has to be computed iteratively." In the age of the computer, this should be no argument anymore, for at least three complementary reasons: a simple, lucid implicit definition (like the one for M-estimators) is preferable to a complicated, inscrutable explicit definition and leads to a simple, elegant theory; implicit equations pose no problem for the computer; and for "hand calculations," there are very simple one-step versions (cf. Andrews et al., 1972) which are asymptotically equivalent and for finite samples virtually equivalent to the fully iterated estimators. Besides, general maximum

likelihood estimators always were defined only implicitly, which did not prevent them from becoming accepted even in precomputer days.

4. "The fraction of gross errors ε has to be known." There are several counterarguments: in moderately large to large samples, one can try to estimate ε from the data (e.g., as in Huber's "Proposal 3"—see Huber, 1964); an experienced data analyst often has some idea about the quality of his or her data; and most importantly, we are lucky numerically since within a wide (and most important) range the choice of ε hardly matters. Thus, according to Huber's (1964) Table I, for all choices of k between 1 and 1.8 for the Huber-estimator (corresponding to ε between 14 and 1.6%) and all true ε between 1 and 10%, the asymptotic variance lies at most 7% above the minimax variance.

5. "Minimax solutions are too pessimistic." In general this may be true, but it is not so in our case. Neither are the least favorable distributions unrealistic, nor are the efficiency losses at the ideal model a reason for concern. The least favorable distributions are rather short-tailed with all moments existing, and for small ε they are very similar to the normal; Huber (1981, p. 91) showed that for ε around a few percent, they are quite good examples for what the distribution of large samples of high-quality data can look like. They are thus shorter-tailed than distributions with distant gross errors, showing again that distant outliers are easier to handle than "doubtful outliers" nearby and contamination in the flanks of the distribution. And again we are lucky with the quantitative aspects: the increase in variance over that of the arithmetic mean under the strict normal model is 11% for the Huber-estimator with $k = 1.0$, 5% for $k = 1.4$, and 2% for $k = 1.8$. If one compares these values with the potential increase of variance of $\bar{X}$ in a small neighborhood of the normal, which is unbounded and which even after rejection of outliers can easily be around 20% (cf. Section 1.4), the choice should be clear.—It should be noted that we are not always so lucky with the quantitative aspects: in robust regression with "leverage points" (highly informative and at the same time highly danger-ous points, cf. Chapter 6), the efficiency losses of decently robust estimators are considerable.

6. "The gross-error model contains too many unrealistic distributions. In particular, for every positive ε it contains distributions without any moments, but in reality all distributions have all moments since the range of their values is always bounded." It does no harm to have too many distributions included as long as the relevant distributions, namely the least favorable ones, are not too unrealistic—and this we have just seen. But the argument about the existence of moments is a rather deep-rooted confusion

between mathematics as a simplified model of reality trying to capture relevant features of it, and mathematics taken literally. Already Tukey, a long time ago, has tried to explain the difference by means of a Cauchy curtailed at $\pm 10^{10}$, thus possessing all moments, and a Normal with 10^{-10} of a Cauchy mixed in, thus possessing no moments. The relevant statistical features for not extremely large sample sizes are still those of a Cauchy and Normal respectively, and not vice versa.

7. "A neighborhood based on the first few moments (say, up to the fourth, or even only the second) is also rather "full," rather simple and at least as reasonable as the gross-error model." Such models are unrealistic, since moments are a straitjacket. A single outlier can make all moments arbitrarily large. If it is objected that all observations are bounded, then "arbitrarily" still has to be replaced by "very." If the bound on the data is so low that the moments cannot be changed by much, then there is hardly any robustness problem; but then there is hardly any statistical problem at all, since the data are known quite well a priori. It may still be argued that moments can be relevant under three circumstances: conditionally, given that there are no outliers, or after rejection of outliers, or if one considers only those underlying distributions which would not be rejected by an appropriate goodness-of-fit test applied to the data. The first situation leaves open what to do in the case of outliers and thus does not lead to fully specified statistical procedures; the second situation looks only at the second part of a statistical procedure and not at it as a whole; only the neighborhood of "distributions indistinguishable from the true underlying one" (or: "compatible with the empirical one") merits some theoretical interest, but in almost all cases, as $n \to \infty$, it will eventually not contain the strict parametric model anymore (contrary to a number of so-called robustness theories based on "shrinking neighborhoods"), hence it is not clear how to treat it. Perhaps Huber's (1964; cf. also Huber, 1974b) idea of considering the least informative distribution in such a neighborhood can lead further, but at any rate it would lead to much more complex procedures which are only marginally better than the simple ones already existing (cf. also Subsection 8.2d).

8. "The gross-error model is too narrow since it does not contain all realistic alternatives to the model." This is true: neither gaps and troughs in the ideal model distribution nor rounding errors and the like are covered. In a heuristic sense, the gross-error model is only "half" of the "total-variation neighborhood" (cf. Hampel, 1968), and we need a neighborhood in the weak topology (cf. Section 2.2). However, this weakness leads to an interesting discussion of the role of mathematical models. On the one hand, the

gross-error model does capture the most important deviations from the parametric model, and its "solution," the Huber-estimators, are also very good in the larger neighborhoods as they are little affected (for k not too small) by the additional types of deviations. On the other hand, Huber (1964, 1981) has also treated a realistic neighborhood, namely that based on the Kolmogorov distance, and the solution turns out to be rather similar, only more complicated. For these reasons, and since the exact quantification of the neighborhood is somewhat arbitrary, we prefer the simpler solution. (It also turns out that the simpler solution is also the solution for a simple *and* realistic model, namely the one based on Taylor expansions around the parametric model; cf. Subsection 1.3d.)

9. "The minimax approach cannot be generalized to arbitrary parametric models." Unfortunately, this seems to be a real and important limitation. There appears to be no simple general procedure to derive the least favorable distribution, and some of Huber's (1981) solutions require already considerable ingenuity. On the other hand, the minimax approach *can* be generalized to the most important and most frequently needed parametric models, including the linear model with normal errors and the covariance matrices of multivariate normal distributions.

10. "The minimax approach works only for symmetric distributions." This is usually to be interpreted that both the model distribution and the contamination have to be symmetric. It is one of the worst prejudices against robust statistics, because obviously the assumption of exact symmetry is as unrealistic as the assumption of an exact parametric model. The prejudice has been reinforced by the Princeton Monte Carlo study published in Andrews et al. (1972) which made many statisticians believe that robust statistics is "the statistics of symmetric long-tailed distributions." An answer can be given on different levels.

Most superficially, it can be said that not a single theorem in Huber (1964) needs the assumption of symmetry or contains the word "symmetry," and in fact Huber also solves the scale problem where neither the model distribution (of the log absolute values) nor the contamination is symmetric; moreover, Andrews et al. (1972) contains also Monte Carlo results for asymmetric distributions, which were included upon insistence by Huber and the present writer. To this can be objected that Huber's (1964) Theorem 1 contains a more general but equally stringent condition; that its corollary (with the most important application) does in fact assume symmetry; that in the scale case Huber seems to waver before settling on the choice of the estimator which is Fisher-consistent at the parametric model; and finally, that Andrews et al. (1972) does not contain an analysis of the asymmetric situations (which came about because the computer outputs arrived in Zurich too late).

It seems, however, often to be overlooked that already Huber (1964) contains quite a remarkable discussion of the bias caused by asymmetric contamination, including its increasing importance with increasing sample size and the role of the median as the best choice for very large sample sizes (which some folklore in applied statistics also seems to have arrived at on intuitive grounds and which is contrary to many "adaptive" and "shrinking neighborhoods" theories). Huber uses an approximation to get bounds for the bias and treats it separately from his other theory; but the asymptotic variance is also only an approximation for finite sample sizes, and about all Huber did not do was to combine variance and bias to an ("asymptotic," but sample-size-dependent) mean-squared error. Also the asymmetric situations in Andrews et al. (1972) can be analyzed in this way, and while providing some interesting additional aspects and broader information, their results fit in excellently with the results for the symmetric situations. (Cf. also the discussions of bias and asymmetry in Huber, 1981).

While in Huber's first approach, the minimax approach, symmetry still plays a certain role, though mainly for mathematical convenience, it seems to be often overlooked that in two other major theories of robustness, namely Huber's (1965, 1968) second approach via robust testing and confidence intervals, and in Hampel's (1968) approach, there is not the least need for or mention of symmetry. It is amazing how a superficial understanding of just one out of three theories (and not even the most general one), and perhaps even more just the existence of a big but incompletely analyzed Monte Carlo study, could lead to such a deep-rooted prejudice.

Probably there is another, perhaps often subconscious reason. Classical estimation theory is based on the mathematical fiction of a uniquely defined estimand, even though for every finite n only some neighborhood of it is identifiable. This paradigm has to be abandoned as soon as an unknown though limited distortion of the parametric model is allowed for, since for every n this distortion can at most be partly identified, leaving an unknown bias which is bounded but in general nonvanishing for $n \to \infty$, as is already shown in Huber (1964). Reluctance to accept this unavoidable fact has resulted in a number of attempts to make the bias disappear by artificial tricks, but it is time to go beyond the debilitating relativism of many mathematical statisticians who are always willing to sacrifice realism of their models for some spurious mathematical simplicity. (For a more detailed discussion, see Subsection 8.2d.)

To summarize the discussion of various objections to Huber's first approach: In the conflict between a mathematically elegant and a practically relevant theory, this approach is a very good compromise, treating the most important robustness aspects in a manageable way.

8.2b.   "Robust Statistics Is Not Necessary, Because..."

We shall not discuss that well-known peculiar humor of some statisticians, who with regard to modern robustness research say something like: "(i) We do not need robust methods at all. (ii) We have always used robust methods. (iii) We have done it much better."

More serious is that widespread pseudologic which says: "If you don't believe that your model (e.g., of normal errors) is correct, choose another one and use maximum likelihood—or Bayesian—methods for the new model." What, if I don't believe in the new model either? It takes a lot of stubbornness to flood the world with a host of rather arbitrary and probably hardly interpretable models and claim they are exactly true. The point of robust statistics is that one may keep a parametric model *although the latter is known to be wrong.* All that is needed is that it is simple and reasonable under the circumstances of the subject matter science, and that it still gives a useful *approximation* for the data (in the sense of: at least for the majority of the data). The robust methods are then designed and known to work well also in those cases where the model is only approximately valid. This property is what is tacitly hoped for when classical methods are used, and it is even more tacitly hoped for (though perhaps for narrower neighborhoods) by those who propose to drop the old models and replace them by collections of new ones.

It might also be noted that the advice regarding change of the model instead of using "robust" methods on p. 344 in the second edition of the well-known book by Draper and Smith (1981) is obviously purely academic, since virtually the whole book is about normal theory methods.

A frequent objection to robust statistics as a whole is that "it is not necessary, since statisticians have always been getting along without it." To take a specific example, there is no need for the median since statisticians have for centuries used the "good old" arithmetic mean and were happy (?) with it. On the surface, the argument is debatable but unrealistic, since if we would follow it, we would still live in the Stone Age, carving and using Stone Age tools, since we have done it successfully for untold millenaries. Or, to consider a more contemporary situation, we would not use high-speed computers at all since we could always do our calculations with a desk calculator or even by hand.

Looked at more closely, the argument is plainly wrong. Good statisticians have always used the median along with the mean; they have switched to the median in view of long-tailed data; they have rejected outliers and considered them separately. They have used and do use at least informal robust methods. It is true that they had no theory and hence no deeper formal understanding of what they were doing, and the consequences can still be seen in the clumsy treatment of formal rejection of outliers in the recent literature (cf. Section 1.4). But the informal intuition of many statisticians was already excellent and included robustness considerations as a matter of fact.

Thus it can only be claimed that the *theory* of robustness is new. Is it necessary? Of course not always. With fairly simple data sets, one can do a very good job without using the additional insight and technology provided by robustness theory. Even with complicated data sets, it is possible to do a fair job with a lot of effort and quite a bit of ingenuity. But already in rather simple situations, robustness theory can at least help by improving understanding, ease, and efficiency. And to a large extent, it is also due to the computer revolution, with the large and highly structured masses of data processed nowadays, that formal robust methods and good understanding of their theoretical properties become more and more necessary, not only convenient to have. It is in multiple regression, high-dimensional covariance matrices, and the like, not in simple location estimates, that formal robust methods become indispensible, at least as a control and, if necessary, corrective for classical methods.

This does not imply that classical methods, like least squares, should not be used anymore. As far as robust substitutes have been developed, it is not necessary to use the former anymore, and they should certainly never be used alone without at least informal outlier checks and the like; but whether one wants to use them in addition along with robust methods, is more a matter of taste.

Thus, it would be foolish to demand that the arithmetic mean should now always be replaced by the median. On the other hand, if one wanted to set up an artificial alternative only between mean and median (e.g., for a very short beginners' book), one would have to weigh between the reliability and rather constant moderate inefficiency of the median and the mathematical simplicity and greatly variable and unreliable behavior of the mean which makes special measures like rejection of outliers not only desirable, but even necessary. It has to be repeated again (because it is so often forgotten or misunderstood) that the mean combined with some reasonable form of rejection of outliers is quite often quite good, and that only the plain mean as done by the computer is highly risky. Those who still believe that the risk is low because outliers were rather rare, should read the corresponding discussions and examples given and cited in this book; probably they, too, have already unknowingly overlooked outliers or deviating substructures, even if they belong to that group of statisticians which deals only with high-quality data and have no understanding for their less fortunate colleagues.

Clearly, classical methods often have a great simplicity and esthetic elegance which is appealing. But many robust methods are not much less elegant, and there are two properties which are often claimed for classical parametric methods but which they do not possess: namely that they are "exact" and "optimal." They are exact and optimal only under strict assumptions which are virtually never fulfilled, and then their alledged properties hold only as approximately as, or rather even "less" approximately than those of robust methods, and their efficiency losses, usually unnoticed, are frequently in the region between 5 and 50%. It is strange that often the same persons who insist on the usual classical methods also denounce efficiency losses of, say, 5% as unacceptable. This attitude is self-contradictory and untenable. Either one prefers classical methods for reasons of simplicity and accepts the resulting avoidable efficiency losses, or one uses the best robust methodology available.

Some other claims that robust statistics be unnecessary stem from misinterpreted mathematical theorems. Thus, it is true that the arithmetic mean is the best estimate of the true expectation of the underlying distribution in a nonparametric sense, when no assumptions (except a moment condition) are made about the shape of the distribution, but who is ever going to be interested in the expectation of the actual underlying distribution (distorted by gross errors, rounding, etc.), except in a truly nonparametric situation? The Gauss–Markov theorem tells us that for linear models the method of least squares gives us the best linear unbiased estimators (BLUEs), but as Fisher has already stressed, *all* linear estimators are good at most in a very small neighborhood of the normal error distribution, and often they are very bad. Rejection of outliers is nonlinear, and so is any decent robust estimator. While it is probably fair to say that most work on BLUEs has been and is being done mainly because of mathematical convenience, the results may still be of broader usefulness, namely as starting points for robustification.

Many statisticians are proud of the so-called robustness of the t-test and, more generally, of the tests in fixed-effects models in the analysis of variance. But this robustness is only a rather moderate and limited robustness of the level ("robustness of validity"); the power ("robustness of efficiency") and hence also the length of confidence intervals and size of standard errors is very nonrobust. Consequently, a significant result can be believed, but nonsignificance may just be due to the inefficiency of least squares. The robustness of validity stems from the use of an asymptotically nonparametric (i.e., valid) measure of variability (of a nonrobust statistic, though), contrary to the tests for variances and in random-effects models in the analysis of variance which make strong use of the fourth moment of the normal distribution and hence cannot even keep the level approximately. Nonparametric tests keep the level exactly; but it is perhaps more a fortunate accident that most rank tests are also stable with regard to their power, while customary nonrobust randomization tests are again badly unsafe with regard to their power.

To summarize, informal robust statistics was always necessary, and formal robust statistics becomes necessary as soon as standards are high or the capacity for informal data analysis is surpassed; this latter situation is the rule in the computer age.

*8.2c. Some Details on Redescending Estimators

There is hardly any group of robust estimators about which opinions are so strongly divided among "robust" statisticians as they are about redescending M-estimators (cf. Section 2.6). Some statisticians enthusiastically advocate them and use them, sometimes even carelessly, on every occasion; others try to downplay them and their qualities, and emphasize their dangers. Since the reasons for this unbalance of judgment are partly historical, and since the thoughts behind the construction of the first fully redescending M-estimators were never put into print before, it may be suitable to discuss the historical background of these estimators in some detail (cf. also Subsection 1.3d).

Estimators with nonmonotone continuous influence functions were discussed by such statisticians as Newcomb (1886) and Jeffreys (1939); furthermore, many maximum likelihood estimators have this property, and for all hard rejection rules the IF even jumps back to zero. Nevertheless, when during the Princeton Monte Carlo study (see Andrews et al., 1972 for the estimators to follow) the first (two-part) redescending M-estimator (HMD) was proposed, with continuous IF becoming exactly zero outside an interval, "in order to do with M-estimators what every decent data analyst does informally" (cf. also Section 1.4), hardly any participant in the study believed it would work at all. There was some fear that something terrible might happen. The programmer even changed a parameter value in the definition (from 6 to 5.5) to give it, in his opinion, a slightly better chance at least to converge. According to the first computer outputs, HMD did in fact work, though it was strangely and mysteriously similar to the mediocre median. In order to play it safe, a flat piece was then added to the influence function, and a range of parameter values tried, defining the three-part redescending M-estimators from 25A to ADA. The next round of computer outputs happened to contain only rather long-tailed distributions, and in each case the whole new group came out in front of all other estimators included till then (which were ordered in the outputs by their variances). This caused an overly strong psychological impact, which later provoked some counterreactions. The mystery about HMD was partly resolved when the first stylized sensitivity curves (a version of empirical influence functions; cf. Subsection 2.1e) were drawn by the computer: the strange behavior was due to a programming error, causing twice two jumps in the IF, and the corrected version of HMD behaved about as it should have, though it never became particularly popular.

Now what are the main dangers of redescending M-estimators? First, that too many data are on the rejecting (zero) part of the IF at the start or sometime during the iterations, causing convergence (if any) to the wrong local solutions of the M-estimator equation; second, that too many data are on the redescending part, causing problems with the iterations and the asymptotic variance. In fact, iterations and variance may explode when the denominator $\int \psi' \, dF$ of the IF becomes zero. It is therefore important to keep $\int \psi'(x - T) \, dF_n(x)$ bounded away from zero also during the iterations for T. A large part of the problem is merely one of the (e.g., Newton–Raphson) algorithm, but it can occur for most underlying distributions. Even choosing a safer algorithm, however, does not help for those specific distributions where the asymptotic variance becomes infinite. The danger is considerably increased in regression, because the effect of the negative slope is multiplied by the influence of position. For example, in simple (Huber-type) robust regression through the origin, a single distant leverage point can cause rather peculiar effects (such as repulsion of the fitted line away from the data) with redescending estimators.

In the location case, the following precautions were taken for the HMD to ADA group: the starting value for the iterations was always the highly reliable median, and the scale was kept fixed at the equally reliable median deviation. The first corner point was always at least one median deviation away from the median, thus ensuring that at the start at least half of the observations were on the ascending part. With the exception of the experimental 22A, all descending slopes were much flatter than the ascending slopes (mostly about $\frac{1}{2}$ to $\frac{1}{4}$ as steep), so it would need a strong preponderance of data on the negative slopes to cause trouble; and the definition of 22A at least ensured that more than half of the data were on the ascending slope when T was in a small neighborhood of the median. With the exception of HMD, all estimators had a "safety zone" of constant influence between the ascending and the descending part. There were still some finer points involved in the construction of the estimators, but the preceding remarks should suffice to show that the construction was not done thoughtlessly.

Subject to the safety constraints given, the parameters were chosen to give good to decent efficiency at and near the normal and decent behavior for long-tailed distributions. The straight-line pieces were chosen just for simplicity, though later they appeared to possess also a (not very important) optimality property of their own (cf. Subsection 1.3d). In retrospect, HMD descended towards zero too early and too steeply to have good efficiency near the normal. It seems easy to underestimate the efficiency loss due to redescending parts of the IF, compared with a monotone IF. The three-part descending estimators have their (mildly) weak spot at their second corner, as became clear rather soon: A relatively large increase of the numerator of the asymptotic variance, linked with a relatively strong decrease of the denominator, could be caused by additional probability mass at that corner. Smoothing the corner led to the discovery of the hyperbolic tangent (tanh) estimators in 1972 (cf. Subsections 1.3d and 2.6). The sine estimator (Example 2 in Subsection 2.6a) in Andrews et al. (1972) smoothed in effect both corners of the three-part descending estimators; this was a deterioration for the first corner and the central part of the IF (which ceased to be an optimal compromise between robustness and efficiency), but it was an improvement for the second corner. Hence the sine estimator should tend to be relatively worse for shorter-tailed distributions and relatively better for longer-tailed distributions (with more mass near the second corner); this delicate effect actually showed up in the Princeton Monte Carlo study and was also noticed empirically by Tukey in Andrews et al. (1972, p. 155).

The main motivation for Andrews's proposal of the sine estimator was Jaeckel's observation that with it the solution of the M-estimator equation can frequently be found in a single step. Apart from this computational aspect, the sine estimator does not seem to have any specific optimality property. On the contrary, if it is combined with the median deviation, as proposed by Andrews, there is, with sufficiently large samples, always the (theoretical) possibility that $\Sigma\psi'(x_i - T) < 0$ in the first iteration step, although this is usually unlikely in practice. (Cf. also the empirical findings of Rey, 1976, about the dangers of careless use of redescending estimators.)

Tukey's biweight (Example 3 in Subsection 2.6a) may be considered an improvement over the sine function, as its steepest descent is only $\frac{4}{5}$ of its steepest ascent; hence $\Sigma\psi' > 0$ in the first iteration step as long as its parameter r (the rejection point) is greater than about 5.4 median deviations when scaling by MAD is used. Its greatest virtue, apart from its nice smoothness, is probably its very elementary and simple analytical form. It does not seem to be distinguished by any optimality property either, and its steepest descending slope may still be considered dangerously steep. See also Table 3 in Subsection 2.6c for a numerical comparison of various redescending estimators.

Actually, also some tanh estimators descend too steeply. If the rejection point approaches the first corner, the estimator approaches a hard rejection rule (this holds correspondingly also for other redescending M-estimators); the properties and dangers of hard rejection rules were discussed at length in Section 1.4. If a tanh estimator descends too steeply or if, in other words, there should be a tighter bound on the change-of-bias sensitivity in addition to the bound on the change-of-variance sensitivity, it seems advisable to replace the last part of the tanh by the tangent at the last tanh point used (cf. Hampel, 1974, p. 393). Perhaps one can also investigate some simple approximations to these combinations of tanh estimators and two-part descending estimators.

There are many more aspects of redescending estimators which have been discussed elsewhere. A detailed study of convergence of the iterations is contained in Clarke (1984a). The breakdown aspects are analyzed in Huber (1984). The gains of good redescenders over Huber-estimators in the presence of distant outliers are often about 5–20% (cf. Table 1 in Subsection 1.4b and Table 3 in Subsection 2.6c; cf. also Hampel, 1983a, p. 227f., and its discussion of Huber, 1981, p. 103), while the losses of the tanh estimators against Huber-estimators in Huber's minimax setup are typically only one or a few percent; the losses of the 25A–12A group as in Huber's (1981, p. 145) Exhibit 6.6.2 are somewhat higher, but they occur only for highly specific contamination and are not representative of the full potential of redescending M-estimators.

It may also be pointed out again that L- and R-estimators are not able to reject outliers properly, that is, depending on their distance from the bulk of the data (R-estimators either "reject" always or never, in the sense of giving zero influence, and "rejection" with L-estimators is rather delicate). An empirical and theoretical discussion of what can and cannot be achieved by R-estimators with redescending IF is to be found in Hampel (1983a).

*8.2d.   What Can Actually Be Estimated?

The question raised in this subsection has already been discussed briefly at the end of item 10 in Subsection 8.2a, but as it seems to be one of the most difficult philosophical questions in the background of robust statistics, and as the fact (some would say claim) of the "unavoidable bias" of robust estimators still met with difficulties of understanding, for example, at the

1984 Oberwolfach meeting on robust statistics, a more detailed discussion should be attempted. The problem can be viewed from different angles, and we shall try to discuss several aspects. However, we shall not discuss nonparametric estimation.

Our basic ingredients in parametric and in robust estimation are data, parametric models, and the parameters of the models. What do we mean by that?

The parametric models, together with the parameters (which are often merely simple and convenient but otherwise arbitrary identifiers of the model distributions), are concise and complete descriptions of stochastic mechanisms or laws. Thus, ordinarily they contain information about arbitrarily large data sets, which is more than the information contained in any fixed finite data set. Hence it is not possible in general to deduce a model, let alone its parameters, exactly from the data alone. It may be possible to deduce a model from a physical or similar law, but, with few exceptions perhaps, such models are only approximations to reality and are only conceived as such. Even given an exact model, its parameters are in general not exactly determined by any finite data set, but only up to the fuzziness described by a standard error, confidence interval, or the like.

We thus have the fuzziness of the model superimposed on the fuzziness of the parameter, for any fixed finite data set. The meaning and also the value of the parameter will change with the model, even though one can sometimes identify some restricted subclasses of models, such as symmetric distributions about some θ, for which the value of the parameter does not change (yet the meaning does; remember, we are not talking about a "nonparametric" center of symmetry, but about the complete description of a stochastic model).

Now it seems to be a standard argument of classical frequentist theory that the fuzziness of the estimated parameter can be made to disappear by letting the number n of observations tend to infinity. Then we "know exactly what we are estimating." But several questions arise. Is it always legitimate to increase the (hypothetical) number of observations, even beyond any limits? For what model do we want to define our parameter? And what do we do if our model turns out to be wrong?

The step of simply and naively letting n tend to infinity has been criticized by de Finetti, Jeffreys, and other Bayesians, and with some deep justification. Even if it is possible, at least in principle, to obtain more observations, the "model"—if we believe in such a hypothetical construction—will change in general for the new data, since the circumstances change; or, if we use a slightly different concept of a "true model," it has to be augmented and to become more complex in order to incorporate the new data; or, in a third variant, we cannot identify the "true model" even approximately with finitely many data. Contrary to what extreme subjectivists seem to say, the idea of a "true model" appears to be a useful concept in science, and we can leave the question open whether such "true models" exist only in the minds of some scientists or in some "objective reality"; but there seem to be relatively few cases (like the Longcor data, cf. Subsection 8.1c) where a simple "true model" appears to continue to hold even for huge data sets. Frequently, theories become more refined when more (and more exact) data become available. To put it sharply, the operational "true model" depends on the sample size.

We have assumed here that the sample size can be increased arbitrarily, as is often the case in the physical sciences. But frequently we have only a limited sample (of experiments or observations), perhaps even a sample of size one, like the natural history of the earth, and moreover it is sometimes not clear at all of which population a given data set can be reasonably assumed to be a sample; or else, to which (larger) population its results can be extrapolated, even if it happens to be a proper random sample from a well-defined population (e.g., of the properties of a certain strain of laboratory mice at a certain time and place). Clearly, the concept of a "true model" becomes even less sharp in such situations.

Another way of talking about estimation starts by talking about "true constants" to be estimated, such as the velocity of light in vacuum (or of neon light in beer, if one prefers). Our belief in such "true constants" is tied to theories and assumptions about the real world. First, one had to discover that there exists something like a finite velocity of light; not long ago, its velocity in "vacuum" would have been considered meaningless, and one expected different velocities in different directions on the earth; and fairly recently there were some doubts whether the speed of light was really constant over time, until another explanation was found for the contradictory experimental results (cf. Subsection 8.1a). But even if one believes in a "true constant," its value is tied to our practical experience only through measurements with their errors (including "pure" measurement errors, gross errors, semi-systematic and systematic errors), and as long as we assume nothing about these errors, we learn nothing about the constant. In fact, the scientist will make a breakup of the error similar to the one given; will estimate the size and distribution of the unavoidable random error from replications; will try to avoid or eliminate gross errors; and will do everything to make the systematic and semi-systematic errors small and will try, by judgment and experience, to assess approximate bounds on them. But the systematic and semi-systematic errors are virtually always there, and this means that in addition to the fuzziness of the parametric model for replications and to the fuzziness of its estimated parameters, there is also the fuzziness of the unknown systematic and semi-systematic errors (which can of course be integrated into the uncertainty of the error distribution by adding an unknown and only approximately bounded shift). Cf. Section 8.1 for more details on semi-systematic errors; they and the systematic errors are another obstacle to a sharp mathematical *and* applicable definition of "what we want to estimate."

Let us now ignore the problems of existence of a "true model," of the possibility of replications, and of systematic and semi-systematic errors, and let us assume we have a data set obtained under well-defined conditions, with the possibility for replications if desired. What does our data set leave us with? It leaves us, to start with, not with a single model distribution, but rather with the totality of all model distributions "compatible" with the data. "Compatible" may mean not rejected by some goodness-of-fit test on some level. We shall ignore the additional fuzziness of "compatible" reflected in the level of the test, which after all may be chosen arbitrarily small. Another, not unimportant question is the choice of the goodness-of-fit test or the metric used for it. A simple choice is the Kolmogorov–Smirnov test: The true distribution ought to lie between two confidence bounds above and below the empirical cumulative distribution function, and the width of the confidence strip shrinks to zero as n tends to infinity. However, the Kolmogorov distance does not generate the weak topology, which is a disadvantage under some robustness aspects, and it is intuitively not optimal in "distinguishing what can be distinguished" (the test is generally considered to have too little power in the tails of the distribution). Locally, and ignoring the rounding and the discreteness problem, the Hellinger distance with its difference between two root densities and its relation to the chi-square test and the information statistic might come closer to measuring discernible differences appropriately, but there is still the question whether a supremum rather than an integral or some combination of both might be more suitable. Other distances might be discussed equally well. In addition, there is the problem of correlations—most tests are valid for i.i.d. observations, but in reality correlations are present, and trying to estimate them and bounds for them blows up the confidence set of "compatible" distributions, up to nonergodicity and complete unidentifiability of parameters if no restrictions are made. These effects are tied to the systematic and semi-systematic errors discussed above, and the restrictions are tied to the simplicity of models discussed below. But if we ignore correlations, we can find to each data set a fairly simple confidence set of compatible distributions, which shrinks with increasing n, and under the usual idealistic assumptions of mathematical statistics these sets even shrink towards a single distribution, the "true model" (leaving aside all the reservations made before).

Given our data and the set of all probability distributions which might have generated them, how do we select a model distribution to describe the data? A parametric model should be "simple," and it should be "useful" for describing the data. Both concepts are ill-defined, but there is some intuition about them which makes them often applicable in specific situations. Now it might be considered a general policy to select a model distribution from the set of compatible distributions—be it any simple one, or be it the "simplest" one in the set if "simplest" can be defined, or be it, for example, the distribution with the smallest Fisher information in this set. The problem with this policy (which clearly contains adaptive and nonparametric features) is that in almost all cases the "true model" will be very complicated, and hence more and more simple models will be excluded as n increases, until only a very complicated and almost useless (and most probably nongeneralizable) description remains. Hence what might have looked as an endorsement of shrinking neighborhood approaches (namely the recognition of shrinking neighborhoods of undiscernible distributions) ultimately shows their unsuitability in parametric and robust (though not in nonparametric) statistics. The point is that the shrinking neighborhoods of parametric models which are considered in such approaches eventually do not overlap anymore with the shrinking neighborhoods consisting of the distributions compatible with the actual observed distributions as n increases. It is quite possible that some extrapolations obtained from these approaches for fixed finite n may be useful in practice, and there is even a subgroup of shrinking neighborhood approaches which recaptures the extrapolations used in this book in a mathematically rigorous way; but frequently the basic philosophy of their limiting statements is that of a simple parametric model which is exactly true, combined with the (more or less realistic) restriction that we are not sure about this for finite n.

We are not discussing the technical details of various shrinking neighborhood approaches, but it should be kept in mind that it would be important to distill the true intuitive meaning from the mathematical formalism if one wanted to pursue such an approach. Thus, during the 1984 Oberwolfach meeting on robust statistics, H. Rieder analyzed the meaning of the LAM (local asymptotic minimax) condition and found it highly restrictive and rather different from the vague intuitive beliefs usually associated with it.

Once we realize that as a rule all simple models will eventually be rejected by the data as more and more data are obtained, we may consider entertaining a *simple model* of which we *know that it is wrong*. This is not as counterintuitive and impractical as it may seem first. Every honest data analyst considers his or her model only as an approximation to truth. The deviations from the model may even be statistically highly significant but not relevant in the given situation. Furthermore, experience shows that simplified models which do not explain all idiosyncrasies of the data, are often much better generalizable than more complicated ones which are not rejected by the data. Of course, some peculiarities of the data may turn out to be most interesting, but then there is a lot of subject matter science experience and judgment involved, and the point is not that the scientist should always discard part of the information of the data, but that he or she should have a chance to do so if he or she deems it reasonable (e.g., sometimes in the case of gross errors), and that the scientist may want to analyze the coarser and the finer features of the data step by step, or layer by layer. We now can contemplate the possibility of keeping the same simple model for all n, although explicitly only as an approximation. But this is what is done in the theories of robust statistics discussed in this book. They do not destroy the beauty of classical parametric models, they rather keep them and only fortify them against mishap. The price is that any model distribution and hence any parameter in some neighborhood of the actual distribution may be chosen to describe the data, but this fuzziness is only one out of several ones existing already for other reasons. And there is no necessity to invent any new supermodels.

Sometimes it is argued that one should use a minimum distance estimator for some distance and minimize the distance between the actual and the chosen model distribution. This, it is said, would give a unique estimand. But choosing a distance is about as arbitrary as choosing an estimator (in fact, many estimators can be viewed as minimum distance estimators for a suitably chosen distance), hence the problem is only shifted from the choice of the estimator (which also defines an estimand) to another arbitrary choice (unless, of course, one can give compelling reasons for choosing one particular distance). If a distance is chosen for the good robustness properties of the estimator it yields, the choice is really one of the estimator and is guided by robustness theory (which leads to reasonable but not unique choices). If the distance is chosen without any regard to robustness properties of the resulting estimator, this estimator may be quite foolish. For any minimum distance estimator considered, one ought to compute the breakdown point and, if possible, the influence function (or some other description of the local properties). Some distances which are rather attractive in robustness theories, such as the Prohorov and the Lévy distance, lead to estimators which are quite difficult to handle both in theory and in practice, even though Kozek (1982) was able to obtain some theoretical results; these estimators therefore do not seem to be very useful in practice.

We can go one step further in our modeling and postulate that the actual distribution has arisen from distortion of an ideal model distribution (e.g., local changes, like rounding, of all data, and replacement of a few data by arbitrary gross errors, which fills out neighborhoods in the weak topology).

This distortion is the same for all sample sizes (unless, of course, we change the accuracy of the data with the sample size; even then, we can imbed our situation for a given n into a sequence with constant distortion). Then we have an ideal parameter to be estimated, but one that is not definable from the observed distribution, since this distribution can have arisen from different model distributions. Our estimand will change in general when we distort the model distribution; it *has* to change at least sometimes in most parametric models, since some distortions (such as the same small shift for all data in location problems) generate neighboring model distributions, and even if we are free in what we want to estimate outside the parametric model, we almost certainly want to estimate the correct parameter if an ideal model distribution is true (i.e., we require Fisher-consistency). Now all we can do is to keep the supremum of the unavoidable change of the estimand small for a small distortion. But in the infinitesimal approximation, this change is bounded by the sum of multiples of the gross-error sensitivity and the local-shift sensitivity. It suffices therefore to put bounds on these two quantities separately. Very often the local-shift sensitivity is already sufficiently low, and then we are back to the problem we have discussed at great length in this book. We know that there is a positive lower bound on the gross-error sensitivity since the estimand (the value of the functional, or the limiting value of the estimator) must be able to follow the changing parameters of the parametric model (i.e., must keep Fisher-consistency). And we know that in general there is a conflict between the requirement of low gross-error sensitivity and the requirement of efficiency or low asymptotic variance; the optimal compromises are a central theme of the book. (Those who like game theory may also consider the game where Nature chooses an ideal model distribution as well as an arbitrary but bounded contamination, and the statistician tries to minimize his or her maximum approximate (extrapolated) mean-squared error, as described in Subsection 1.3e.)

We may expand the present theory and include consideration of the local-shift sensitivity in the search for optimal compromises. In doing this, we may also replace the local-shift sensitivity by a lower bound which will be a better approximation but also more complicated, namely the "mean shift sensitivity" $\nu(T, F) := \int |\mathrm{IF}'| \, dF + \sum |h(x_i)| f(x_i)$, where the x_i and the $h(x_i)$ are the places and the heights of the jumps of the IF of T at F [and IF$'$ is the derivative of the IF with respect to x, and $f(x_i)$ is the density of F in x_i assumed to exist, otherwise we may put $\nu = \infty$].

Whichever variant we choose (and the one chosen in this book is the simplest one, at the same time being already quite useful), the idea is to choose an estimator which estimates the correct quantity if the model were exactly true, and which changes as little as possible in neighborhoods of the model, subject to the constraints of Fisher-consistency as mentioned and of a certain asymptotic efficiency under the ideal model. We believe and assume that we are in a certain neighborhood of an ideal parametric model distribution (as long as we retain the model, of course), a neighborhood which does not change with n and into which we can extrapolate from the model as long as its maximum radius (e.g., in Prohorov distance) is clearly and sufficiently smaller than the breakdown point. We assume our actual underlying distribution has been generated from a model distribution with "ideal" but unknown parameter by some unknown but bounded distortion; this distortion causes a bounded but unknown and unavoidable bias. In practice, we often have some idea about the likely percentage of gross errors and the size of rounding and grouping effects, and hence also about the size of the neighborhood and that of the bias. We then project our actual underlying distribution back onto the parametric model, in trying to fit a well-fitting model distribution from which the actual one also could have arisen by a small distortion, ending up, at any rate, in a small neighborhood of the original (ideal or fictive) model distribution, which is still augmented somewhat by the random variability. Since the "projection" is done in terms of the locally approximately linear functional used as estimator, hence is an approximation to a simple linear projection, we also might consider our procedure as an approximate minimum distance estimator, thus finding a tie to this general idea discussed above.

There is one finer additional point that may be made to our approach: In practice one will often *want* to use a more complex parametric model when more data are available. This is connected with the fact that a larger number of meaningful parameters may become estimable; but they usually do so in a more complex design. Thus, in most cases we shall not want to estimate a many-parameter distribution for a location problem, but rather augment the location problem to one of a many-factor analysis of variance or regression (with a simple model error distribution). In order to apply our extrapolations, we may then start with the sample size and design complexity and contemplated model complexity at hand and from there on increase the fictitious number of observations for this fixed design and model until we are in the limit of local properties of functionals from which we can extrapolate back to our actual finite sample size.

8.3. ROBUSTNESS IN TIME SERIES

8.3a. Introduction

Robustness for dependent data has been a neglected area for many years although it has been known for a long time that a few outliers or transient phenomena can have disastrous effects on the classical procedures. This neglect is not surprising in view of the many new difficulties that arise in this area. To name just a few of these: (1) Many more types of deviations from a model have to be considered. (2) The pattern of time points where the contaminations occur becomes very important, for instance, scattered and patchy outliers can have quite different effects and both occur in reality. (3) From a theoretical point of view limit theorems become much more delicate. Moreover, in time-series analysis the nonparametric spectrum estimators play a prominent role, but the classical robustness theory was not developed for such a problem.

Nevertheless, in the last years the subject developed quickly and it is not possible to give an overview of the whole field on a few pages. We concentrate here on the influence function for time series defined in Künsch (1984) and Martin and Yohai (1984a) since this provides much insight in the problem and it is the main topic of the book. Some other areas are mentioned in the last section, but for more details and problems not treated here, like the different proposals of estimators for ARMA processes, we refer the reader to Kleiner et al. (1979), Martin (1979, 1980, 1981), Martin and Yohai (1984b).

8.3b. The Influence Function for Time Series

To begin with a relatively simple case consider an estimator which is obtained from

$$\sum_{i=m}^{n} \psi(X_i, X_{i-1}, \ldots, X_{i-m+1}; T_n) = 0. \qquad (8.3.1)$$

X_i, ψ, and T_n can all be vector valued. In analogy to (2.3.3) we call such an estimator an M-estimator although often this term is reserved for a much smaller class. If the observations $(X_1, X_2, \ldots, X_n, \ldots)$ are distributed according to a stationary ergodic distribution F, then under mild regularity conditions T_n converges a.s. to $T(F^{(m)})$. Here $F^{(m)}$ denotes the m-dimensional marginal of F and the functional T acts on probability measures on the m-fold product space and is defined by

$$\int \psi(x_m, \ldots, x_1; T(G)) \, dG(x_1, \ldots, x_m) = 0. \qquad (8.3.2)$$

Formally, we can define like in the i.i.d. case

$$\mathrm{IF}\left(x_m, \ldots, x_1; T, F^{(m)}\right) = \lim_{h \downarrow 0} \frac{T\left((1 - h)F^{(m)} + h\Delta_{(x_1, \ldots, x_m)}\right) - T\left(F^{(m)}\right)}{h}$$

$$= M\left(\psi, F^{(m)}\right)\psi\left(x_m, \ldots, x_1; T(F^{(m)})\right), \quad (8.3.3)$$

where

$$M\left(\psi, F^{(m)}\right) = - \int \frac{\delta}{\delta t}\psi(x_m, \ldots, x_1; t)\bigg|_{t = T(F^{(m)})} dF^{(m)}(x_1, \ldots, x_m).$$

$$(8.3.4)$$

However, the mixture $(1 - h)F^{(m)} + h\Delta_{(x_1, \ldots, x_m)}$ is not directly related to any contamination of the original series, for example, substituting some values by gross errors. We neither can add an additional m-tuple of observations to a series because it is important where to add them. Nevertheless, we are going to show that the function defined in Eq. (8.3.3) contains all the relevant information on the asymptotic covariance, on the asymptotic bias, and on the effect of changing a finite number of observations in a large series.

Let us first treat the asymptotic covariance. By a Taylor-series expansion similar to the derivation of Eq. (2.3.5) we obtain

$$T_n - T\left(F^{(m)}\right) = \frac{1}{n}\sum_{i=m}^{n} \mathrm{IF}\left(X_i, \ldots, X_{i-m+1}; T, F^{(m)}\right) + \text{remainder}.$$

$$(8.3.5)$$

Under regularity conditions the remainder is of order $o_p(n^{-1/2})$. Hence under a condition of sufficiently weak dependence

$$\sqrt{n}\left(T_n - T\left(F^{(m)}\right)\right) \overset{\text{weak}}{\to} \mathcal{N}\left(0, V(T, F)\right),$$

where

$$V(T, F) = \sum_{k=-\infty}^{\infty} V_k(T, F),$$

$$V_k(T, F) = \int \mathrm{IF}\left(x_m, \ldots, x_1; T, F^{(m)}\right)\mathrm{IF}\left(x_{k+m}, \ldots, x_{k+1}; T, F^{(m)}\right)^T dF.$$

$$(8.3.6)$$

This complicated expression simplifies in an important special case. If

$$E_F\left[\psi\left(X_i,\ldots,X_{i-m+1};T(F^{(m)})\right)\big|X_{i-1},X_{i-2},\ldots\right]\equiv 0, \qquad (8.3.7)$$

then $V_k = 0$ for all $k \neq 0$ whence

$$V(T,F)=\int \mathrm{IF}\left(x_m,\ldots,x_1;T,F^{(m)}\right)\mathrm{IF}\left(x_m,\ldots,x_1;T,F^{(m)}\right)^T dF$$

like in the i.i.d. case. Moreover, under (8.3.7) no assumption of weak dependence is needed because $\mathrm{IF}(X_i,\ldots,X_{i-m+1};T,F^{(m)})$ is then a so-called martingale difference (cf. Billingsley, 1961). If the X_i form a higher-order Markov process, then it is no essential restriction to assume (8.3.7) (cf. Künsch, 1984, Theorem 1.3). Moreover, (8.3.7) holds for many of the proposed choices of ψ for ARMA processes.

In order to examine the asymptotic bias, consider the contamination model

$$X_i^h = \left(1 - V_i^h\right)Y_i + V_i^h Z_i, \qquad (8.3.8)$$

where V_i^h is a stationary ergodic 0–1 process independent of all Y_i, Z_i with $P[V_i^h = 1] = h$. Such a model contains many different types of contaminations since we can choose the dependence of the V_i^h and the dependence of Z_i on the clean process Y_i arbitrarily. If the V_i^h are independent we have for small h isolated outliers, while patchy outliers can be modeled by a strong dependence of the V_i^h. Z_i independent of Y_i leads to so-called substitution outliers, but with a suitable Z_i we can also model additive outliers.

Denoting by F_h the distribution of X_i^h it is easily seen that $F_h^{(m)}$ is not a mixture of $F_0^{(m)}$ with some contamination H, but we have the following expansion of the asymptotic bias:

$$T\left(F_h^{(m)}\right) - T\left(F_0^{(m)}\right) \approx \int \mathrm{IF}\left(x_m,\ldots,x_1;T,F_0^{(m)}\right) d\left(F_h^{(m)} - F_0^{(m)}\right)$$

$$\approx h\int \mathrm{IF}\left(x_m,\ldots,x_1;T,F_0^{(m)}\right)c_m\, d\left(G^{(m)} - F_0^{(m)}\right)$$

$$= h\int \mathrm{IF}\left(x_m,\ldots,x_1;T,F_0^{(m)}\right)c_m\, dG^{(m)}. \qquad (8.3.9)$$

Here c_m is a positive constant depending only on the distribution of the V_i^h and $G^{(m)}$ is a probability measure depending on V_i^h, Y_i, Z_i but not on T.

The calculation of $G^{(m)}$ is usually straightforward; for example, for independent V_i^h, $c_m = m$, $G^{(m)} = (1/m)\sum_{k=1}^{m} G_{k,m}$, where $G_{k,m}$ is the joint distribution of $(Y_1, \ldots, Y_{k-1}, Z_k, Y_{k+1}, \ldots, Y_m)$. Equation (8.3.9) says that the "arc" $(F_h^{(m)})_{h \geq 0}$ in measure space is differentiable with derivative $c_m(G^{(m)} - F_0^{(m)})$, and that for the infinitesimal bias we may replace the "arc" $(F_h^{(m)})_{h \geq 0}$ by the segment $(1 - c_m h)F_0^{(m)} + c_m h G^{(m)}$. Note that such an argument is possible also for other contamination models instead of (8.3.8), for example, for innovation outliers in a linear time-series model or for small changes in the correlation structure of a Gaussian model.

Finally, consider a large series of observations $(X_1, \ldots, X_n)$ with distribution F and let $(Y_1, \ldots, Y_n)$ be another series with $Y_i = X_i$ for all i except $i_1, \ldots, i_r$. Letting n tend to infinity while keeping r and $i_1, \ldots, i_r$ fixed, we obtain by the usual Taylor expansion:

$$\lim_n \left\{ n \left(T_n(X_1, \ldots, X_n) - T_n(Y_1, \ldots, Y_n) \right) \right\}$$

$$= \sum_{k \in I} \left(\mathrm{IF}\left(X_k, \ldots, X_{k-m+1}; T, F^{(m)} \right) - \mathrm{IF}\left(Y_k, \ldots, Y_{k-m+1}; T, F^{(m)} \right) \right),$$

$$(8.3.10)$$

where $I = \{ k; k = i_j + q \text{ for some } j, 1 \leq j \leq r \text{ and some } q, 0 \leq q < m \}$. Thus the influence of an observation depends on its neighboring values on both sides and this "conditional influence" is not additive.

Equations (8.3.6), (8.3.9), and (8.3.10) show that the function IF defined by (8.3.3) is also in the time-series context a fundamental object, and for this reason Künsch (1984) calls IF the influence function. Because of Eq. (8.3.9), Martin and Yohai (1984a) propose to call $c_m \int \mathrm{IF}(x_m, \ldots, x_1; T, F^{(m)}) \, dG^{(m)}$ the influence function which is then a function of the contamination model. There is a similar controversy over the definition of the gross-error sensitivity: By (8.3.9) it has to be some supremum of $\| \int \mathrm{IF}(x_m, \ldots, x_1; T, F^{(m)}) \, dG^{(m)} \|$, but it is not clear over what kind of $G^{(m)}$'s this supremum should be taken. Martin and Yohai suggest to take a supremum separately over the different subclasses of the contamination model (8.3.8), that is, one would have a gross-error sensitivity for isolated substitution outliers, another one for patchy substitution outliers, and so on. On the other hand, Künsch argues that even the supremum over all models (8.3.8) is not sufficient because this gives no bound for the right-hand side of (8.3.10). He proposes therefore to bound $\sup_{x_1, \ldots x_m} \| \mathrm{IF}(x_m, \ldots, x_1; T, F^{(m)}) \|$ or some standardized version of this. In doing so some care is needed because different ψ's can give rise to asymptotically equivalent estimators, but this difficulty can be overcome and the same optimality results as in Chapter 4 are obtained for autoregressions.

The class of estimators defined by (8.3.1) can be extended to functionals on the m-dimensional marginal distributions (see Künsch, 1984), but a far more severe restriction is the finite range m of the function ψ. For instance, in ARMA models not even the classical efficient estimators can be put in this form, and also for AR models an infinite argument of ψ might be helpful for robustification. Martin and Yohai (1984a) introduce the class of estimators defined by

$$\sum_{i=1}^{n} \psi_i(X_i, X_{i-1}, \ldots, X_1; T_n) = 0, \qquad (8.3.11)$$

where the sequence of functions $\psi_i(x_0, x_{-1}, \ldots, x_{-i+1}; t)$ converges to some $\psi(x_0, x_{-1}, \ldots; t)$. The asymptotic functional is then given by

$$\int \psi(x_0, x_{-1}, \ldots; T(F)) \, dF = 0 \qquad (8.3.12)$$

and formally

$$\mathrm{IF}(x_0, x_{-1}, \ldots; T, F) = M(\psi, F)^{-1} \psi(x_0, x_{-1}, \ldots; T(F))$$

$$M(\psi, F) = - \int \frac{\delta}{\delta t} \psi(x_0, x_{-1}, \ldots; t) \bigg|_{t=T(F)} dF.$$

$$(8.3.13)$$

The formula (8.3.6) for the asymptotic covariance still holds as well as (8.3.10), but the analog of (8.3.9) becomes more delicate. They show that there is a linear functional D defined on a certain class of functions such that

$$T(F_h) - T(F_0) \approx hD(\mathrm{IF}(x_0, x_{-1}, \ldots; T, F_0)). \qquad (8.3.14)$$

However, D is not given by a measure like in (8.3.9). Even for bounded ψ, the supremum of $\|D(\mathrm{IF}(x_0, x_{-1}, \ldots; T, F_0))\|$ over different Y_i, Z_i and fixed V_i^h can be unbounded. This means that for an infinite range boundedness of ψ is not sufficient for robustness.

Staab (1985) discusses also influences for time series. He uses shrinking contaminations of the transition probabilities.

8.3c.   Other Robustness Problems in Time Series

Generalizations of qualitative robustness and of the breakdown point have been discussed by Papantoni-Kazakos and Gray (1979), Cox (1981), Boente et al. (1982). The problem consists of choosing a metric on the set of distributions of stochastic processes and has not yet been settled completely.

An entirely different approach to robust inference is the so-called robust filter of Kleiner et al. (1979) (see also Martin and Thomson, 1982). The idea consists in cleaning the possibly contaminated data X_i. If X_i differs too much from a robust prediction $\hat{X}_i$ based on the other values, then X_i is replaced by a value closer to $\hat{X}_i$. This procedure is expected to have a high breakdown point and is intuitively very appealing. It provides a simple robust nonparametric spectrum estimate by the application of the classical procedure to the cleaned data. However, there are also some problems. The spectrum of the cleaned data seems to be different from the spectrum of the original data also when the latter are uncontaminated. (In fact, it has to this date not even been proved that the cleaned data are stationary.) Moreover, for the prediction $\hat{X}_i$ we have to fit first some AR model. In order to overcome these problems Martin and Yohai (oral communication) suggest to determine all necessary constants for the filter by minimizing a robust scale estimate of $X_i - \hat{X}_i$, but there still remain many gaps to close.

Another problem is the effect of slowly decaying correlations mentioned in Section 8.1. For the M-estimators defined by formula (8.3.1) the asymptotic value depends only on some finite-dimensional marginal, so it is robust against such departures. However, the asymptotic covariance $V(T, F)$ of Eq. (8.3.6) depends on the whole distribution of the process and is necessarily very sensitive to deviations in the dependence structure of F. $V(T, F)$ will tend to infinity if we come closer to a process with long-range dependence and in this sense there can be no V-robustness in time series. The problem of constructing robust confidence intervals for parameter estimates in time series is a difficult one and has not yet received sufficient attention.

*8.4.   SOME SPECIAL TOPICS RELATED TO THE BREAKDOWN POINT

8.4a.   Most Robust (Median-Type) Estimators on the Real Line

This section discusses two parts (2.2 and 3.1) of Hampel (1975) which are concerned with structural and breakdown-point aspects of three- and more-parameter models on the real line and of the analysis of variance, respectively. On the one hand, these structural aspects of statistics have been unduly neglected, although some growing awareness for them can be found in the recent interest in exploratory data analysis, probability-free data analysis, projection-pursuit-related work, and a theory of data analysis. On the other hand, the paper cited has proven to be rather inaccessible. Furthermore, because of the imposed page limit, it was condensed so much that it was mistaken for a program for future research instead of for a summary of past research. Some of the material can also be found briefly in Hampel (1980, 1.4.2-4, 1.5.4).

The aim of the present subsection is to find median-like estimators for three- and four-parameter models on the real line (and for more-parameter models if desired). Some interesting parametric families of this type are the three-parameter power transformations of the normal distribution, the three-parameter Gamma distributions (with shift), and the four-parameter Pearson curves. Clearly, there are many others. The optimal robustifications of the maximum likelihood estimators can be found by the theory in this book (cf. Section 4.3). (It may be noted that the partial superefficiency of the maximum likelihood estimator for the shift parameter of the Gamma distribution disappears again with the robustification, as for the rectangular distribution—cf. Subsections 1.2b and 1.3f. The same is true for the Pearson curves.) In addition, it is important to have a simple family of estimators with high breakdown point for all parameters. They may be used as a starting point for the iterations for the robustified maximum likelihood estimator; as crude but reliable estimators themselves if the sample is large enough and no higher accuracy is required; as a check for more refined estimators; and possibly as a basis for rejection of outliers if desired.

Let a data set $x_1, \ldots, x_n$ on the real line be given. Two simple "most robust" estimators of location and scale with joint breakdown point $\frac{1}{2}$ are known to be the median $m = \text{med}(x_i)$ and the median deviation $d = \text{med}(|x_i - m|)$. (We remember that the interquartile range has only breakdown point $\frac{1}{4}$.) An equivalent but perhaps more natural choice for scale would be $\log d = \text{med}(\log|x_i - m|)$.

It might be tempting to use the fraction Q of data between m and $m + d$ (which is related to the quadrant correlation) as the basis for a measure of skewness. However, when Q_0 denotes the "true" or basic fraction of mass between m and $m + d$, the gross-error breakdown point (or its finite-sample variant with replacement, cf. Subsection 2.2a) is only $\frac{1}{2} - 2|Q_0 - \frac{1}{4}| = \min(2Q_0, 1 - 2Q_0)$, since a global shift of such a data mass reaches an extreme limit of skewness.

A better choice is possible in the case of the power transformations of the normal: determine that power transformation which yields $Q = \frac{1}{4}$ for the transformed data and use it as estimate, together with median and median deviation of the transformed data. We can also test the null hypothesis of the identity transformation (symmetry, as measured by Q: equal masses in the four intervals determined by m and $m \pm d$) with an approximate sign test. It is clear that these methods are not very efficient and make sense mainly in large samples, but they are very robust and fairly simple and complement the nonrobust classical methods and their less extreme, moderately robust, and more efficient optimal robustifications.

We now devise a general sequence of median-like analogs of moment estimators for higher moments. Naturally, in order to apply them to any specific three- or more-parameter model, one has to gauge them specifically for that model, that is, one has to find a one-to-one correspondence between the median-like parameters (estimators, functionals) and whatever parametrization one uses otherwise. The most natural correspondence is again the one at the model distributions themselves, that is, we require Fisher-consistency (and not, e.g., unbiasedness). But apart from their use for specific parametric models, the new statistics can also be used as general descriptive statistics.

We start again with our data set of x's on the real line and split the line by the median m which serves as our location measure.

Then we map each piece of the real line on the full real line again by taking $\log|x - m|$. (Other choices are possible, of course, but this one appears to be especially simple.) We obtain a certain superposition of the two pieces, and now we can make another split by taking the median again to obtain $\log d = \text{med}(\log|x_i - m|)$ as our scale measure. For brevity, we put $y_i = (x_i - m)/d$.

We now have (by m and $m \pm d$) four quarter pieces on the original real line, and we can again map both original halves on the full line, but this time such that the first quarter is mapped on the third (and not the fourth) quarter, and the second on the fourth one. This is achieved by taking $\text{sgn}(x - m)$ $(\log|x - m| - \log d)$, and we obtain a skewness measure by taking $\text{med}(\text{sgn}(y_i)\log|y_i|)$.

For obtaining a measure of kurtosis, we have to assign the data near m and $\pm \infty$ large positive values and those near $m \pm d$ large negative values. This is achieved by mapping each quarter piece on the whole real line with the function $\log|\log|x - m| - \log d|$, and our kurtosis measure becomes $\text{med}(\log|\log|y_i||)$ or else $k = \text{med}(|\log|y_i||)$ if we want it to be nonnegative.

If desired, we can continue the construction to the analogs of higher-moment measures. As examples, we give the analogs for the fifth to seventh moment. We note that k together with m and d divides the real line into eight pieces (by the four points which yield k). These eight pieces can again be mapped to yield the desired pattern. The estimators then are:

$$\text{med}(\text{sgn}(y_i)(\log|\log|y_i\|| - \log k))$$
$$\text{med}(\text{sgn}(\log|y_i|)(\log|\log|y_i\|| - \log k))$$
$$\text{med}(\text{sgn}(y_i)\text{sgn}(\log|y_i|)(\log|\log|y_i\|| - \log k))$$

or, to write the last formula in more detail,

$$\text{med}(\text{sgn}(x_i - m)\text{sgn}(\log|x_i - m| - \log d)$$
$$\times (\log|\log|x_i - m| - \log d| - \log k)).$$

We note that we have to imitate the sign pattern of a polynomial of the corresponding degree with positive leading coefficient and maximum number of zeros. The sign patterns for the zeroth- to seventh-moment estimators in our eight pieces of the real line determined by m, d, and k are:

$$
\begin{array}{cccccccc}
+ & + & + & + & + & + & + & + \\
- & - & - & - & + & + & + & + \\
+ & + & - & - & - & - & + & + \\
- & - & + & + & - & - & + & + \\
+ & - & - & + & + & - & - & + \\
- & + & + & - & + & - & - & + \\
+ & - & + & - & - & + & - & + \\
- & + & - & + & - & + & - & +
\end{array}
$$

These orthogonal patterns are familiar in the analysis of variance.

8.4b. Special Structural Aspects of the Analysis of Variance

We shall now discuss various points connected with the question of how one can establish the presence of outliers in the analysis of variance, and how one can identify them. One reason for considering this question separately is that typically in ANOVA there are many parameters compared with the number of observations, such that often only a few degrees of freedom (if any) are left for error, and the problem of redundance necessary for being able to identify outliers becomes rather acute. Another reason is that there are some interesting structural aspects in ANOVA connected with the intuitive meaning of models and parameters. We shall mainly think of a balanced fixed-effects two-way layout without replicates as a prototype analysis of variance, although the points made are of course more general. (The usual model is written $Y_{ij} = \mu + A_i + B_j + \varepsilon_{ij}$.)

First, we have to recall some general points of data analysis. Every design (in the narrow sense, as the structure of the data) is associated with a partially ordered class of estimable models. For an unreplicated two-way layout, the most obvious models are: a constant (no effects), A-effects only, B-effects only, A- and B-effects, and A- and B-effects with AB-interactions (and, alas, no error estimate and no redundancy); but there are some other, less important ones. The partial order is given by the terms in the model. Clearly, if some factor (on more than two levels) is quantitative, there are still more possibilities. The data themselves can be used to define new models, such as done with Tukey's one degree of freedom for nonadditivity (odoffna) or, more generally, with Tukey's vacuum cleaner (cf. Tukey, 1962). General statistical experience puts a high plausibility on some models and a very low one on others (e.g., a pure interaction model). The general prior knowledge of the subject matter science and the specific prior knowledge about the data and their immediate background can make many more models identifiable.

For example, the error may be known a priori, allowing tests even in a fully saturated design. Prior assumptions and beliefs, if only tentative, allow even "super-saturated" designs. Thus, in the "desperado designs" considered by Hampel in consulting more than a decade ago, one puts $2^{2^{k-1}} - 1$ factors on two levels each into a 2^k-factorial design—for example, 127 factors on 8 runs (!)—by identifying each main effect with a nontrivial split of the data, in the desperate hope of finding that one main effect (if existent) that clearly dominates all the rest.

It often happens that not a single cell but rather some bigger substructure of the design such as a row or a group of rows behaves like an outlier, that is, shows an unusual behavior (cf. Daniel, 1976 for examples). In such a case the absolute residuals in this substructure tend to be large, especially the residuals from a robust fit (which brings out outliers much more clearly than least squares). If a majority of the residuals in the substructure are large, this larger variability might be found by a robust analysis of variance of the absolute residuals from a robust fit. In some cases perhaps a least-squares analysis of the absolute residuals from a robust fit might be more informative (namely if only a decided minority of the substructure, but no single cells, show outlying behavior).—It is well known that in the analysis of residuals there are many regressions based on residuals, fitted values, and estimated effects which may produce useful information on improving the model. We have already mentioned odoffna; another method would be, for example, to regress the absolute residuals on the fitted values to detect some systematic heteroscedasticity, and the data analyst is encouraged to find other sensible plots and relationships.

The desirable structure of an ANOVA is usually that apart from factors with no effects at all one would like some clear main effects. Mathematically, main effects are defined after determination of the grand mean, which involves all data, and with a somewhat arbitrary side condition, but the operational meaning of a clear main effect is a practically constant difference between two rows, say (not involving any other rows). We therefore call any difference $Y_{ij} - Y_{i'j}$ an "element" of the main effect between rows i and i'. These elements are very useful in practical work; an example is given in Section 4.7 of Daniel (1976). If the elements $Y_{ij} - Y_{i'j}$ for all j are about constant, we have a clear main-effects difference $A_i - A_{i'}$; if at least the majority is constant, we still can speak of a well-defined main-effects difference, but only for part of the data, and we know that the remaining pairs of data contain outliers, though we still do not know which value in each pair is the outlier. We may be able to tell this by comparison with other rows. We thus have found a useful tool for the identification of outliers in a main-effects model which we shall discuss in more detail below.

The "elements" of first-order interactions are clearly the tetrad differences $Y_{ij} - Y_{ij'} - Y_{i'j} + Y_{i'j'}$, while the "elements" of the grand mean are the Y_{ij} themselves. When we want to determine a grand mean, we first have to fix the main effects if there are any (e.g., decide whether their mean or median should be zero) and take them out of the data to obtain a near-constant set of "adjusted elements" for the grand mean. Similarly, if there are interactions, we first have to take them out of the data before we can determine the main effects. The order of operations is thus the reverse of the usual mathematical one where main effects are differences against the grand mean, and so on.

Disadvantages of the elements are that there usually are too many (many more than parameters, yielding a lot of redundancy), and that they are not independent (e.g., $Y_{ij} - Y_{i'j}$, $Y_{i'j} - Y_{i''j}$, and $Y_{i''j} - Y_{ij}$). The main advantages are that they describe directly the most relevant features of the model, and that they are purely local, depending only on the minimum number of data necessary for defining the features. One is also reminded of the role of cross-product ratios in contingency table analysis.

An example showing the properties mentioned is a main-effects model with a big outlier put into one cell. Then all and only all tetrad differences containing this cell will be big in absolute value, while all others will be practically zero, thus showing that part of the data still obeys a main-effects model. The task is then to find the outlier or outliers. By contrast, the least-squares analysis of variance distributes the outlier over the whole design, causing "interactions" everywhere, and blows up the interaction mean squares of all orders. It is an important experience that isolated outliers or local outlying substructures (such as single rows) are the most common source for high-order interactions (cf. Daniel, 1976).

We now discuss the identification of outliers (such as big gross errors) in detail. We shall think only of clear, distant outliers, not of marginal cases. Outliers exist relative to a model: an outlier in a main-effects model is merely a big interaction in a saturated model. The treatment of outliers is more difficult than that of missing values because first the cell containing the outlier has to be identified before it can be treated like in the case of a missing value and fitted with a fitted value not influenced by the outlier.

A principle which we shall follow is that we try to get along with the smallest possible number of outliers. Clearly, if we find one outlier, it could be theoretically that this is the only "good" value and all the others are "wrong." If there is little redundance (e.g., only three values), our outlier identification may be shaky, but it is the best we can do (apart perhaps from pointing out some less likely alternative solutions).

Sometimes one is able to say that at least one outlier exists in a data set without being able to pinpoint its location. An extreme example is a 2×3 design with a main-effects model where the last column does not fit the pattern (with one vanishing tetrad difference) set by the first two columns. Either or both values in the last column are outliers, since the unknown last column effect can accommodate any one value, but not both. If, however, the column effects are block effects or other random effects of which we can assume that they have the same order of magnitude, then we have one degree of freedom for estimating their variability and may be able to exclude extreme values of the third column effect and thus sometimes exclude just one data point.

If we have several outliers, we may be able to say something about their structure, and assumptions about their structure may make a big difference in what outliers we can identify. The two extreme cases are those of "wild outliers" where every outlier is totally unrelated to every other one and to the "good" data, best modeled by occasional independent random errors with huge (possibly infinite) variance; and "mean contamination" where we assume Nature tries to be mean to us and play tricks on us, usually by building a piece of another nice model elsewhere with suitably chosen outliers, so that we have problems deciding which piece of a nice model we shall believe in. Overall, mean contamination will be rarer, but there may be specific reasons for it, such as a temporarily shifted origin, a part of the data measuring a different population, or just an instrument getting stuck repeatedly at the same value. Mean contamination is also more difficult to deal with. Take, for example, a 4 × 4 design (with main effects) and two outliers in the last column. If they are wild outliers, they will in all probability not form any vanishing tetrad difference (or a "good" main-effects element) and hence can be easily identified. But if they are "mean," they behave like good data belonging to another column effect (i.e., they equal the original good data shifted by the same amount), and without additional outside knowledge we have no means to identify the true column effect, both possible values being indicated by two observations.

In situations with little redundance, as often in the analysis of variance, the role of prior information may be crucial. We mean the vague prior information which tells us whether some effect or error is quite possible or is very surprising and unlikely. For example, sometimes an effect should be positive, or in some region of plausible values; sometimes the error should be small compared with the absolute values of the observations; or the error is quite well determined or at least bounded by a single observation or even without any observation because of some model assumptions, as in some models for counting data (binomial, Poisson, root Poisson, etc.). Various levels of prior information together with various levels of assumptions about the meanness of contamination can lead to various stages in the assessment of outliers. We shall exemplify this by considering the simplest cases of very few observations for a location parameter.

If we have a single observation, we need prior information both on the possible range of the true parameter value and on the possible size of the random error in order to be able to identify at least sometimes a distant outlier as an outlier. There is no redundance, hence no possibility to use other data to correct the value, and the latter is lost (apart from the possibility of checking on its origin, or on the likely source of a gross error —such as a wrong decimal point—and correcting it in this way).

If we have two observations, we may be able to reject one or both with prior knowledge about parameter and error. Moreover, it now suffices to have prior knowledge about the possible size of the error in order to be able to say that at least one value is an outlier if they are improbably far apart; but we cannot say which one. The practice of dissolving the dilemma by taking a third observation in this situation has been discussed in Subsection 1.4a.

Three observations provide minimum redundance both about the parameter (or fit) and the error. If two values are close together and the third one is very far away, then the third one is likely to be an outlier (except in the improbable case of "mean" contamination by two instead of one outliers, where the third one may be the only "good" value). Of course, the one degree of freedom for error in the closest pair of data yields very meager information, and the information given by the data alone can easily be reinforced or superseded by prior information on parameter and/or error. Thus, it could be that all three values are rejected as implausible a priori, or that only one in the closest pair is retained if the error is known to be much smaller than the span of the closest pair and if the range of plausible parameters also excludes the other value. Clearly, there are already many possible combinations of prior knowledge and information given by the data themselves.

While three observations are the minimum number to provide shaky evidence against a single outlier by the data themselves (shaky both because of the weak 2:1 majority and the extremely badly determined error scale), four observations can provide a fairly comfortable 3:1 majority with a considerably better (though still not good) error estimate against a single outlier. They can also provide very shaky evidence against two "wild" outliers (when only one pair is very close together), but they can only suggest the existence of two "mean" outliers in the case of two very close pairs of data far apart, without being able to tell which pair (if any) is the "proper" one. Various prior information can again help and confirm or correct the indications by the data alone, especially in the cases of two outliers (and prior information may even identify three or four outliers).

Five observations can provide comfortable evidence against a single outlier, good evidence against two wild outliers, shaky evidence against two mean outliers (which are barely in the minority), and even very shaky evidence against three wild outliers; all this without using prior knowledge (which again could lead to much more information).

These considerations, which clearly can be worked out and extended, are equally valid for parts of an ANOVA design, for example, the three observations in a row of a $k \times 3$ design which besides other effects determine the row mean just by themselves, or the three elements of a main-effects difference which may describe a full or partial or absent "main-effects property" for the two rows concerned. An important additional feature is that one can "borrow strength," to use Tukey's terms, from the other parts of the data: usually they provide an error estimate, and sometimes also information on the effects (as in the 2×3 example above with the block effects). This collateral information takes the place of (or complements) prior information.

The elements can be used to identify parts of the data which obey the same simple model. For example, several row differences which are equal within random error identify a subset of data points which can be described by a main-effects model. The redundance can also be supplied by a closed chain of elements: thus, if $(Y_{11} - Y_{21}) + (Y_{22} - Y_{32}) + (Y_{33} - Y_{13}) \approx 0$ within random error of such an algebraic sum, these six values are close to a nice simple main-effects fit. If we believe only in wild outliers, any such piece of a nice fit is very unlikely to have arisen by chance from some outliers and hence identifies part of the true model. In principle, we need only two equal elements or one closed chain as above to identify part of the fit, although we will be happier if we have more redundance. If we fear mean outliers, however, we have to check many more elements and closed chains of elements, since it may easily happen that we get several contradictory parts of nice fits. (The outliers can "cooperate" to deceive or at least confuse the statistician, as in the 4×4 example above.) If there is a clear majority of data or elements in one nice fit, we shall tentatively believe in this one; but there are easily cases (e.g., with two mean outliers in a 3×3) with several equivalent-looking fits.

It is very instructive to study how many wild outliers and how many mean outliers in various positions of various designs can be identified or at least indicate the presence of outliers somewhere. These investigations are important if one fears a certain percentage of gross errors and wants to build a design sufficiently large and redundant so that certain effects can still be safely identified. As mentioned (cf. also the 2×3 above), we need more redundance than for estimation of missing values, even if we fear only wild outliers, and still more if we fear also mean outliers.—

To round off this section, we may add a few words about the remaining contents of Hampel (1975) for those who have read the paper. While "most robust" estimation of single correlations is treated, there is still the question of how to achieve a positive semidefinite correlation matrix. The claim for the breakdown point of the "rather strange-looking method" for covariance matrix estimation (Part 2.3) is wrong, as was found by Stahel (1981a); but the method itself, also according to Stahel, still deserves closer consideration. The "most robust" regression estimator has meanwhile been worked out in detail, generalized and programmed by Rousseeuw (cf. Subsection 6.4a). The infinite sequence of estimators containing the "shordth" still merits a closer look. In particular, the minimum shordth deviation estimator can be used to describe data containing two different systematic errors (in all data) and being contaminated by up to $n/2$ gross errors, and the following members of the sequence describe contaminated data containing an additive overlay of several systematic errors, as described in Youden (1972). Most other points in the above-mentioned article are discussed extensively in this book.

*8.5.   SMALL-SAMPLE ASYMPTOTICS

8.5a.   Introduction

Very often it is an intractable problem to find the exact finite-sample distribution of robust estimators and test statistics and approximations must be used. If the statistic is asymptotically normal, the first natural candidate is the normal approximation which can be carried out using the influence function [cf. (2.1.8), and (4.2.2)]. Although this approximation is often sufficient for simple robust estimators and simple models (e.g., M-estimators for location), it deteriorates quickly when accuracy is required for small-sample sizes and more complex models. In particular, numerical work by Maronna (1980, personal communication) for regression through the origin indicates that the density of M-estimators converges slowly to the asymptotic normal density especially when the design is unbalanced.

One might try to improve the normal approximation by using the first few terms of an Edgeworth expansion (cf. Feller, 1971, Chapter 16). This is an expansion in powers of $n^{-1/2}$, where the constant term is the normal density. It turns out in general that the Edgeworth expansion provides a good approximation in the center of the density, but can be inaccurate in the tails where it can even become negative. Thus it can be unreliable for calculating tail probabilities (the values usually of interest) when the sample size is small. In this section we discuss techniques that overcome this problem.

Saddlepoint techniques, already widely used in statistical mechanics (cf. Khinchin, 1949), were used by H. E. Daniels in a pioneering paper in 1954 to derive an approximation to p_n, the density of the mean of n independent identically distributed observations $x_1, \ldots, x_n$ with the underlying density f. The key idea is as follows. The density p_n can be written as an integral on the complex plane by means of a Fourier transform. Since the integrand is of the form $\exp(n \cdot w(z))$, the major contribution to this integral for large n will come from a neighborhood of the saddlepoint z_0, a zero of $w'(z)$. By means of the method of steepest descent, one can then derive a complete expansion for p_n with terms in powers of n^{-1}. Daniels (1954) also showed that this expansion is equivalent to that obtained using the idea of the conjugate density (see Cramer, 1938; Khinchin, 1949). The key point can be summarized as follows. First, recenter the original underlying distribution f at the point t where p_n is to be approximated, that is, to f define its conjugate (or associate) density h_t. Then use the Edgeworth expansion locally at t with respect to h_t and transform the results back in terms of the original density f. Since t is the mean of the conjugate density h_t the

Edgeworth expansion at t with respect to h_t is in fact of order n^{-1} and provides a good approximation locally at that point.

Other "small-sample asymptotic techniques," which are closely related to saddlepoint techniques, were introduced independently by Hampel (1973b) and are explicitly based on the idea of recentering the original distribution. The key difference is the expansion of the logarithmic derivative p'_n/p_n instead of p_n. A consequence of this is that the normalizing constant, that is, the constant that makes the total mass equal 1, must be determined numerically. This turns out to be an advantage since this rescaling greatly improves the approximation. The striking characteristic of these approximations is that the first few terms (or even just the leading term) of the expansion give very accurate approximation even in the extreme tails and down to small-sample sizes ($n = 3, 4$).

8.5b. Small-Sample Asymptotics for M-Estimators

In this subsection we derive briefly the small-sample asymptotic approximation to the density of a location M-estimator T_n defined by

$$\sum_{i=1}^{n} \psi(x_i - T_n) = 0$$

[cf. (2.3.3) and (2.3.9)]. We follow the presentation in Field and Hampel (1982). For details, numerical results and a deeper discussion we refer to that paper.

To begin, assume ψ is strictly monotone. Then

$$P(T_n \le t) = P\left\{ \sum_{i=1}^{n} \psi(x_i - t) \le 0 \right\} = \int \cdots \int \prod_{i=1}^{n} g_t(y_i)\, dy_1 \cdots dy_n,$$

$$p_n(t) = n \int \cdots \int \prod_{i=1}^{n-1} g_t(y_i)\, \partial/\partial t \{ g_t(y_n) \}\, dy_1 \cdots dy_n, \qquad (8.5.1)$$

where the integrals are over the range

$$\left\{ \sum_{i=1}^{n} y_i \le 0 \right\}$$

and g_t is the density of $\psi(x_1 - t)$. Since $\int \partial/\partial t g_t(y)\, dy = f\{\psi^{-1}(y) + t\}$, we get from (8.5.1)

$$p_n(t) = n \int \cdots \int f \left\{ \psi^{-1} \left(- \sum_{i=1}^{n-1} y_i \right) + t \right\} \prod_{i=1}^{n-1} g_t(y_i)\, dy_1 \cdots dy_{n-1}.$$

$$(8.5.2)$$

The key step is to recenter g_t about t by replacing g_t with the conjugate or associate density h_t.

Let

$$h_t(y) = c_t \exp(\alpha_t y) g_t(y),$$

where the constants c_t and α_t are determined by

$$\int h_t(y)\, dy = 1 \quad \text{and} \quad \int y h_t(y)\, dy = 0,$$

and denote by $j_{\nu,t}$ the density of the sum of ν independent random variables each with density h_t. Then, after some calculations, we obtain from (8.5.2)

$$p_n(t) = nc_t^{-n} \int j_{n-1,t}(s) \psi' \{ \psi^{-1}(-s) \} h_t(-s)\, ds, \qquad (8.5.3)$$

and similarly

$$p_n'(t) = n(n-1)c_t^{-n} \int\int j_{n-2,t}(r-s)\psi'\{\psi^{-1}(-r)\}$$

$$\times f'/f\{\psi^{-1}(s) + t\} h_t(-r)h_t(s)\, dr\, ds$$

$$+ nc_t^{-n} \int j_{n-1,t}(s)\psi'\{\psi^{-1}(-s)\} f'/f\{\psi^{-1}(-s) + t\} h_t(-s)\, ds;$$

$$(8.5.4)$$

whence $p_n'(t)/p_n(t)$ is obtained as a fraction not involving c_t.

The last step is to approximate $j_{n,t}(s)/j_{n,t}(0)$ by a series with the coefficients determined by the Edgeworth expansion of $j_{n,t}(s)$ at the origin. This enables us to use locally the good properties of the Edgeworth expansion at the origin. We finally obtain the following expansion:

$$p_n'(t)/p_n(t) = \left(n - \tfrac{1}{2}\right) A_{2,t} + \lambda_{3,t} A_{4,t}/(2\sigma_t) - A_{4,t} A_{3,t}/\left(\sigma_t^2 A_{1,t}\right)$$

$$- A_{6,t}/\left(2\sigma_t^2\right) + A_{5,t}/A_{1,t} + O(1/n), \qquad (8.5.5)$$

where

$$A_{1,t} = \int \psi'(x - t) c_t \exp\{\alpha_t \psi(x - t)\} f(x)\, dx,$$

$$A_{2,t} = \int c_t \exp\{\alpha_t \psi(x - t)\} f'(x)\, dx,$$

$$A_{3,t} = \int \psi(x - t) \psi'(x - t) c_t \exp\{\alpha_t \psi(x - t)\} f(x)\, dx,$$

$$A_{4,t} = \int \psi(x - t) c_t \exp\{\alpha_t \psi(x - t)\} f'(x)\, dx,$$

$$A_{5,t} = \int \psi'(x - t) c_t \exp\{\alpha_t \psi(x - t)\} f'(x)\, dx,$$

$$A_{6,t} = \int \psi^2(x - t) c_t \exp\{\alpha_t \psi(x - t)\} f'(x)\, dx,$$

c_t and α_t satisfy the equations

$$\int c_t \exp\{\alpha_t \psi(x - t)\} f(x)\, dx = 1,$$

$$\int \psi(x - t) c_t \exp\{\alpha_t \psi(x - t)\} f(x)\, dx = 0,$$

$$\sigma_t^2 = \int \psi^2(x - t) c_t \exp\{\alpha_t \psi(x - t)\} f(x)\, dx,$$

$$\lambda_{3,t} = \int \psi^3(x - t) c_t \exp\{\alpha_t \psi(x - t)\} f(x)\, dx/\sigma_t^3,$$

where all the integrals are over the range $(-\infty, \infty)$.

Remark 1. Although it is theoretically less satisfactory, it is technically simpler to derive an expansion directly for $p_n(t)$. To second order this is

$$p_n(t) = (2\pi)^{-1/2} n^{1/2} c_t^{-n} \sigma_t^{-1} A_{1,t} \{1 + O(1/n)\}. \qquad (8.5.6)$$

This expression coincides with the saddlepoint approximation (the leading term of the saddlepoint expansion). For higher accuracy in numerical computations one will determine the normalizing constant empirically by numerical integration, and thus one needs only the approximation

$$p_n(t) \propto c_t^{-n} \sigma_t^{-1} A_{1,t}. \qquad (8.5.7)$$

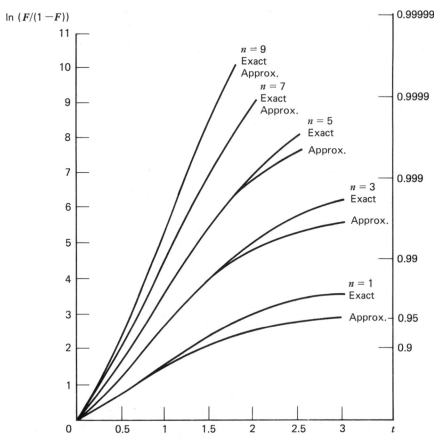

Figure 1. Exact and approximate cumulative [using (8.5.5) or (8.5.7)] of Huber-estimator ($k = 1.4$) under 5%-contaminated normal (with contamination at $\pm\infty$) in logistic scale.

Remark 2. The formulas (8.5.5) and (8.5.6) remain true if the condition of strict monotonicity of ψ is replaced by that ψ is a bounded monotone increasing continuous function.

Remark 3. Direct saddlepoint techniques can be used to derive small-sample asymptotic approximations to p_n (cf. Daniels, 1983). It turns out that the second-order formula (8.5.5) and the saddlepoint approximation *with renormalization* give identical results (see Field and Hampel, 1982; Daniels, 1983).

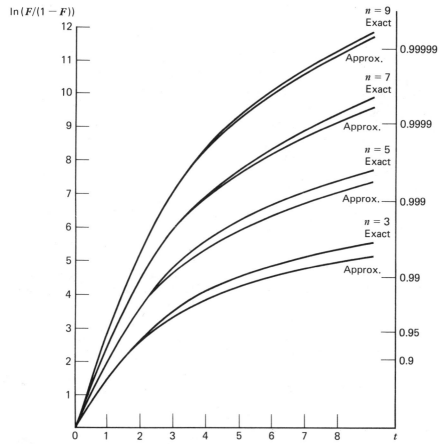

Figure 2. Exact and approximate cumulative [using (8.5.5) or (8.5.7)] of Huber-estimator ($k = 1.5$) under Cauchy in logistic scale.

Figures 1 and 2 show the excellent approximations to the density of the Huber-estimator down to very small-sample sizes.

8.5c. Further Applications

In the last few years there has been a revival of interest in this area. Daniels (1980) showed that the normal, Gamma, and inverse normal are the only possible densities for which the renormalized saddlepoint approximation reproduces exactly the density of the mean. Field (1982) extended the results presented in Subsection 8.5b to multivariate M-estimators (as defined in Subsection 4.2c). Durbin (1980) applied similar techniques to derive approximations of the density of sufficient estimators. Field and Ronchetti (1985) derived small-sample asymptotic approximations to the tail area of M-statistics and used them in robust testing (see subsection 3.2c). Easton and Ronchetti (1984) extended saddlepoint approximations to general statistics and applied them to approximate the density of linear combinations of order statistics.

We refer to the papers by Barndorff-Nielsen and Cox (1979) and Field and Hampel (1982) for an overview and comparison between these new techniques and the classical ones.

Some subjects of ongoing research include the application of these techniques to robust regression and to the problem of finding optimal robust estimators for fixed finite n as mentioned in Field and Hampel, (1978, p. 11).

EXERCISES AND PROBLEMS

The attentive reader who has worked through the book so far will have noticed that the main text of Chapter 8 is full of potential exercises and problems, and he will not need a separate list of them. Some major types of exercises are the following:

(i)   Work out the details of the proofs or arguments only sketched in the text.

(ii) Work out numerical examples for statements in the text.

(iii) Find real life examples illuminating the text.

(iv) Work on the open research problems.

References

Sections in which the reference is cited are given in brackets.

Abramowitz, M. and Stegun, I. A. (eds.) (1972). *Handbook of Mathematical Functions* (9th printing). Dover, New York. [5.4b]

Ahnert, P. (1961). *Kalender für Sternfreunde*. Johann Ambrosius Barth-Verlag, Leipzig. [8.1a]

Akaike H. (1973). Information theory and an extension of the maximum likelihood principle. In: 2nd International Symposium on Information Theory, B. N. Petrov and F. Csaki (eds.). Academiai Kiado, Budapest, pp. 267–281. [7.1b, 7.3d]

Andersen, A. H., Jensen, E. B., and Schou, G. (1981). Two-way analysis of variance with correlated errors. *Int. Statist. Rev.* **49**, 153–167. [7.1a]

Anderson, T. W. (1958). *An Introduction to Multivariate Statistical Analysis*. Wiley, New York. [2.0]

Andrews, D. F. (1974). A robust method for multiple linear regression. *Technometrics* **16**, 523–531. [6.4a]

Andrews, D. F., Bickel, P. J., Hampel, F. R., Huber, P. J., Rogers, W. H., and Tukey, J. W. (1972). *Robust Estimates of Location: Survey and Advances*. Princeton University Press, Princeton, N.J. [1.2b, 1.2e, 1.3d, 1.3e, 1.4b, 1.ex, 2.1e, 2.3a, 2.6a, 2.6b, 4.4a, 8.2a, 8.2c]

Anonymous (1821). Dissertation sur la recherche du milieu le plus probable, entre les résultats de plusieurs observations ou expériences. *Ann. Math. Pures Appl.* **12**, 181–204. [1.3a]

Anscombe, F. J. (1960). Rejection of outliers. *Technometrics* **2**, 123–147. [1.1b, 1.4b]

Anscombe, F. J. and Tukey, J. W. (1963). The examination and analysis of residuals. *Technometrics* **5**, 141–160. [1.1b]

Atkinson, A. C. (1982). Robust and diagnostic regression analysis. *Commun. Statist. A* **11**, 2559–2571. [6.4d]

Bahadur, R. R. (1960). Stochastic comparison of tests. *Ann. Math. Statist.* **31**, 276–295. [3.6a]

Bahadur, R. R. (1971). *Some Limit Theorems in Statistics.* SIAM, Philadelphia. [3.6a]

Barndorff-Nielsen, O. and Cox, D. R. (1979). Edgeworth and saddle-point approximations with statistical applications (with discussion). *J. Roy. Statist. Soc. Ser. B* **41**, 279–312. [8.5c]

Barndorff-Nielsen, O., Blæsild, P., Jensen, J. L., and Jørgensen, B. (1982). Exponential transformation models. *Proc. Roy. Soc. London Ser. A* **379**, 41–65. [4.5a]

Barnett, V. and Lewis, T. (1978). *Outliers in Statistical Data.* Wiley, New York. [1.1b, 1.3a, 1.4a]

Bartlett, M. S. (1935). The effect of non-normality on the t-distribution. *Proc. Cambridge Philos. Soc.* **31**, 223–231. [1.3a]

Beaton, A. E. and Tukey, J. W. (1974). The fitting of power series, meaning polynomials, illustrated on band-spectroscopic data. *Technometrics* **16**, 147–185. [2.6a]

Beckman, R. J. and Cook, R. D. (1983). Outlier..........s (with discussion). *Technometrics* **25**, 119–163. [1.4a]

Bednarek-Kozek, B. and Kozek, A. (1977). Stability, sensitivity and sensitivity of characterizations. Preprint 99. Institute of Mathematics, Polish Academy of Sciences, Warsaw. [1.1b]

Bednarski, T. (1980). Solution of minimax test problems for special capacities. Preprint 230. Institute of Mathematics, Polish Academy of Sciences, Warsaw. [1.1b, 3.7]

Bednarski, T. (1982). Binary experiments, minimax tests and 2-alternating capacities. *Ann. Statist.* **10**, 226–232. [1.1b, 3.7]

Belsley, D. A., Kuh, E., and Welsch, R. E. (1980). *Regression Diagnostics: Identifying Influential Data and Sources of Collinearity.* Wiley, New York. [1.1c, 6.4d]

Beran, J. (1984). Maximum likelihood estimation for Gaussian processes with long-range dependence. Abstracts Book, 16th European Meeting of Statisticians, Marburg, West Germany, p. 71. [8.1d]

Beran, J. and Künsch, H. (1985). Location estimators for processes with long range dependence. Research Report 40. Seminar für Statistik, ETH, Zurich. [8.1c]

Beran, R. (1972). Upper and lower risks and minimax procedures. Proceedings of the Sixth Berkeley Symposium on Mathematical Statistics and Probability, Vol. 1, pp. 1–16. [1.1b]

Beran, R. (1977a). Robust location estimates. *Ann. Statist.* **5**, 431–444. [1.1b]

Beran, R. (1977b). Minimum Hellinger distance estimates for parametric models. *Ann. Statist.* **5**, 445–463. [1.1b, 2.3d]

Beran, R. (1979). Testing for ellipsoidal symmetry of a multivariate density. *Ann. Statist.* **7**, 150–162. [1.1b]

Beran, R. (1980). Asymptotic lower bounds for risk in robust estimation. *Ann. Statist.* **8**, 1252–1264. [1.1b]

Beran, R. (1982). Robust estimation in models for independent non-identically distributed data. *Ann. Statist.* **10**, 415–428. [1.1b, 6.4d]

Berger, J. (1980). A robust generalized Bayes estimator and confidence region for a multivariate normal mean. *Ann. Statist.* **8**, 716–761. [1.1b]

Berger, J. (1982). Bayesian robustness and the Stein effect. *J. Am. Statist. Assoc.* **77**, 358–368. [1.1b]

Berkson, J. (1938). Some difficulties of interpretation encountered in the application of the chi-square test. *J. Am. Statist. Assoc.* **33**, 526–536. [8.1c]

Bernoulli, D. (1777). Dijudicatio maxime probabilis plurium observationum discrepantium atque verisimillima inductio inde formanda. *Acta Acad. Sci. Petropolit.* **1**, 3–33. (English translation by C. G. Allen (1961). *Biometrika* **48**, 3–13.) [1.3a]

Bessel, F. W. (1818). *Fundamenta Astronomiae.* Nicolovius, Königsberg. [1.2b]

Bessel, F. W. and Baeyer, J. J. (1838). Gradmessung in Ostpreussen und ihre Verbindung mit Preussischen und Russischen Dreiecksketten. Druckerei der Königlichen Akademie der Wissenschaften Berlin. Reprinted in part in *Abhandlungen von F. W. Bessel*, R. Engelmann (ed.). W. Engelmann, Leipzig, 1876, Vol. 3, pp. 62–138. [1.2b, 1.3a]

Bhansali, R. J. and Downham, D. Y. (1977). Some properties of the order of an autoregressive model selected by a generalization of Akaike's FPE criterion. *Biometrika* **64**, 547–551. [7.3d]

Bickel, P. J. (1975). One-step Huber estimates in the linear model. *J. Am. Statist. Assoc.* **70**, 428–434. [1.1b, 2.3a, 6.4c]

Bickel, P. J. (1976). Another look at robustness: A review of reviews and some new developments. *Scand. J. Statist.* **3**, 145–168. [1.1b, 7.3a]

Bickel, P. J. (1981). Quelques aspects de la statistique robuste. Lecture Notes in Mathematics 876, Springer, Berlin. [1.1b, 4.3c]

Bickel, P. J. (1984a). Parametric robustness: Small biases can be worthwhile. *Ann. Statist.* **12**, 864–879. [1.1b]

Bickel, P. J. (1984b). Robust regression based on infinitesimal neighbourhoods. *Ann. Statist.* **12**, 1349–1368. [1.1b, 6.3b]

Bickel, P. J. and Collins, J. R. (1983). Minimizing Fisher information over mixtures of distributions. *Sankhyā A* **45**, 1–19. [1.1b]

Bickel, P. J. and Herzberg, A. M. (1979). Robustness of design against autocorrelation in time I: Asymptotic theory, optimality for location and linear regression. *Ann. Statist.* **7**, 77–95. [1.1b]

Bickel, P. J. and Lehmann, E. L. (1975). Descriptive statistics for nonparametric models. I. Introduction. II. Location. *Ann. Statist.* **3**, 1038–1044. 3, 1045–1069. [1.1b]

442

Billingsley, P. (1961). The Lindeberg-Lévy theorem for martingales. *Proc. Am. Math. Soc.* **12**, 788–792. [8.3b]

Bloomfield, P. and Steiger, W. L. (1983). *Least Absolute Deviations: Theory, Applications, and Algorithms.* Birkhäuser, Boston. [6.2]

Boente, G., Fraiman, R., and Yohai, V. J. (1982). Qualitative robustness for general stochastic processes. Technical Report 26. Department of Statistics, University of Washington, Seattle, Wash. [8.3c]

Boos, D. D. (1981). Minimum distance estimators for location and goodness of fit. *J. Am. Statist. Assoc.* **76**, 663–670. [2.3d]

Boos, D. D. and Serfling, R. J. (1976). Development and comparison of *M*-estimators for location on the basis of the asymptotic variance functional. Statistics Report M380. Florida State University, Tallahassee. [2.5a]

Boos, D. D. and Serfling, R. J. (1980). A note on differentials and the CLT and LIL for statistical functions, with application to *M*-estimates. *Ann. Statist.* **8**, 618–624. [1.3f, 2.1b, 2.3a]

Boscovich, R. J. (1757). De litteraria expeditione per pontificiam ditionem, et synopsis amplioris operis, ac habentur plura eius ex exemplaria etiam sensorum impressa. *Bononiensi Scientiarum et Artium Instituto Atque Academia Commentarii* **4**, 353–396. [1.3a]

Box, G. E. P. (1953). Non-normality and tests on variances. *Biometrika* **40**, 318–335. [1.2e, 1.3a, 3.1]

Box, G. E. P. and Andersen, S. L. (1955). Permutation theory in the derivation of robust criteria and the study of departures from assumption. *J. Roy. Statist. Soc. Ser. B* **17**, 1–34. [1.2e, 1.3a]

Box, G. E. P., Hunter, W. G., and Hunter, J. S. (1978). *Statistics for Experimenters.* Wiley, New York. [8.1a]

Box, G. E. P. and Tiao, G. C. (1962). A further look at robustness via Bayes's theorem. *Biometrika* **49**, 419–432. [1.1b]

Box, G. E. P. and Tiao, G. C. (1968). A Bayesian approach to some outlier problems. *Biometrika* **55**, 119–129. [1.1b]

Box, G. E. P. and Tiao, G. C. (1973). *Bayesian Inference in Statistical Analysis.* Addison-Wesley, Reading, Mass. [1.1b]

Breuer, P. and Major, P. (1983). Central limit theorems for non-linear functionals of Gaussian fields. *J. Mult. Anal.* **13**, 425–441. [8.16]

Broffitt, B., Clarke, W. R., and Lachenbruch, P. A. (1980). The effect of Huberizing and trimming on the quadratic discriminant function. *Commun. Statist. A* **9**, 13–25. [5.1]

Brown, G. W. and Mood, A. M. (1951). On median tests for linear hypotheses. Proceedings of the Second Berkeley Symposium on Mathematical Statistics and Probability, University of California Press, Berkeley, Calif., pp. 159–166. [6.4a]

Brownlee, K. A. (1965). *Statistical Theory and Methodology in Science and Engineering,* 2nd ed. Wiley, New York. [6.4a]

Bühler, W. J. and Puri, P. S. (1966). On optimal asymptotic tests of composite hypotheses with several constraints. *Z. Wahrsch. verw. Geb.* **5**, 71–88. [7.4a]

Buja, A. (1980). Sufficiency, least favorable experiments and robust tests. Ph.D. thesis. ETH, Zurich. [1.1b]

Cambanis S., Huang S., and Simons G. (1981). On the theory of elliptically contoured distributions. *J. Mult. Anal.* **11**, 368–385. [5.2b]

Campbell, N. A. (1978). The influence function as an aid in outlier detection in discriminant analysis. *Appl. Statist.* **27**, 251–258. [1.1b]

Campbell, N. A. (1980). Robust procedures in multivariate analysis I: Robust covariance estimation. *Appl. Statist.* **29**, 231–237. [1.1b, 5.1]

Campbell, N. A. (1982). Robust procedures in multivariate analysis II: Robust canonical variate analysis. *Appl. Statist.* **31**, 1–8. [1.1b, 5.1]

Carroll, R. J. (1979). On estimating variances of robust estimators when the errors are asymmetric. *J. Am. Statist. Assoc.* **74**, 674–679. [6.4d]

Carroll, R. J. (1980). A robust method for testing transformations to achieve approximate normality. *J. Roy. Statist. Soc. Ser. B* **42**, 71–78. [7.3a]

Carroll, R. J. and Ruppert, D. (1980). A comparison between maximum likelihood and generalized least squares in a heteroscedastic linear model. *J. Am. Statist. Assoc.* **77**, 878–882. [6.4d]

Carroll, R. J. and Ruppert, D. (1982). Robust estimation in heteroscedastic linear models. *Ann. Statist.* **10**, 429–441. [6.4d]

Chauvenet, W. (1863). Method of Least Squares. Appendix to *Manual of Spherical and Practical Astronomy 2*. pp. 469–566, tables pp. 593–599. Reprinted (1960) (5th ed.). Dover, New York. [1.3a]

Chen, Z. and Li, G. (1981). Robust principal components and dispersion matrices via projection pursuit. Research Report. Department of Statistics, Harvard University, Cambridge, Mass. [5.1]

Clarke, B. R. (1983). Uniqueness and Fréchet differentiability of functional solutions to maximum likelihood type equations. *Ann. Statist.* **11**, 1196–1205. [1.3f, 2.3a]

Clarke, B. R. (1984a). Asymptotic theory for description of regions in which Newton-Raphson iterations converge to location *M*-estimators. Research Report. Murdoch University, Murdoch, Western Australia. [1.3f, 8.2c]

Clarke, B. R. (1984b). Nonsmooth analysis and Fréchet differentiability of *M*-functionals. Research Report. Murdoch University, Murdoch, Western Australia. [1.3f, 2.3a]

Coale, A. J. and Stephan, F. F. (1962). The case of the Indians and the teen-age widows. *J. Am. Statist. Assoc.* **57**, 338–347. [1.2c]

Cochran, W. G. (1947). Some consequences when the assumptions for the analysis of variance are not satisfied. *Biometrics* **3**, 22–38. [1.2c]

Cochran, W. G. and Cox, G. M. ((1950) 1957). *Experimental Designs.* Wiley, New York. [1.1d]

Collins, J. R. (1976). Robust estimation of a location parameter in the presence of asymmetry. *Ann. Statist.* **4**, 68–85. [2.3a, 2.6c, 2.7]

Collins, J. R. (1977). Upper bounds on asymptotic variances of M-estimators of location. *Ann. Statist.* **5**, 646–657. [2.3a, 2.7]

Collins, J. R. (1982). Robust M-estimators of location vectors. *J. Mult. Anal.* **12**, 480–492. [5.1]

Collins, J. R. and Portnoy, S. L. (1981). Maximizing the variance of M-estimators using the generalized method of moment spaces. *Ann. Statist.* **9**, 567–577. [2.5a, 2.7]

Collins, J. R., Sheahan, J. N., and Zheng, Z. (1985). Robust estimation in the linear model with asymmetric error distributions. *J. Mult. Anal.* **15** (to appear). [6.4d]

Cook, R. D. and Weisberg, S. (1980). Characterization of an empirical influence function for detecting influential cases in regression. *Technometrics* **22**, 495–508. [6.4d]

Cook, R. D. and Weisberg, S. (1982). *Residuals and Influence in Regression.* Chapman and Hall, New York. [1.1c, 6.4d]

Cox, D. (1981). Metrics on stochastic processes and qualitative robustness. Technical Report 3. Department of Statistics, University of Washington, Seattle, Wash. [8.3c]

Cox, D. R. and Hinkley, D. V. (1974). *Theoretical Statistics.* Chapman and Hall, London. [4.6b]

Cramèr, H. (1938). Sur un nouveau théorème-limite de la théorie des probabilités. *Actualités scientifiques et industrielles* **736**, Hermann, Paris. [8.5a]

Cushny, A. R. and Peebles, A. R. (1905). The action of optical isomers. II. Hyoscines. *J. Physiol.* **32**, 501–510. [2.0]

Daniel, C. (1976). *Applications of Statistics to Industrial Experimentation.* Wiley, New York. [1.1c, 1.1d, 1.2c, 1.4a, 8.4b]

Daniel, C. and Wood, F. S. ((1971) 1980). *Fitting Equations to Data.* Wiley, New York. [1.2c, 1.4a, 6.4a, 6.4c, 8.1a]

Daniels, H. E. (1954). Saddlepoint approximations in statistics. *Ann. Math. Statist.* **25**, 631–650. [1.3d, 3.2c, 7.5b, 8.5a]

Daniels, H. E. (1980). Exact saddlepoint approximations. *Biometrika* **67**, 59–63. [8.5c]

Daniels, H. E. (1983). Saddlepoint approximations for estimating equations. *Biometrika* **70**, 89–96. [8.5b]

de Finetti, B. (1961). The Bayesian approach to the rejection of outliers. Proceedings of the Fourth Berkeley Symposium on Mathematical Statistics and Probability, Vol. 1, University of California Press, Berkeley, Calif., pp. 199–210. [1.1b]

Dempster, A. P. (1968). A generalization of Bayesian inference (with discussion). *J. Roy. Statist. Soc. Ser. B* **30**, 205–247. [1.1b]

Dempster, A. P. (1975). A subjectivist look at robustness. Proceedings of the 40th Session of the ISI, Warsaw, Vol. XLVI, Book 1, pp. 349–374. [1.1b, 1.3f, 1.4a]

Denby, L. and Mallows, C. L. (1977). Two diagnostic displays for robust regression analysis. Technometrics 19, 1–13. [1.1b]

Devlin, S. J., Gnanadesikan, R., and Kettenring, J. R. (1975). Robust estimation and outlier detection with correlation coefficients. Biometrika 62, 531–545. [5.1]

Devlin, S. J., Gnanadesikan, R., and Kettenring, J. R. (1981). Robust estimation of dispersion matrices and principal components. J. Am. Statist. Assoc. 76, 354–362. [5.1]

Dobrushin, R. L. (1979). Gaussian and their subordinated self-similar random generalized fields. Ann. Prob. 7, 1–28. [8.1b]

Dobrushin, R. L. and Major, P. (1979). Non-central limit theorems for non-linear functionals of Gaussian fields. Z. Wahrsch. verw. Geb. 50, 27–52. [8.1b]

Donoho, D. L. (1982). Breakdown properties of multivariate location estimators. Ph.D. qualifying paper. Department of Statistics, Harvard University, Cambridge, Mass. [2.3d, 5.1, 5.5c]

Donoho, D. L. and Huber, P. J. (1983). The notion of breakdown point. In A Festschrift for Erich L. Lehmann, P. J. Bickel, K. A. Doksum, J. L. Hodges Jr. (eds.). Wadsworth, Belmont, Calif., pp. 157–184. [1.4b, 2.2a, 5.1, 5.5c]

Donoho, D. L., Johnstone, I., Rousseeuw, P. J., and Stahel, W. A. (1985). Comments on 'Projection Pursuit' by P. J. Huber. Ann. Statist. 13, 496–500. [2.3d]

Draper, N. R. and Smith, H. ((1966) 1981). Applied Regression Analysis. Wiley, New York. [6.4a, 7.5d, 8.2b]

Durbin, J. (1980). Approximations for densities of sufficient estimators. Biometrika 67, 311–333. [8.5c]

Dutter, R. (1976). Computer linear robust curve fitting program LINWDR. Research Report 10. Fachgruppe für Statistik, ETH, Zurich. [6.4c]

Dutter, R. (1977). Numerical solution of robust regression problems: Computational aspects, a comparison. J. Statist. Comput. Simul. 5, 207–238. [6.4c]

Dutter, R. (1979). Robuste Regression. Bericht Nr. 135 (1980). Institut für Statistik, Technische Universität, Graz, Austria. [6.4c]

Dutter, R. (1983). Computer program BLINWDR for robust and bounded influence regression. Research Report 8. Institut für Statistik, Technische Universität, Graz, Austria. [6.4c]

Dwyer, P. S. (1967). Some applications of matrix derivatives in multivariate analysis. J. Am. Statist. Assoc. 62, 607–625. [5.2b]

Easton, G. S. and Ronchetti, E. (1984). General saddlepoint approximations with applications to L-statistics. Technical Report 274, Series 2. Department of Statistics, Princeton University, Princeton, N.J. (to appear in J. Am. Statist. Assoc.). [8.5c]

Edgeworth, F. Y. (1883). The method of least squares. Philos. Mag. 23, 364–375. [1.3a]

Efron, B. (1979). Bootstrap methods: Another look at the jackknife. *Ann. Statist.* **7**, 1–26. [2.1e]

Eplett, W. J. R. (1980). An influence curve for two-sample rank tests. *J. Roy. Statist. Soc. Ser. B* **42**, 64–70. [3.1, 3.6b]

Ezekiel, M. and Fox, K. A. (1959). *Methods of Correlation and Regression Analysis.* Wiley, New York. [6.1b]

Feller, W. (1951). The asymptotic distribution of the range of sums of independent random variables. *Ann. Math. Statist.* **22**, 427–432. [8.1b]

Feller, W. ((1966) 1971). *An Introduction to Probability Theory and Its Applications*, Vol. II. Wiley, New York. [8.5a]

Fernholz, L. T. (1983). *Von Mises Calculus for Statistical Functionals.* Lecture Notes in Statistics 19. Springer, New York. [1.3f, 2.1b, 2.3a]

Field, C. A. (1982). Small sample asymptotic expansions for multivariate *M*-estimates. *Ann. Statist.* **10**, 672–689. [7.5b, 8.5c]

Field, C. A. and Hampel, F. R. (1978). Small-sample asymptotic distributions of *M*-estimators of location. Research Report 17. Fachgruppe für Statistik, ETH, Zurich. [1.3d, 3.2c, 7.5b, 8.2a, 8.5b, 8.5c]

Field, C. A. and Hampel, F. R. (1982). Small-sample asymptotic distributions of *M*-estimators of location. *Biometrika* **69**, 29–46. [1.3d, 3.2c, 7.5b, 8.2a, 8.5b, 8.5c]

Field, C. A. and Ronchetti, E. (1985). A tail area influence function and its application to testing. *Commun. Statist. Ser. C.* **4** (19–41). [3.2c, 8.5c]

Filippova, A. A. (1961). Mises' theorem on the asymptotic behavior of functionals of empirical distribution functions and its statistical applications. *Theory Prob. Appl.* **7**, 24–57. [2.1d]

Fisher, R. A. (1920). A mathematical examination of the methods of determining the accuracy of an observation by the mean error and by the mean square error. *Mon. Not. Roy. Astr. Soc.* **80**, 758–770. [1.2d]

Fisher, R. A. (1922). On the mathematical foundations of theoretical statistics. *Philos. Trans. Roy. Soc. London Ser. A* **222**, 309–368. [1.2d, 1.2e, 1.3f, 8.2a]

Fisher, R. A. ((1925)1970). *Statistical Methods for Research Workers.* Oliver & Boyd, Edinburgh. [1.2a, 2.0]

Fisher, R. A. and Mackenzie, W. A. (1922). The correlation of weekly rainfall. *Quart. J. Roy. Meteorol. Soc.* **48**, 234–242. [8.1a]

Foutz, R. V. and Srivastava, R. C. (1977). The performance of the likelihood ratio test when the model is incorrect. *Ann. Statist.* **5**, 1183–1194. [7.2b]

Fox, R. and Taqqu, M. S. (1984). Maximum likelihood type estimation for the self-similarity parameter in Gaussian sequences. Technical Report 590. Operations Research, Cornell University, Ithaca, N.Y. [8.1b, 8.1d]

Fraser, D. A. S. (1968). *The Structure of Inference.* Wiley, New York. [4.5a]

Freedman, D., Pisani, R., and Purves, R. (1978). *Statistics.* Norton, New York. [1.2c, 8.1a]

Freedman, H. W. (1966). The "little variable factor". A statistical discussion of the reading of seismograms. *Bull. Seismol. Soc. Am.* **56**, 593–604. [1.2c]

Gastwirth, J. L. (1966). On robust procedures. *J. Am. Statist. Assoc.* **61**, 929–948. [2.3b]

Gastwirth, J. L. and Rubin, H. (1975). The behavior of robust estimators on dependent data. *Ann. Statist.* **3**, 1070–1100. [1.1b]

Gauss, C. F. (1816). Bestimmung der Genauigkeit der Beobachtungen. *Z. Astr. verw. Wiss.* In *Werke 4*, Dieterichsche Universitäts-Druckerei, Göttingen (1880), pp. 109–117. [1.4b]

Gauss, C. F. (1823). Theoria combinationis observationum erroribus minimis obnoxiae (pars prior), presented 2/15/1821. Commentationes societatis regiae scientiarum Gottingensis recentiores, Vol. 5, Göttingen. In *Werke 4*, Dieterichsche Universitäts-Druckerei, Göttingen (1880), pp. 1–108. [1.3a]

Gayen, A. K. (1950). The distribution of the variance ratio in random samples of any size drawn from non-normal universes. *Biometrika* **37**, 236–255. [1.3a]

Geary, R. C. (1936). The distribution of 'Student's' ratio for non-normal samples. *J. Roy. Statist. Soc. Suppl.* **3**, 178–184. [1.3a]

Geary, R. C. (1947). Testing for normality. *Biometrika* **34**, 209–242. [1.3a]

Gentleman, W. M. (1965). Robust estimation of multivariate location by minimizing p-th power deviations. Ph.D. thesis. Princeton University, Princeton, N.J. [2.3d]

Glaisher, J. W. L. (1872-73). On the rejection of discordant observations. *Mon. Not. Roy. Astr. Soc.* **33**, 391–402. [1.3a]

Gnanadesikan, R. (1977). *Methods for Statistical Data Analysis of Multivariate Observations.* Wiley, New York. [5.1]

Gnanadesikan, R. and Kettenring, J. R. (1972). Robust estimates, residuals, and outlier detection with multiresponse data. *Biometrics* **28**, 81–124. [5.1]

Graf, H. (1983). Long-range correlations and estimation of the self-similarity parameter. Ph.D. thesis. ETH, Zurich. [8.1a, 8.1b, 8.1d]

Graf, H., Hampel, F. R., and Tacier, J. (1984). The problem of unsuspected serial correlations. In *Robust and Nonlinear Time Series Analysis*. J. Franke, W. Härdle, R. D. Martin (eds.), Lecture Notes in Statistics 26. Springer, New York, pp. 127–145. [8.1a, 8.1b, 8.1c, 8.1d]

Grenander, U., Pollak, H. O., and Slepian, D. (1959). The distribution of quadratic forms in normal variates: A small sample theory with applications in spectral analysis. *J. Soc. Ind. Appl. Math.* **7**, 374–401. [7.5b]

Gross, A. M. and Tukey, J. W. (1973). The estimators of the Princeton robustness study. Technical Report 38, Series 2. Department of Statistics, Princeton University, Princeton, N.J. [1.4b]

Grubbs, F. E. (1969). Procedures for detecting outlying observations in samples. *Technometrics* **11**, 1–21. [1.4b]

Hájek, J. and Šidák, Z. (1967). *Theory of Rank Tests.* Academic Press, New York. [3.3a, 3.3b]

Hampel, F. R. (1968). Contributions to the theory of robust estimation. Ph.D. thesis. University of California, Berkeley. [1.1a, 1.1c, 1.3a, 1.3d, 1.3f, 2.1b, 2.1c, 2.2a, 2.4a, 2.7, 8.2a]

Hampel, F. R. (1971). A general qualitative definition of robustness. *Ann. Math. Statist.* **42**, 1887–1896. [1.1c, 1.3d, 2.2a, 2.2b, 2.3a, 2.3c]

Hampel, F. R. (1973a). Robust estimation: A condensed partial survey. *Z. Wahrsch. verw. Geb.* **27**, 87–104. [1.1a, 1.2e, 1.3d, 1.4a, 2.5a, 5.1, 6.1b, 6.2, 7.1b, 8.2a]

Hampel, F. R. (1973b). Some small sample asymptotics. Proceedings of the Prague Symposium on Asymptotic Statistics, Prague, pp. 109–126. [1.3d, 3.2c, 7.5b, 8.5a]

Hampel, F. R. (1974). The influence curve and its role in robust estimation. *J. Am. Statist. Assoc.* **69**, 383–393. [1.1c, 1.1d, 1.3d, 1.4b, 2.1b, 2.1c, 2.2b, 2.5a, 2.ex, 6.2, 8.2c]

Hampel, F. R. (1975). Beyond location parameters: Robust concepts and methods (with discussion). Proceedings of the 40th Session of the ISI, Vol. XLVI, Book 1, pp. 375–391. [1.1d, 1.2b, 1.3d, 2.3d, 4.6b, 5.1, 5.4b, 6.4a, 8.4a, 8.4b]

Hampel, F. R. (1976). On the breakdown points of some rejection rules with mean. Research Report 11. Fachgruppe für Statistik, ETH, Zurich. [1.2c, 1.2d, 1.4b, 2.2b]

Hampel, F. R. (1977). Rejection rules and robust estimates of location: An analysis of some Monte Carlo results. Transactions of the Seventh Prague Conference on Information Theory, Statistical Decision Functions, Random Processes, 1974 European Meeting of Statisticians, Vol. A, pp. 187–194. [1.2d, 1.4b]

Hampel, F. R. (1978a). Modern trends in the theory of robustness. *Math. Operationsforschung Statist. Ser. Statist.* **9**, 425–442. [1.2d, 1.3d, 1.4a, 1.4b, 6.1b, 8.2a]

Hampel, F. R. (1978b). Optimally bounding the gross-error-sensitivity and the influence of position in factor space. Proceedings of the ASA Statistical Computing Section, ASA, Washington, D.C., pp. 59–64. [6.2, 6.3a, 6.3b]

Hampel, F. R. (1979). Discussion of the meeting on robustness. Proceedings of the 42nd Session of the ISI, Manila, Vol. XLVIII, Book 2, pp. 100–102. [1.1c, 8.1c]

Hampel, F. R. (1980). Robuste Schätzungen: Ein anwendungsorientierter Überblick. *Biom. J.* **22**, 3–21. [1.1d, 1.2d, 1.3d, 1.4a, 1.4b, 6.1b, 7.1b, 8.2a, 8.4a]

Hampel, F. R. (1983a). The robustness of some nonparametric procedures. In *A Festschrift for Erich L. Lehmann*, P. J. Bickel, K. A. Doksum, and J. L. Hodges Jr. (eds.). Wadsworth, Belmont, Calif., pp. 209–238. [1.1b, 1.3d, 1.3e, 8.2c]

Hampel, F. R. (1983b). Some aspects of model choice in robust statistics. Proceedings of the 44th Session of the ISI, Madrid, Book 2, pp. 767–771. [1.3f, 7.3d]

Hampel, F. R. (1985). The breakdown points of the mean combined with some rejection rules. *Technometrics* **27**, 95–107. [1.2c, 1.4b, 8.2a]

Hampel, F. R., Rousseeuw, P. J., and Ronchetti, E. (1981). The change-of-variance curve and optimal redescending *M*-estimators. *J. Am. Statist. Assoc.* **76**, 643–648. [1.3d, 1.3e, 2.5a, 2.6a, 2.6b, 2.6c]

Hampel, F. R., Marazzi, A., Ronchetti, E., Rousseeuw, P. J., Stahel, W. A., and Welsch, R. E. (1982). Handouts for the instructional meeting on robust statistical methods. 15th European Meeting of Statisticians, Palermo, Italy. Fachgruppe für Statistik, ETH, Zurich. [1.3d, 1.4b, 2.1e, 2.2a, 6.4d, 8.1c, 8.1d]

Hampel, F. R., Schweingruber, M., and Stahel, W. (1983). Das Ergebnis des konfirmatorischen Tests im Grossversuch IV zur Hagelabwehr. Research Report 35. Fachgruppe für Statistik, ETH, Zurich. [7.3c]

Handschin, E., Schweppe, F. C., Kohlas, J., and Fiechter, A. (1975). Bad data analysis for power system state estimation. *IEEE Trans. Power Appar. Sys.* **PAS-94**, 329–337. [6.3a]

Harter, H. L. (1974–1976). The method of least squares and some alternatives, Parts I-VI. *Rev. Int. Inst. Statist.* **42**, 147–174 (Part I). **42**, 235–264 (Part II). **43**, 1–44 (Part III). **43**, 125–190 (Part IV). **43**, 269–278 (Part V). **44**, 113–159 (Part VI). [1.3a]

Hawkins, D. M. (1980). *Identification of Outliers.* Chapman and Hall, London. [1.4a]

Helmers, R. (1978). Edgeworth expansions for linear combinations of order statistics. Ph.D. thesis. Rijksuniversiteit Leiden, The Netherlands. [2.3b]

Helmers, R. (1980). Edgeworth expansions for linear combinations of order statistics with smooth weight function. *Ann. Statist.* **8**, 1361–1374. [2.3b]

Henderson, H. V. and Searle, S. R. (1979). Vec and Vech operators for matrices, with some uses in Jacobians and multivariate statistics. *Canad. J. Statist.* **7**, 65–81. [5.2a]

Hettmansperger, T. P. and Utts, J. M. (1977). Robustness properties for a simple class of rank estimates. *Commun. Statist. A* **6**, 855–868. [3.2d]

Hill, M. and Dixon, W. J. (1982). Robustness in real life: A study of clinical laboratory data. *Biometrics* **38**, 377–396. [1.2e]

Hill, R. W. (1977). Robust regression when there are outliers in the carriers. Ph.D. thesis. Harvard University, Cambridge, Mass. [6.3a, 6.4b]

Hill, R. W. and Holland, P. W. (1977). Two robust alternatives to least squares regression. *J. Am. Statist. Assoc.* **72**, 828–833. [6.4d]

Hoaglin, D. C., Mosteller, F., and Tukey, J. W. (eds.) (1983). *Understanding Robust and Exploratory Data Analysis.* Wiley, New York. [1.1b]

Hoaglin, D. C. and Welsch, R. E. (1978). The hat matrix in regression and ANOVA. *Am. Statist.* **32**, 17–22. [1.1b, 7.3c]

Hodges, J. L. Jr. (1967). Efficiency in normal samples and tolerance of extreme values for some estimates of location. Proceedings of the Fifth Berkeley Symposium on Mathematical Statistics and Probability, Vol. 1, University of California Press, Berkeley, Calif., pp. 163–186. [2.2a]

Hodges, J. L. Jr. and Lehmann, E. L. (1956). The efficiency of some nonparametric competitors of the *t*-test. *Ann. Math. Statist.* **27**, 324–335. [1.3a, 3.2d]

Hodges, J. L. Jr. and Lehmann, E. L. (1963). Estimates of location based on rank tests. *Ann. Math. Statist.* **34**, 598–611. [1.3a, 2.0, 2.3c, 3.2d]

Hogg, R. V. (1974). Adaptive robust procedures: A partial review and some suggestions for future applications and theory (with discussion). *J. Am. Statist. Assoc.* **69**, 909–927. [1.1b]

Hogg, R. V. (1977). An introduction to robust procedures. *Commun. Statist. A* **6**, 789–794. [1.1b]

Holland, P. W. and Welsch, R. E. (1977). Robust regression using iteratively reweighted least-squares. *Commun. Statist. A* **6**, 813–888. [2.6a, 6.4c]

Huber, P. J. (1964). Robust estimation of a location parameter. *Ann. Math. Statist.* **35**, 73–101. [1.1a, 1.2b, 1.3a, 1.3b, 1.3d, 1.3e, 1.4a, 2.0, 2.3a, 2.4b, 2.5a, 2.5c, 2.5d, 2.5f, 2.7, 2.ex, 4.2d, 8.2a]

Huber, P. J. (1965). A robust version of the probability ratio test. *Ann. Math. Statist.* **36**, 1753–1758. [1.1a, 1.3a, 1.3c, 3.1, 3.7, 8.2a]

Huber, P. J. (1967). The behavior of maximum likelihood estimates under nonstandard conditions. Proceedings of the Fifth Berkeley Symposium on Mathematical Statistics and Probability, Vol. 1, pp. 221–233. [1.3b, 2.3a, 2.5a, 4.2c]

Huber, P. J. (1968). Robust confidence limits. *Z. Wahrsch. verw. Geb.* **10**, 269–278. [1.3a, 1.3c, 1.ex, 3.1, 3.7, 8.2a]

Huber, P. J. (1970). Studentizing robust estimates. In *Nonparametric Techniques in Statistical Inference*, M. L. Puri (ed.). Cambridge University Press, Cambridge, England, pp. 453–463. [1.1b]

Huber, P. J. (1972). Robust statistics: A review. *Ann. Math. Statist.* **43**, 1041–1067. [1.3a, 1.3b, 1.3d]

Huber, P. J. (1973a). Robust regression: Asymptotics, conjectures, and Monte Carlo. *Ann. Statist.* **1**, 799–821. [1.1c, 1.3b, 6.1b, 6.2, 6.3a]

Huber, P. J. (1973b). The use of Choquet capacities in statistics. Proceedings of the 39th Session of the ISI, Vol. 45, pp. 181–188 (discussion: pp. 189–191). [1.1b]

Huber, P. J. (1974a). Early cuneiform evidence for the planet Venus. AAAS Annual Meeting, San Francisco, Calif. [1.2c]

Huber, P. J. (1974b). Fisher information and spline interpolation. *Ann. Statist.* **2**, 1029–1033. [8.2a]

Huber, P. J. (1975a). Robustness and designs. In *A Survey of Statistical Design and Linear Models*, J. N. Srivastava (ed.). North Holland, Amsterdam, pp. 287–301. [1.2b]

Huber, P. J. (1975b). Application vs. abstraction: The selling out of mathematical statistics? *Suppl. Adv. Appl. Prob.* **7**, 84–89. [1.1a]

Huber, P. J. (1977a). *Robust Statistical Procedures.* SIAM, Philadelphia. [1.3b, 1.4b, 2.1b, 2.3b, 6.4c]

Huber, P. J. (1977b). Robust covariances. In *Statistical Decision Theory and Related Topics*, Vol. 2, S. S. Gupta and D. S. Moore (eds.). Academic Press, New York, pp. 165–191. [1.3b, 5.1, 5.3c, 5.5a]

Huber, P. J. (1977c). Robust methods of estimation of regression coefficients. *Math. Operationsforschung Statist. Ser. Statist.* **8**, 41–53. [2.7, 6.1b, 6.2]

Huber, P. J. (1979). Robust smoothing. In *Robustness in Statistics*, R. L. Launer and G. N. Wilkinson (eds.). Academic Press, New York, pp. 33–47. [1.1b]

Huber, P. J. (1981). *Robust Statistics*. Wiley, New York. [1.1b, 1.2d, 1.3b, 1.3d, 1.4b, 2.1b, 2.2a, 2.2b, 2.3a, 2.3c, 2.7, 2.ex, 4.2c, 4.2d, 4.3c, 5.1, 5.3c, 6.2, 7.3a, 8.2a, 8.2c]

Huber, P. J. (1983). Minimax aspects of bounded-influence regression (with discussion). *J. Am. Statist. Assoc.* **78**, 66–80. [6.2, 6.3a, 6.4d]

Huber, P. J. (1984). Finite sample breakdown of M- and P-estimators. *Ann. Statist.* **12**, 119–126. [2.2b, 2.3d, 8.2c]

Huber, P. J. (1985). Projection pursuit (with discussion). *Ann. Statist.* **13**, 435–525. [5.5c]

Huber, P. J. and Dutter, R. (1974). Numerical solution of robust regression problems. COMPSTAT 1974, Proceedings in Computational Statistics, G. Bruckmann, F. Ferschl, and L. Schmetterer (eds.). Physica, Vienna, pp. 165–172. [6.4c]

Huber, P. J. and Strassen, V. (1973). Minimax tests and the Neyman-Pearson lemma for capacities. *Ann. Statist.* **1**, 251–263; **2**, 223–224. [1.3c, 3.1, 3.7]

Huber-Carol, C. (1970). Etude asymptotique de tests robustes. Ph.D. thesis. ETH, Zurich. [3.1, 3.2c, 3.7]

Hunt, G. and Moore, P. (1982). *The Planet Venus*. Faber and Faber, London. [1.4a]

Hurst, H. E. (1951). Long term storage capacity of reservoirs. *Trans. Am. Soc. Civil Engineers* **116**, 770–808. [8.1b]

Hurst, H. E. (1956). Methods of using long-term storage in reservoirs. *Proc. Inst. Civil Engs.*, Part 1, pp. 519ff. [8.1b]

Hurst, H. E., Black, R. P., and Simaika, Y. M. (1965). *Long-Term Storage, An Experimental Study*. Constable, London. [8.1b]

Iglewicz, B. (1983). Robust scale estimators and confidence intervals for location. In *Understanding Robust and Exploratory Data Analysis*, D. C. Hoaglin, F. Mosteller, and J. W. Tukey (eds.). Wiley, New York. [2.3d]

Irwin, J. O. (1925). On a criterion for the rejection of outlying observations. *Biometrika* **17**, 238–250. [1.3a]

Iversen, G. R., Longcor, W. H., Mosteller, F., Gilbert, J. P., and Youtz, C. (1971). Bias and runs in dice throwing and recording: A few million throws. *Psychometrika* **36**, 1–19. [8.1c]

Jaeckel, L. A. (1971). Robust estimates of location: Symmetry and asymmetric contamination. *Ann. Math. Statist.* **42**, 1020–1034. [2.7]

Jaeckel, L. A. (1972). Estimating regression coefficients by minimizing the dispersion of residuals. *Ann. Math. Statist.* **43**, 1449–1458. [2.3d, 6.4a, 6.4c]

Jeffreys, H. (1932). An alternative to the rejection of outliers. *Proc. Roy. Soc. London Ser. A* **137**, 78–87. [1.3a]

Jeffreys, H. (1939, 1948, 1961). *Theory of Probability*. Clarendon Press, Oxford. [1.2b, 1.3a, 8.1a, 8.1c, 8.2c]

Johns, M. V. (1979). Robust Pitman-like estimators. In *Robustness in Statistics*, R. L. Launer and G. N. Wilkinson (eds.). Academic Press, New York, pp. 49–60. [2.3d]

Johnson, N. L. and Kotz, S. (1970). *Continuous Univariate Distributions* 2. Houghton Mifflin, Boston. [7.5b]

Johnstone, I. and Velleman, P. (1985). The resistant line and related regression methods. *J. Am. Statist. Assoc.* (to appear). [6.4a, 6.ex]

Jurečková, J. (1971). Nonparametric estimate of regression coefficients. *Ann. Math. Statist.* **42**, 1328–1338. [1.1b, 6.4d]

Jurečková, J. (1977). Asymptotic relations of *M*-estimates and *R*-estimates in linear regression model. *Ann. Statist.* **5**, 464–472. [1.1b, 6.4d]

Jurečková, J. (1981). Tail-behaviour of location estimators. *Ann. Statist.* **9**, 578–585. [1.1b]

Kallianpur, G. (1963). Von Mises functions and maximum likelihood estimation. *Sankhyā A* **23**, 149–158. [2.3a]

Kallianpur, G. and Rao, C. R. (1955). On Fisher's lower bound to asymptotic variance of a consistent estimate. *Sankhyā A* **15**, 331–342. [2.1a]

Kariya, T. (1980). Locally robust tests for serial correlation in least squares regression. *Ann. Statist.* **8**, 1065–1070. [1.1b]

Kelly, G. (1984). The influence function in the errors in variables problem. *Ann. Statist.* **12**, 87–100. [6.4d]

Kempthorne, O. (1966). Some aspects of experimental inference. *J. Am. Statist. Assoc.* **61**, 11–34. [1.2b]

Kent, J. T. (1982). Robust properties of likelihood ratio tests. *Biometrika* **69**, 19–27. [7.2b]

Khinchin, A. I. (1949). *Mathematical Foundations of Statistical Mechanics*. Dover, New York. [8.5a]

Kleiner, B., Martin, R. D., and Thomson, D. J. (1979). Robust estimation of power spectra. *J. Roy. Statist. Soc. Ser. B* **41**, 313–351. [1.2c, 1.2e, 8.3a, 8.3c]

Knüsel, L. F. (1969). Über Minimum-Distance-Schätzungen. Ph.D. thesis. ETH, Zurich. [2.3d]

Kolmogorov, A. N. (1940). Wienersche Spiralen und einige andere interessante Kurven im Hilbertschen Raum. *C.R.* (*Doklady*), *Acad. Sci. URSS* (*N.S.*) **26**, 115–118. [8.1b]

Kolmogorov, A. N. (1941). Local structure of turbulence in an incompressible liquid for very large Reynolds numbers. *C.R.* (*Doklady*), *Acad. Sci. URSS* (*N.S.*) **30**, 299–303. In *Turbulence: Classic Papers on Statistical Theory*, S. K. Friedlander and L. Topper (eds.) (1961), pp. 151–155. [8.1b]

Kozek, A. (1982). Minimum Lévy distance estimation of a translation parameter. Preprints in Statistics 70. University of Cologne, W. Germany. [8.2d]

Krasker, W. S. (1977). Parametric estimation in approximate parametric models. Unpublished manuscript. [4.3a, 6.3b]

Krasker, W. S. (1980). Estimation in linear regression models with disparate data points. *Econometrica* **48**, 1333–1346. [1.3d, 6.3b, 7.3c]

Krasker, W. S. and Welsch, R. E. (1982). Efficient bounded-influence regression estimation. *J. Am. Statist. Assoc.* **77**, 595–604. [4.3c, 6.3a, 6.3b]

Kuhn, T. S. ((1962) 1970). *The Structure of Scientific Revolutions.* University of Chicago Press, Chicago, Ill. [1.1a]

Künsch, H. (1981). Thermodynamics and statistical analysis of Gaussian random fields. *Z. Wahrsch. verw. Geb.* **58**, 407–421. [8.1d]

Künsch, H. (1984). Infinitesimal robustness for autoregressive processes. *Ann. Statist.* **12**, 843–863. [8.3a, 8.3b]

Lambert, D. (1981). Influence functions for testing. *J. Am. Statist. Assoc.* **76**, 649–657. [3.1, 3.6a]

Lambert, D. (1985). Robust two-sample permutation tests. *Ann. Statist.* **13**, 606–625. [7.3c]

Lambert, D. and Hall, W. J. (1982). Asymptotic lognormality of *P*-values. *Ann. Statist.* **10**, 44–64. [3.6a]

Lawson, C. L. and Hanson, R. J. (1974). *Solving Least Squares Problems.* Prentice Hall, Englewood Cliffs, N.J. [6.4c]

Lax, D. A. (1975). An interim report of a Monte Carlo study of robust estimators of width. Technical Report 93, Series 2. Department of Statistics, Princeton University, Princeton, N.J. [2.3d]

Lehmann, E. L. (1959). *Testing Statistical Hypotheses.* Wiley, New York. [7.1b]

Leroy, A. and Rousseeuw, P. J. (1984). PROGRES: A program for robust regression. Technical Report 201. Free University, Brussels, Belgium. [6.4a, 6.4c]

Li, G. and Chen, Z. (1981). Robust projection pursuit estimator for dispersion matrices and principal components. Research Report PJH-11. Department of Statistics, Harvard University, Cambridge, Mass. [5.1]

Magnus, J. R. and Neudecker, H. (1979). The commutation matrix: Some properties and applications. *Ann. Statist.* **7**, 381–394. [5.2a]

Major, P. (1981a). *Multiple Wiener-Itô Integrals.* Lecture Notes in Mathematics 849. Springer, Berlin. [8.1b]

Major, P. (1981b). Limit theorems for non-linear functionals of Gaussian sequences. *Z. Wahrsch. verw. Geb.* **57**, 129–158. [8.1b]

Mallows, C. L. (1971). Hoeffding, Tukey, Hájek, and von Mises. Unpublished talk at the Princeton Robustness Seminar, May 10, 1971. [1.3d]

Mallows, C. L. (1973). Some comments on C_p. *Technometrics* **15**, 661–675. [2.1b, 7.3d]

Mallows, C. L. (1975). On some topics in robustness. Technical Memorandum. Bell Telephone Laboratories, Murray Hill, N.J. [2.1b]

Mallows, C. L. (1979). Robust methods—some examples of their use. *Am. Statist.* **33**, 179–184. [1.2c, 1.2e]

Mallows, C. L. (1983). Discussion of Huber: Minimax aspects of bounded-influence regression. *J. Am. Statist. Assoc.* **78**, 77. [6.3b]

Mandelbrot, B. B. (1965). Une classe de processus homothétiques à soi: Application à la loi climatologique de H. E. Hurst. *C.R. Acad. Sci. Paris* **260**, 3274–3277. [8.1b]

Mandelbrot, B. B. (1972). Statistical methodology for nonperiodic cycles: From the covariance to R/S analysis. *Ann. Econ. Social Meas.* **1**, 259–290. [8.1d]

Mandelbrot, B. B. ((1977) 1983). *The Fractal Geometry of Nature.* Freeman, New York. [8.1b]

Mandelbrot, B. B. and Taqqu, M. S. (1979). Robust R/S analysis of long run serial correlation. Proceedings of the 42nd Session of the ISI, Manila, Vol. XLVIII, Book 2, pp. 69–99. [8.1b, 8.1d]

Mandelbrot, B. B. and van Ness, J. W. (1968). Fractional Brownian motions, fractional noises and applications. *SIAM Rev.* **10**, 422–437. [8.1b]

Mandelbrot, B. B. and Wallis, J. R. (1968). Noah, Joseph, and operational hydrology. *Water Res. Res.* **4**, 909–918. [8.1b]

Mandelbrot, B. B. and Wallis, J. R. (1969a). Computer experiments with fractional Gaussian noises. *Water Res. Res.* **5**, 228–267. [8.1b]

Mandelbrot, B. B. and Wallis, J. R. (1969b). Some long-run properties of geophysical records. *Water Res. Res.* **5**, 321–340. [8.1b]

Mandelbrot, B. B. and Wallis, J. R. (1969c). Robustness of the rescaled range R/S in the measurement of noncyclic long run statistical dependence. *Water Res. Res.* **5**, 967–988. [8.1d]

Marazzi, A. (1980a). Robust Bayesian estimation for the linear model. Research Report 27. Fachgruppe für Statistik, ETH, Zurich. [6.4d]

Marazzi, A. (1980b). ROBETH, a subroutine library for robust statistical procedures. COMPSTAT 1980, Proceedings in Computational Statistics. Physica, Vienna. [6.4c]

Marazzi, A. (1980c). Robust linear regression programs in ROBETH. ROBETH Document No. 2, Research Report 23. Fachgruppe für Statistik, ETH, Zurich. [6.4c]

Marazzi, A. (1980d). Robust affine invariant covariances in ROBETH. ROBETH Document No. 3, Research Report 24. Fachgruppe für Statistik, ETH, Zurich. [6.4c]

Maronna, R. A. (1976). Robust M-estimators of multivariate location and scatter. *Ann. Statist.* **4**, 51–67. [6.4a]

Maronna, R. A., Bustos, O. H., and Yohai, V. J. (1979). Bias- and efficiency-robustness of general M-estimators for regression with random carriers. In *Smoothing Techniques for Curve Estimation*, T. Gasser and M. Rosenblatt (eds.). Lecture Notes in Mathematics 757. Springer, Berlin, pp. 91–116. [6.4a, 6.4b, 7.5c]

Maronna, R. A. and Yohai, V. J. (1981). Asymptotic behaviour of general M-estimates for regression and scale with random carriers. *Z. Wahrsch. verw. Geb.* **58**, 7–20. [4.3c, 5.1, 5.3c, 5.5, 6.3a]

Martin, R. D. (1979). Robust estimation for time series autoregressions. In *Robustness in Statistics*, R. L. Launer and G. N. Wilkinson (eds.). Academic Press, New York, pp. 147–176. [8.3a]

Martin. R. D. (1980). Robust estimation of autoregressive models. In *Directions in Time Series*, D. R. Brillinger and G. C. Tiao (eds.). Institute of Mathematical Statistics, Hayward, Calif., pp. 228–254. [7.3d, 8.3a]

Martin, R. D. (1981). Robust methods for time series. In *Applied Time Series II*, D. F. Findley (ed.). Academic Press, New York, pp. 683–759. [8.3a]

Martin, R. D. and Thomson, D. J. (1982). Robust-resistant spectrum estimation. *Proc. IEEE* **70**, 1097–1115. [8.3c]

Martin, R. D. and Yohai, V. J. (1984a). Influence curves for time series. Technical Report 51. Department of Statistics, University of Washington, Seattle, Wash. [8.3a, 8.3b]

Martin, R. D. and Yohai, V. J. (1984b). Robustness in time series and estimating ARMA models. In *Handbook of Statistics, Vol. 4: Nonparametric Methods*, D.R. Brillinger and P. R. Krishnaiah (eds.). Elsevier Science, Amsterdam and New York. [8.3a]

McCulloch, C. E. (1982). Symmetric matrix derivatives with applications. *J. Am. Statist. Assoc.* **77**, 679–682. [5.2a, 5.2b]

McKean, J. W. and Schrader, R. M. (1980). The geometry of robust procedures in linear models. *J. Roy. Statist. Soc. Ser. B* **42**, 366–371. [6.2]

McLeod, A. I. and Hipel, K. W. (1978). Preservation of the rescaled range R/S in the measurement of noncyclic long run statistical dependence. *Water Res. Res.* **5**, 967–988. [8.1d]

Mendeleev, D. I. (1895). Course of work on the renewal of prototypes or standard measures of lengths and weights (Russian). *Vremennik Glavnoi Palaty Mer i Vesov* **2**, 157–185. Reprinted 1950: *Collected Writings (Socheneniya)*, Izdat. Akad. Nauk, SSSR, Leningrad-Moscow, Vol. 22, pp. 175–213. [1.3a]

Merrill, H. M. and Schweppe, F. C. (1971). Bad data suppression in power system static state estimation. *IEEE Trans. Power App. Syst.* **PAS-90**, 2718–2725. [6.3a]

Meshalkin, L. D. (1968). On the robustness of some characterizations of the normal distribution. *Ann. Math. Statist.* **39**, 1747–1750. [1.1b]

Michelson, A. A., Pease, F. G., and Pearson, F. (1935). Measurement of the velocity of light in a partial vacuum. Contributions from the Mount Wilson Observatory, Carnegie Institution of Washington, Vol. XXII, No. 522, pp. 259–294. [8.1a]

Millar, P. W. (1981). Robust estimation via minimum distance methods. *Z. Wahrsch. verw. Geb.* **55**, 73–89. [1.1b]

Mohr, D. L. (1981). Modeling data as a fractional Gaussian noise. Ph.D. thesis. Princeton University, Princeton, N.J. [8.1d]

Morgenthaler S. (1979). Untersuchung und Vergleich robuster Regressions-schätzungen. Diploma thesis. Fachgruppe für Statistik, ETH, Zurich. [6.4d]

Mosteller, F. and Tukey, J. W. (1977). *Data Analysis and Regression*. Addison-Wesley, Reading, Mass. [1.1b, 8.1a]

Muirhead, R. J. and Waternaux, C. M. (1980). Asymptotic distributions in canonical correlation analysis and other multivariate procedures for nonnormal populations. *Biometrika* **67**, 31–43. [5.1]

Newcomb, S. (1886). A generalized theory of the combination of observations so as to obtain the best result. *Am. J. Math.* **8**, 343–366. [1.3a, 8.2c]

Newcomb, S. (1895). Astronomical Constants (The Elements of the Four Inner Planets and the Fundamental Constants of Astronomy). Supplement to the American Ephemeris and Nautical Almanac for 1897. U.S. GPO, Washington, D.C. [8.1a]

Neyman, J. (1958). Optimal asymptotic tests of composite statistical hypotheses. In *Probability and Statistics: The Cramèr Volume*, U. Grenander (ed.). Almquist and Wiksells, Uppsala, Sweden, pp. 213–234. [7.4a]

Neyman, J. (1979). $C(\alpha)$-tests and their use. *Sankhyā A* **41**, 1–21. [7.4a]

Noether, G. E. (1955). On a theorem of Pitman. *Ann. Math. Statist.* **26**, 64–68. [2.5f, 3.2b, 3.2c, 3.6a, 3.6b]

O'Brien, F. L. (1984). Polyefficient and polyeffective simple linear regression estimators and the absolute polyefficiency of the biweight regression estimator. Ph.D. thesis. Department of Statistics, Princeton University, Princeton, N.J. [6.4d]

Papantoni-Kazakos, P. and Gray, R. M. (1979). Robustness of estimators on stationary observations. *Ann. Prob.* **7**, 989–1002. [8.3c]

Parr, W. C. and Schucany, W. R. (1980). Minimum distance and robust estimation. *J. Am. Statist. Assoc.* **75**, 616–624. [2.3d]

Pearson, E. S. (1929). The distribution of frequency constants in small samples from non-normal symmetrical and skew populations. *Biometrika* **21**, 259–286. [1.3a]

Pearson, E. S. (1931). The analysis of variance in cases of non-normal variation. *Biometrika* **23**, 114–133. [1.2e, 1.3a]

Pearson, E. S. and Chandra Sekar, C. (1936). The efficiency of statistical tools and a criterion for the rejection of outlying observations. *Biometrika* **28**, 308–320. [1.3a]

Pearson, K. (1902). On the mathematical theory of errors of judgement, with special reference to the personal equation. *Philos. Trans. Roy. Soc. Ser. A* **198**, 235–299. [1.2b, 8.1a]

Peirce, B. (1852). Criterion for the rejection of doubtful observations. *Astr. J.* **2**, 161–163. [1.3a]

Peters, S. C., Samarov, A., and Welsch, R. E. (1982). Computational procedures for bounded-influence and robust regression (TROLL: BIF and BIFMOD). Technical Report 30. Center for Computational Research in Economics and Management Science, Massachusetts Institute of Technology, Cambridge, Mass. [6.4c]

Pfanzagl, J. (1968). *Allgemeine Methodenlehre der Statistik, Band II.* De Gruyter, Berlin. [2.0]

Portnoy, S. L. (1977). Robust estimation in dependent situations. *Ann. Statist.* **5**, 22–43. [1.1b]

Portnoy, S. L. (1979). Further remarks on robust estimation in dependent situations. *Ann. Statist.* **7**, 224–231. [1.1b]

Pregibon, D. (1981). Logistic regression diagnostics. *Ann. Statist.* **9**, 705–724. [1.1b, 6.4d]

Pregibon, D. and Tukey, J. W. (1981). Assessing the behaviour of robust estimates of location in small samples: Introduction to configural polysampling. Technical Report 185, Series 2. Department of Statistics, Princeton University, Princeton, N.J. [1.1b, 6.4d]

Prohorov, Y. V. (1956). Convergence of random processes and limit theorems in probability theory. *Theor. Prob. Appl.* **1**, 157–214. [2.2a]

Quenouille, M. H. (1956). Notes on bias in estimation. *Biometrika* **43**, 353–360. [2.1e]

Ramsay, J. O. (1977). A comparative study of several robust estimates of slope, intercept, and scale in linear regression. *J. Am. Statist. Assoc.* **72**, 608–615. [1.1b]

Ramsay, J. O. and Novick, M. R. (1980). PLU robust Bayesian decision theory: Point estimation. *J. Am. Statist. Assoc.* **75**, 901–907. [1.1b]

Randles, R. H., Broffitt, J. D., Ramberg, J. S., and Hogg, R. V. (1978). Generalized linear and quadratic discriminant functions using robust estimates. *J. Am. Statist. Assoc.* **73**, 564–568. [5.1]

Reeds, J. A. (1976). On the definition of von Mises functionals. Research Report S 44. Department of Statistics, Harvard University, Cambridge, Mass. [1.3f, 2.1b, 2.3a]

Relles, D. A. and Rogers, W. H. (1977). Statisticians are fairly robust estimators of location. *J. Am. Statist. Assoc.* **72**, 107–111. [1.2d, 1.4b]

Rey, W. J. J. (1976). *M*-estimators in robust regression, a case study. European Meeting of Statisticians, Grenoble. [8.2c]

Rey, W. J. J. (1983). *Introduction to Robust and Quasi-Robust Statistical Methods.* Universitext, Springer, Heidelberg. [1.1b, 1.3a]

Rieder, H. (1978). A robust asymptotic testing model. *Ann. Statist.* **6**, 1080–1094. [3.1, 3.2c, 3.3a, 3.7, 7.4d]

Rieder, H. (1980). Estimates derived from robust tests. *Ann. Statist.* **8**, 106–115. [3.1, 3.7]

Rieder, H. (1981). Robustness of one- and two-sample rank tests against gross errors. *Ann. Statist.* **9**, 245–265. [3.1, 3.2c, 3.7]

Rieder, H. (1982). Qualitative robustness of rank tests. *Ann. Statist.* **10**, 205–211. [3.1]

Rivest, L. P. (1984). Bartlett's, Cochran's and Hartley's tests on variances are liberal when the underlying distribution is long-tailed. Submitted to *J. Am. Statist. Assoc.* [3.1]

Rocke, D. M. (1983). Robust statistical analysis of interlaboratory studies. *Biometrika* **70**, 421–431. [6.4d]

Rocke, D. M., Downs, G. W., and Rocke, A. J. (1982). Are robust estimators really necessary? *Technometrics* **24**, 95–101. [1.2e]

Romanowski, M. (1970). *The theory of random errors based on the concept of modulated normal distributions.* National Research Council of Canada (NRC-11432), Division of Physics, Ottawa, Canada. [1.2b]

Romanowski, M. and Green, E. (1965). Practical applications of the modified normal distribution. *Bull. Géodésique* **76**, 1–20. [1.2b]

Ronchetti, E. (1979). Robustheitseigenschaften von Tests. Diploma thesis. ETH, Zurich. [1.3d, 2.5f, 3.1, 3.3a]

Ronchetti, E. (1982a). Robust alternatives to the *F*-test for the linear model. In *Probability and Statistical Inference*, W. Grossmann, C. Pflug, and W. Wertz (eds.). Reidel, Dortrecht, pp. 329–342. [6.1b, 7.1a, 7.1b, 7.3b]

Ronchetti, E. (1982b). Robust testing in linear models: The infinitesimal approach. Ph.D. thesis. ETH, Zurich. [1.3f, 6.1b, 6.3c, 7.1a, 7.1b, 7.2b, 7.3b, 7.3c, 7.3d, 7.4a, 7.4d, 7.5c]

Ronchetti, E. (1985). Robust model selection in regression. *Statist. Prob. Let.* **3**, 21–23. [7.3d].

Ronchetti, E. and Rousseeuw, P. J. (1980). A robust *F*-test for the linear model. Abstracts Book, 13th European Meeting of Statisticians, Brighton, England, pp. 210–211. [7.2b, 7.3a]

Ronchetti, E. and Rousseeuw, P. J. (1985). Change-of-variance sensitivities in regression analysis. *Z. Wahrsch. verw. Geb.* **68**, 503–519. [6.3b, 6.3c]

Ronchetti, E. and Yen, J. H. (1985). Variance-stable *R*-estimators. *Math. Operationsforsch. Statist., Ser. Statist.* (to appear). [2.5f]

Rosenthal, R. (1978). How often are our numbers wrong? *Am. Psychol.* **33**(11), 1005–1008. [1.2c]

Rousseeuw, P. J. (1979). Optimally robust procedures in the infinitesimal sense. Proceedings of the 42nd Session of the ISI, Manila, pp. 467–470. [1.3d, 2.4c, 3.1, 3.4]

Rousseeuw, P. J. (1981a). A new infinitesimal approach to robust estimation. *Z. Wahrsch. verw. Geb.* **56**, 127–132. [1.3d, 1.3e, 2.1c, 2.5a, 2.5b, 2.7]

Rousseeuw, P. J. (1981b). New infinitesimal methods in robust statistics. Ph.D. thesis. Vrije Universiteit, Brussels, Belgium. [2.5a, 2.5c, 2.5d, 2.5f, 2.6a, 2.6b, 2.6c]

Rousseeuw, P. J. (1981c). Infinitesimal criteria in robust estimation of location. *Rev. Belge Statist. Infor. Recherche Operat.* 21 (4), 24–42. [2.5a, 2.6a]

Rousseeuw, P. J. (1982a). Most robust M-estimators in the infinitesimal sense. *Z. Wahrsch. verw. Geb.* 61, 541–555. [2.5a, 2.6a, 2.6b, 2.6c]

Rousseeuw, P. J. (1982b). Estimation and testing by means of optimally robust statistics. *Rev. Belge Statist. Infor. Recherche Operat.* 22 (3), 3–19. [2.4c, 3.1, 3.4]

Rousseeuw, P. J. (1983a). Location M-estimators are characterized by the infinitesimal behaviour of their asymptotic variance. *Bull. Soc. Math. Belgique* 35B, 167–176. [2.5a, 2.5f]

Rousseeuw, P. J. (1983b). Multivariate estimation with high breakdown point. Fourth Pannonian Symposium on Mathematical Statistics, Bad Tatzmannsdorf, Austria, September 4–9, 1983. [2.3d, 5.5c]

Rousseeuw, P. J. (1984). Least median of squares regression. *J. Am. Statist. Assoc.* 79, 871–880. [2.3d, 6.4a]

Rousseeuw, P. J. and Ronchetti, E. (1979). The influence curve for tests. Research Report 21. Fachgruppe für Statistik, ETH, Zurich. [3.1, 3.2, 3.6a]

Rousseeuw, P. J. and Ronchetti, E. (1981). Influence curves for general statistics. *J. Comput. Appl. Math.* 7, 161–166. [1.3d, 3.1, 3.2, 3.6a]

Rousseeuw, P. J. and Yohai, V. (1984). Robust regression by means of S-estimators. In *Robust and Nonlinear Time Series Analysis*, J. Franke, W. Härdle, and R. D. Martin (eds.), Lecture Notes in Statistics 26, Springer, New York, pp. 256–272. [2.3d, 6.4a]

Ruppert, D. (1985). On the bounded-influence regression estimator of Krasker and Welsch. *J. Am. Statist. Assoc.* 80, 205–208. [6.3b]

Ruppert, D. and Carroll, R. J. (1980). Trimmed least squares estimation in the linear model. *J. Am. Statist. Assoc.* 75, 828–838. [6.4d]

Ruymgaart, F. H. (1981). A robust principal component analysis. *J. Mult. Anal.* 11, 485–497. [5.1]

Sacks, J. and Ylvisaker, D. (1978). Linear estimation for approximately linear models. *Ann. Statist.* 6, 1122–1137. [1.1b]

Samarov, A. (1983). Bounded-influence regression via local minimax mean squared error. Technical Report 40. Center for Computational Research in Economics and Management Science, Massachusetts Institute of Technology, Cambridge, Mass. [6.4d]

Samarov, A. and Welsch, R. E. (1982). Computational procedures for bounded-influence regression. COMPSTAT 1982, Proceedings in Computational Statistics. Physica, Vienna. [6.4c]

Scheffé, H. (1959). *The Analysis of Variance*. Wiley, New York. [1.2b, 1.2c, 1.3a, 6.1b, 7.1b, 8.1a]

Schönholzer, H. (1979). Robuste Kovarianz. Ph.D. thesis. ETH, Zurich. [5.1, 5.3c]

Schrader, R. M. and Hettmansperger, T. P. (1980). Robust analysis of variance based upon a likelihood ratio criterion. *Biometrika* **67**, 93–101. [6.1b, 7.1b, 7.2c, 7.3a]

Schrader, R. M. and McKean, J. W. (1977). Robust analysis of variance. *Commun. Statist. A* **6**, 879–894. [7.1b, 7.2c]

Schweingruber, M. (1980). Das Monte Carlo Verhalten einiger Verwerfungsregeln. Diploma thesis. Fachgruppe für Statistik, ETH, Zurich. [1.4b]

Searle, S. R. (1978). A univariate formulation of the multivariate linear model. In *Contributions to Survey Sampling and Applied Statistics, Papers in Honor of H. O. Hartley*, H. A. David (ed.). Academic Press, New York. [5.2a]

Seber, G. A. F. ((1966) 1980). *The Linear Hypothesis: A General Theory*. Griffin, London. [7.1b]

Sen, P. K. (1968). Estimates of the regression coefficient based on Kendall's Tau. *J. Am. Statist. Assoc.* **63**, 1379–1389. [6.4a]

Sen, P. K. (1982). On *M* tests in linear models. *Biometrika* **69**, 245–248. [3.3a]

Serfling, R. J. (1980). *Approximation Theorems of Mathematical Statistics*. Wiley, New York. [1.1b]

Shapiro, S. S. and Wilk, M. B. (1965). An analysis of variance test for normality (complete samples). *Biometrika* **52**, 591–611. [1.4b]

Shoemaker, L. H. (1982). Robust estimation and testing in random effects models. Unpublished manuscript. [6.4d]

Shoemaker, L. H. (1984). Robustness properties for a class of scale estimators. *Commun. Statist. A* **13**, 15–28. [2.3d]

Shoemaker, L. H. and Hettmansperger, T. P. (1982). Robust estimates and tests for the one- and two-sample scale models. *Biometrika* **69**, 47–53. [2.3d]

Siegel, A. F. (1982). Robust regression using repeated medians. *Biometrika* **69**, 242–244. [4.6b, 5.5c, 6.4a]

Smith, A. F. M. (1981). Bayesian approaches to outliers and robustness. Research Report 07-81. Nottingham Statistics Group, University of Nottingham, Nottingham, England. [1.1b]

Smith, H. Fairfield (1938). An empirical law describing heterogeneity in the yields of agricultural crops. *J. Agric. Sci.* **28**, 1–23. [8.1a]

Solomon, H. and Stephens, M. A. (1977). Distribution of a sum of weighted chi-square variables. *J. Am. Statist. Assoc.* **72**, 881–885. [7.5b]

Spjøtvoll, E. and Aastveit, A. H. (1980). Comparison of robust estimators on data from field experiments. *Scand. J. Statist.* **7**, 1–13. [1.2e]

Spjøtvoll, E. and Aastveit, A. H. (1983). Robust estimators on laboratory measurements of fat and protein in milk. *Biom. J.* **25**, 627–639. [1.2e]

Staab, M. (1985). Robust parameter estimation for ARMA models. Ph.D. thesis. University of Bayreuth, Bayreuth, W. Germany. [8.3b]

Stahel, W. A. (1981a). Robust estimation: Infinitesimal optimality and covariance matrix estimators (in German). Ph.D. thesis. ETH, Zurich. [1.3d, 4.3a, 4.6a, 4.6b, 5.1, 5.3c, 5.4a, 6.4c, 7.2b, 8.4b]

Stahel, W. A. (1981b). Breakdown of covariance estimators. Research Report 31. Fachgruppe für Statistik, ETH, Zurich. [5.1, 5.5c, 6.4c]

Staudte, R. G. (1980). *Robust Estimation*. Queen's Papers in Pure and Applied Mathematics 53. Queen's University, Kingston, Canada. [1.1b]

Steele, J. M. and Steiger, W. L. (1984). Algorithms and complexity for least median of squares regression. Technical Report 286, Series 2. Department of Statistics, Princeton University, Princeton, N.J. [6.4a]

Stefanski, L. A., Carroll, R. J., and Ruppert D. (1984). Bounding influence and leverage in logistic regression. Mimeo Series No. 1554. Department of Statistics, University of North Carolina, Chapel Hill, N.C. [4.1]

Stefansky, W. (1971). Rejecting outliers by maximum normed residual. *Ann. Math. Statist.* **42**, 35–45. [1.4b]

Stefansky, W. (1972). Rejecting outliers in factorial designs. *Technometrics* **14**, 469–479. [1.4b]

Stigler, S. M. (1973). Simon Newcomb, Percy Daniell, and the history of robust estimation 1885–1920. *J. Am. Statist. Assoc.* **68**, 872–879. [1.3a]

Stigler, S. M. (1975). Contribution to the discussion of the meeting on robust statistics. Proceedings of the 40th Session of the ISI, Warsaw, Vol. XLVI, Book 1, pp. 383–384. [1.2b]

Stigler, S. M. (1976). The anonymous Professor Gergonne. *Hist. Math.* **3**, 71–74. [1.3a]

Stigler, S. M. (1977). Do robust estimators work with real data? *Ann. Statist.* **5**, 1055–1098. [1.2e]

Stigler, S. M. (1980). Studies in the history of probability and statistics XXXVIII: R. H. Smith, a Victorian interested in robustness. *Biometrika* **67**, 217–221. [2.1c, 2.6a]

Stigler, S. M. (1981). Gauss and the invention of least squares. *Ann. Statist.* **9**, 465–474. [6.1b]

Stone, E. J. (1868). On the rejection of discordant observations. *Mon. Not. Roy. Astr. Soc.* **28**, 165–168. [1.3a]

Stone, E. J. (1873). On the rejection of discordant observations. *Mon. Not. Roy. Astr. Soc.* **34**, 9–15. [1.3a]

Stone, M. (1977). An asymptotic equivalence of choice of model by cross-validation and Akaike's criterion. *J. Roy. Statist. Soc. Ser. B* **39**, 44–47. [7.3d]

Strassen, V. (1964). Messfehler und Information. *Z. Wahrsch. verw. Geb.* **2**, 273–305. [1.1b]

Student (1908). The probable error of a mean. *Biometrika* **6**, 1–25. [2.0]

Student (1927). Errors of routine analysis. *Biometrika* **19**, 151–164. [1.2b, 1.3a, 8.1a, 8.1c]

Takeuchi, K. (1971). A uniformly asymptotically efficient estimator of a location parameter. *J. Am. Statist. Assoc.* **66**, 292–301. [1.1b]

Taqqu, M. S. (1975). Weak convergence to fractional Brownian motion and to the Rosenblatt process. *Z. Wahrsch. verw. Geb.* **31**, 287–302. [8.1b]

Taqqu, M. S. (1979a). Self-similar processes and related ultraviolet and infrared catastrophes. Technical Report 423. Operations Research, Cornell University, Ithaca, N.Y. [8.1b]

Taqqu, M. S. (1979b). Convergence of iterated processes of arbitrary Hermite rank. *Z. Wahrsch. verw. Geb.* **50**, 53–83. [8.1b]

Theil, H. (1950). A rank-invariant method of linear and polynomial regression analysis, I, II, III. *Nederland. Akad. Wetensch. Proc. Ser. A* **53**, 386–92, 521–525, 1397–1412. [6.4a]

Thompson, W. R. (1935). On a criterion for the rejection of observations and the distribution of the ratio of deviation to sample standard deviation. *Ann. Math. Statist.* **6**, 214–219. [1.3a]

Tracy, D. S. and Dwyer, P. S. (1969). Multivariate maxima and minima with matrix derivatives. *J. Am. Statist. Assoc.* **64**, 1576–1594. [5.2b]

Tukey, J. W. (1958). Bias and confidence in not-quite large samples (Abstract). *Ann. Math. Statist.* **29**, 614. [1.1b, 2.1e]

Tukey, J. W. (1960). A survey of sampling from contaminated distributions. In *Contributions to Probability and Statistics*, I. Olkin (ed.). Stanford University Press, Stanford, Calif., pp. 448–485. [1.1b, 1.2b, 1.2d, 1.3a, 1.3b, 2.1e]

Tukey, J. W. (1962). The future of data analysis. *Ann. Math. Statist.* **33**, 1–67. [1.1b, 1.2c, 1.4b, 8.4b]

Tukey, J. W. ((1970–71) 1977). *Exploratory Data Analysis* (1970–71: preliminary edition). Addison-Wesley, Reading, Mass. [1.1b, 1.2c, 2.0, 2.1e, 2.3d, 6.4a, 6.ex]

Tukey, J. W. (1981). Some advanced thoughts on the data analysis involved in configural polysampling directed toward high performance estimates. Technical Report 189, Series 2. Department of Statistics, Princeton University, Princeton, N.J. [1.1b, 6.4d]

Tukey, J. W. and McLaughlin, D. H. (1963). Less vulnerable confidence and significance procedures for location based on a single sample: Trimming/Winsorization I. *Sankhyā A* **25**, 331–352. [1.1b]

Tyler, D. E. (1981). Asymptotic inference for eigenvectors. *Ann. Statist.* **9**, 725–736. [5.1]

Tyler, D. E. (1982). Radial estimates and the test for sphericity. *Biometrika* **69**, 429–436. [5.1, 5.2a]

Tyler, D. E. (1983a). A class of asymptotic tests for principal component vectors. *Ann. Statist.* **11**, 1243–1250. [5.1]

Tyler, D. E. (1983b). Robustness and efficiency properties of scatter matrices. *Biometrika* **70**, 411–420. [5.2a]

von Mises, R. (1937). Sur les fonctions statistiques. In: Conf. de la Réunion Internationale des Math., Gauthier-Villars, Paris, pp. 1–8. [2.1b]

von Mises, R. (1947). On the asymptotic distribution of differentiable statistical functions. *Ann. Math. Statist.* **18**, 309–348. [2.1b, 7.2b]

Wainer, H. (1983). Pyramid power: Searching for an error in test scoring with 830,000 helpers. *Am. Statist.* **37**, 87–91. [1.4a]

Wang, P. C. C. (1981). Robust asymptotic tests of statistical hypotheses involving nuisance parameters. *Ann. Statist.* **9**, 1096–1106. [3.7, 7.4a, 7.4b, 7.4d]

Watson, G. S. (1983). Optimal and robust estimation on the sphere. Proceedings of the 44th Session of the ISI, Madrid, Book 2, pp. 816–818. [4.1]

Welsch, R. E. (1979). Robust and bounded-influence regression. Proceedings of the 42nd Session of the ISI, Manila, Vol. XLVIII, Book 2, pp. 59–68. [6.3a]

Welsch, R. E. (1981). Notes for the Princeton robustness reunion. Unpublished manuscript. [6.3b, 7.3c]

Welsch, R. E. (1982). Influence functions and regression diagnostics. In *Modern Data Analysis*, R. L. Launer and A. F. Siegel (eds.). Academic Press, New York, pp. 149–169. [6.4d]

Whittle, P. (1956). On the variation of yield variance with plot size. *Biometrika* **43**, 337–343. [8.1a, 8.1b]

Wieand, H. S. (1976). A condition under which the Pitman and Bahadur approaches to efficiency coincide. *Ann. Statist.* **4**, 1003–1011. [3.6a]

Wolf, G. (1977). Obere und untere Wahrscheinlichkeiten. Ph.D. thesis. ETH, Zurich. [1.1b]

Wolfowitz, J. (1957). The minimum distance method. *Ann. Math. Statist.* **28**, 75–88. [2.3d]

Wright, T. W. (1884). *A Treatise on the Adjustment of Observations by the Method of Least Squares*. Van Nostrand, New York. [1.3a]

Ylvisaker, D. (1977). Test resistance. *J. Am. Statist. Assoc.* **72**, 551–556. [1.1c]

Yates, F. ((1949) 1981). *Sampling Methods for Censuses & Surveys*. Griffin, London. [8.1a]

Yates, F. and Finney, D. J. (1942). Statistical problems in field sampling for wireworms. *Ann. Appl. Biol.* **29**, 156–167. [8.1a]

Yohai, V. J. and Maronna, R. A. (1979). Asymptotic behavior of M-estimators for the linear model. *Ann. Statist.* **7**, 258–268. [6.3a]

Youden, W. J. (1972). Enduring values. *Technometrics* **14**, 1–11. [8.1a, 8.4b]

Zabell, S. (1976). Arbuthnot, Heberden and the bills of mortality. Technical Report 40. Department of Statistics, University of Chicago, Chicago, Ill. [1.2c]

Zieliński, R. (1977). Robustness: A quantitative approach. *Bull. Acad. Polon. Sci. Ser. Math., Astr. Phys.* **25**, 1281–1286. [1.1b]

Zieliński, R. (1981). Robust statistical procedures: A general approach. Preprint No. 254. Institute of Mathematics, Polish Academy of Sciences, Warsaw, Poland. [1.1b]

Zolotarev, V. M. (1977). General problems of the stability of mathematical models. Proceedings of the 41st Session of the ISI, New Delhi, Vol. XLVII, Book 2, pp. 382–401. [1.1b]

Author Index

Subject Index